Control Systems Theory

Control Systems Theory

N. Sukavanam

CAMBRIDGE
UNIVERSITY PRESS

CAMBRIDGE
UNIVERSITY PRESS

Shaftesbury Road, Cambridge CB2 8EA, United Kingdom

One Liberty Plaza, 20th Floor, New York, NY 10006, USA

477 Williamstown Road, Port Melbourne, VIC 3207, Australia

314 to 321, 3rd Floor, Plot No.3, Splendor Forum, Jasola District Centre, New Delhi 110025, India

Cambridge University Press is part of Cambridge University Press & Assessment, a department of the University of Cambridge.

We share the University's mission to contribute to society through the pursuit of education, learning and research at the highest international levels of excellence.

www.cambridge.org
Information on this title: www.cambridge.org/9781009633475

First published 2026

Printed in India by Manipal Technologies Limited, Manipal

A catalogue record for this publication is available from the British Library

A Cataloging-in-Publication data record for this book is available from the Library of Congress

ISBN 978-1-009-63347-5 Paperback

To my parents

Contents

Preface

The aim of this book is to comprehensively present the mathematical aspects of control systems for graduate-level courses in control theory and optimal control theory. Since the subject requires foundational knowledge in linear algebra, functional analysis, and ordinary differential equations, three chapters are devoted to concisely covering the necessary topics from these areas in order to make the book as self-contained as possible.

In Chapter 1, simple mathematical models of various dynamical and control systems are presented to illustrate key concepts of control systems theory, including controllability, observability, stability, and optimal control.

In Chapter 2, essential results from matrix theory that are fundamental to understanding control theory are presented. Specifically, sections on eigenvalues and eigenvectors, canonical forms, matrix functions, and special matrices establish a strong foundation for finite dimensional control systems theory.

In Chapter 3, definitions, theorems, and techniques from linear algebra and functional analysis are briefly reviewed through examples. The primary purpose of this chapter is to introduce the notations and concepts that will be used in subsequent chapters. The main result is the Banach fixed point theorem, presented in Section 3.5 and applied in Chapter 4 to establish the existence and uniqueness theorem. Section 3.6 on compact operators is included primarily to support Chapter 10 on infinite dimensional systems.

Systems of first-order ordinary differential equations play a vital role in control systems theory. In Chapter 4, key results from theory of differential equations, essential for developing control theory, are presented. Theorem 4.5.1 on the existence and uniqueness of solutions to initial value problems, Section 4.6 on systems of linear ODEs, Section 4.8 on fundamental matrices and the state transition matrix, and Theorem 4.10.1 on Gronwall's inequality are key focal points for better understanding the material in Chapters 5–8. Controllability and observability are desirable properties of a control system. Controllability implies that a control input can steer the system to any arbitrary state, while observability means that the system's state over a period of time can be uniquely determined from given observation data. Chapter 5 presents results related to the controllability and observability of a system.

In Chapter 6, optimal control problems focused on obtaining optimal solutions (both state and control) that minimize or maximize a given objective functional are discussed. Using the Euler–Lagrange equations, the state and co-state canonical matrix is formed to derive the optimal state and co-state functions. The optimal control is then expressed in terms of the optimal co-state. In Chapter 7, the matrix Riccati equation is used to determine the optimal state and control. Section 7.3 discusses the Pontryagin principle for finding the optimal control for systems with constraints on the control function. Applications are presented in Section 7.4.

In Chapter 8, the stability properties of linear and nonlinear dynamical systems are considered. For autonomous systems, the stability of an equilibrium is studied using methods based on (i) the eigenvalues of the system matrix, (ii) linearization, and (iii) Lyapunov theorems. For nonautonomous systems, the eigenvalue method and Lyapunov theorems are discussed for stability analysis. In addition, the stabilizability of dynamical systems is discussed in Section 8.5.

For the computational study of a continuous-time system, it is necessary to discretize the system equations and analyze the equivalent discrete-time system. Thus, the study of discrete-time systems becomes important in applications. In Chapter 9, results on controllability, observability, stability, stabilizability, and optimal control for discrete-time systems are discussed. These results are analogous to those of continuous-time systems presented in Chapters 5–8, with appropriate modifications.

In Chapter 10, the concept of controllability for control systems defined in infinite dimensional spaces is discussed. In particular, the controlled heat equation is studied within the framework of semigroup theory.

Readers familiar with undergraduate mathematics courses may quickly glance through Chapters 2–4 for certain notations and proceed directly to Chapters 5–9. Readers without a background in basic functional analysis may skip both Section 3.6 and Chapter 10.

Having taught courses on control theory and optimal control theory for many years at the Indian Institute of Technology Roorkee, this book is based on the notes I prepared during that time. I am sincerely grateful to my teachers, students, and the many authors of outstanding books on control theory and related topics.

I am grateful to my student, Kiran, for his constant support over the past three years in the technical aspects of manuscript typesetting and figure creation. His positive comments greatly helped improve the manuscript. I also thank all my students for their support at various points. The Department of Mathematics, IIT

Roorkee, has provided a peaceful atmosphere for writing this book, for which I am thankful. Thanks are due to Ankush Kumar, Karan Gupta, Simran Meena, and other editorial members at Cambridge University Press, for their patience and positive interactions throughout the writing process. Finally, I thank my wife, Anuradha, and daughter, Sindhuja, for their understanding and support, and I extend special thanks to my granddaughter, Krishika, for being a great source of happiness.

Any comments and corrections from the readers of this book are most welcome.

1
Introduction

A collection or network of objects or entities is called a **system** if they are interconnected or can interact to produce outputs or results according to given inputs. **Dynamical system** is a natural or man-made system that evolves with time. In other words, the instantaneous description of the system, known as the **state** of the system, changes with respect to time. **Control systems** are those dynamical systems that can be controlled using suitable inputs, called **control variables**. So, a control system can be considered a particular type of dynamical system.

The motion of a planet can be described as a dynamical system, while the motion of a vehicle driven by a driver can be described as a control system. Dynamical systems, whether natural or man-made, are present in every facet of our day-to-day life. The solar system, wind motion, epidemics, population dynamics, economy, human metabolism, missiles, and automobiles are some examples of dynamical or control systems.

Mathematical modeling is a procedure for translating a real-world problem into a mathematical form and obtaining information about the problem using the solution of the mathematical model. If a proper procedure is followed in formulating a mathematical model, one can expect reliable information about the real-life problem from the model.

In Section 1.1, general mathematical forms of continuous-time and discrete-time dynamical systems are introduced. Simple mathematical models of population dynamics and prey–predator population models are presented in Section 1.2. Mathematical models of some mechanical systems, namely simple harmonic motion and the simple pendulum, are discussed in Section 1.3. The concentration dynamics of a fluid while mixing a substance into it is discussed in Section 1.4. A brief discussion on mathematical models of decay of mass of a substance and temperature changes in a substance are given in Sections 1.5 and 1.6, respectively. In Section 1.7, the inverted pendulum model is discussed in detail.

1.1 Dynamical Systems

In this section, continuous-time and discrete-time dynamical systems are introduced.

Continuous-time Dynamical Systems

The mathematical formulation of a continuous-time dynamical system can be expressed as the following system of first order ordinary differential equations:

$$\begin{cases} \dfrac{dx_1(t)}{dt} = f_1(t, x_1(t), x_2(t), \ldots, x_n(t)), \\ \dfrac{dx_2(t)}{dt} = f_2(t, x_1(t), x_2(t), \ldots, x_n(t)), \\ \vdots \\ \dfrac{dx_n(t)}{dt} = f_n(t, x_1(t), x_2(t), \ldots, x_n(t)), \end{cases} \tag{1.1}$$

where $(x_1(t), x_2(t), \ldots, x_n(t))$ represents the **state** of the system at time t, $0 \leq t_0 \leq t \leq \tau < \infty$, and $f_i(t, x_1(t), x_2(t), \ldots, x_n(t)) : i = 1, 2, \ldots, n$ are real valued functions.

If one (or more) of the function(s) $f_i : i = 1, 2, \ldots, n$ is (are) nonlinear in the state variables, then the system (1.1) is called a **nonlinear dynamical system**. If all $f_i : i = 1, 2, \ldots, n$ are linear functions of the state variables, then the system (1.1) is called a **linear dynamical system**.

If $f_i : i = 1, 2, \ldots, n$ are functions only of the state variables (t does not appear explicitly in f_i), then (1.1) is called an **autonomous** or a **time-invariant system**. If any f_i is an explicit function of t, then (1.1) is called a **nonautonomous** or **time-varying system**.

The dynamical system (1.1) can be written in the vector form as

$$\frac{dx(t)}{dt} = f(t, x(t)), \tag{1.2}$$

where $x(t) = \begin{bmatrix} x_1(t) \\ x_2(t) \\ \vdots \\ x_n(t) \end{bmatrix} \in \mathbb{R}^n$ for each $t \in [t_0, \tau] =: J$ and $f : J \times \mathbb{R}^n \to \mathbb{R}^n$ is a vector function given by $f(t,x(t)) = \begin{bmatrix} f_1(t, x_1(t), x_2(t), \ldots, x_n(t)) \\ f_2(t, x_1(t), x_2(t), \ldots, x_n(t)) \\ \vdots \\ f_n(t, x_1(t), x_2(t), \ldots, x_n(t)) \end{bmatrix}$.

Here, $\mathbb{R}$ denotes the set of all real numbers, and $\mathbb{R}^n$ denotes the set of all vectors with n real components, i.e., $\mathbb{R}^n = \{(a_1, a_2, \ldots, a_n)' : a_i \in \mathbb{R}, i = 1, 2, \ldots, n\}$, where $'$ represents the transpose of a vector.

A linear dynamical system is of the form

$$\frac{dx_i}{dt} = a_{i1}(t)x_1 + a_{i2}(t)x_2 + \cdots + a_{in}(t)x_n + b_i(t), \quad i = 1, 2, \ldots, n. \tag{1.3}$$

In vector form, the linear system (1.3) can be written as

$$\frac{dx}{dt} = A(t)x + b(t),$$

where

$$A(t) = \begin{bmatrix} a_{11}(t) & a_{12}(t) & \cdots & a_{1n}(t) \\ a_{21}(t) & a_{22}(t) & \cdots & a_{2n}(t) \\ \vdots & & & \\ a_{n1}(t) & a_{n2}(t) & \cdots & a_{nn}(t) \end{bmatrix} \text{ and } b(t) = \begin{bmatrix} b_1(t) \\ b_2(t) \\ \vdots \\ b_n(t) \end{bmatrix}.$$

***Example* 1.1.1** (i) The following differential equation

$$\frac{dx}{dt} = 3x, \quad 0 \le t < \infty$$

is a linear autonomous system with one state variable x.

(ii) The following is a nonlinear autonomous system with two state variables x_1 and x_2

$$\frac{dx_1}{dt} = 2x_1 - x_1x_2,$$
$$\frac{dx_2}{dt} = -x_2 + x_1x_2.$$

(iii) A nonautonomous nonlinear system is given below

$$\frac{dx_1}{dt} = 3x_2,$$
$$\frac{dx_2}{dt} = tx_1 - x_1x_2^2.$$

Continuous-time Control System

A continuous-time control system can be represented in the form

$$\frac{dx_i}{dt} = f_i(t, x_1, x_2, \ldots, x_n, u_1, u_2, \ldots, u_m), \quad i = 1, 2, \ldots, n, \tag{1.4}$$

where $x_i : i = 1, 2, \ldots, n$ are state variables and $u_i : i = 1, 2, \ldots, m$ are **control variables.**

In vector form, the control system (1.4) can be written as

$$\frac{dx(t)}{dt} = f(t, x(t), u(t)), \tag{1.5}$$

where $x(t) \in \mathbb{R}^n$ and $u(t) \in \mathbb{R}^m$ for each $t \in J$, and $f : J \times \mathbb{R}^n \times \mathbb{R}^m \to \mathbb{R}^n$ is a nonlinear map. If the function $f(t, x, u)$ is linear in x and u, then the system (1.5) is called a linear control system. For example,

$$\frac{dx(t)}{dt} = A(t)x(t) + B(t)u(t)$$

is a linear time-varying control system, where

$$A(t) = \begin{bmatrix} a_{11}(t) & a_{12}(t) & \cdots & a_{1n}(t) \\ a_{21}(t) & a_{22}(t) & \cdots & a_{2n}(t) \\ \vdots & & & \\ a_{n1}(t) & a_{n2}(t) & \cdots & a_{nn}(t) \end{bmatrix} \text{ and } B(t) = \begin{bmatrix} b_{11}(t) & b_{12}(t) & \cdots & b_{1m}(t) \\ b_{21}(t) & b_{22}(t) & \cdots & b_{2m}(t) \\ \vdots & & & \\ b_{n1}(t) & b_{n2}(t) & \cdots & b_{nm}(t) \end{bmatrix}.$$

If A and B are constant matrices, then the above system is called a linear time-invariant (autonomous) control system.

Converting a Differential Equation into a Dynamical System

Consider an nth order ordinary differential equation

$$a_0(t)\frac{d^n x}{dt^n} + a_1(t)\frac{d^{n-1}x}{dt^{n-1}} + \cdots + a_n(t)x = b(t), \quad t \in J, \tag{1.6}$$

where $a_i(t) : i = 0,1,2,\ldots,n$ and $b(t)$ are real valued functions and $a_0(t) \neq 0$ for $t \in J$.

The above equation can be expressed as a system of first order differential equations by defining state variables as

$$x_1 = x,\ x_2 = \frac{dx}{dt},\ x_3 = \frac{d^2x}{dt^2},\cdots,\ x_n = \frac{d^{n-1}x}{dt^{n-1}}.$$

Hereafter, we shall denote $\frac{dx}{dt}$ by $\dot{x}$. Then

$$\begin{cases} \dot{x}_1 = x_2, \\ \dot{x}_2 = x_3, \\ \vdots \\ \dot{x}_{n-1} = x_n, \\ \dot{x}_n = -\dfrac{a_n(t)}{a_0(t)}x_1 - \dfrac{a_{n-1}(t)}{a_0(t)}x_2 - \cdots - \dfrac{a_1(t)}{a_0(t)}x_n + \dfrac{b(t)}{a_0(t)} \end{cases}$$

represents a dynamical system corresponding to (1.6). In vector form, the above system can be written as

$$\dot{x} = A(t)x + B(t),$$

where

$$x = \begin{bmatrix} x_1 \\ x_2 \\ \vdots \\ x_n \end{bmatrix},\ A(t) = \begin{bmatrix} 0 & 1 & 0 & \cdots & 0 \\ 0 & 0 & 1 & \cdots & 0 \\ \vdots & & & & \\ 0 & 0 & 0 & \cdots & 1 \\ -\frac{a_n(t)}{a_0(t)} & -\frac{a_{n-1}(t)}{a_0(t)} & -\frac{a_{n-2}(t)}{a_0(t)} & \cdots & -\frac{a_1(t)}{a_0(t)} \end{bmatrix},\ \text{and } B(t) = \begin{bmatrix} 0 \\ 0 \\ \vdots \\ \frac{b(t)}{a_0(t)} \end{bmatrix}.$$

Discrete-time Dynamical Systems

If the state $x(t)$ of a dynamical system is considered at $N+1$ discrete-time instances, say, $k_0, k_0+1, k_0+2, \ldots, k_0+N$, then the discrete-time system can be expressed as

$$x_i(k+1) = f_i\left(k, x_1(k), x_2(k), \ldots, x_n(k)\right), \quad i = 1,2,\ldots,n,$$

$$k = k_0, k_0+1, k_0+2, \ldots, k_0+N-1.$$

A discrete-time control system is of the form

$$x_i(k+1) = f_i\left(k, x_1(k), x_2(k), \ldots, x_n(k), u_1(k), u_2(k), \ldots, u_m(k)\right), \quad i = 1, 2, \ldots, n,$$
$$k = k_0, k_0+1, k_0+2, \ldots, k_0+N-1.$$

Analogous to continuous-time systems, the above system can be written in vector form as

$$x(k+1) = f\left(k, x(k), u(k)\right),$$

where $x(k) \in \mathbb{R}^n$ and $u(k) \in \mathbb{R}^m$ for each $k \in S$, and $f : S \times \mathbb{R}^n \times \mathbb{R}^m \to \mathbb{R}^n$ with $S = \{k_0, k_0+1, k_0+2, \ldots, k_0+N\}$.

***Example* 1.1.2** The following linear discrete-time control system

$$\begin{cases} x_1(k+1) = 2x_1(k) + x_2(k), \\ x_2(k+1) = -x_1(k) + 3x_2(k) + u(k) \end{cases}$$

can be written as $x(k+1) = Ax(k) + Bu(k)$, $k = 0, 1, 2, \ldots$, where

$$x(k) = \begin{bmatrix} x_1(k) \\ x_2(k) \end{bmatrix}, \; A = \begin{bmatrix} 2 & 1 \\ -1 & 3 \end{bmatrix}, \; \text{and } B = \begin{bmatrix} 0 \\ 1 \end{bmatrix}.$$

In the following sections (Sections 1.2–1.7), mathematical models representing various real-life problems, such as population dynamics, motion of a mass, oscillation of a simple pendulum, dynamics of an inverted pendulum, the concentration of a substance in a mixture, etc., are presented. To model these problems mathematically, certain ideal conditions are assumed. Therefore, these models can be thought of only as a preliminary step toward a more accurate mathematical modeling of real-life problems.

1.2 Population Dynamics

Continuous-time Model

Let $x(t)$ be the population size of a species at time t. If b and d denote the rates of birth and death, respectively, of the species, then, by assuming that the population changes at a rate proportional to its current size, one can write a simple mathematical model as

$$\begin{cases} \dot{x} = (b-d)x = \alpha x \text{ (say)}, \quad t > t_0, \\ x(t_0) = x_0, \end{cases} \tag{1.7}$$

where x_0 is the population size at an initial time t_0.

The above model yields $x(t) = e^{\alpha(t-t_0)}x_0$. If $\alpha > 0$, then $x(t)$ increases exponentially as t increases, and if $\alpha < 0$, then $x(t) \to 0$. Thus, the model is not realistic for long time durations.

Discrete-time Model

Let $x(k)$ be the population size of a species at time instant k. If the change in the population size during the period $[k, k+1)$ is proportional to $x(k)$, then

$$x(k+1) - x(k) = \alpha(k)x(k), \tag{1.8}$$

where $\alpha(k) = b(k) - d(k)$ is the proportionality constant during $[k, k+1)$. Here $b(k)$ and $d(k)$ are the birth rate and death rate, respectively, during the period $[k, k+1)$. $\alpha(k)$ can be positive or negative, depending on whether the population size increases or decreases. From (1.8), the discrete-time model can be written as

$$x(k+1) = \beta(k)x(k), \quad k = k_0, k_0+1, k_0+2, \ldots,$$

where $\beta(k) = 1 + \alpha(k)$.

A more realistic model can be obtained using appropriate birth and death rates, as shown in the **logistic model** in the next subsection.

Logistic Model

If a population increases beyond a certain level, then due to overcrowding and limitations of resources, the birth rate decreases and the death rate increases proportionally to the population size x. This leads to the following expressions for birth and death rates

$$\begin{aligned} b &= b_1 - b_2 x, \\ d &= d_1 + d_2 x, \quad b_1, b_2, d_1, d_2 > 0. \end{aligned}$$

Equation (1.7) becomes

$$\begin{cases} \dot{x} = x((b_1 - d_1) - (b_2 + d_2)x) = x(a - cx), \quad a > 0, \quad c > 0, \quad t > t_0, \\ x(t_0) = x_0, \end{cases} \tag{1.9}$$

where $a = b_1 - d_1$ and $c = b_2 + d_2$

In the discrete-time case, a similar equation can be obtained as shown below.

In (1.8), let $b(k) = b_1 - b_2 x(k)$ and $d(k) = d_1 + d_2 x(k)$. Then we obtain the following logistic model.

$$\begin{cases} x(k+1) = x(k)\,(\alpha - \beta x(k))\,, & k = k_0, k_0+1, k_0+2, \ldots, \\ x(k_0) = x_0, \end{cases} \tag{1.10}$$

where $\alpha = 1 + (b_1 - d_1)$ and $\beta = b_2 + d_2$.

Prey–Predator Model

Let us consider a system with two species: prey and predator. Let $x(t)$ and $y(t)$ denote the number of preys and predators at time t, respectively. The interaction between these two species is described by the following system of differential equations. The rate of change of the prey population is given by

$$\frac{d}{dt}x(t) = \alpha x(t) - \beta x(t)y(t), \quad \alpha > 0, \beta > 0, \tag{1.11}$$

where the term $\alpha x(t)$ represents the growth of the prey population, and the term $-\beta x(t)y(t)$ represents the decrease in the prey population due to predation. The loss due to predation is proportional to both the number of predators and prey.

For the predator, the rate of change of its population is given by

$$\frac{d}{dt}y(t) = -\gamma y(t) + \delta x(t)y(t), \quad \gamma > 0, \delta > 0, \tag{1.12}$$

where $-\gamma y(t)$ is the effect of mortality in the absence of prey, and $\delta x(t)y(t)$ is the growth from predation. Now, (1.11) and (1.12) can be expressed as

$$\begin{cases} \dot{x} = x(\alpha - \beta y), \\ \dot{y} = y(-\gamma + \delta x). \end{cases}$$

These equations are known as the Lotka–Volterra system. Analysis of its solution will provide insight into how predator and prey populations fluctuate over a period of time. Note that the system is nonlinear and autonomous. In Chapter 8, the solution and stability analysis of this model will be discussed.

1.3 Mechanical Systems

A particle of mass m moves in a straight line on a smooth horizontal plane (without friction) under the influence of a force f. If $x(t)$ is the distance of m from a fixed point O at time t, then, according to *Newton's law of motion* (mass $\times$ acceleration $=$ force), the equation of motion of the particle is $m\ddot{x} = f$ (see Figure 1.1).

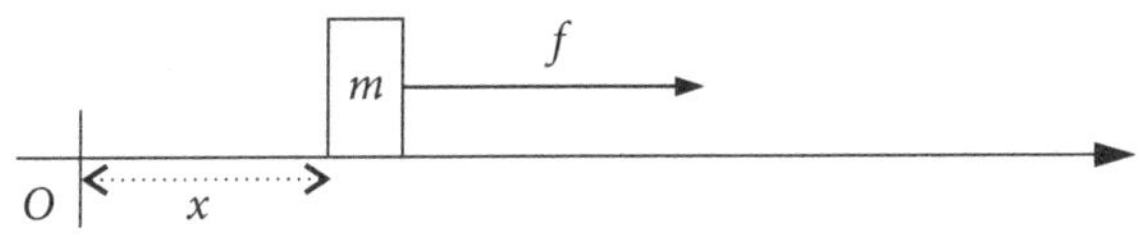

Figure 1.1 Object moving in a straight line

If friction force is present, it will act on the body in the direction opposite to the velocity. Hence, friction can be modeled as $-\alpha\dot{x}$, where $\alpha > 0$ is a constant. Thus, the equation of motion can be written as $m\ddot{x} = f - \alpha\dot{x}$.

By defining $x_1 = x$ and $x_2 = \dot{x}$, the above equation can be written in the form of a control system

$$\dot{x} = Ax + Bu, \tag{1.13}$$

where

$$x = \begin{bmatrix} x_1 \\ x_2 \end{bmatrix}, A = \begin{bmatrix} 0 & 1 \\ 0 & -\frac{\alpha}{m} \end{bmatrix}, B = \begin{bmatrix} 0 \\ \frac{1}{m} \end{bmatrix}, \text{ and } u = f.$$

In practical situations, it may be required to determine a control function $f(t)$ (force) such that $x(t)$ (position and velocity) satisfies preassigned initial and final conditions, i.e., $x(t_0) = x_0$ and $x(t_1) = x_1$. This leads to the question of **controllability** of the system (1.13), which will be discussed in Chapter 5.

Simple Harmonic Motion

A particle is said to be in simple harmonic motion if it oscillates about a mean position O along a straight line under the influence of a force that acts toward O and is proportional to the distance from O. The mathematical equation describing the simple harmonic motion can be obtained from *Newton's law of motion*.

Let O represent the origin on the real axis (Figure 1.2). If $x(t)$ is the displacement of the mass m from O at time t under a force f, then the equation of motion is $m\ddot{x} = f = -\alpha x(t)$, where $\alpha > 0$. Let $x(0) = l$ and $\dot{x}(0) = 0$, where $l > 0$ is the initial displacement.

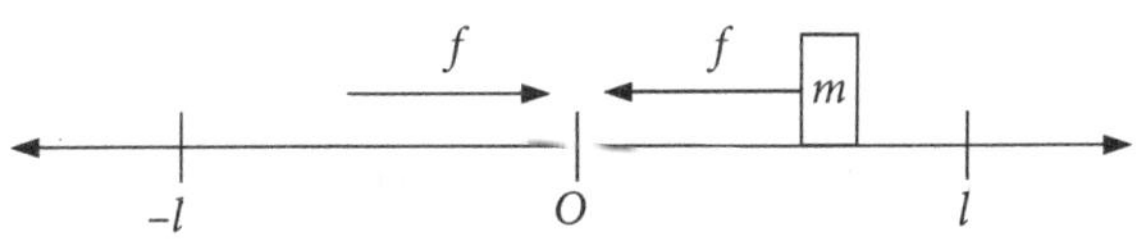

Figure 1.2 Simple harmonic motion

The negative sign is due to the fact that f is always directed toward the origin O. If a damping force proportional to the velocity is present in the system, then the equation becomes $m\ddot{x} = -\alpha x - \beta\dot{x}$, where $\beta > 0$.

By defining $x_1 = x$ and $x_2 = \dot{x}$, the above equation can be written as

$$\begin{cases} \dot{X} = AX, \\ X(0) = X_0, \end{cases}$$

where

$$X = \begin{bmatrix} x_1 \\ x_2 \end{bmatrix}, A = \begin{bmatrix} 0 & 1 \\ -\frac{\alpha}{m} & -\frac{\beta}{m} \end{bmatrix}, \text{ and } X_0 = \begin{bmatrix} l \\ 0 \end{bmatrix}.$$

Simple Pendulum

Consider a simple pendulum OA hanging from O, with a mass m attached at the end point A, as shown in Figure 1.3. Assume that the mass of the portion OA of the pendulum is negligible compared to m.

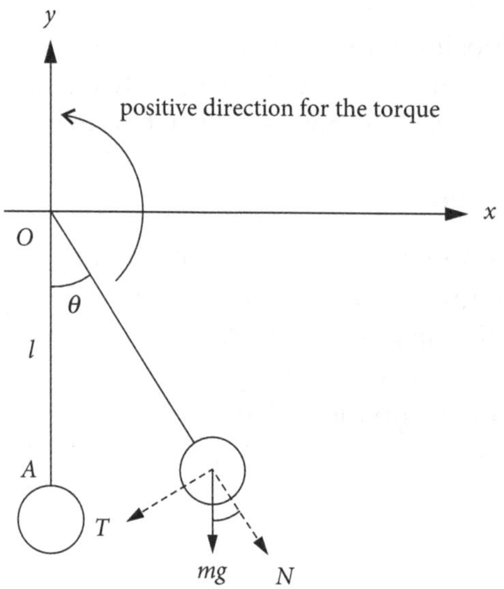

Figure 1.3 Controlled pendulum

Let O be the origin of the xy-plane, as shown in Figure 1.3. Let $OA = l$. During the motion, since gravity is the only force acting on m, the tangential component T and the normal component N of the force mg at A are $mg \sin\theta$ and $mg \cos\theta$, repetitively.

The torque at O due to the tangential force at A is $-lmg\sin\theta$. The negative sign is due to the following fact: the torque will be positive or negative depending on whether it induces a rotation in the anti-clockwise direction (from x to y) or in the clockwise direction (from y to x).

The inertia of the mass m with respect to O is ml^2. According to *Newton's law for rotating motion* (inertia $\times$ angular acceleration $=$ torque), the equation of motion of the simple pendulum $ml^2\ddot{\theta} = -mgl\sin\theta$. In case damping forces (such as air resistance) are included in the model, the mathematical model becomes

$$ml^2\ddot{\theta} = -mgl\sin\theta - k\dot{\theta}, \tag{1.14}$$

where k is a suitable positive constant. Air resistance acts in the opposite direction of the velocity of the mass. Hence, it is modeled as $-k\dot{\theta}$. Note that (1.14) is a nonlinear equation.

By defining state variables $x_1 = \theta$ and $x_2 = \dot{\theta}$, the above equation can be written as $\dot{x} = f(x)$, where

$$x = \begin{bmatrix} x_1 \\ x_2 \end{bmatrix} \text{ and } f(x) = \begin{bmatrix} x_2 \\ -\frac{g}{l}\sin x_1 - \frac{k}{ml^2}x_2 \end{bmatrix}.$$

If we assume θ to be small, we can replace $\sin x_1$ by x_1. Then, the system becomes linear, and $f(x) = Ax$, where

$$A = \begin{bmatrix} 0 & 1 \\ -\frac{g}{l} & -\frac{k}{ml^2} \end{bmatrix}.$$

A detailed discussion on the **stability analysis** of the above system can be found in Chapter 8.

1.4 Mixing of Substances

When a soluble substance is mixed into a liquid continuously, its concentration in the fluid changes continuously. Finding the concentration at any given time is essential in certain practical problems. As concentration is the amount of substance present in unit volume of the liquid, one can determine the concentration by finding the amount of substance in the entire volume of the liquid. The problem is illustrated with two examples below.

***Example* 1.4.1 (Mixing in one tank)** Consider a tank containing 100 liters of salt water, with an initial amount of salt in the mixture being 2 kg. The tank has an inlet and an outlet, as shown in Figure 1.4. If 4 liters per second of salt water containing 100 grams of salt per liter is flowing in, and 4 liters per second of the mixture is flowing out of the tank, write a mathematical model for determining the amount of salt in the tank at any time $t \in [0, \tau]$.

Solution Let $x(t)$ be the amount of salt at time t.

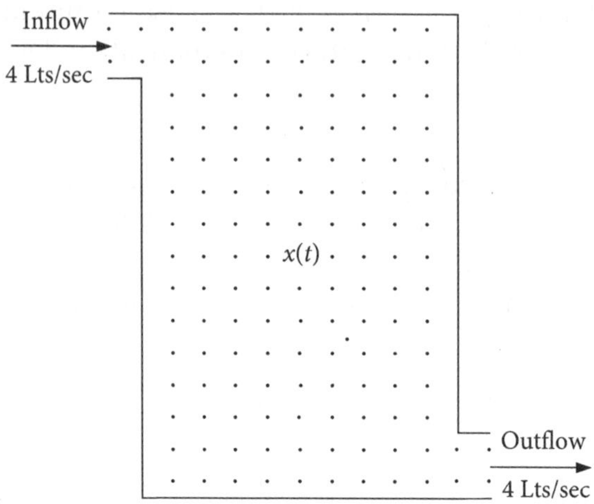

Figure 1.4 A tank with inflow and outflow

It is given that $x(0) = 2$. The inflow adds 400 grams of salt per second, and the outflow removes $\frac{4}{100}x(t) = \frac{1}{25}x(t)$ amount of salt per second. So, $\dot{x}$, the rate of change of the amount of salt, is given by

$$\dot{x}(t) = 0.4 - \frac{x(t)}{25}, \quad 0 \le t \le \tau, \tag{1.15}$$

with the initial condition $x(0) = 2$. $x(t)$ can be obtained by solving the above model. Hence, the concentration of salt in the mixture at time t is $\frac{x(t)}{100}$.

Note that (1.15) is a continuous-time system, while the data given in the problem is in discrete-time (per second). Thus, the solution $x(t)$ of (1.15) can only be considered as an approximation to the amount of salt at any time t.

In the above problem, if the salt concentration in the inflow is a function of time, $u(t)$ grams per liter, then the mathematical model becomes a control system as follows

$$\begin{cases} \dot{x} = -\dfrac{x}{25} + 4u, & 0 \le t \le \tau, \\ x(0) = 2. \end{cases} \tag{1.16}$$

Here, $x(t)$ can be controlled by suitably choosing the control function $u(t)$.

A discrete-time model for the problem is

$$\begin{cases} x((k+1)h) = e^{-\frac{1}{25}h}x(kh) + 4u(kh), & k = 0,1,2,\dots,N-1, \\ x(0) = 2, \end{cases}$$

where $h = \frac{\tau}{N}$ is the partition length in the time interval $[0,\tau]$. The method for obtaining a discrete-time system corresponding to a continuous-time system is discussed in detail in Chapter 9.

***Example* 1.4.2 (Mixing in two tanks)** Each of the two tanks (Figure 1.5) contains 250 gallons of salt water, with 15 kg of salt in tank-1 and 25 kg of salt in tank-2, respectively. The inflow into tank-1 is 15 gal/min, with $(1+\cos t)$ kg/gal of salt present. The outflow from tank-2 is 15 gal/min. The mixtures are circulated between the tanks at a rate of 20 gal/min from tank-1 to tank-2 and 5 gal/min from tank-2 to tank-1. Find the amount of salt in each tank at any time t.

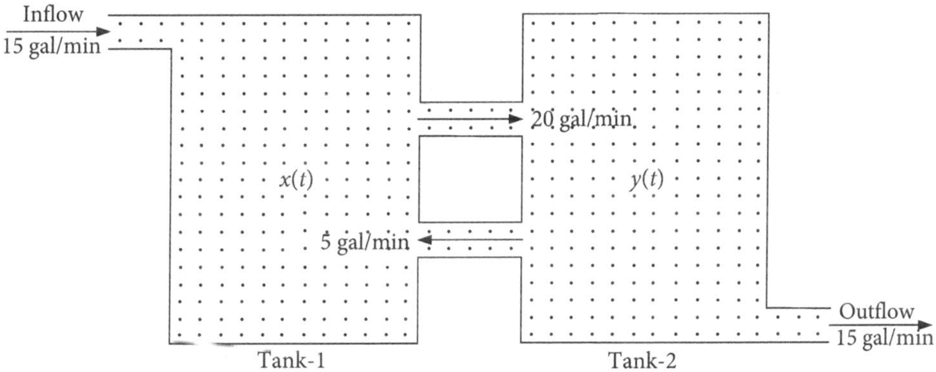

Figure 1.5 Mixing in two tanks

Solution Let $x_1(t)$ and $x_2(t)$ denote the amounts in kg of salt in tank-1 and tank-2, respectively, at time t. The rate at which salt is added in tank-1 from the inflow $= 15(1+\cos t)$, the rate at which salt is added in tank-1 from tank-2 $= 5 \times \frac{x_2(t)}{250} = \frac{x_2(t)}{50}$, and the rate at which salt is removed from tank-1 $= 20 \times \frac{x_1(t)}{250} = \frac{2}{25}x_1(t)$. Hence, the rate of change of the amount of salt in tank-1 is

$$\dot{x}_1 = 15(1+\cos t) + \frac{x_2}{50} - \frac{2x_1}{25}.$$

Similarly, the rate of change of the amount of salt in tank-2 is

$$\dot{x}_2 = \frac{2x_1}{25} - \frac{2x_2}{25}.$$

The initial conditions are $x_1(0) = 15$ and $x_2(0) = 25$. The above equations can be written in matrix form as $\dot{x} = Ax + Bu$ with $x(0) = x_0$, where $x = \begin{bmatrix} x_1 \\ x_2 \end{bmatrix}$, $A = \frac{1}{25}\begin{bmatrix} -1 & \frac{1}{2} \\ 2 & -2 \end{bmatrix}$, $B = \begin{bmatrix} 1 \\ 0 \end{bmatrix}$, and $u(t) = 15(1+\cos t)$.

1.5 Exponential Decay of Masses

Experiments show that radioactive substances decompose at a rate proportional to the present amount of the substance. For example, Lead-210 reduces to half its mass in 22 years. The time taken by a substance to loose half its mass is called the half-life of that substance.

A simple mathematical model of this physical process can be obtained by utilizing the given property of the natural phenomenon. We denote by $x(t)$ the amount of substance present at time t. The rate of change $\dot{x}$, according to the physical law governing the process of radiation, is proportional to x, i.e., $\dot{x} = \alpha x$, where α is a constant. Clearly, since the amount of substance is positive and decreases with time, $\dot{x}$ is negative and so is α. The constant α is a definite physical constant whose numerical value is known for various radioactive substances. α can be determined by measuring $x(t)$ at two instances of time, say at t_0 and t_1. Then, by solving the differential equation, we obtain

$$x(t_1) = e^{\alpha(t_1-t_0)}x(t_0) \implies \alpha = \frac{1}{t_1 - t_0}\log\frac{x(t_1)}{x(t_0)}.$$

If α is known for a substance, then its half-life t_2 can be determined as follows:

$$x(t_2) = \frac{x(t_0)}{2} \implies \frac{x(t_0)}{2} = e^{\alpha(t_2-t_0)}x(t_0).$$

Hence, $t_2 = t_0 + \frac{1}{\alpha}\log\frac{1}{2}$, which implies that the mass m at time t_0 will be reduced to $m/2$ in $\frac{1}{\alpha}\log\frac{1}{2}$ time units.

1.6 Propagation of Heat

Let $h(t)$ represent the temperature of a substance at time t, and let c be the constant room temperature. It is natural to assume that the rate of change of temperature of the substance is directly proportional to the difference between the current temperature $h(t)$ and the room temperature c. Therefore, a simple mathematical model can be written as

$$\dot{h} = \alpha\,(h(t) - c), \tag{1.17}$$

where α is a suitable constant (depending on the substance).

Let h_0 denote the temperature at the initial time t_0. Then the solution of (1.17) is given by

$$h(t) = c + (h_0 - c)e^{\alpha(t-t_0)}. \tag{1.18}$$

If the constant α in (1.17) is unknown, it can be determined from the values of $h(t)$ at two distinct instants t_0 and t_1. Let $h(t_1) = h_1$. Then, using (1.18), we obtain

$$\frac{h_1 - c}{h_0 - c} = e^{\alpha(t_1 - t_0)},$$

which yields

$$\alpha = \frac{1}{t_1 - t_0}\log\left(\frac{h_1 - c}{h_0 - c}\right).$$

If $h_0 > c$, then $h(t)$ will decrease for $t > t_0$ and stabilize at c; and if $h_0 < c$, then $h(t)$ will increase for $t > t_0$ and stabilize at c. Hence, in both the cases $\alpha < 0$ in equation (1.17).

Heat Control Problem

Instead of allowing the substance to stabilize at room temperature, if one is interested in controlling its temperature using an external source of heat that produces a rate of change of temperature $u(t)$, the model becomes

$$\begin{cases} \dot{h} = \alpha\,(h(t) - c) + u(t), \\ h(0) = h_0. \end{cases} \tag{1.18}$$

In a practical problem, it may be necessary to attain the temperature h_1 at time $t = \tau$. In other words, we aim to find a control function $u(t)$ such that the solution $h(t)$ of (1.18) satisfies $h(\tau) = h_1$.

In this problem, it will be easy to find at least one control function $u(t)$ such that the solution $h(t)$ of (1.18) satisfies $h(\tau) = h_1$. This can be achieved by directly solving (see (4.14) in Chapter 4) equation (1.18). For any $u(t)$, the solution of (1.18) is (see (4.14) in Chapter 4)

$$h(t) - c = e^{\alpha t}(h_0 - c) + \int_0^t e^{\alpha(t-s)}u(s)ds.$$

At time $t = \tau$, we get

$$e^{-\alpha\tau}\left[h(\tau) - c - e^{\alpha\tau}(h_0 - c)\right] = \int_0^\tau e^{-\alpha s}u(s)ds. \tag{1.19}$$

Since the LHS is a constant, let it be K. Then, one can choose $u(s) = \frac{e^{\alpha s}K}{\tau}$ to satisfy (1.19) among infinitely many possible control functions $u(s)$.

But, in a general control system with multiple state variables, finding such a control function is more complicated. This issue will be discussed when we introduce the concept of **controllability** in Chapter 5.

1.7 Inverted Pendulum Control

Consider an inverted pendulum *OA* (Figure 1.6) of mass m, which is mounted vertically on a wheeled cart. The point O of the pendulum is attached to the cart in such a way that *OA* can oscillate in the xy-plane while the cart moves back and forth along the x-direction under the influence of a force $f(t)$. The problem is to determine a suitable force function $f(t)$ (control) such that the pendulum can be maintained in the upright vertical position without falling.

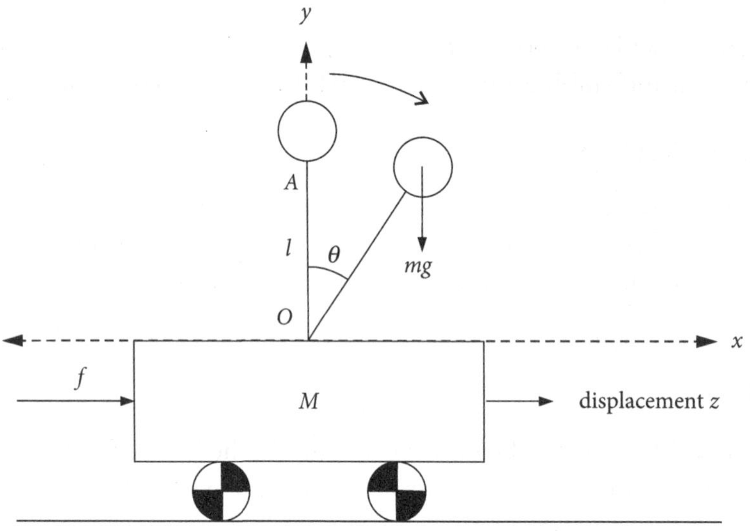

Figure 1.6 Inverted pendulum

We can derive a mathematical model for the problem under the following assumptions:

(i) The pendulum is a thin rod *OA* with a mass m attached at A. It is assumed that mass of the rod *OA* is negligible compared to m.

(*ii*) The wheeled cart is placed on a horizontal surface with no friction.

Let $OA = L$ and the mass of the cart be M. Let z be the displacement of the cart in x direction due to the force $f(t)$. Let $p(t)$ and $q(t)$ be the horizontal (x-direction) and vertical (y-direction) forces, respectively, acting on the mass m at A (Figure 1.7).

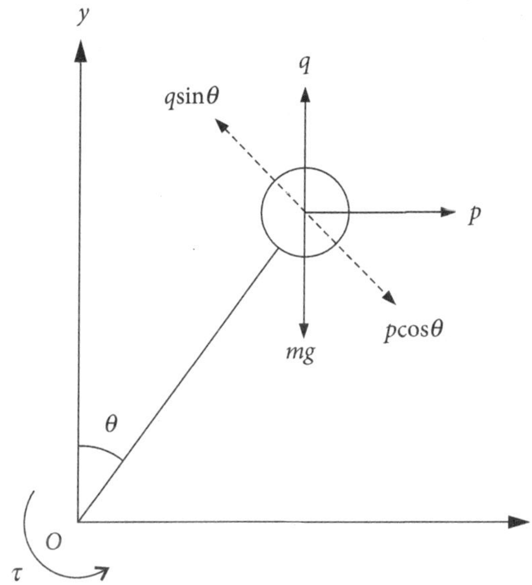

Figure 1.7 Forces acting on m

The displacement of A in x-direction due to the motion of the pendulum is $L\sin\theta$, and the displacement of A in the x-direction due to the motion of the cart is z. Hence,

$$p(t) = m\frac{d^2}{dt^2}(L\sin\theta) + m\ddot{z} = mL(\cos\theta \cdot \ddot{\theta} - \sin\theta \cdot \dot{\theta}^2) + m\ddot{z}. \tag{1.20}$$

The displacement of A in the vertical direction is $-(L - L\cos\theta)$. Therefore,

$$q(t) = -mg + m\frac{d^2}{dt^2}(L\cos\theta - L) = mg - mL(\sin\theta \cdot \ddot{\theta} + \cos\theta \cdot \dot{\theta}^2). \tag{1.21}$$

The total torque at the point O due to the forces $p(t)$ and $q(t)$ is

$$\tau = Lq(t)\sin\theta - Lp(t)\cos\theta. \tag{1.22}$$

According to *Newton's law of rotational motion* (inertia $\times$ angular acceleration $=$ total torque),

$$mL^2\ddot{\theta} = -Lp\cos\theta + Lq\sin\theta. \tag{1.23}$$

As the force $f(t)$ compensates for the forces in the x-direction, we have

$$f(t) = p + M\ddot{z}. \tag{1.24}$$

From equations (1.20)–(1.23), we get the equation of motion of the system as

$$\begin{aligned} mL^2\ddot{\theta} &= q(t)L\sin\theta - p(t)L\cos\theta \\ &= L\sin\theta\left[-mg - mL(\sin\theta\cdot\ddot{\theta} + \cos\theta\cdot\dot{\theta}^2)\right] \\ &\quad - L\cos\theta\left[mL(\cos\theta\cdot\ddot{\theta} - \sin\theta\cdot\dot{\theta}^2) + m\ddot{z}\right] \\ \implies mL\ddot{\theta} &= -mg\sin\theta - mL\sin^2\theta\cdot\ddot{\theta} - mL\sin\theta\cos\theta\cdot\dot{\theta}^2 \\ &\quad - mL\cos^2\theta\cdot\ddot{\theta} + mL\cos\theta\sin\theta\cdot\dot{\theta}^2 - m\cos\theta\cdot\ddot{z} \\ \implies 2mL\ddot{\theta} &= -mg\sin\theta - m\cos\theta\cdot\ddot{z}. \end{aligned} \tag{1.25}$$

From equations (1.20) and (1.24), we obtain

$$f(t) = mL(\cos\theta\cdot\ddot{\theta} - \sin\theta\cdot\dot{\theta}^2) + m\ddot{z} + M\ddot{z}. \tag{1.26}$$

Equations (1.25) and (1.26) together provide a mathematical model for the motion of the inverted pendulum.

As $\theta(t)$ and $\dot{\theta}(t)$ are expected to be very small during the motion, we can approximate $\sin\theta \approx \theta$, $\cos\theta \approx 1$, $\dot{\theta}^2 \approx 0$. Hence, equations (1.25) and (1.26), respectively, become $2mL\ddot{\theta} = -mg\theta - m\ddot{z}$ and $mL\ddot{\theta} + (m+M)\ddot{z} = f$. On solving the above two equations for $\ddot{\theta}$ and $\ddot{z}$, for particular values of the parameters, say $m = 1$, $L = 1, M = 9$, and $g = 10$, we get

$$\begin{aligned} \ddot{\theta} &= -\frac{100}{19}\theta - \frac{1}{19}f, \\ \ddot{z} &= \frac{2}{19}f + \frac{10}{19}\theta. \end{aligned}$$

By defining new variables (states of the system) as $x_1 = \theta, x_2 = \dot{\theta}, x_3 = z$, and $x_4 = \dot{z}$, we obtain the control system as

$$\begin{bmatrix} \dot{x}_1 \\ \dot{x}_2 \\ \dot{x}_3 \\ \dot{x}_4 \end{bmatrix} = \begin{bmatrix} 0 & 1 & 0 & 0 \\ -\frac{100}{19} & 0 & 1 & 0 \\ 0 & 0 & 0 & 1 \\ \frac{10}{19} & 0 & 0 & 0 \end{bmatrix} \begin{bmatrix} x_1 \\ x_2 \\ x_3 \\ x_4 \end{bmatrix} + \begin{bmatrix} 0 \\ -\frac{1}{19} \\ 0 \\ \frac{2}{19} \end{bmatrix} f. \tag{1.27}$$

Equation (1.27) represents a linear autonomous control system in which $f(t)$ is the control function.

In this problem, it is required to find a stabilizing control force $f(t)$ such that $x_1(t)$ and $x_2(t)$ remain very close to 0, ensuring that pendulum is kept in the upward vertical position. Such problems will be discussed in Chapter 8, which focuses on the stability and stabilizability of dynamical systems.

In a practical situation, it is possible to measure (observe) the angle $x_1(t)$, from which we can define an observation function $Y(t)$ as $Y(t) = x_1(t)$. One can ask the following question: Using this observation, is it possible to obtain information about all the state variables of the system uniquely? This important question leads to the concept of **observability** of a system, which will be discussed in detail in Chapter 5.

1.8 Exercise

1. At time $t = 0$, the temperature of a cup of tea is 75°C. After two minutes, the temperature drops to 70°C. If the room temperature is 30°C, find the time t at which the temperature becomes 60°C. (Assume that the rate of change in temperature of the tea is proportional to the difference between its current temperature and the room temperature.)
2. In the year 2022, an archaeologist excavates a wooden object, measures the radioactive carbon content, and finds that it contains 30% of the carbon present in living wood (initial value). After one year, the carbon content of the same object is found to be 29.6% of the initial amount. Find the approximate year in which the object was made. (Assume that the carbon radiation follows a simple decay model $\frac{dc}{dt} = -kc$.)
3. The mass of a radioactive substance was 100 g in the year 1990, and it decreased to 99 g by the year 2024. Find the half-life of the substance.
4. An object of mass m moves along a straight line on a horizontal plane under the continuous influence of a force $u(t) = 15 - t$. Assume that the friction is $1/10$th of the velocity. If $x(t)$ and $\dot{x}(t)$, respectively, denote the displacement and velocity of the object at time t, with initial conditions $x(0) = 0$ and $\dot{x}(0) = 0$, then write a mathematical model for the motion of the object and find $x(t)$ and $\dot{x}(t)$ at $t = 30$.
5. A vessel contains a volume V of a solution with concentration c_0 at time $t = 0$ (concentration is defined as the amount of the substance present in unit volume of the solution). A solution with concentration c_1 enters the vessel from an overhead tank at a constant rate R and leaves the vessel at the same rate. Write a mathematical model for the concentration $c(t)$ for $t > 0$ and derive the expression for $c(t)$.
6. At time $t = 0$, each of the three tanks contains 1000 liters of salt water, in which 15, 20, and 25 kgs of salt, respectively, are dissolved in tank-1, tank-2, and tank-

3. The inflow of pure water into tank-1 is 15 lt/sec, the outflow from tank-2 is 10 lt/sec, and the outflow from tank-3 is 5 lt/sec. The mixtures in the three tanks are circulated as follows: 20 lt/sec from tank-1 to tank-2 and 5 lt/sec from tank-2 to tank-1; 10 lt/sec from tank-2 to tank-3 and 5 lt/sec from tank-3 to tank-2. Write a mathematical model for the amount of salt $x(t), y(t)$, and $z(t)$ in tanks-1, 2, and 3, respectively, at any time t.

7. Convert the difference equation $x(k+3) - 2x(k+2) + 4x(k+1) - x(k) = u(k)$ into a discrete-time dynamical system.

8. Let $x(t)$ and $y(t)$ denote the populations of deer and tigers at time t (in months), respectively. In the absence of tigers, the deer population increases at a rate of 2 percent per month, and the predators consume prey at a rate of 5 percent of the prey population per 100 predators. In the absence of deer, the tiger population decreases at a rate of 3 percent per month. In the presence of deer, the tiger population increases at a rate of 1 percent of its population per 100 prey. Write a mathematical model with $x(t)$ and $y(t)$ as state variables.

9. Let $x_1(t)$ and $x_2(t)$ denote the populations of two species, S_1 and S_2, respectively, which depend on a common prey S_3, whose population is denoted by $x_3(t)$. Assuming that S_1 and S_2 are not prey to each other, write a mathematical model in the form of a system of ordinary differential equations for $x_i : i = 1, 2, 3$.

10. Let A and B be two vehicles moving on an xy-plane. At time $t = 0$, A is at $(x_0, 0)$ and B is at the origin $(0, 0)$. Assume that B moves along the y-axis with a constant velocity v and A always moves toward B with a constant velocity u. Write a differential equation describing the motion of A.

2

Basics from Matrix Theory

2.1 Introduction

Results from matrix theory and linear algebra play an important role in the theory of finite dimensional control systems. This chapter presents key results from matrix theory that are essential for understanding control theory concepts.

In Section 2.2, linear dependence and independence of vectors in $\mathbb{R}^n$ and basis and dimension of a subspace are discussed. In Section 2.3, elementary row operation, row reduced echelon form, range and rank of a matrix, and inverse and determinant of a square matrix are presented. Section 2.4 is on solving a system of linear algebraic equations. In Section 2.5, null space and nullity are defined. Eigenvalues and eigenvectors of a matrix and related results are discussed in Section 2.6. Canonical forms (diagonal and Jordan forms) are explained in Section 2.7. Function of a matrix and computation of exponential of a matrix are presented in Section 2.8. In Section 2.9, special matrices and their properties are discussed.

2.2 The $\mathbb{R}^n$ and $\mathbb{C}^n$ Spaces

Let $\mathbb{R}$ and $\mathbb{C}$ be the set of all real and complex numbers, respectively. The n-dimensional Euclidean space, denoted by $\mathbb{R}^n$, is defined as $\mathbb{R}^n = \{(x_1, x_2, \ldots, x_n) : x_i \in \mathbb{R} : i = 1, 2, \ldots, n\}$. Each element of $\mathbb{R}^n$ can be considered as a vector with n-components, and $\mathbb{R}^n$ is called the n-dimensional vector space over $\mathbb{R}$ (see Definition 2.2.12).

Note that the sum of two vectors is a vector, i.e., if $x = (x_1, x_2, \ldots, x_n) \in \mathbb{R}^n$ and $y = (y_1, y_2, \ldots, y_n) \in \mathbb{R}^n$, then $x + y = (x_1 + y_1, x_2 + y_2, \ldots, x_n + y_n) \in \mathbb{R}^n$, and the product of a vector with a real number is also a vector in $\mathbb{R}^n$, i.e., if $\alpha \in \mathbb{R}$, $x = (x_1, x_2, \ldots, x_n) \in \mathbb{R}^n$, then $\alpha x = (\alpha x_1, \alpha x_2, \ldots, \alpha x_n) \in \mathbb{R}^n$.

If $\alpha_i : i = 1, 2, \ldots, k$ are scalars (real numbers) and $v_i : i = 1, 2, \ldots, k$ are vectors from $\mathbb{R}^n$, then the **sum** $\alpha_1 v_1 + \alpha_2 v_2 + \cdots + \alpha_k v_k$ is a vector in $\mathbb{R}^n$.

We denote by Θ the zero vector $(0, 0, \ldots, 0) \in \mathbb{R}^n$.

A subset S of $\mathbb{R}^n$ is called a **subspace** of $\mathbb{R}^n$ if (i) for all $x, y \in S$, $x + y \in S$, and (ii) for all $\alpha \in \mathbb{R}$ and $x \in S$, $\alpha x \in S$. The empty set and the whole space $\mathbb{R}^n$ are trivial subspaces of $\mathbb{R}^n$.

The space $\mathbb{C}^n$ is defined as $\mathbb{C}^n = \{(x_1, x_2, \ldots, x_n) : x_i \in \mathbb{C}, i = 1, 2, \ldots, n\}$. $\mathbb{C}^n$ is an n-dimensional vector space over $\mathbb{C}$ and is a $2n$-dimensional vector space over $\mathbb{R}$.

***Example* 2.2.1** In $\mathbb{R}^2$, $S = \{(x_1, x_2) : 3x_1 + 2x_2 = 0\}$ is a subspace because (i) $(x_1, x_2) \in S$ and $(y_1, y_2) \in S \implies (x_1 + y_1, x_2 + y_2) \in S$, as $3(x_1 + y_1) + 2(x_2 + y_2) = 0$, and (ii) $\alpha \in \mathbb{R}$ and $(x_1, x_2) \in S \implies \alpha(x_1, x_2) \in S$, as $3(\alpha x_1) + 2(\alpha x_2) = 0$.

***Example* 2.2.2** $S = \{(x_1, x_2) : 3x_1 + 2x_2 = 1\}$ is not a subspace R^2 because $(1, -1) \in S$ but $2(1, -1) \notin S$.

Linear Dependence of Vectors

Definition 2.2.3 Let $\{v_1, v_2, \ldots, v_k\} \subset \mathbb{R}^n$ and $\alpha_1, \alpha_2, \ldots, \alpha_k \in \mathbb{R}$. Then $\alpha_1 v_1 + \alpha_2 v_2 + \cdots + \alpha_k v_k$ is called a **linear combination** of the vectors $v_i : i = 1, 2, \ldots, k$.

Definition 2.2.4 A set of vectors $\{v_1, v_2, \ldots, v_k\} \subset \mathbb{R}^n$ is said to be **linearly dependent** if a linear combination of them yields the zero vector for some nonzero scalars, i.e., if

$$\alpha_1 v_1 + \alpha_2 v_2 + \cdots + \alpha_k v_k = \Theta \tag{2.1}$$

for at least one of the scalars α_i is nonzero.

From the above definition, it is clear that

- any set containing the zero vector is linearly dependent, as $\alpha\Theta = \Theta$ for any nonzero scalar α.
- if a set of nonzero vectors is linearly dependent, then at least one of them can be written as a linear combination of the remaining vectors.

Therefore, if $\{v_1, v_2, \ldots, v_k\}$ is linearly dependent, then $\alpha_1 v_1 + \alpha_2 v_2 + \cdots + \alpha_k v_k = \Theta$ for some nonzero scalars α_i. In this combination, if $\alpha_1 \neq 0$, say, then

$$v_1 = -\frac{\alpha_2}{\alpha_1} v_2 - \frac{\alpha_3}{\alpha_1} v_3 - \cdots - \frac{\alpha_k}{\alpha_1} v_k.$$

Similarly if $\alpha_r \neq 0$, then v_r can be written as a linear combination of $v_i : i = 1, 2, \cdots, r-1, r+1, \cdots, k$.

Linear Independence of Vectors

If a set of vectors $\{v_1, v_2, \ldots, v_k\}$ is not linearly dependent, then it is called **linearly independent**. This means that

$$\alpha_1 v_1 + \alpha_2 v_2 + \cdots + \alpha_k v_k = \Theta \tag{2.2}$$

if and only if **all** the scalars $\alpha_i : i = 1, 2, \ldots, k$ are zero. So, if a set $\{v_1, v_2, \ldots, v_k\}$ is linearly independent, then none of the vectors v_i can be expressed as a linear combination of the remaining vectors.

***Example* 2.2.5** Check if the set of vectors $\{(1,1,0), (1,2,1), (3,5,2)\}$ is linearly independent or dependent.

According to the definition, the set is linearly dependent if there exists at least one nonzero scalar $\alpha_i : i = 1, 2, 3$ such that

$$\alpha_1(1,1,0) + \alpha_2(1,2,1) + \alpha_3(3,5,2) = (0,0,0).$$

Equating the LHS and RHS element-wise, we obtain

$$\begin{cases} \alpha_1 + \alpha_2 + 3\alpha_3 = 0, \\ \alpha_1 + 2\alpha_2 + 5\alpha_3 = 0, \\ \alpha_2 + 2\alpha_3 = 0. \end{cases}$$

Solving the above system of equations, we find $\alpha_2 = -2\alpha_3$ and $\alpha_1 = -\alpha_3$. Taking $\alpha_3 = 1$, we get $\alpha_1 = -1$ and $\alpha_2 = -2$. Hence, the given set is linearly dependent. Since all the coefficients are nonzero, any vector in the given set can be expressed as a linear combination of the other two.

***Example* 2.2.6** The set $\{(1,0), (2,0), (3,1)\}$ is clearly linearly dependent. A linear combination $\alpha_1 v_1 + \alpha_2 v_2 + \alpha_3 v_3 = \Theta$ gives $\alpha_1 = -2\alpha_2$ and $\alpha_3 = 0$. As $\alpha_3 = 0$, v_3 cannot be expressed as a linear combination of v_1 and v_2.

Maximal Linearly Independent Subset

Let $A = \{v_1, v_2, \ldots, v_k\}$ be a linearly dependent set of nonzero vectors. By solving

$$\alpha_1 v_1 + \alpha_2 v_2 + \cdots + \alpha_k v_k = \Theta,$$

one can obtain some nonzero scalars. If $\alpha_r \neq 0$, say, then v_r can be expressed in terms of the remaining vectors. Removing the vector v_r from the set A, we obtain the set

$$A_1 = \{v_1, v_2, \ldots, v_{r-1}, v_{r+1}, \ldots, v_k\}.$$

If A_1 is linearly dependent, then by following the same procedure, we can remove one vector v_l from A_1 to obtain the set

$$A_2 = \{v_1, v_2, \ldots, v_{r-1}, v_{r+1}, \ldots, v_{l-1}, v_{l+1}, \ldots, v_k\}.$$

If A_2 is linearly independent, then A_2 is called a maximal linearly independent subset of A. If A_2 is linearly dependent, we can proceed further and remove one vector at a time. After a few steps, we will arrive at a set that is linearly independent, which is called a **maximal linearly independent set**.

Note that from a linearly dependent set of nonzero vectors, we may obtain more than one maximal linearly independent set. For a given set, **all maximal linearly independent sets contain the same number of elements.** (Prove!)

Example 2.2.7 $A = \{(1,0),(2,0),(1,1)\}$ is linearly dependent in $\mathbb{R}^2$. Here, both $\{(1,0),(1,1)\}$ and $\{(2,0),(1,1)\}$ are maximal linearly independent sets.

Span of Vectors $\{v_1, v_2, \ldots, v_k\} \subset \mathbb{R}^n$

The set S of all linear combinations of $v_1, v_2, \ldots, v_k$ is called the **span** of these vectors and is defined as

$$S = \{\alpha_1 v_1 + \alpha_2 v_2 + \cdots + \alpha_k v_k : \alpha_i \in \mathbb{R}, i = 1, 2, \ldots, k\}.$$

Note that the span S is a **subspace** of $\mathbb{R}^n$ and the set $v_1, v_2, \cdots, v_k$ is called a generator of S.

Example 2.2.8 In $\mathbb{R}^3$, the xy-plane consists of vectors of the form $(x,y,0)$, which can be written as $x(1,0,0)+y(0,1,0)$. Therefore, the set $\{(1,0,0),(0,1,0)\}$ spans the xy-plane. Note that the xy-plane is also spanned (generated) by the set $\{(1,0,0),(1,1,0)\}$.

Example 2.2.9 In $\mathbb{R}^3$, the subspace $S = \{(x,y,z) : 2x - y + z = 0\}$ is spanned by $\{(1,2,0),(0,1,1)\}$ because if $(x,y,z) \in S$, then $y = 2x + z$, and $(x,y,z) = (x, 2x+z, z) = x(1,2,0) + z(0,1,1)$, where x and z are arbitrary real numbers.

Lemma 2.2.10 Let $B_1 = \{v_1, v_2, \cdots, v_m\}$ and $B_2 = \{w_1, w_2, \cdots, w_k\}$ be sets in a subspace S. If B_1 is a generator of S and B_2 is linearly independent, then $k \leq m$.

Proof. Since B_1 spans S and $w_1 \in S$, the set $C_1 = \{w_1, v_1, \ldots, v_m\}$ also spans S and it is linearly dependent. If $\alpha_1 w_1 + \beta_1 v_1 + \beta_2 v_2 + \cdots. + \beta_m v_m = \theta$, then at least one β_i must be nonzero (If all β_i are zeros then α_1 must be nonzero. Hence w_1 is zero vector which is a contradiction.) Let $\beta_r \neq 0$. Then v_r can be eliminated from the set C_1 to obtain the set $D_1 = \{w_1, v_1, \ldots, v_{r-1}, v_{r+1}, \ldots, v_m\}$ which also spans S. Now consider the set $C_2 = \{w_2, w_1, v_1, \ldots, v_{r-1}, v_{r+1}, \ldots, v_m\}$. It is clear that this set is linearly dependent and spans S. If we equate to θ a linear combination of vectors in C_2 then at least one of the coefficients of $v_i's$ must be nonzero (if not then w_1, w_2 become linearly dependent, which is a contradiction). Let $\beta_l \neq 0$. Then v_l can be eliminated from the set C_2 to obtain the set $D_2 = \{w_2, w_1, v_1, \ldots, v_{r-1}, v_{r+1}, \ldots, v_{l-1}, v_{l+1}, \ldots, v_m\}$ which also spans S. Now consider the set $C_3 = \{w_3, w_2, w_1, v_1, \ldots, v_{r-1}, v_{r+1}, \ldots, v_{l-1}, v_{l+1}, \ldots, v_m\}$ which is linearly dependent and generates S. Proceeding as above we obtain sets $D_1, D_2, \ldots, D_h$, respectively, by replacing one vector of B_1 by one vector of B_2 in the sets $C_1, C_2, \ldots, C_h$. At every step these sets span S. This procedure will be stopped when B_2 is exhausted or there is no v_i available in D_h. Now three cases arise: (i) If $h = k$ then D_h contains all vectors of B_2 along with some remaining vectors of B_1. In this case $k < m$. (ii) If $h = k$ and $D_h = B_2$ then $k = m$. (iii) $h < k$ and there is no vector v_i in D_h. This is not possible because this will imply that D_h contains h elements of B_2 and spans the remaining vectors of B_2. Since B_2 is linearly independent, case(iii) is not possible. Hence $k \leq m$. □

Basis and Dimension

A subspace S is said to be generated (spanned) by a set $\mathscr{B} = \{v_1, v_2, \ldots, v_m\}$ if the span of the set $\mathscr{B}$ is S. Here, $\mathscr{B}$ may or may not be linearly independent.

Definition 2.2.11 If a set $\mathscr{B} = \{v_1, v_2, \ldots, v_m\}$ is linearly independent and spans S, then $\mathscr{B}$ is called a **basis** of S.

It can be shown that any two bases of S have the same number of vectors. Let S_1 and S_2 be two bases of S, containing m and k elements, respectively. In Lemma 2.2.10, replace B_1 and B_2 by S_1 and S_2, respectively. Then $k \leq m$. Now, replace B_1 and B_2 by S_2 and S_1 respectively. Then $m \leq k$. Hence $m = k$.

Definition 2.2.12 The **dimension** of a subspace S is defined as the number of vectors in a basis of S.

***Example* 2.2.13** The set $\{e_1, e_2, \ldots, e_n\}$ forms a basis of the space $\mathbb{R}^n$, where $e_i = (0, 0, \ldots, 1, 0, \ldots, 0)$ with 1 in the ith position and 0 in all other coordinates. This set $\{e_1, e_2, \ldots, e_n\}$ is called the **standard basis** of $\mathbb{R}^n$.

***Example* 2.2.14** The set $B = \{(1,1,0), (1,0,0), (0,1,0)\}$ spans the xy-plane in $\mathbb{R}^3$, as $(x, y, 0)$ can be expressed as $(x, y, 0) = \alpha(1,1,0) + (x - \alpha)(1,0,0) + (y - \alpha)(0,1,0)$, where α is any arbitrary real number. Note that the set B is linearly dependent, and the maximal linearly independent subset $\{(1,0,0), (0,1,0)\}$ also spans the xy-plane in $\mathbb{R}^3$. Other maximal subsets $\{(1,1,0), (1,0,0)\}$ and $\{(1,1,0),\ (0,1,0)\}$ also span the xy-plane, since any element $(a, b, 0)$ on the xy-plane can be expressed as $(a, b, 0) = \alpha(1,1,0) + \beta(1,0,0)$, where $\alpha = b$ and $\beta = a - b$ or $(a, b, 0) = \gamma(1,1,0) + \delta(0,1,0)$, where $\gamma = a$ and $\delta = b - a$.

A maximal linearly independent set in a subspace is a basis for that subspace. In Example 2.2.14, the *xy*-plane is a subspace of $\mathbb{R}^3$, and the maximal linearly independent sets are the bases of the *xy*-plane among its infinitely many possible bases.

Exercise 2.2.1 Let A be a real matrix of order n and $x \in \mathbb{R}^n$ be such that the set $S = \{x, Ax, \ldots, A^{k-1}x\}$ is linearly independent. If $A^k x$ is linearly dependent on the vectors in S, show that $A^r x$ for all $r > k$ is linearly dependent on the vectors in S.

Exercise 2.2.2 Show that $\mathbb{C}^n$ is an n-dimensional vector space over $\mathbb{C}$ and is a $2n$-dimensional vector space over $\mathbb{R}$.

2.3 Range and Rank of a Matrix

Definition 2.3.1 The **column rank** of a matrix A is the number of linearly independent columns of A. The **row rank** of a matrix A is the number of linearly independent rows of A.

In the next section, we shall show that for any matrix A, the column rank $=$ the row rank. This number is called the **rank of** A. An equivalent definition of rank is given below using the range of A. Let

$$A = \begin{bmatrix} a_{11} & a_{12} & \cdots & a_{1n} \\ a_{21} & a_{22} & \cdots & a_{2n} \\ \vdots & & & \\ a_{m1} & a_{m2} & \cdots & a_{mn} \end{bmatrix}$$

be an $m \times n$ matrix, which can be denoted by $\{a_{ij}\}_{i=1,2,\ldots,m;j=1,2,\ldots,n}$. For each

$$x = \begin{bmatrix} x_1 \\ x_2 \\ \vdots \\ x_n \end{bmatrix} \in \mathbb{R}^n,$$

$$Ax = \begin{bmatrix} a_{11}x_1 + a_{12}x_2 + \cdots + a_{1n}x_n \\ a_{21}x_1 + a_{22}x_2 + \cdots + a_{2n}x_n \\ \vdots \\ a_{m1}x_1 + a_{m2}x_2 + \cdots + a_{mn}x_n \end{bmatrix} = \begin{bmatrix} y_1 \\ y_2 \\ \vdots \\ y_m \end{bmatrix} = y \in \mathbb{R}^m. \tag{2.3}$$

So, the matrix A can be considered as a mapping from $\mathbb{R}^n$ to $\mathbb{R}^m$, which can be represented as $A : \mathbb{R}^n \to \mathbb{R}^m$. The element $y \in \mathbb{R}^m$ is called the image of $x \in \mathbb{R}^n$.

Definition 2.3.2 (Range of a matrix) The **range** of a matrix A, denoted by Range(A), is the set of all vectors in $\mathbb{R}^m$ that are images of vectors in $\mathbb{R}^n$ under the mapping defined by A. That is,

$$\text{Range}(A) = \{y \in \mathbb{R}^m : y = Ax, x \in \mathbb{R}^n\}.$$

Now, we show that the range of A is generated by its columns. Let $x = \begin{bmatrix} x_1 \\ x_2 \\ \vdots \\ x_n \end{bmatrix}$ be an arbitrary element in $\mathbb{R}^n$. Then from (2.3),

$$Ax = \begin{bmatrix} a_{11}x_1 + a_{12}x_2 + \cdots + a_{1n}x_n \\ a_{21}x_1 + a_{22}x_2 + \cdots + a_{2n}x_n \\ \vdots \\ a_{m1}x_1 + a_{m2}x_2 + \cdots + a_{mn}x_n \end{bmatrix} = y \in \mathbb{R}^m,$$

which can be written as

$$= x_1 \begin{bmatrix} a_{11} \\ a_{21} \\ \vdots \\ a_{m1} \end{bmatrix} + x_2 \begin{bmatrix} a_{12} \\ a_{22} \\ \vdots \\ a_{m2} \end{bmatrix} + \cdots + x_n \begin{bmatrix} a_{1n} \\ a_{2n} \\ \vdots \\ a_{mn} \end{bmatrix}$$

$$= x_1c_1 + x_2c_2 + \cdots + x_nc_n, \tag{2.4}$$

where $c_i : i = 1, 2, \ldots, n$ are the columns of A.

From the above, it is clear that the range of A is a subspace of $\mathbb{R}^m$, spanned by the columns of A. Any maximal linearly independent subset of $\{c_1, c_2, \ldots, c_n\}$ forms a basis for the range of A.

Definition 2.3.3 (Rank of a matrix) The **rank** of a matrix A is the dimension of its range.

From the above discussion, it can be concluded that rank(A) = the maximum number of linearly independent columns of A. Later, we will show that this number also equals the number of linearly independent rows of A.

Elementary Row Operations on a Matrix

Let A be an $m \times n$ matrix with rows $r_1, r_2, \ldots, r_m$ which can be considered as vectors in $\mathbb{R}^n$. Consider the following three operations on the rows of A:

(i) interchanging two rows (vectors),

(ii) multiplying a row (vector) by a nonzero scalar,

(iii) adding any two rows (vectors).

Theorem 2.3.4 Let $S = \{r_1, r_2, \ldots, r_m\}$ be a set of vectors in $\mathbb{R}^n$ (the rows of A). Let k be the maximum number of linearly independent vectors in S. If we perform any of the elementary operations on the set S, the resulting set, say S_1, will also have the same number of linearly independent vectors.

Proof. (i) Let k (where $k \leq m$) be the maximum number of linearly independent vectors in S. Interchanging any two vectors in S does not affect the set, except for the order of the vectors. Hence, the maximum number of linearly independent vectors remains the same.

(ii) Let $S = \{r_1, r_2, \ldots, r_m\}$. Multiply r_1 by a nonzero scalar α. Let $S_1 = \{\alpha r_1, r_2, \ldots, r_m\}$. If $x \in$ span of S, then

$$
\begin{aligned}
x &= \beta_1 r_1 + \beta_2 r_2 + \cdots + \beta_m r_m \quad \text{for some scalars } \beta_i, \\
&= \frac{\beta_1}{\alpha}(\alpha r_1) + \beta_2 r_2 + \cdots + \beta_m r_m \\
&= \gamma_1 (\alpha r_1) + \beta_2 r_2 + \cdots + \beta_m r_m,
\end{aligned}
$$

where $\gamma_1 = \frac{\beta_1}{\alpha}$. This implies that both S and S_1 span the same subspace and hence the maximum number of linearly independent vectors in S and S_1 are the same. In the above proof, r_1 can be replaced by any $r_i : i = 1, 2, \ldots, m$.

(*iii*) Let $S = \{r_1, r_2, \dots, r_m\}$. Add the first and second vectors to form a new set $S_1 = \{r_1 + r_2, r_2, \dots, r_m\}$. Let $x \in$ span of S and $x = \beta_1 r_1 + \beta_2 r_2 + \cdots + \beta_m r_m$ for some scalars β_i. Then

$$\begin{aligned} x &= \beta_1 (r_1 + r_2) + (\beta_2 - \beta_1) r_2 + \beta_3 r_3 + \cdots + \beta_m r_m \\ &= \gamma_1 (r_1 + r_2) + \gamma_2 r_2 + \beta_3 r_3 + \cdots + \beta_m r_m, \end{aligned}$$

where $\gamma_1 = \beta_1$ and $\gamma_2 = \beta_2 - \beta_1$. Thus, $x \in$ the span of S_1, which implies that span of S is a subspace of span of S_1. Similarly it can be shown that span of S_1 is a subspace of S. Hence, S and S_1 span the same subspace and have the same number of linearly independent vectors. For the general case, r_1 and r_2 can be replaced by any two vectors in S.

From the above results, it is clear that the maximum number of linearly independent rows of a matrix will not change if we perform row operations on the matrix. Hence, the row rank of a matrix remains the same after row operations. □

The next theorem shows that the column rank of a matrix remains the same after **row operations**.

Theorem 2.3.5 Let A be an $m \times n$ matrix and A_1 be the matrix obtained from A by performing an elementary row operation. Let $\{c_1, c_2, \dots, c_n\}$ and $\{d_1, d_2, \dots, d_n\}$ be the columns of A and A_1, respectively. Then, the maximum number of linearly independent columns of A and A_1 are equal, i.e., the column rank of $A =$ the column rank of A_1.

Proof. Let $A = \{c_{ij}\}_{i=1,2,\dots,m; j=1,2,\dots,n}$ and $c_j = \begin{bmatrix} c_{1j} \\ c_{2j} \\ \vdots \\ c_{mj} \end{bmatrix}$ be the jth column of A. Let $A_1 = \{d_{ij}\}_{i=1,2,\dots,m; j=1,2,\dots,n}$ and $d_j = \begin{bmatrix} d_{1j} \\ d_{2j} \\ \vdots \\ d_{mj} \end{bmatrix}$ be the jth column of A_1.

(i) If A_1 is obtained from A by interchanging the first and second rows, then $d_j = \begin{bmatrix} c_{2j} \\ c_{1j} \\ c_{3j} \\ \vdots \\ c_{mj} \end{bmatrix}$ for $j = 1,2,\ldots,n$. If $v = \begin{bmatrix} x_1 \\ x_2 \\ \vdots \\ x_m \end{bmatrix}$ is obtained by a linear combination $\alpha_1 c_1 + \alpha_2 c_2 + \cdots + \alpha_n c_n$, then $v_1 = \begin{bmatrix} x_2 \\ x_1 \\ \vdots \\ x_m \end{bmatrix}$ is obtained by $\alpha_1 d_1 + \alpha_2 d_2 + \cdots + \alpha_n d_n$ for the same set of scalars $\alpha_1, \alpha_2, \ldots, \alpha_n$, which shows that the sets $\{c_1, c_2, \ldots, c_n\}$ and $\{d_1, d_2, \ldots, d_n\}$ are either linearly independent or dependent together. Hence, the maximum number of linearly independent vectors in both sets are the same, and hence, the column ranks of A and A_1 are equal.

(ii) If the first row of A is multiplied by a nonzero constant α, then $d_j = \begin{bmatrix} \alpha c_{1j} \\ c_{2j} \\ \vdots \\ c_{mj} \end{bmatrix}$ for $\alpha \neq 0$ and $j = 1,2,\ldots,n$. If $v = \begin{bmatrix} x_1 \\ x_2 \\ \vdots \\ x_m \end{bmatrix} = \alpha_1 c_1 + \alpha_2 c_2 + \cdots + \alpha_n c_n$ for some scalars, then $v_1 = \begin{bmatrix} \alpha x_1 \\ x_2 \\ \vdots \\ x_m \end{bmatrix} = \alpha_1 d_1 + \alpha_2 d_2 + \cdots + \alpha_n d_n$ for the same set of scalars. Hence, the column rank of $A =$ the column rank of A_1. A similar argument can be extended to the third row operation. □

Row Reduced Echelon Form

Definition 2.3.6 A matrix $E = \{a_{ij}\}_{i=1,2,\ldots,m; j=1,2,\ldots,n}$ is said to be in **row reduced echelon form** if it satisfies the following:

(i) In each row, the first nonzero element is 1.

(ii) If a_{ij} is the first nonzero element in the ith row and $a_{i+1,l}$ is the first nonzero element in the $(i+1)^{\text{th}}$ row, then $j < l$.

(iii) If a_{ij} is the first nonzero entry in the ith row, then all elements except a_{ij} in the jth column are zero.

(iv) Zero rows (rows having only zeros), if any, appear below all nonzero rows.

Definition 2.3.7 If a matrix satisfies the conditions (ii) and (iv) in the definition above, then it is said to be in **row echelon form.**

Example **2.3.8** The following matrices are in row reduced echelon form

$$\begin{bmatrix} 1 & 0 & 0 & 0 \\ 0 & 1 & 0 & 0 \\ 0 & 0 & 1 & 2 \end{bmatrix}_{3\times 4} \quad \text{and} \quad \begin{bmatrix} 1 & 2 & 0 & 0 & 7 \\ 0 & 0 & 1 & 0 & 0 \\ 0 & 0 & 0 & 1 & 5 \\ 0 & 0 & 0 & 0 & 0 \end{bmatrix}_{4\times 5}.$$

Example **2.3.9** The following matrix is in row echelon form

$$\begin{bmatrix} 4 & 3 & 0 & 3 & 0 \\ 0 & 0 & 1 & 0 & 7 \\ 0 & 0 & 0 & 2 & 5 \\ 0 & 0 & 0 & 0 & 1 \end{bmatrix}_{4\times 5}.$$

Note that in the above examples, we can easily find the rank of these echelon matrices just by counting the number of nonzero rows. Using row operations, any given $m \times n$ matrix A can be transformed into a **unique row reduced echelon form** E. (Why?)

Theorem 2.3.10 For any matrix, the row rank = the column rank.

Proof. Since the row reduced echelon form E is obtained from A by performing a series of row operations, from Theorem 2.3.4, one can conclude that row rank(A) = row rank(E). It follows from Theorem 2.3.5 that column rank(A) = column rank(E). From the structure of E, it can be easily seen that row rank(E) = column rank(E). Hence, row rank(A) = column rank(A), which is simply called the rank of A and is denoted by rank(A). □

Remark 2.3.11 From Theorems 2.3.4 and 2.3.5, we infer the following: If A is an $n \times n$ matrix with rank n, then the row reduced echelon form of A is the $n \times n$ identity matrix.

***Example* 2.3.12** Find the rank of the following matrix

$$A = \begin{bmatrix} 1 & 2 & -1 & 0 \\ 2 & 1 & 3 & -2 \\ 1 & 1 & 1 & -1 \\ -1 & 3 & 0 & -4 \end{bmatrix}.$$

Solution $R_i : i = 1,2,3,4$ denotes the ith row. The row operations $R_2 \to R_2 - 2R_1$, $R_3 \to R_3 - R_1$, and $R_4 \to R_4 + R_1$ yield

$$\begin{bmatrix} 1 & 2 & -1 & 0 \\ 0 & -3 & 5 & -2 \\ 0 & -1 & 2 & -1 \\ 0 & 5 & -1 & -4 \end{bmatrix}.$$

Interchanging R_2 and R_3, we get

$$\begin{bmatrix} 1 & 2 & -1 & 0 \\ 0 & -1 & 2 & -1 \\ 0 & -3 & 5 & -2 \\ 0 & 5 & -1 & -4 \end{bmatrix}.$$

Performing $R_3 \to R_3 - 3R_2$ and $R_4 \to R_4 + 5R_2$, we obtain

$$\begin{bmatrix} 1 & 2 & -1 & 0 \\ 0 & -1 & 2 & -1 \\ 0 & 0 & -1 & 1 \\ 0 & 0 & 9 & -9 \end{bmatrix}.$$

The operation $R_4 \to R_4 + 9R_3$ give the following row echelon form:

$$\begin{bmatrix} 1 & 2 & -1 & 0 \\ 0 & -1 & 2 & -1 \\ 0 & 0 & -1 & 1 \\ 0 & 0 & 0 & 0 \end{bmatrix}.$$

As the row echelon matrix above has three nonzero rows, $\text{rank}(A) = 3$.

The row reduced echelon form of the above matrix is

$$\begin{bmatrix} 1 & 0 & 0 & 0 \\ 0 & 1 & 0 & 0 \\ 0 & 0 & 1 & 0 \\ 0 & 0 & 0 & 0 \end{bmatrix}.$$

Finding Inverse Using Row Operators

Let A be an $m \times n$ matrix and I the $m \times m$ identity matrix. Let P_1 be the matrix obtained from I by interchanging the ith and j^{th} rows. Then $A_1 = P_1 A$ is the matrix obtained from A by interchanging its ith and j^{th} rows. Similarly, if P is a matrix obtained from I by performing any elementary row operation, then PA provides the matrix resulting from performing the same elementary row operation on A. For example, let $A = \begin{bmatrix} 2 & 1 & 4 & 0 \\ 0 & 3 & 5 & 1 \\ 1 & 5 & 7 & 2 \end{bmatrix}$. If we perform the following row operations on A: (i) interchange the rows R_2 and R_1, and (ii) replace the row R_3 by $2R_2+R_3$, we obtain $B = \begin{bmatrix} 0 & 3 & 5 & 1 \\ 2 & 1 & 4 & 0 \\ 5 & 7 & 15 & 2 \end{bmatrix}$.

Now, let us perform the same elementary row operations, (i) and (ii), on the 3×3 identity matrix. Then we obtain $P = \begin{bmatrix} 0 & 1 & 0 \\ 1 & 0 & 0 \\ 2 & 0 & 1 \end{bmatrix}$. It can now be verified that $PA = B$.

Thus, if P is the matrix obtained from I by performing k elementary row operations, namely $P_1, P_2, \ldots, P_k$, in that order, then $P = P_k P_{k-1} \cdots P_1 I$. Therefore, $PA = B$ is the matrix obtained from A by performing the elementary row operations on A in the same order.

Since the row reduced echelon form E of A is obtained by performing some elementary row operations Q_i (say) for $i = 1, 2, \ldots, l$ successively, we have

$$E = Q_l Q_{l-1} \cdots Q_1 A. \tag{2.5}$$

If A is an $n \times n$ matrix with rank n, then its row reduced echelon form is $I_{n\times n}$. Hence, from (2.5), we get $I = Q_l Q_{l-1} \cdots Q_1 A$. Letting $Q = Q_l Q_{l-1} \cdots Q_1$, we get $I = QA$, which implies that $Q = A^{-1}$. This procedure provides a method for computing the inverse of a matrix, if it exists. If we simultaneously perform the operations $Q_i : i = 1, 2, \cdots, l$ on A and I, then A is transformed to I and I is transformed to A^{-1}.

***Example* 2.3.13** Find the inverse of $A = \begin{bmatrix} 3 & 0 & 2 \\ 1 & 0 & 1 \\ 0 & 1 & 1 \end{bmatrix}$ using elementary row operations.

Solution Consider the augmented matrix $\left[\begin{array}{ccc|ccc} 3 & 0 & 2 & 1 & 0 & 0 \\ 1 & 0 & 1 & 0 & 1 & 0 \\ 0 & 1 & 1 & 0 & 0 & 1 \end{array}\right]$. Now

$$R_1 \to \frac{1}{3}R_1 \implies \left[\begin{array}{ccc|ccc} 1 & 0 & \frac{2}{3} & \frac{1}{3} & 0 & 0 \\ 1 & 0 & 1 & 0 & 1 & 0 \\ 0 & 1 & 1 & 0 & 0 & 1 \end{array}\right]$$

$$R_2 \to R_2 - R_1 \implies \left[\begin{array}{ccc|ccc} 1 & 0 & \frac{2}{3} & \frac{1}{3} & 0 & 0 \\ 0 & 0 & \frac{1}{3} & -\frac{1}{3} & 1 & 0 \\ 0 & 1 & 1 & 0 & 0 & 1 \end{array}\right]$$

$$R_2 \to R_3 \implies \left[\begin{array}{ccc|ccc} 1 & 0 & \frac{2}{3} & \frac{1}{3} & 0 & 0 \\ 0 & 1 & 1 & 0 & 0 & 1 \\ 0 & 0 & \frac{1}{3} & -\frac{1}{3} & 1 & 0 \end{array}\right]$$

$$R_3 \to 3R_3 \implies \left[\begin{array}{ccc|ccc} 1 & 0 & \frac{2}{3} & \frac{1}{3} & 0 & 0 \\ 0 & 1 & 1 & 0 & 0 & 1 \\ 0 & 0 & 1 & -1 & 3 & 0 \end{array}\right]$$

$$R_2 \to R_2 - R_3 \implies \left[\begin{array}{ccc|ccc} 1 & 0 & \frac{2}{3} & \frac{1}{3} & 0 & 0 \\ 0 & 1 & 0 & 1 & -3 & 1 \\ 0 & 0 & 1 & -1 & 3 & 0 \end{array}\right]$$

$$R_1 \to R_1 - \frac{2}{3}R_3 \implies \left[\begin{array}{ccc|ccc} 1 & 0 & 0 & 1 & -2 & 0 \\ 0 & 1 & 0 & 1 & -3 & 1 \\ 0 & 0 & 1 & -1 & 3 & 0 \end{array}\right].$$

Hence, $A^{-1} = \begin{bmatrix} 1 & -2 & 0 \\ 1 & -3 & 1 \\ -1 & 3 & 0 \end{bmatrix}.$

Determinant

Let $A = \{a_{ij}\}_{i,j=1,2,\ldots,n}$ be a **square** matrix of order n. The determinant of A is defined as

$$\det(A) = a_{i1}M_{i1} + a_{i2}M_{i2} + \cdots + a_{in}M_{in} \quad \text{for any } i \in \{1,2,\ldots,n\}, \tag{2.6}$$

where

$$M_{ij} = (-1)^{i+j} \text{ minor of } a_{ij},$$

where minor of a_{ij} = the determinant of the submatrix of A obtained by deleting the ith row and j^{th} column of A.

The formula (2.6) provides a method for finding the determinant of A using the ith row, $i = 1, 2, \ldots, n$. From the properties of the determinant of a matrix A of order n, it is known that:

(i) If i^{th} and k^{th} rows are interchanged to obtain a matrix B, then $\det(B) = (-1)^{i+k}\det(A)$.

(ii) If a nonzero scalar α is multiplied with each element of any row of A to obtain B, then $\det(B) = \alpha\det(A)$.

(iii) If the ith row of A is replaced by the sum of the ith and k^{th} rows to obtain B, then $\det(B) = \det(A)$.

From (i) and (ii), it follows that if any row of A is linearly dependent on another row(s), then $\det(A) = 0$. Hence,

$$a_{j1}M_{i1} + a_{j2}M_{i2} + \cdots + a_{jn}M_{in} = 0, \text{ if } i \neq j. \tag{2.7}$$

From (2.6) and (2.7), we obtain $A(\text{adj}(A))' = \det(A)I$, where $\text{adj}(A) = \{M_{ij}\}_{i,j=1,2,\ldots,n}$. This implies that $A^{-1} = \frac{(\text{adj}(A))'}{\det(A)}$ if $\det(A) \neq 0$.

From the above properties, one can infer that if $\det(A) \neq 0$, then after performing elementary row operations, the determinant of the resulting matrix remains nonzero. Since the row reduced echelon form E of A is obtained through elementary row operations, $\det(A)$ **and** $\det(E)$ **are both either zero or nonzero**. Thus, if $\text{rank}(A) = n$, then $\det(A) \neq 0$ because the row reduced echelon form of A is the identity matrix I, whose determinant is 1. If $\text{rank}(A) < n$, then $\det(A) = 0$ because $\det(E) = 0$.

From the above discussion, we can conclude the following:

Theorem 2.3.14 The following statements are equivalent for a matrix A of order n:

(i) $\text{rank}(A) = n$

(ii) $\det(A) \neq 0$

(iii) A^{-1} exists

Exercise 2.3.1 Let A be an $n \times n$ matrix and $M_{ij} = (-1)^{i+j}$minor of a_{ij} for $i, j = 1, 2, \ldots, n$. Prove that $a_{11}M_{11} + a_{12}M_{12} + \cdots + a_{1n}M_{1n} = a_{21}M_{21} + a_{22}M_{22} + \cdots + a_{2n}M_{2n}$.

Exercise 2.3.2 Using the row reduction method, find the determinant of the matrix $\begin{bmatrix} 1 & 0 & 5 \\ 2 & 4 & 6 \\ 6 & 1 & 3 \end{bmatrix}$

2.4 Solving System of Equations

A system of linear algebraic equations can be written as

$$\begin{cases} a_{11}x_1 + a_{12}x_2 + \cdots + a_{1n}x_n = b_1, \\ a_{21}x_1 + a_{22}x_2 + \cdots + a_{2n}x_n = b_2, \\ \vdots \\ a_{m1}x_1 + a_{m2}x_2 + \cdots + a_{mn}x_n = b_m. \end{cases} \tag{2.8}$$

The above system can be expressed in matrix form as

$$Ax = b, \tag{2.9}$$

where

$$A = \begin{bmatrix} a_{11} & a_{12} & \cdots & a_{1n} \\ a_{21} & a_{22} & \cdots & a_{2n} \\ \vdots & & & \\ a_{m1} & a_{m2} & \cdots & a_{mn} \end{bmatrix}, x = \begin{bmatrix} x_1 \\ x_2 \\ \vdots \\ x_n \end{bmatrix} \in \mathbb{R}^n \text{ and } b = \begin{bmatrix} b_1 \\ b_2 \\ \vdots \\ b_m \end{bmatrix} \in \mathbb{R}^m.$$

Hence, A is a mapping from $\mathbb{R}^n$ to $\mathbb{R}^m$. The problem of solving the system (2.8) is equivalent to finding the vector x, given A and b in (2.9). If $m = n$ and A is nonsingular, i.e., A^{-1} exists, then from (2.9), we get

$$x = A^{-1}b.$$

If A is either singular or $m \neq n$, then we encounter three cases for the equation (2.9):

(*i*) there exists a unique solution,

(*ii*) there exist infinitely many solutions,

(*iii*) no solution.

Since the system of equations (2.8) is represented in the form $Ax = b$ as in (2.9), we observe that if the vector b is in the range of A, then there exists a solution $x \in \mathbb{R}^n$ (as b is the image of x). If b is not in the range of A, then equation (2.9) has no solution. Since the Range(A) is generated by the columns of A, equation (2.9) has a solution if $b \in \mathbb{R}^m$ is a linear combination of the columns of A.

If we denote by (A,b) the $m \times (n+1)$ augmented matrix obtained by appending b as the $(n+1)^{\text{th}}$ column to A, then the system (2.9) has a solution if

$$\text{rank}(A) = \text{rank}(A,b), \tag{2.10}$$

since b is linearly dependent on the columns of A.

By converting (A,b) to its row echelon form, condition (2.10) can be verified and the solution (if it exists) can be obtained. From the echelon form, it can also be seen that system (2.9) has

(*i*) infinitely many solutions if $\text{rank}(A) = \text{rank}(A,b) < n$,

(*ii*) a unique solution if $\text{rank}(A) = \text{rank}(A,b) = n$,

(*iii*) no solution if $\text{rank}(A) < \text{rank}(A,b)$.

***Example* 2.4.1** Solve the following system of linear equations

$$\begin{cases} 3x - y + 2z + 5p = -1, \\ 2x + y + z + p = 8, \\ 4x - 3y - 4z - 4p = -2. \end{cases}$$

Solution The augmented matrix of the system is

$$(A,b) = \begin{bmatrix} 3 & -1 & 2 & 5 & -1 \\ 2 & 1 & 1 & 1 & 8 \\ 4 & -3 & -4 & -4 & -2 \end{bmatrix}.$$

By performing row operations successively, we obtain the following matrices

$$(A,b) \sim \begin{bmatrix} 1 & -\frac{1}{3} & \frac{2}{3} & \frac{5}{3} & -\frac{1}{3} \\ 1 & \frac{1}{2} & \frac{1}{2} & \frac{1}{2} & 4 \\ 1 & -\frac{3}{4} & -1 & -1 & -\frac{2}{4} \end{bmatrix} \qquad (R_1 \to \frac{1}{3}R_1, R_2 \to \frac{1}{2}R_2, R_3 \to \frac{1}{4}R_3)$$

$$\sim \begin{bmatrix} 1 & -\frac{1}{3} & \frac{2}{3} & \frac{5}{3} & -\frac{1}{3} \\ 0 & \frac{5}{6} & -\frac{1}{6} & -\frac{7}{6} & \frac{13}{3} \\ 0 & -\frac{5}{12} & -\frac{5}{3} & -\frac{8}{3} & -\frac{1}{6} \end{bmatrix} \quad (R_2 \to R_2 - R_1, R_3 \to R_3 - R_1)$$

$$\sim \begin{bmatrix} 1 & -\frac{1}{3} & \frac{2}{3} & \frac{5}{3} & -\frac{1}{3} \\ 0 & 1 & -\frac{1}{5} & -\frac{7}{5} & \frac{26}{5} \\ 0 & 1 & 4 & -\frac{32}{5} & -\frac{2}{5} \end{bmatrix} \quad (R_2 \to \frac{6}{5}R_2, R_3 \to -3R_3)$$

$$\sim \begin{bmatrix} 1 & -\frac{1}{3} & \frac{2}{3} & \frac{5}{3} & -\frac{1}{3} \\ 0 & 1 & -\frac{1}{5} & -\frac{7}{5} & \frac{26}{5} \\ 0 & 0 & \frac{21}{5} & \frac{39}{5} & -\frac{24}{5} \end{bmatrix} \quad (R_3 \to R_3 - R_2)$$

$$\sim \begin{bmatrix} 1 & -\frac{1}{3} & \frac{2}{3} & \frac{5}{3} & -\frac{1}{3} \\ 0 & 1 & -\frac{1}{5} & -\frac{7}{5} & \frac{26}{5} \\ 0 & 0 & 1 & \frac{39}{21} & -\frac{24}{21} \end{bmatrix} \quad (R_3 \to \frac{5}{21}R_3),$$

which is a row echelon form. Then, we infer that

$$z = -\frac{39}{21}p - \frac{24}{21}, y = \frac{1}{5}z + \frac{7}{5}p + \frac{26}{5}, \text{ and } x = \frac{1}{3}y - \frac{2}{3}z - \frac{5}{3}p - \frac{1}{3}.$$

Setting $p = 1$, we obtain $z = -3, y = 6$, and $x = 2$, which provides one solution of the system. However, there are infinitely many solutions of the system as $\text{rank}(A) = 3$ and the number of unknowns is 4. These solutions can be obtained by solving the homogeneous system as shown in the following.

Consider a system of equations given by $Ax = b$. If b is a nonzero vector, the system is called non-homogeneous, and $Ax = \Theta$ (in the context of matrix equations we use the symbol Θ to denote the zero column matrix while the symbol is also used to denote the zero vector in a vector space) is called the corresponding homogeneous system.

If x_1 is a solution of the non-homogeneous system and x_h is a solution of the homogeneous system, then $Ax_1 = b$ and $Ax_h = \Theta$. Adding these equations, we obtain

$$A(x_1 + x_h) = b, \tag{2.11}$$

which implies that $x_1 + x_h$ is also a solution of the non-homogeneous system.

Exercise 2.4.1 Find all possible values of a and b for which the system of equations

$$\begin{cases} 2x+y+az=0, \\ y+2z=4, \\ x+by=1 \end{cases}$$

has (i) a unique solution, (ii) infinitely many solutions, and (iii) no solution.

2.5 Null Space of a Matrix

Let A be an $m \times n$ matrix. Then, A is a mapping from $\mathbb{R}^n$ into $\mathbb{R}^m$, denoted by $A : \mathbb{R}^n \to \mathbb{R}^m$.

Definition 2.5.1 The **null space** of an $m \times n$ matrix A, denoted by $\text{Null}(A)$, is a subset of $\mathbb{R}^n$ defined as

$$\text{Null}(A) = \{x \in \mathbb{R}^n : Ax = \Theta\},$$

which is the set of all $x \in \mathbb{R}^n$ for which the image under A is $\Theta \in \mathbb{R}^m$.

It can be observed that $\text{Null}(A)$ is a **subspace** of $\mathbb{R}^n$ because if $x, y \in \text{Null}(A)$ and $\alpha \in \mathbb{R}$, then $x+y \in \text{Null}(A)$ and $\alpha x \in \text{Null}(A)$ as $A(x+y) = Ax + Ay = \Theta$ and $A(\alpha x) = \alpha(Ax) = \Theta$.

Definition 2.5.2 (Nullity of a matrix) The dimension of the null space of a matrix A is called the **nullity** of A.

***Example* 2.5.3** Find the null space, $\text{Null}(A)$ and the nullity of the matrix

$$\begin{bmatrix} 1 & 2 & -2 & -1 \\ 3 & 0 & 1 & 6 \\ 2 & -3 & 1 & -1 \end{bmatrix}.$$

Solution By the definition, $\text{Null}(A) = \left\{x = (x_1, x_2, x_3, x_4)' \in \mathbb{R}^4 : Ax = \Theta\right\}$. Therefore, we have the following system of equations $Ax = \Theta$, which is equivalent to

$$\begin{cases} x_1 + 2x_2 - 2x_3 - x_4 = 0, \\ 3x_1 + x_3 + 6x_4 = 0, \\ 2x_1 - 3x_2 + x_3 - x_4 = 0. \end{cases}$$

The row echelon form of the coefficient matrix is

$$\begin{bmatrix} 1 & 0 & 0 & 1 \\ 0 & 1 & 0 & 2 \\ 0 & 0 & 1 & 3 \end{bmatrix}.$$

Thus, $x_1 + x_4 = 0$, $x_2 + 2x_4 = 0$, and $x_3 + 3x_4 = 0$, which imply $x_1 = -x_4$, $x_2 = -2x_4$, and $x_3 = -3x_4$. Hence, $\text{Null}(A) = \{(x_1, x_2, x_3, x_4) : x_1 = -x_4, x_2 = -2x_4, x_3 = -3x_4\} = \{(-x_4, -2x_4, -3x_4, x_4)' : x_4 \in \mathbb{R}\}$, which is spanned by the vector $(-1, -2, -3, 1)'$. Therefore, the nullity is 1.

Now, consider a non-homogeneous system of equations $Ax = b$. From equation (2.11), it is clear that if x_1 is a solution of $Ax = b$, then the set of all solutions is given by $\{x = x_1 + x_h ; x_h \in \text{Null}(A)\}$.

An important relation between the rank and nullity of a matrix is given in the following theorem (stated without proof).

Theorem 2.5.4 (The rank-nullity theorem) If A is an $m \times n$ matrix, then

$$\text{rank}(A) + \text{nullity of } A = n.$$

Exercise 2.5.1 Let A and B be $m \times n$ and $n \times p$ matrices with ranks r_1 and r_2, respectively. Then show that

(*i*) if $n < p$, then $\text{Null}(AB) > 0$.

(*ii*) if $r_2 < p$, then $\text{Null}(AB) > 0$.

(*iii*) if $r_2 = p$ and $m < p$, then $\text{Null}(AB) > 0$.

(*iv*) if $p = r_2 \leq n$ and $r_1 = n$, then $\text{Null}(AB) = 0$.

(*v*) if $m = p$ and $n < p$, then $\text{Null}(AB) > 0$.

Exercise 2.5.2 Let A be a matrix of order n. Show that:

(*i*) $\text{Range}(A^k) \supset \text{Range}(A^{k+1})$ and $\text{Null}(A^k) \subset \text{Null}(A^{k+1})$ for $k = 1, 2, \ldots, r$, where $r < n$.

(*ii*) If $\text{Range}(A^r) = \text{Range}(A^{r+1})$ for some r, then $\text{Null}(A^r) = \text{Null}(A^{r+1})$.

2.6 Eigenvalues and Eigenvectors of a Matrix

Let A be an $n \times n$ real matrix. Then A maps $\mathbb{R}^n$ to $\mathbb{R}^n$ and can also be considered as a mapping from $\mathbb{C}^n$ to $\mathbb{C}^n$.

Definition 2.6.1 A scalar λ (real or complex) is called an **eigenvalue** of A if there exists a **nonzero** vector $X \in \mathbb{C}^n$ such that

$$AX = \lambda X. \tag{2.12}$$

From (2.12), it follows that

$$(A - \lambda I)X = \Theta, \tag{2.13}$$

where I is the $n \times n$ identity matrix. Equation (2.13) implies that the matrix $(A - \lambda I)$ is not invertible since $X \neq \Theta$. (If $(A - \lambda I)^{-1}$ exists, multiplying both sides of (2.13) by $(A - \lambda I)^{-1}$ gives $X = \Theta$.) By Theorem 2.3.14,

$$\det(A - \lambda I) = 0. \tag{2.14}$$

Equation (2.14) is known as the **characteristic equation** of A. The expression $\det(A - \lambda I)$ is an nth degree polynomial in λ, known as the **characteristic polynomial** of A. Equation (2.14) can be written as

$$\lambda^n + a_1\lambda^{n-1} + a_2\lambda^{n-2} + \cdots + a_n\lambda = 0. \tag{2.15}$$

From (2.13) and (2.15), it is clear that any scalar λ (real or complex) satisfying (2.15) is an eigenvalue of A. Since there are n roots (distinct or repeated) for any n^{th} degree polynomial, we can conclude that any $n \times n$ matrix A will have n eigenvalues, which are the roots of its characteristic polynomial.

Let $\lambda_1, \lambda_2, \ldots, \lambda_n$ be the eigenvalues of A. For each eigenvalue λ_i, there exists an eigenvector X_i satisfying $AX_i = \lambda_i X_i$, which can be obtained by solving the system $(A - \lambda_i I)X_i = \Theta$.

Since $\det(A - \lambda_i I) = 0$, it follows from Theorem 2.3.14 that $\text{rank}(A - \lambda_i I) < n$ and from Theorem 2.5.4 that the nullity of $A - \lambda_i I$ is greater than or equal to 1. Hence, there exist nonzero vectors $X_i \in$ the null space of $(A - \lambda_i I)$. Any such nonzero vector X_i is called an **eigenvector** of A corresponding to λ_i.

It can be observed from (2.12) that if X is an eigenvector of a matrix A, then for any nonzero scalar α, αX is also an eigenvector of A.

***Example* 2.6.2** Find the eigenvalues and eigenvectors of the matrix $A = \begin{bmatrix} 0 & 3 \\ 2 & 1 \end{bmatrix}$.

Solution The characteristic equation of A is given by

$$\det(A-\lambda I)=\begin{vmatrix}-\lambda & 3\\ 2 & 1-\lambda\end{vmatrix}=0 \implies (\lambda-3)(\lambda+2)=0.$$

Thus, the eigenvalues of A are $\lambda_1=3$ and $\lambda_2=-2$.

To find an eigenvector $X_1=\begin{bmatrix}x_{11}\\ x_{12}\end{bmatrix}$ corresponding to $\lambda_1=3$, solve the equation $(A-3I)X_1=\Theta$, which yields $\begin{bmatrix}-3 & 3\\ 2 & -2\end{bmatrix}\begin{bmatrix}x_{11}\\ x_{12}\end{bmatrix}=\begin{bmatrix}0\\ 0\end{bmatrix}$. Hence, $x_{11}=x_{12}$. Taking $x_{12}=1$, we get $X_1=\begin{bmatrix}1\\ 1\end{bmatrix}$ as one of the eigenvectors corresponding to $\lambda_1=3$.

To find an eigenvector $X_2=\begin{bmatrix}x_{21}\\ x_{22}\end{bmatrix}$ corresponding to $\lambda_2=-2$, solve the equation $(A+2I)X_2=\Theta$, which yields $\begin{bmatrix}2 & 3\\ 2 & 3\end{bmatrix}\begin{bmatrix}x_{21}\\ x_{22}\end{bmatrix}=\begin{bmatrix}0\\ 0\end{bmatrix}$. Hence, $x_{22}=-\frac{2}{3}x_{21}$. Taking $x_1=1$, we get $X_2=\begin{bmatrix}1\\ -\frac{2}{3}\end{bmatrix}$ as an eigenvector corresponding to $\lambda_2=-2$.

***Example* 2.6.3** Find the eigenpairs of the matrix $A=\begin{bmatrix}3 & 2 & 4\\ 2 & 0 & 2\\ 4 & 2 & 3\end{bmatrix}$.

Solution The characteristic equation of A is given by

$$\begin{vmatrix}3-\lambda & 2 & 4\\ 2 & -\lambda & 2\\ 4 & 2 & 3-\lambda\end{vmatrix}=0 \implies \lambda^3-6\lambda^2-15\lambda-8=0.$$

The eigenvalues of A are $\lambda_1 = \lambda_2 = -1$ and $\lambda_3 = 8$. To find eigenvectors corresponding to -1, solve the equation

$$(A+I)X = \Theta \implies \begin{bmatrix} 4 & 2 & 4 \\ 2 & 1 & 2 \\ 4 & 2 & 4 \end{bmatrix} \begin{bmatrix} x_1 \\ x_2 \\ x_3 \end{bmatrix} = \begin{bmatrix} 0 \\ 0 \\ 0 \end{bmatrix} \implies 2x_1 + x_2 + 2x_3 = 0.$$

Since there is one equation with three unknowns, we can write $x_1 = -\frac{1}{2}x_2 - x_3$, with x_2 and x_3 as arbitrary variables. Taking $x_2 = 0$ and $x_3 = 1$, we get the eigenvector $X_1 = \begin{bmatrix} -1 \\ 0 \\ 1 \end{bmatrix}$. Taking $x_2 = 1$ and $x_3 = 0$, we get the eigenvector $X_2 = \begin{bmatrix} -\frac{1}{2} \\ 1 \\ 0 \end{bmatrix}$. These are the two linearly independent eigenvectors corresponding to $\lambda = -1$.

To find an eigenvector corresponding to 8, solve the equation

$$(A-8I)X_3 = \Theta \implies \begin{bmatrix} -5 & 2 & 4 \\ 2 & 8 & 2 \\ 4 & 2 & -5 \end{bmatrix} \begin{bmatrix} x_1 \\ x_2 \\ x_3 \end{bmatrix} = \begin{bmatrix} 0 \\ 0 \\ 0 \end{bmatrix} \implies 2x_2 = \frac{x_1}{2} \text{ and } x_3 = x_1.$$

Taking $x_1 = 1$, we get the eigenvector $X_3 = \begin{bmatrix} 1 \\ \frac{1}{2} \\ 1 \end{bmatrix}$.

Sum and Product of Eigenvalues

Let the characteristic polynomial of an $n \times n$ matrix A be

$$\det(A - \lambda I) = \begin{vmatrix} a_{11} - \lambda & a_{12} & \cdots & a_{1n} \\ a_{21} & a_{22} - \lambda & \cdots & a_{2n} \\ \vdots & & & \\ a_{n1} & a_{n2} & \cdots & a_{nn} - \lambda \end{vmatrix} = \alpha_0 \lambda^n + \alpha_1 \lambda^{n-1} + \cdots + \alpha_n. \tag{2.16}$$

Note that

$$\begin{cases} \alpha_0 = (-1)^n, \\ \alpha_1 = (-1)^{n-1}(a_{11} + a_{22} + \cdots + a_{nn}). \end{cases} \tag{2.17}$$

Other coefficients can be obtained by actually computing the determinant. Equation (2.16) is valid for any value of λ (real or complex). Setting $\lambda = 0$ yields

$$\det(A) = \alpha_n, \tag{2.18}$$

which indicates that the constant α_n in the characteristic polynomial is nothing but the determinant of A.

Let $\lambda_1, \lambda_2, \ldots, \lambda_n$ be the eigenvalues of A, which are the roots of the characteristic equation $\alpha_0\lambda^n + \alpha_1\lambda^{n-1} + \cdots + \alpha_n = 0$. Then, from (2.17) and (2.18),

$$\begin{aligned} \text{the sum of the roots} = \lambda_1 + \lambda_2 + \cdots + \lambda_n = -\frac{\alpha_1}{\alpha_0} = a_{11} + a_{22} + \cdots + a_{nn} \\ = \text{the sum of the diagonal elements of } A = \text{the trace of } A \end{aligned} \tag{2.19}$$

and

$$\text{the product of the roots } = \lambda_1\lambda_2\cdots\lambda_n = (-1)^n\frac{\alpha_n}{\alpha_0} = \alpha_n = \det(A). \tag{2.20}$$

From (2.20), it is clear that all eigenvalues are nonzero if and only if $\det(A) \neq 0$, and A has a zero eigenvalue if and only if $\det(A) = 0$.

Combining this information with Theorem 2.3.14, we obtain the following result:

Theorem 2.6.4 The following statements are equivalent:

(*i*) A is invertible,

(*ii*) $\det(A) \neq 0$,

(*iii*) $\operatorname{rank}(A) = n$,

(*iv*) all eigenvalues of A are nonzero.

Remark 2.6.5 A matrix that satisfies any one of the conditions stated above is called **nonsingular**. If A is an $n \times n$ singular matrix, then the equation $Ax = 0$ has nonzero solutions.

Algebraic Multiplicity and Geometric Multiplicity of an Eigenvalue

If A is an $n \times n$ matrix, then its characteristic equation has n roots, which are the eigenvalues of A. Some of these roots may be repeated. Let $\lambda_1, \lambda_2, \ldots, \lambda_k$ be the distinct eigenvalues of A, where λ_i is repeated m_i times. Thus, $m_1 + m_2 + \cdots + m_k = n$. The number m_i is called the **algebraic multiplicity** of $\lambda_i : i = 1, 2, \ldots, k$.

To find the eigenvectors corresponding to λ_i, we solve the system of equations

$$(A - \lambda_i I)X = \Theta. \tag{2.21}$$

If equation (2.21) has r_i linearly independent solutions $X_1^i, X_2^i, \ldots, X_{r_i}^i$, then r_i is called the **geometric multiplicity** of λ_i.

Note that $r_i \leq m_i$ for all $i = 1, 2, \ldots, k$.

***Example* 2.6.6** Find the algebraic multiplicity and geometric multiplicity of the eigenvalues of the matrix $A = \begin{bmatrix} 2 & 1 & 0 \\ 1 & 2 & 0 \\ 0 & 0 & 3 \end{bmatrix}$.

Solution The eigenvalues of A are 3, 3, and 1. Hence, the algebraic multiplicity of the eigenvalue 3 is 2. The linearly independent eigenvectors corresponding to 3 are $\begin{bmatrix} 1 \\ 1 \\ 0 \end{bmatrix}$ and $\begin{bmatrix} 1 \\ 1 \\ 1 \end{bmatrix}$. Therefore, the geometric multiplicity of the eigenvalue 3 is 2. Since the algebraic multiplicity of the eigenvalue 1 is 1, its geometric multiplicity must also be 1.

***Example* 2.6.7** Find the algebraic and geometric multiplicities of the matrix $A = \begin{bmatrix} 2 & 1 & 0 \\ 0 & 2 & 0 \\ 1 & 1 & 3 \end{bmatrix}$.

Solution The eigenvalues of A are 2 (with algebraic multiplicity 2) and 3 (with algebraic multiplicity 1). Eigenvectors corresponding to the eigenvalue 2 can be obtained by solving $\begin{bmatrix} 0 & 1 & 0 \\ 0 & 0 & 0 \\ 1 & 1 & 1 \end{bmatrix} \begin{bmatrix} x_1 \\ x_2 \\ x_3 \end{bmatrix} = \begin{bmatrix} 0 \\ 0 \\ 0 \end{bmatrix}$. This yields $x_2 = 0$ and $x_1 + x_3 = 0$. By taking $x_3 = 1$, we obtain $X = \begin{bmatrix} -1 \\ 0 \\ 1 \end{bmatrix}$ as a linearly independent eigenvector. Therefore, the geometric multiplicity of the eigenvalue 2 is 1. As the algebraic multiplicity of the eigenvalue 3 is 1, its geometric multiplicity must also be 1.

Exercise 2.6.1 Let A be an $n \times n$ matrix, with its kth row given by $k\begin{bmatrix} 1 & 2 & \cdots & n \end{bmatrix}$ for $k = 1, 2, \ldots, n$. Find the eigenpairs of A.

2.7 Canonical Forms of a Matrix

A matrix of order n is said to be in **diagonal form** if all its off-diagonal elements are zeros.

Definition 2.7.1 A matrix A is called **diagonalizable** if there exists a nonsingular matrix P such that $P^{-1}AP = D$, where D is a diagonal matrix.

It can be shown that a matrix can be diagonalized if, for every eigenvalue, the *algebraic multiplicity* = the *geometric multiplicity*. In other words, any matrix of order n is diagonalizable if it has n linearly independent eigenvectors.

Theorem 2.7.2 Let $\lambda_i : i = 1, 2, \ldots, n$ be the eigenvalues of an $n \times n$ matrix A with corresponding linearly independent eigenvectors $X_i : i = 1, 2, \ldots, n$. Let P be the matrix with X_i as its ith column, and let D be the diagonal matrix with λ_i as its ith diagonal element. Then $P^{-1}AP = D$.

Proof. Since $X_i : i = 1, 2, \ldots, n$ are eigenvectors corresponding to $\lambda_i : i = 1, 2, \ldots, n$,

$$AX_i = \lambda_i X_i, \quad i = 1, 2, \ldots, n.$$

If

$$P = \begin{bmatrix} X_1 & X_2 & \cdots & X_n \\ \downarrow & \downarrow & \cdots & \downarrow \end{bmatrix},$$

then $\text{rank}(P) = n$ since the columns of P are linearly independent. Therefore, P is invertible. Now,

$$AP = A \begin{bmatrix} X_1 & X_2 & \cdots & X_n \\ \downarrow & \downarrow & \cdots & \downarrow \end{bmatrix} = \begin{bmatrix} \lambda_1 X_1 & \lambda_2 X_2 & \cdots & \lambda_n X_n \\ \downarrow & \downarrow & \cdots & \downarrow \end{bmatrix}$$

$$= \begin{bmatrix} X_1 & X_2 & \cdots & X_n \\ \downarrow & \downarrow & \cdots & \downarrow \end{bmatrix} \begin{bmatrix} \lambda_1 & & & 0 \\ & \lambda_2 & & \\ & & \ddots & \\ 0 & & & \lambda_n \end{bmatrix} = PD,$$

which yields $P^{-1}AP = D$. □

Example **2.7.3** Diagonalize the matrix $A = \begin{bmatrix} 2 & 1 & 0 \\ 1 & 2 & 0 \\ 0 & 0 & 3 \end{bmatrix}$.

Solution The eigenvalues of A are $\lambda_1 = 1$ and $\lambda_2 = 3$ with the algebraic multiplicities 1 and 2, respectively. An eigenvector corresponding to λ_1 is $\begin{bmatrix} -1 \\ 1 \\ 0 \end{bmatrix}$, and two linearly independent eigenvectors corresponding to λ_2 are $\begin{bmatrix} 1 \\ 1 \\ 0 \end{bmatrix}$ and $\begin{bmatrix} 1 \\ 1 \\ 1 \end{bmatrix}$. For both the eigenvalues of A, the algebraic multiplicity = the geometric multiplicity. Hence, A is diagonalizable. If $P = \begin{bmatrix} -1 & 1 & 1 \\ 1 & 1 & 1 \\ 0 & 0 & 1 \end{bmatrix}$, then, from Theorem 2.7.2, $P^{-1}AP = D = \begin{bmatrix} 1 & 0 & 0 \\ 0 & 3 & 0 \\ 0 & 0 & 3 \end{bmatrix}$ is the diagonal form of A.

Jordan Canonical Form

If a matrix A is not diagonalizable, then the other canonical form of A is the **Jordan canonical form**. A matrix A cannot be diagonalized if any of its eigenvalues has the *geometric multiplicity* < the *algebraic multiplicity*. Let A be an $n \times n$ matrix with distinct eigenvalues $\lambda_1, \lambda_2, \ldots, \lambda_k$, where $k < n$. Let m_i and l_i denote the algebraic and geometric multiplicities of $\lambda_i : i = 1, 2, \ldots, k$, respectively. Then, the Jordan canonical form can be described as:

$$J = \begin{bmatrix} J_1 & & & \\ & J_2 & & \\ & & \ddots & \\ & & & J_k \end{bmatrix}_{n \times n}, \tag{2.22}$$

where $J_i : i = 1, 2, \ldots, k$ are called **Jordan blocks** corresponding to $\lambda_i : i = 1, 2, \ldots, k$, respectively, and each J_i is of order m_i. All non-diagonal blocks contain zeros. Now, let

$$J_i = \begin{bmatrix} J_{i1} & & & \\ & J_{i2} & & \\ & & \ddots & \\ & & & J_{il_i} \end{bmatrix}_{m_i \times m_i}. \tag{2.23}$$

J_i contains l_i Jordan subblocks, each corresponding to a distinct linearly independent eigenvector. Each Jordan subblock J_{ir} is of the form:

$$J_{ir} = \begin{bmatrix} \lambda_i & 1 & & & 0 \\ & \lambda_i & 1 & & \\ & & \ddots & \ddots & \\ & & & \ddots & 1 \\ 0 & & & & \lambda_i \end{bmatrix}, \quad r = 1, 2, \ldots, l_i. \tag{2.24}$$

The order of each J_{ir} depends on the matrix A and its eigenpairs. This is explained as follows:

First, consider λ_i with algebraic multiplicity m_i and geometric multiplicity l_i. Let $X_1^i, X_2^i, \ldots, X_{l_i}^i$ be linearly independent eigenvectors corresponding to λ_i. For each eigenvector $X_j^i : j = 1, 2, \ldots, l_i$, we define generalized eigenvectors $Y_{j1}^i, Y_{j2}^i, \ldots, Y_{jr_{ij}}^i$ as solutions of the following equations. Since X_j^1 is an eigenvector for λ_1, it satisfies

$$AX_j^i = \lambda_i X_j^i. \tag{2.25}$$

To find generalized eigenvectors, solve $(A-\lambda_i I)Y_{j1}^i = X_j^i$. After finding Y_{j1}^i, we can find Y_{j2}^i by solving

$$(A-\lambda_i I)Y_{j2}^i = Y_{j1}^i. \tag{2.26}$$

Proceed in this manner, we obtain

$$\begin{cases} (A-\lambda_i I)Y_{j3}^i = Y_{j2}^i, \\ \vdots \\ (A-\lambda_i I)Y_{jr_{ij}}^i = Y_{j(r_{ij}-1)}^i. \end{cases} \tag{2.27}$$

If the system $(A-\lambda_i I)Y = Y_{jr_{ij}}^i$ is inconsistent, we conclude that there are r_{ij} generalized eigenvectors corresponding to X_j^i. Thus, X_j^i along with its generalized eigenvectors $Y_{j1}^i, Y_{j2}^i, \ldots, Y_{jr_{ij}}^i$ form $r_{ij}+1$ linearly independent vectors. Consequently, the Jordan subblock J_{ij} is a $(r_{ij}+1)\times(r_{ij}+1)$ matrix, which is of the form

$$J_{ij} = \begin{bmatrix} \lambda_i & 1 & & & 0 \\ & \lambda_i & 1 & & \\ & & \ddots & \ddots & \\ & & & \ddots & 1 \\ 0 & & & & \lambda_i \end{bmatrix}_{(r_{ij}+1)\times(r_{ij}+1)}, \quad i = 1,2,\ldots,k, j = 1,2,\ldots,l_i.$$

The n linearly independent vectors

$$\bigcup_{i=1}^{k} \{X_1^i, Y_{11}^i, Y_{12}^i, \ldots, Y_{1r_{i1}}^i, X_2^i, Y_{21}^i, Y_{22}^i, \ldots, Y_{2r_{i2}}^i, \ldots, X_{l_i}^i, Y_{l_i1}^i, Y_{l_i2}^i, \ldots, Y_{l_ir_{il_i}}^i\} \tag{2.28}$$

form a basis of $\mathbb{R}^n$. Let P be the nonsingular matrix whose columns are given in (2.28), written in the same order. Then, using (2.25)–(2.27), it can be shown that $AP = PJ$. Hence, we obtain the relation

$$P^{-1}AP = J. \tag{2.29}$$

***Example* 2.7.4** Let the matrix J given below be the Jordan canonical form of an 8×8 matrix

$$J = \left[\begin{array}{cc|c|ccc|cc} 2 & 1 & 0 & & & & & \\ 0 & 2 & 0 & & & & & \\ \hline 0 & 0 & 2 & 0 & 0 & 0 & & \\ \hline & & 0 & 4 & 1 & 0 & & \\ & & 0 & 0 & 4 & 1 & & \\ & & 0 & 0 & 0 & 4 & & \\ \hline & & & 0 & 0 & 0 & 4 & 1 \\ & & & 0 & 0 & 0 & 0 & 4 \end{array}\right]_{8\times 8}.$$

Then,the eigenvalues of J are $\lambda_1 = 2$ and $\lambda_2 = 4$. λ_1 has algebraic multiplicity $m_1 = 3$ and geometric multiplicity $l_1 = 2$. X_1^1 and X_2^1 are linearly independent eigenvectors corresponding to λ_1. X_1^1 has one generalized eigenvector Y_{11}. So, $r_{11} = 1$, and hence, the subblock $J_{11} = \begin{bmatrix} 2 & 1 \\ 0 & 2 \end{bmatrix}_{2\times 2}$. X_2^1 has no generalized eigenvector. So, $r_{12} = 0$, and hence, the subblock $J_{12} = \begin{bmatrix} 2 \end{bmatrix}$.

For $\lambda_2 = 4$, we have $m_2 = 5$ and $l_2 = 2$. Let X_1^2 and X_2^2 be linearly independent eigenvectors corresponding to λ_2. X_1^2 has generalized eigenvectors Y_{11}^2 and Y_{12}^2, implying $r_{21} = 2$, and hence, the subblock $J_{21} = \begin{bmatrix} 4 & 1 & 0 \\ 0 & 4 & 1 \\ 0 & 0 & 4 \end{bmatrix}_{3\times 3}$. X_2^2 has one generalized eigenvector Y_{21}^2. So, $r_{22} = 1$, and hence, the subblock $J_{22} = \begin{bmatrix} 4 & 1 \\ 0 & 4 \end{bmatrix}_{2\times 2}$.

Example **2.7.5** For the matrix $A = \begin{bmatrix} 2 & 1 & 0 \\ 0 & 2 & 0 \\ 1 & 1 & 3 \end{bmatrix}$ given in Example 2.6.7, the eigenvalues are 2, 2, and 3. The eigenvalue 2 has one linearly independent eigenvector $\begin{bmatrix} -1 \\ 0 \\ 1 \end{bmatrix}$ and a generalized eigenvector $\begin{bmatrix} 1 \\ -1 \\ 1 \end{bmatrix}$. The eigenvalue 3 has eigenvector $\begin{bmatrix} 0 \\ 0 \\ 1 \end{bmatrix}$. Hence,

$$P = \begin{bmatrix} -1 & 1 & 0 \\ 0 & -1 & 0 \\ 1 & 1 & 1 \end{bmatrix} \text{ and } J = \begin{bmatrix} 2 & 1 & 0 \\ 0 & 2 & 0 \\ 0 & 0 & 3 \end{bmatrix}.$$

Jordan-type Canonical Form

Let $J(\alpha)$ be the Jordan canonical form with a given scalar $\alpha \neq 0$ on the super diagonal. For example, a 3×3 Jordan form $J(\alpha)$ is given by $J(\alpha) = \begin{bmatrix} \lambda & \alpha & 0 \\ 0 & \lambda & \alpha \\ 0 & 0 & \lambda \end{bmatrix}$. To obtain $J(\alpha)$ using the eigenvector x corresponding to the eigenvalue λ, and the generalized eigenvectors y_1 and y_2 (as shown in (2.25)–(2.27)), let $z_1 = \alpha y_1$ and $z_2 = \alpha^2 y_2$, and $P = \begin{bmatrix} x & z_1 & z_2 \\ \downarrow & \downarrow & \downarrow \end{bmatrix}$. Then, $P^{-1}AP = J(\alpha)$. In general, if the eigenvalue λ has eigenvector x and $r-1$ generalized eigenvectors denoted by $y_1, y_2, \dots, y_{r-1}$ as defined in (2.25)–(2.27), then for a given nonzero scalar α, define $P = \begin{bmatrix} x & z_1 & z_2 & \cdots & z_{r-1} \\ \downarrow & \downarrow & \downarrow & \downarrow & \downarrow \end{bmatrix}$, where

$z_i = \alpha^i y_i : i = 1,2,\ldots,r-1$. Now it can be seen that $AP = PJ(\alpha)$, implying $P^{-1}AP =$

$$J(\alpha), \text{where } J(\alpha) = \begin{bmatrix} \lambda & \alpha & 0 & \cdots & & 0 \\ 0 & \lambda & \alpha & \cdots & & 0 \\ \vdots & & & & & \\ 0 & 0 & & \cdots & \lambda & \alpha \\ 0 & 0 & & \cdots & 0 & \lambda \end{bmatrix}.$$

***Example* 2.7.6** For the matrix $A = \begin{bmatrix} 4 & 0 & 0 \\ 1 & 4 & 0 \\ 0 & 1 & 4 \end{bmatrix}$, find P such that $P^{-1}AP = J(0.5)$, where

$$J(0.5) = \begin{bmatrix} 4 & 0.5 & 0 \\ 0 & 4 & 0.5 \\ 0 & 0 & 4 \end{bmatrix}.$$

Solution The eigenvalue of A is 4 (algebraic multiplicity 3 and geometric multiplicity 1) with eigenvector $x = \begin{bmatrix} 0 \\ 0 \\ 1 \end{bmatrix}$. The two generalized eigenvectors are $y_1 =$ $\begin{bmatrix} 0 \\ 1 \\ 0 \end{bmatrix}$ and $y_2 = \begin{bmatrix} 1 \\ 0 \\ 0 \end{bmatrix}$. Let $z_1 = 0.5y_1$ and $z_2 = 0.5^2 y_2$. Now define $P = \begin{bmatrix} 0 & 0 & 0.5^2 \\ 0 & 0.5 & 0 \\ 1 & 0 & 0 \end{bmatrix}$. Then $P^{-1}AP = J(0.5)$.

Exercise 2.7.1 Let A be an $n \times n$ matrix with the first row $r_1 = \begin{bmatrix} 1 & 2 & \cdots & n \end{bmatrix}$, the second row $r_2 = \begin{bmatrix} n & n-1 & \cdots & 1 \end{bmatrix}$, and the kth row $r_k = \sum_{i=1}^{k-1} r_i$ for $k = 3,4,\ldots,n$. Prove that the eigenvalues of A are $2^{n-2}(n+1), -1$, and 0. Also, show that A is diagonalizable. (**Hint:** $\begin{bmatrix} 1 & -1 & 0 & \cdots & 0 \end{bmatrix}'$ is an eigenvector.)

Exercise 2.7.2 Let $A = \begin{bmatrix} A_1 & & \\ & A_2 & \\ & & A_3 \end{bmatrix}_{6\times 6}$, where the block matrices $A_i : i = 1,2,3$ are given by $A_1 = \begin{bmatrix} 2 & 1 & 0 \\ 0 & 2 & 3 \\ 0 & 0 & 2 \end{bmatrix}$, $A_2 = \begin{bmatrix} 2 \end{bmatrix}$, and $A_3 = \begin{bmatrix} 3 & 0 \\ 1 & 3 \end{bmatrix}$. Find P such that $P^{-1}AP = J(3)$, the Jordan-type canonical form.

Exercise 2.7.3 Let A be an $n \times n$ matrix, with its kth row given by $\begin{bmatrix} 1 & 2 & \cdots & n \end{bmatrix}$ for $k = 1,2,\cdots,n$. Is A diagonalizable?

2.8 Functions of a Matrix

If f is a real valued function of a real variable x, it is denoted by $f : \mathbb{R} \to \mathbb{R}$. For example, $f(x) = x^n$, $g(x) = a_0 + a_1x + a_2x^2 + \cdots + a_nx^n$, and $h(x) = e^{ax}$ are some real valued functions on $\mathbb{R}$. Similarly, one can define a function of an $n \times n$ matrix as $f : \mathbb{R}^{n\times n} \to \mathbb{R}^{n\times n}$, where $\mathbb{R}^{n\times n}$ denotes the space of all $n \times n$ real matrices. Eigenvalues and eigenvectors play an important role in evaluating functions $f(A)$ of a matrix A.

Diagonalizable Matrix

First, let us consider matrices that are diagonalizable. In this case, there exists a nonsingular matrix P such that $P^{-1}AP = D$, which gives $A = PDP^{-1}$. Then, $A^2 = \left(PDP^{-1}\right)\left(PDP^{-1}\right) = PD^2P^{-1}$. Similarly, we can show that

$$A^k = PD^kP^{-1}. \tag{2.30}$$

If $f(A) = a_0I + a_1A + a_2A^2 + \cdots + a_kA^k$, then using (2.30),

$$\begin{aligned} f(A) &= a_0PP^{-1} + a_1PDP^{-1} + \cdots + a_kPD^kP^{-1} \\ &= P(a_0I + a_1D + \cdots + a_kD^k)P^{-1} = Pf(D)P^{-1}. \end{aligned} \tag{2.31}$$

Let

$$D = \begin{bmatrix} \lambda_1 & & & 0 \\ & \lambda_2 & & \\ & & \ddots & \\ 0 & & & \lambda_n \end{bmatrix}$$

be the diagonal form of A with eigenvalues λ_i, $i = 1, 2, \ldots, n$. Then,

$$D^m = \begin{bmatrix} \lambda_1^m & & & 0 \\ & \lambda_2^m & & \\ & & \ddots & \\ 0 & & & \lambda_n^m \end{bmatrix},$$

and from equation (2.31), we obtain

$$f(D) = \begin{bmatrix} f(\lambda_1) & & & 0 \\ & f(\lambda_2) & & \\ & & \ddots & \\ 0 & & & f(\lambda_n) \end{bmatrix} \text{ and } f(A) = P \begin{bmatrix} f(\lambda_1) & & & 0 \\ & f(\lambda_2) & & \\ & & \ddots & \\ 0 & & & f(\lambda_n) \end{bmatrix} P^{-1}. \tag{2.32}$$

Now, let $f(A) = e^{\alpha A}$ for $\alpha \in \mathbb{R}$. By writing $e^{\alpha A} = I + \alpha A + \frac{\alpha^2 A^2}{2!} + \cdots + \frac{\alpha^n A^n}{n!} + \cdots$ and replacing A^m *by* PD^mP^{-1}, we get $e^{\alpha A} = P(I + \alpha D + \frac{\alpha^2}{2!}D^2 + \frac{\alpha^n}{n!}D^n + \cdots)P^{-1} = Pe^{\alpha D}P^{-1}$, where

$$e^{\alpha D} = \begin{bmatrix} e^{\alpha\lambda_1} & & & 0 \\ & e^{\alpha\lambda_2} & & \\ & & \ddots & \\ 0 & & & e^{\alpha\lambda_n} \end{bmatrix}.$$

Example **2.8.1** For the matrix $A = \begin{bmatrix} 2 & 1 \\ 1 & 2 \end{bmatrix}$, find A^{100} and e^{3A}.

Solution The eigenpairs of A are $\left(3, \begin{bmatrix} 1 \\ 1 \end{bmatrix}\right)$ and $\left(1, \begin{bmatrix} 1 \\ -1 \end{bmatrix}\right)$. Hence, $P = \begin{bmatrix} 1 & 1 \\ 1 & -1 \end{bmatrix}$ and $D = \begin{bmatrix} 3 & 0 \\ 0 & 1 \end{bmatrix}$. From (2.32), $A^{100} = PD^{100}P^{-1} = \frac{1}{2}\begin{bmatrix} 3^{100}+1 & 3^{100}-1 \\ 3^{100}-1 & 3^{100}+1 \end{bmatrix}$ and $e^{3A} = Pe^{3D}P^{-1} = \frac{e^3}{2}\begin{bmatrix} e^6+1 & e^6-1 \\ e^6-1 & e^6+1 \end{bmatrix}$.

Non-diagonalizable Matrix

If A is not diagonalizable, then it has a Jordan canonical form J, given by the relation $P^{-1}AP = J$, as in (2.29). Following the steps of the previous discussion, we obtain

$$A^m = PJ^mP^{-1}, f(A) = Pf(J)P^{-1} \tag{2.33}$$

where $f(A)$ is a polynomial in A, and $e^{\alpha A} = Pe^{\alpha J}P^{-1}$.

Now, let us consider an $r \times r$ Jordan subblock J_s corresponding to an eigenvalue λ as follows:

$$J_s = \begin{bmatrix} \lambda & 1 & 0 & \cdots & 0 \\ & \lambda & 1 & & \\ & & \ddots & \ddots & \\ & & & \ddots & 1 \\ 0 & & & & \lambda \end{bmatrix}_{r\times r}.$$

Then

$$J_s^2 = \begin{bmatrix} \lambda^2 & 2\lambda & 1 & 0 & 0 & \cdots & 0 \\ 0 & \lambda^2 & 2\lambda & 1 & 0 & \cdots & 0 \\ \cdots & \cdots & \cdots & \cdots & \cdots & \cdots & \cdots \\ 0 & 0 & 0 & 0 & \cdots & \cdots & \lambda^2 \end{bmatrix}.$$

It can be shown by induction that

$$J_s^m = \begin{bmatrix} \lambda^m & \binom{m}{1}\lambda^{m-1} & \binom{m}{2}\lambda^{m-2} & \cdots & \binom{m}{r-1}\lambda^{m-r+1} \\ & \lambda^m & \binom{m}{1}\lambda^{m-1} & \cdots & \binom{m}{r-2}\lambda^{m-r+2} \\ & & \ddots & \ddots & \\ & & & \ddots & \binom{m}{1}\lambda^{m-1} \\ 0 & & & & \lambda^m \end{bmatrix}. \tag{2.34}$$

Note that if $\lambda = 0$, then $J_s^m = \Theta$ for $m \geq r$.

Now, let J be the Jordan canonical form of an $n \times n$ matrix A. Then, from (2.22), we have

$$J = \begin{bmatrix} J_1 & & & \\ & J_2 & & \\ & & \ddots & \\ & & & J_k \end{bmatrix}_{n \times n}.$$

We know from (2.23) that J_i is an $m_i \times m_i$ matrix for $i = 1, 2, \cdots, k$ and $J_i - \lambda_i I_i$ is an upper triangular matrix with zeros on the diagonal. Here, I_i is an $m_i \times m_i$ identity matrix. Hence, from (2.34), $(J_i - \lambda_i I_i)^{m_i}$ is zero matrix of order m_i. Now,

$$(J - \lambda_1 I)^{m_1} = \begin{bmatrix} J_1 - \lambda_1 I_1 & & & \\ & J_2 - \lambda_1 I_2 & & \\ & & \ddots & \\ & & & J_k - \lambda_1 I_k \end{bmatrix}^{m_1}$$

$$= \begin{bmatrix} (J_1 - \lambda_1 I_1)^{m_1} & & & \\ & (J_2 - \lambda_1 I_2)^{m_1} & & \\ & & \ddots & \\ & & & (J_k - \lambda_1 I_k)^{m_1} \end{bmatrix},$$

which implies that the first m_1 rows are zero. Similarly, $(J-\lambda_2 I)^{m_2}$ has the next m_2 rows as zero rows, and $(J-\lambda_i I)^{m_i}$ has i^{th} block of m_i zero rows. Hence,

$$(J-\lambda_1 I)^{m_1}(J-\lambda_2 I)^{m_2}\cdots(J-\lambda_k I)^{m_k}=\Theta. \tag{2.35}$$

Replacing J with $P^{-1}AP$ and I with $P^{-1}P$, we get

$$(J-\lambda_i I)^{m_i}=P^{-1}(A-\lambda_i I)^{m_i}P, \quad i=1,2,\ldots,k.$$

Substituting the above in (2.35), we get

$$(A-\lambda_1 I)^{m_1}(A-\lambda_2 I)^{m_2}\cdots(A-\lambda_k I)^{m_k}=\Theta.$$

In the above equation if A is replaced by λ and Θ is replaced by 0 we obtain the characteristic equation of the matrix. Thus we arrive at the following result.

Theorem 2.8.2 (Cayley–Hamilton theorem) Every matrix satisfies its own characteristic equation.

***Example* 2.8.3** If A is the matrix

$$\begin{bmatrix} 2 & 1 \\ -1 & 3 \end{bmatrix}$$

find A^4 using Cayley–Hamilton theorem.

Solution From the characteristic equation of A and from Theorem 2.8 we obtain $A^2=5A+I$. Multiplying both sides by A we get $A^3=5A^2+A=26A+5I$. Then $A^4=26A^2+5A=135A+26I$.

Computation of $e^{\alpha J}$

Let J be a Jordan canonical matrix of order n and J_s be a Jordan block. We know that

$$e^{J_s}=I+J_s+\frac{1}{2!}J_s^2+\cdots+\frac{1}{n!}J_s^n+\cdots.$$

Substituting $J_s^m : m=1,2,\ldots$ as given in (2.34) into the above series and adding element-wise, we obtain

$$e^{J_s}=\begin{bmatrix} e^{\lambda} & e^{\lambda} & \frac{1}{2!}e^{\lambda} & \cdots & \frac{1}{(r-1)!}e^{\lambda} \\ & e^{\lambda} & e^{\lambda} & \cdots & \frac{1}{(r-2)!}e^{\lambda} \\ & & \ddots & \ddots & \\ & & & \ddots & e^{\lambda} \\ 0 & & & & e^{\lambda} \end{bmatrix}=e^{\lambda}\begin{bmatrix} 1 & 1 & \frac{1}{2!} & \cdots & & \frac{1}{(r-1)!} \\ 0 & 1 & 1 & \cdots & & \frac{1}{(r-2)!} \\ \vdots & & & & & \\ 0 & 0 & & \cdots & 1 & 1 \\ 0 & 0 & & \cdots & 0 & 1 \end{bmatrix}.$$

Similarly,

$$e^{\alpha J_s} = e^{\alpha\lambda} \begin{bmatrix} 1 & \alpha & \frac{\alpha^2}{2!} & \cdots & \frac{\alpha^{r-1}}{(r-1)!} \\ & 1 & \alpha & \cdots & \frac{\alpha^{r-2}}{(r-2)!} \\ & & \ddots & \ddots & \\ & & & \ddots & \alpha \\ 0 & & & & 1 \end{bmatrix}. \tag{2.36}$$

If J is a Jordan canonical form containing Jordan subblocks as

$$J = \begin{bmatrix} \boxed{J_1} & & & \\ & \boxed{J_2} & & \\ & & \ddots & \\ & & & \boxed{J_k} \end{bmatrix},$$

then

$$J^m = \begin{bmatrix} \boxed{J_1^m} & & & \\ & \boxed{J_2^m} & & \\ & & \ddots & \\ & & & \boxed{J_k^m} \end{bmatrix} \tag{2.37}$$

and

$$e^{\alpha J} = \begin{bmatrix} \boxed{e^{\alpha J_1}} & & & \\ & \boxed{e^{\alpha J_2}} & & \\ & & \ddots & \\ & & & \boxed{e^{\alpha J_k}} \end{bmatrix}. \tag{2.38}$$

Now it follows from (2.29) that $e^{\alpha A} = Pe^{\alpha J}P^{-1}$.

***Example* 2.8.4** For the Jordan canonical form $J = \begin{bmatrix} 4 & 1 & 0 \\ 0 & 4 & 1 \\ 0 & 0 & 4 \end{bmatrix}$, find J^5 and e^{3J}.

Solution From (2.34) and (2.36), we directly obtain

$$J^5 = \begin{bmatrix} 4^5 & 5.4^4 & 10.4^3 \\ 0 & 4^5 & 5.4^4 \\ 0 & 0 & 4^5 \end{bmatrix} \text{ and } e^{3J} = e^{12} \begin{bmatrix} 1 & 3 & \frac{9}{2} \\ 0 & 1 & 3 \\ 0 & 0 & 1 \end{bmatrix}.$$

***Example* 2.8.5** Find J^5 and e^{7J} for the Jordan canonical form

$$J = \begin{bmatrix} 4 & 1 & 0 & 0 & 0 & 0 \\ 0 & 4 & 1 & 0 & 0 & 0 \\ 0 & 0 & 4 & 0 & 0 & 0 \\ 0 & 0 & 0 & 4 & 0 & 0 \\ 0 & 0 & 0 & 0 & 3 & 1 \\ 0 & 0 & 0 & 0 & 0 & 3 \end{bmatrix}.$$

Solution From (2.37) and (2.38), we obtain

$$J^5 = \begin{bmatrix} 4^5 & 5.4^4 & 10.4^3 & 0 & 0 & 0 \\ 0 & 4^5 & 5.4^4 & 0 & 0 & 0 \\ 0 & 0 & 4^5 & 0 & 0 & 0 \\ 0 & 0 & 0 & 4^5 & 0 & 0 \\ 0 & 0 & 0 & 0 & 3^5 & 5.3^4 \\ 0 & 0 & 0 & 0 & 0 & 3^5 \end{bmatrix}$$

and

$$e^{7J} = \begin{bmatrix} e^{28} & 7.e^{28} & \frac{49}{2}.e^{28} & 0 & 0 & 0 \\ 0 & e^{28} & 7.e^{28} & 0 & 0 & 0 \\ 0 & 0 & e^{28} & 0 & 0 & 0 \\ 0 & 0 & 0 & e^{28} & 0 & 0 \\ 0 & 0 & 0 & 0 & e^{21} & 7.e^{21} \\ 0 & 0 & 0 & 0 & 0 & e^{21} \end{bmatrix}.$$

Example **2.8.6** For the matrix $A = \begin{bmatrix} 2 & 1 & 0 \\ 0 & 2 & 0 \\ 1 & 1 & 3 \end{bmatrix}$, find $e^{\alpha A}$.

Solution The eigenvalues of A are 2 (with algebraic multiplicity 2) and 3 (with algebraic multiplicity 1). The eigenvalue 2 has only one linearly independent eigenvector

$$X_1 = \begin{bmatrix} -1 \\ 0 \\ 1 \end{bmatrix}.$$

Hence, the geometric multiplicity is 1. So, we need to find one generalized eigenvector Y satisfying the equation $(A - 2I)Y = X_1$, which is equivalent to

$$\begin{bmatrix} 0 & 1 & 0 \\ 0 & 0 & 0 \\ 1 & 1 & 1 \end{bmatrix} \begin{bmatrix} x_1 \\ x_2 \\ x_3 \end{bmatrix} = \begin{bmatrix} -1 \\ 0 \\ 1 \end{bmatrix}.$$

The above system yields $x_2 = -1$ and $x_1 = 2 - x_3$. By taking $x_3 = 1$, we get the following generalized eigenvector $Y = \begin{bmatrix} 1 \\ -1 \\ 1 \end{bmatrix}$. Now, an eigenvector corresponding to the eigenvalue 3 is obtained by solving

$$\begin{bmatrix} -1 & 1 & 0 \\ 0 & -1 & 0 \\ 1 & 1 & 0 \end{bmatrix} \begin{bmatrix} x_1 \\ x_2 \\ x_3 \end{bmatrix} = \begin{bmatrix} 0 \\ 0 \\ 0 \end{bmatrix},$$

which gives $x_1 = x_2 = 0$. Hence $\begin{bmatrix} 0 \\ 0 \\ 1 \end{bmatrix}$ is an eigenvector for the eigenvalue 3. Let

$$P = \begin{bmatrix} -1 & 1 & 0 \\ 0 & -1 & 0 \\ 1 & 1 & 1 \end{bmatrix}.$$

Then

$$P^{-1}AP = J = \begin{bmatrix} 2 & 1 & 0 \\ 0 & 2 & 0 \\ 0 & 0 & 3 \end{bmatrix} \text{ and } e^{\alpha J} = \left[\begin{array}{cc|c} e^{2\alpha} & \alpha e^{2\alpha} & 0 \\ 0 & e^{2\alpha} & 0 \\ \hline 0 & 0 & e^{3\alpha} \end{array}\right].$$

Hence, from (2.32), we obtain

$$e^{\alpha A} = Pe^{\alpha J}P^{-1} = e^{2\alpha} \begin{bmatrix} -1 & \alpha - 2 & 0 \\ 0 & 1 & 0 \\ e^{\alpha} - 1 & 2e^{\alpha} - 2 - \alpha & e^{\alpha} \end{bmatrix}.$$

Finding e^{At} Using the Laplace Transform

Definition 2.8.7 (Laplace transform) Let $x : [0, \infty) \to \mathbb{R}$ be a real valued function. The Laplace transform of $x(t)$, denoted by $\mathscr{L}(x(t))$, is defined as

$$\mathscr{L}(x(t)) = \int_0^{\infty} e^{-st} x(t) dt, \quad s > 0, \tag{2.39}$$

if the integral exists.

As $\mathscr{L}(x(t))$ is a function of s, it is represented as $X(s)$.
(Inverse Laplace transform) If $\mathscr{L}(x(t)) = X(s)$, then the inverse Laplace transform is defined as $\mathscr{L}^{-1}(X(s)) = x(t)$.

Example **2.8.8** The Laplace and inverse Laplace transforms of some standard functions are provided below. For $n \geq 0$ and $a \in \mathbb{R}$;

(*i*) $\mathscr{L}(t^n) = \frac{n!}{s^{n+1}}$ and $\mathscr{L}^{-1}\left(\frac{1}{s^{n+1}}\right) = \frac{t^n}{n!}$.

(ii) $\mathscr{L}(e^{at}) = \frac{1}{s-a}$ and $\mathscr{L}^{-1}\left(\frac{1}{s-a}\right) = e^{at}$ for $s > a$.

(iii) $\mathscr{L}(\sin at) = \frac{a}{s^2+a^2}$ and $\mathscr{L}^{-1}\left(\frac{1}{s^2+a^2}\right) = \frac{1}{a}\sin at$.

(iv) $\mathscr{L}(\cos at) = \frac{s}{s^2+a^2}$ and $\mathscr{L}^{-1}\left(\frac{s}{s^2+a^2}\right) = \cos at$.

Let $x(t) = \begin{bmatrix} x_1(t) & x_2(t) & \cdots & x_n(t) \end{bmatrix}' \in \mathbb{R}^n$ for $t > 0$. Then

$$X(s) = \mathscr{L}(x(t)) = \begin{bmatrix} X_1(s) & X_2(s) & \cdots & X_n(s) \end{bmatrix}',$$

where $X_i(s) = \mathscr{L}(x_i(t))$ for $i = 1, 2, \ldots, n$.

Let $A(t) = \{a_{ij}(t)\}_{i,j=1,2,\ldots,n}$ be an $n \times n$ matrix function of t. Then

$$\bar{A}(s) = \mathscr{L}(A(t)) = \{\bar{a}_{ij}(s)\}_{i,j=1,2,\ldots,n},$$

where $\bar{a}_{ij}(s) = \mathscr{L}(a_{ij}(t))$ for $i, j = 1, 2, \ldots, n$.

***Example* 2.8.9** For the matrix $A(t) = \begin{bmatrix} 1 & t \\ t^2 & e^t \end{bmatrix}$, $\bar{A}(s) = \mathscr{L}(A(t)) = \begin{bmatrix} \frac{1}{s} & \frac{1}{s^2} \\ \frac{2}{s^3} & \frac{1}{s-1} \end{bmatrix}$ for $s > 1$.

Clearly, for $s > 1$, $\mathscr{L}^{-1}\left(\begin{bmatrix} \frac{1}{s} & \frac{1}{s^2} \\ \frac{2}{s^3} & \frac{1}{s-1} \end{bmatrix}\right) = \begin{bmatrix} 1 & t \\ t^2 & e^t \end{bmatrix}$.

Lemma 2.8.10 (i) If $\mathscr{L}(x(t)) = X(s)$, then $\mathscr{L}(x^{(k)}(t)) = s^k X(s) - s^{k-1}x(0) - s^{k-2}x^{(1)}(0) - \cdots - x^{(k-1)}(0)$, where $x^{(k)}(t)$ denotes the kth derivate of x with respect to t.

(ii) $\mathscr{L}(tx(t)) = -X^{(1)}(s)$ and $\mathscr{L}(t^k x(t)) = (-1)^k X^{(k)}(s)$, where $X^{(k)}s)$ denotes the kth derivative of X with respect to s.

Proof. The desired identities can be easily derived from Definition 2.8.7. □

Consider the system of first order differential equations

$$\dot{x} = Ax, \quad t \geq 0, \tag{2.40}$$

where A is an $n \times n$ matrix and $x(t) \in \mathbb{R}^n$ for each $t \geq 0$. Applying the Laplace transform on both sides of (2.40), we get $sX(s) - x(0) = AX(s)$. Hence, we have $(sI - A)X(s) = x(0)$ and $X(s) = (sI - A)^{-1}x(0)$. Thus,

$$x(t) = \mathscr{L}^{-1}(X(s)) = \mathscr{L}^{-1}((sI - A)^{-1})x(0). \tag{2.41}$$

It can be shown that (see Exercise 2.8.4) that for any matrix A of order n, $\mathscr{L}(e^{At}) = (sI-A)^{-1}$. Hence from (2.41) we obtain.

$$x(t) = e^{At}x(0). \tag{2.42}$$

Since

$$e^{At} = \mathscr{L}^{-1}((sI-A)^{-1}), \tag{2.43}$$

equation (2.43) provides an alternative method for computing e^{At} for any square matrix A.

***Example* 2.8.11** Using the Laplace transform, find e^{At}, $t > 0$, for the matrix $A = \begin{bmatrix} 1 & 4 \\ 2 & 3 \end{bmatrix}$.

Solution From (2.43), we have $e^{At} = \mathscr{L}^{-1}((sI-A)^{-1})$. As

$$(sI-A)^{-1} = \begin{bmatrix} s-1 & -4 \\ -2 & s-3 \end{bmatrix}^{-1} = \frac{1}{(s-5)(s+1)} \begin{bmatrix} s-3 & 4 \\ 2 & s-1 \end{bmatrix},$$

$$\mathscr{L}^{-1}((sI-A)^{-1}) = \mathscr{L}^{-1} \begin{bmatrix} \frac{s-3}{(s-5)(s+1)} & \frac{4}{(s-5)(s+1)} \\ \frac{2}{(s-5)(s+1)} & \frac{s-1}{(s-5)(s+1)} \end{bmatrix}.$$

Now

$$\mathscr{L}^{-1}\left(\frac{s-3}{(s-5)(s+1)}\right) = \mathscr{L}^{-1}\left(\frac{2}{s-5} - \frac{1}{s+1}\right) = 2e^{5t} - e^{-t},$$

$$\mathscr{L}^{-1}\left(\frac{s-1}{(s-5)(s+1)}\right) = \mathscr{L}^{-1}\left(\frac{2/3}{s-5} + \frac{1/3}{s+1}\right) = \frac{2}{3}e^{5t} + \frac{1}{3}e^{-t},$$

$$\mathscr{L}^{-1}\left(\frac{1}{(s-5)(s+1)}\right) = \mathscr{L}^{-1}\left(\frac{1/6}{s-5} - \frac{1/6}{s+1}\right) = \frac{1}{6}e^{5t} - \frac{1}{6}e^{-t}.$$

Therefore,

$$e^{At} = \begin{bmatrix} 2e^{5t} - e^{-t} & \frac{2}{3}(e^{5t} - e^{-t}) \\ \frac{1}{3}(e^{5t} - e^{-t}) & \frac{2}{3}e^{5t} + \frac{1}{3}e^{-t} \end{bmatrix}.$$

Exercise 2.8.1 If $J(\alpha) = \begin{bmatrix} \lambda & \alpha & 0 & 0 & \cdots & 0 \\ 0 & \lambda & \alpha & 0 & \cdots & 0 \\ \vdots & & & & & \\ 0 & 0 & & \cdots & \lambda & \alpha \\ 0 & 0 & & \cdots & 0 & \lambda \end{bmatrix}_{n\times n}$, find $(J(\alpha))^k : k = 2, 3, \ldots$, and show that $e^{J(\alpha)} = e^{\lambda} \begin{bmatrix} 1 & \alpha & \frac{\alpha^2}{2!} & \cdots & & \frac{\alpha^{n-1}}{(n-1)!} \\ 0 & 1 & \alpha & \cdots & & \frac{\alpha^{n-2}}{(n-2)!} \\ \vdots & & & & & \\ 0 & 0 & & \cdots & 1 & \alpha \\ 0 & 0 & & \cdots & 0 & 1 \end{bmatrix}$. Hence, show that $e^{-\lambda} e^{J(\alpha)} = e^{-\alpha\lambda} e^{\alpha J(1)}$.

Exercise 2.8.2 For $A = \begin{bmatrix} 3 & 0 \\ 2 & 3 \end{bmatrix}$, find e^{5A} using (i) eigenpairs of A and (ii) the Laplace transform method.

Exercise 2.8.3 Let $A = \begin{bmatrix} A_1 & & 0 \\ & A_2 & \\ 0 & & A_3 \end{bmatrix}$ be a 6×6 matrix, where the block matrices $A_1 = \begin{bmatrix} 2 & 1 \\ 1 & 2 \end{bmatrix}$, $A_2 = \begin{bmatrix} 3 \end{bmatrix}$, and $A_3 = \begin{bmatrix} 1 & 0 & 0 \\ 3 & 1 & 5 \\ 0 & 0 & 1 \end{bmatrix}$. Find ($i$) e^{2A}, (ii) $\det(e^{2A})$, and (iii) $\det f(A)$, where $f(t)$ is a polynomial with a root at $t = 1$.

Exercise 2.8.4 (*i*) Let $J_s = \begin{bmatrix} \lambda & 1 & & & 0 \\ & \lambda & 1 & & \\ & & \ddots & \ddots & \\ & & & \ddots & 1 \\ 0 & & & & \lambda \end{bmatrix}$ be a Jordan subblock of order r of a matrix A. Find $\mathscr{L}(e^{J_s t})$ and show that it is the same as $(sI - J_s)^{-1}$.

(*ii*) Show that $\mathscr{L}(e^{At}) = (sI - A)^{-1}$ for any square matrix A.

2.9 Dot Product of Vectors in $\mathbb{R}^n$ and $\mathbb{C}^n$

A vector $X \in \mathbb{R}^n$ is represented by the column

$$X = \begin{bmatrix} x_1 \\ x_2 \\ \vdots \\ x_n \end{bmatrix}.$$

The **transpose of** X is denoted by X' and is defined as

$$X' = \begin{bmatrix} x_1 & x_2 & \cdots & x_n \end{bmatrix}.$$

The **dot product** between two vectors X and Y in $\mathbb{R}^n$ is defined by

$$X \cdot Y = x_1 y_1 + x_2 y_2 + \cdots + x_n y_n,$$

which can also be written as $Y'X$.
The **Euclidean norm** of a vector X in $\mathbb{R}^n$ is defined as

$$\|X\| = \sqrt{X \cdot X} = \sqrt{x_1^2 + x_2^2 + \cdots + x_n^2} \text{ and } \|X\|^2 = X'X.$$

The space $\mathbb{C}^n = \{(x_1, x_2, \ldots, x_n) : x_i \in \mathbb{C}, i = 1, 2, \ldots, n\}$. If $X = \begin{bmatrix} x_1 \\ x_2 \\ \vdots \\ x_n \end{bmatrix}$ and $Y = \begin{bmatrix} y_1 \\ y_2 \\ \vdots \\ y_n \end{bmatrix}$ are in $\mathbb{C}^n$, then the dot product is defined as $X \cdot Y = x_1\bar{y}_1 + x_2\bar{y}_2 + \cdots + x_n\bar{y}_n$, where $\bar{y}_i$ is the complex conjugate of y_i. Hence,

$$X \cdot Y = \overline{Y}'X \text{ and } \|X\|^2 = \overline{X}'X = |x_1|^2 + |x_2|^2 + \cdots + |x_n|^2.$$

Orthogonal Vectors: Two vectors $X, Y \in \mathbb{C}^n$ are orthogonal if $X \cdot Y = 0$, i.e., $\sum_{i=1}^{n} x_i\bar{y}_i = 0$.

Orthogonal Set: A set $\{X_1, X_2, \ldots, X_m\} \subset \mathbb{C}^n$ is called an orthonormal set if

$$X_i \cdot X_j = \begin{cases} 0, & \text{if } i \neq j, \\ \text{nonzero}, & \text{if } i = j. \end{cases}$$

Orthonormal Set: A set $\{X_1, X_2, \ldots, X_m\}$ in $\mathbb{C}^n$ is said to be an orthonormal set if

$$X_i \cdot X_j = \begin{cases} 0, & \text{if } i \neq j, \\ 1, & \text{if } i = j. \end{cases}$$

If a set $\{X_1, X_2, \ldots, X_m\}$ is orthogonal, then $\left\{\frac{X_1}{\|X_1\|}, \frac{X_2}{\|X_2\|}, \ldots, \frac{X_n}{\|X_n\|}\right\}$ is orthonormal.

Example **2.9.1** The set $\{(1,1,0), (1,-1,0), (0,0,1)\}$ is orthogonal in $\mathbb{R}^3$. Dividing each vector by its norm, we obtain the corresponding orthonormal set $\{(\frac{1}{\sqrt{2}}, \frac{1}{\sqrt{2}}, 0), (\frac{1}{\sqrt{2}}, -\frac{1}{\sqrt{2}}, 0), (0,0,1)\}$.

Example **2.9.2** The set $\{(1, 1+i, 2), (0, 1+i, -1), (-3-3i, i, 1+i)\}$ is orthogonal in $\mathbb{C}^3$. The corresponding orthonormal set is $\{(\frac{1}{\sqrt{7}}, \frac{1+i}{\sqrt{7}}, \frac{2}{\sqrt{7}}),\ (0, \frac{1+i}{\sqrt{3}}, -\frac{1}{\sqrt{3}}),\ (-\frac{3+3i}{\sqrt{21}}, \frac{i}{\sqrt{21}}, \frac{1+i}{\sqrt{21}})\}$.

The Gram–Schmidt Process of Orthogonalization

From a linearly independent set in $\mathbb{R}^n$, one can obtain an orthogonal set by applying the Gram–Schmidt process, which is given below.

Theorem 2.9.3 Let $S = \{v_1, v_2, \ldots, v_k\}$ be a linearly independent set in $\mathbb{R}^n$. Then the set $\{u_1, u_2, \ldots, u_k\}$, obtained from S, is orthogonal, where

$$
\begin{aligned}
u_1 &= v_1,\\
u_2 &= v_2 - (v_2 \cdot u_1)\frac{u_1}{\|u_1\|^2},\\
u_3 &= v_3 - (v_3 \cdot u_2)\frac{u_2}{\|u_2\|^2} - (v_3 \cdot u_1)\frac{u_1}{\|u_1\|^2},\\
&\vdots\\
u_k &= v_k - (v_k \cdot u_{k-1})\frac{u_{k-1}}{\|u_{k-1}\|^2} - \cdots - (v_k \cdot u_1)\frac{u_1}{\|u_1\|^2}.
\end{aligned}
$$

Proof. The proof can be established by induction by proceeding with the following steps. Step I: Verify that $u_1 \cdot u_2 = 0$. Step II: Assume that $\{u_1, u_2, \ldots, u_r\}$ is an orthogonal set for some $r < k$. Step III: Show that $u_i \cdot u_{r+1} = 0$ for $i = 1, 2, \ldots, r$. Hence, $\{u_1, u_2, \ldots, u_r, u_{r+1}\}$ is orthogonal. Since the set $\{u_1, u_2\}$ is orthogonal, using step III, we conclude that the set $u_1, u_2, \cdots, u_r$ is orthogonal for $r = 3, 4, \cdots, k$. This completes the proof. The details are left to the reader. □

***Example* 2.9.4** Find an orthonormal set from the linearly independent set $\{(1,1,0), (1,0,1), (1,1,1)\}$.

Solution Let the given set be denoted by $\{v_1, v_2, v_3\}$. From the Gram–Schmidt process, we obtain an orthogonal set $\{u_1, u_2, u_3\}$, where

$$
\begin{aligned}
u_1 &= v_1 = (1,1,0),\\
u_2 &= v_2 - (v_2 \cdot u_1)\frac{u_1}{\|u_1\|^2} = (1,0,1) - \frac{(1,1,0)}{2} = (\frac{1}{2}, -\frac{1}{2}, \frac{1}{2}),\\
u_3 &= v_3 - (v_3 \cdot u_2)\frac{u_2}{\|u_2\|^2} - (v_3 \cdot u_1)\frac{u_1}{\|u_1\|^2} = (1,1,1) - \frac{\frac{1}{2}(\frac{1}{2}, -\frac{1}{2}, \frac{1}{2})}{\frac{3}{4}} - \frac{2(1,1,0)}{2}\\
&= (1,1,1) - (\frac{1}{3}, -\frac{1}{3}, \frac{1}{3}) - (1,1,0) = (-\frac{1}{3}, \frac{1}{3}, \frac{2}{3}).
\end{aligned}
$$

Therefore, $\left\{(\frac{1}{\sqrt{2}}, \frac{1}{\sqrt{2}}, 0), (\frac{1}{\sqrt{3}}, -\frac{1}{\sqrt{3}}, \frac{1}{\sqrt{3}}), (-\frac{1}{\sqrt{6}}, \frac{1}{\sqrt{6}}, \frac{\sqrt{2}}{\sqrt{3}})\right\}$ is an orthonormal set, as required.

Cauchy–Schwarz inequality: The following theorem provides a very important inequality called the Cauchy–Schwarz inequality which is a useful tool in analytical and numerical studies in many branches of mathematics. Here, the proof is given for vectors in $\mathbb{C}^n$. However, the inequality is valid in any inner product space, as will be discussed in Chapter 3, Section 3.2.

Theorem 2.9.5 For any two vectors u and v in $\mathbb{C}^n$, $|u \cdot v| \leq \|u\| \|v\|$. In other words, if $u = (u_1, u_2, \ldots, u_n)$ and $v = (v_1, v_2, \ldots, v_n)$, then $\left|\sum_{i=1}^{n} u_i \bar{v}_i\right| \leq \sqrt{\sum_{i=1}^{n} |u_i|^2} \sqrt{\sum_{i=1}^{n} |v_i|^2}$.

Proof. If $u = \Theta$, then $\|u\| = 0$ and $u \cdot v = 0$ for all $v \in \mathbb{C}^n$; hence, the inequality holds trivially. Assume now that $u \neq \Theta$. Let $w = v - \frac{v \cdot u}{\|u\|^2} u = v - \alpha u$, where $\alpha = \frac{v \cdot u}{\|u\|^2}$. Then it is easy to verify that $w \cdot u = 0$. Since $w \cdot w = \|w\|^2 \geq 0$, we obtain $0 \leq (v - \alpha u) \cdot (v - \alpha u) = \|v\|^2 - \bar{\alpha}(v \cdot u) - \alpha(u \cdot v) + |\alpha|^2 \|u\|^2 = \|v\|^2 - \frac{|u \cdot v|^2}{\|u\|^2}$. Hence, $|u \cdot v|^2 \leq \|u\|^2 \|v\|^2$, which implies $|u \cdot v| \leq \|u\| \|v\|$. This completes the proof. □

Some Special Matrices

Let A be an $m \times n$ matrix. Then the **transpose of** A, denoted by A', is obtained by interchanging the rows and columns of A. If $A = \{a_{ij}\}_{i=1,2,\ldots,m; j=1,2,\ldots,n}$, then $A' = \{a_{ji}\}_{j=1,2,\ldots,n; i=1,2,\ldots,m}$.

If A has complex entries, then the conjugate $\bar{A}$ of A is defined by $\bar{A} = \{\bar{a}_{ij}\}_{i=1,2,\ldots,m; j=1,2,\ldots,n}$. The adjoint A^* of A is defined as

$$A^* = \bar{A}' = \{\bar{a}_{ji}\}_{j=1,2,\ldots,n; i=1,2,\ldots,m}.$$

If A is a real matrix (i.e., all $a_{ij} \in \mathbb{R}$), then $A^* = A'$.

Definition 2.9.5

Symmetric matrix: A **real** matrix A of order n is called symmetric if $A = A'$.

Skew-symmetric matrix: A **real** matrix A of order n is called skew-symmetric if $A = -A'$.

Hermitian matrix: An $n \times n$ matrix A is called Hermitian if $A = A^*$.

If A is Hermitian, then $a_{ij} = \bar{a}_{ji}$ for all i, j. In particular, for all i, $a_{ii} = \bar{a}_{ii}$, implying that the diagonal entries are real.

Skew-Hermitian matrix: An $n \times n$ matrix A with complex entries is called skew-Hermitian if $A = -A^*$.

If A is skew-Hermitian, then $a_{ij} = -\bar{a}_{ji}$ for all i, j. In particular, for all i, $a_{ii} = -\bar{a}_{ii}$, implying that the diagonal entries are purely imaginary.

Orthogonal matrix: A real matrix A is called orthogonal if $AA' = A'A = I$, where I is the identity matrix.

Thus for an orthogonal matrix A, $A' = A^{-1}$.

Unitary matrix: A complex matrix A is called unitary if $AA^* = A^*A = I$, where I is the identity matrix.

Hence if A is unitary A, $A^* = A^{-1}$.

Positive definite: A real symmetric (Hermitian) matrix A is called positive definite if $AX \cdot X > 0$, i.e., $X'AX > 0$ for all nonzero $X \in \mathbb{R}^n$ ($\overline{X}'AX > 0$ for all nonzero $X \in \mathbb{C}^n$).
Positive semi-definite: A real symmetric (Hermitian) matrix A is called positive semi-definite if $AX \cdot X \geq 0$, i.e., $X'AX \geq 0$ for all $X \in \mathbb{R}^n$ ($\overline{X}'AX \geq 0$ for all $X \in \mathbb{C}^n$).
Negative definite and **negative semi-definite** matrices are defined analogously by reversing the inequality signs.

Theorem 2.9.6 The eigenvalues of Hermitian matrices are real.

Proof. Let A be an $n \times n$ Hermitian matrix, and let λ be an eigenvalue with corresponding eigenvector X. Then $AX = \lambda X$. We aim to prove that $\lambda \in \mathbb{R}$. If not, let λ be a complex number. Then,

$$AX \cdot X = \lambda X \cdot X = \lambda(X \cdot X) = \lambda \|X\|^2. \tag{2.44}$$

Moreover,

$$\begin{aligned} AX \cdot X &= \overline{X}'(AX) = (\overline{X}'A)X = (\overline{X}'\,\overline{A}')X \quad (\text{since } A = \overline{A}') \\ &= (\overline{A}\,\overline{X})'X = \overline{\lambda}(\overline{X}'X) = \overline{\lambda}(X \cdot X) = \overline{\lambda}\|X\|^2. \end{aligned} \tag{2.45}$$

From (2.44) and (2.45), we get $(\lambda - \overline{\lambda})\|X\|^2 = 0$. Since $X \neq \Theta$, it follows that $\lambda = \overline{\lambda}$. Therefore, λ is a real number. □

Theorem 2.9.7 If A is Hermitian, then the eigenvectors corresponding to distinct eigenvalues are orthogonal.

Proof. Let λ_1 and λ_2 be distinct eigenvalues of A with corresponding eigenvectors X_1 and X_2. Then

$$AX_1 \cdot X_2 = \lambda_1 X_1 \cdot X_2 = \lambda_1 (X_1 \cdot X_2). \tag{2.46}$$

Moreover,

$$\begin{aligned} AX_1 \cdot X_2 &= \overline{X_2}'(AX_1) = (\overline{X_2}'\,\overline{A}')X_1 \quad (\text{since } A = \overline{A}') \\ &= (\overline{A}\,\overline{X_2})'X_1 = \overline{\lambda_2}(\overline{X_2}'X_1) = \lambda_2(X_1 \cdot X_2) \quad (\text{since } \lambda_2 \text{ is real}). \end{aligned} \tag{2.47}$$

From (2.46) and (2.47), we get $(\lambda_1 - \lambda_2)(X_1 \cdot X_2) = 0$. Since $\lambda_1 \neq \lambda_2$, $X_1 \cdot X_2 = 0$, as required. □

Remark 2.9.8 Theorems 2.9.6 and 2.9.7 also hold for real symmetric matrices.

In the following, we state without proof an important property of symmetric (Hermitian) matrices.

Theorem 2.9.9 If A is a symmetric (Hermitian) matrix, then there exists an orthogonal (unitary) matrix P such that $P^*AP = D$, where D is a diagonal matrix having the eigenvalues of A on the diagonal. The columns of P are the corresponding orthonormal eigenvectors of A.

Theorem 2.9.10 If A is skew-symmetric (skew-Hermitian), then

(i) the eigenvalues of A are purely imaginary,

(ii) the eigenvectors corresponding to distinct eigenvalues of A are orthogonal.

The proof of the above theorem is similar to that of Theorems 2.9.6 and 2.9.7.

Theorem 2.9.11

(i) If A is positive definite, then all its eigenvalues are positive.

(ii) If A is positive semi-definite, then all its eigenvalues are nonnegative.

(iii) If A is negative definite, then all its eigenvalues are negative.

(iv) If A is negative semi-definite, then all its eigenvalues are non-positive.

Proof. (i) Let A be an $n \times n$ positive definite matrix. Then A is symmetric (Hermitian), and $AX \cdot X > 0$ for all $X \in \mathbb{R}^n$ ($\mathbb{C}^n$). Let λ be an eigenvalue of A with eigenvector X. Then $AX \cdot X = \lambda(X \cdot X) > 0 \implies \lambda \|X\|^2 > 0 \implies \lambda > 0$.

The proofs of the other parts are similar. □

Exercise 2.9.1 If A is an $n \times n$ symmetric matrix, prove that there exists non trivial upper triangular matrix B such that AB is symmetric.

Exercise 2.9.2 Show that $I + A^*A$ is invertible for any $n \times n$ matrix A and that $I - A^*A$ is invertible if the eigenvalues of A lie in the interior of the unit circle in the complex plane.

2.10 Summary

- For an $m \times n$ matrix A, the column rank is the number of linearly independent columns of A, and the row rank is the number of linearly independent rows of A (Definition 2.3.1).
- For any matrix, the column rank = the row rank, which is called the rank of the matrix (Theorem 2.3.10).

- The rank of a matrix A is the dimension of its range (Section 2.3).
- For an $n \times n$ matrix A, $\text{rank}(A) = n \iff \det(A) \neq 0 \iff A^{-1}$ exists $\iff$ the row reduced echelon form of A is the identity matrix $\iff$ nullity of A is 0 $\iff$ all the eigenvalues of A are nonzero (see Theorem 2.6.4 and Remark 2.3.11).
- A square matrix is diagonalizable if and only if the algebraic multiplicity = the geometric multiplicity for every eigenvalue of A (see Theorem 2.7.2 and equations (2.22)–(2.24)).
- For any eigenvalue of A, if the algebraic multiplicity > the geometric multiplicity, then A has a Jordan canonical form J, and there exists a nonsingular matrix P such that $P^{-1}AP = J$ (see equations (2.25)–(2.29)).
- If any eigenvalue of A has l linearly independent eigenvectors, then there will be l Jordan blocks corresponding to that eigenvalue in the Jordan canonical form of A.
- $e^A = Pe^J P^{-1}$ (see (2.33) and (2.38)).
- Orthogonal set, the Gram–Schmidt orthogonalization, special matrices, and their spectral properties are discussed in Section 2.9.

2.11 Exercise

1. Check the linear dependence/independence of the following sets:

   (i) $\{(2,1,1),(3,0,2),(8,1,5),(1,1,1)\}$,
   (ii) $\{(1,1,-1,1),(2,0,1,0),(3,1,0,1)\}$.

2. Find the row reduced echelon form of the following matrices:

$$(i)\ \begin{bmatrix} 1 & 2 & 3 \\ -1 & 3 & 2 \\ 1 & 4 & 6 \\ 1 & 1 & 0 \end{bmatrix}; \quad (ii)\ \begin{bmatrix} 2 & 2 & 4 & 0 & -8 \\ 1 & 2 & 0 & -4 & 1 \\ 0 & -4 & 1 & 1 & 2 \\ 2 & 0 & -4 & 1 & 1 \\ -4 & 1 & 1 & 2 & 0 \end{bmatrix}.$$

3. Let $\{r_1, r_2, \ldots, r_m\}$ and $\{c_1, c_2, \ldots, c_n\}$ be the rows and columns of an $m \times n$ matrix A. Show that the column rank of A remains the same if

(*i*) two rows are interchanged,

(*ii*) any row is multiplied by a nonzero constant,

(*iii*) the ith row is replaced by the sum of the ith and j^{th} rows.

4. Show that for a row reduced echelon form matrix, the row rank equals the column rank.

5. Determine the constants a and b such that the matrix $\begin{bmatrix} a & 1 & 2 \\ 1 & 3 & 6 \\ 0 & 2 & b \end{bmatrix}$ has (*i*) rank 2 and (*ii*) rank 3.

6. Using row operations, find the inverse of the matrix $A = \begin{bmatrix} 3 & 1 & 0 \\ 1 & 3 & 0 \\ 0 & 4 & 2 \end{bmatrix}$.

7. Let $\{v_1, v_2, v_3\}$ be a basis of a subspace S of $\mathbb{R}^n$ and S_1 a subspace of S. If $v_1, v_2 \notin S_1$ and $v_3 \in S_1$, then which of the following statements is correct?

   (*i*) The dimension of S_1 is 1.

   (*ii*) The dimension of S_1 can be 1 or 2.

8. State with justification whether the following statements are true:

   (*i*) If $\{v_1, v_2, v_3\}$ is linearly dependent and the subsets $\{v_1, v_2\}$ and $\{v_2, v_3\}$ are linearly independent, then $\{v_1, v_3\}$ is linearly dependent.

   (*ii*) If $\{v_1, v_2, v_3\}$ is a basis for a subspace S of $\mathbb{R}^n$, then $\{v_1 + v_2, v_1 - v_2, v_2 + v_3\}$ is also a basis for S.

9. Show that the following sets are subspaces of their respective vector spaces, and find a basis for each subspace:

   (*i*) $S_1 = \{(x_1, x_2, x_3) \in \mathbb{R}^3 : x_1 + x_2 = 0\}$.

   (*ii*) $S_2 = \{a_0 + a_1 x + a_2 x^2 + a_3 x^3 \in P_3(0,1) : a_0 = a_2, a_0 + a_1 = a_3\}$, where $P_3(0,1)$ is the set of all polynomials of degree ≤ 3 defined on $(0,1)$.

   (*iii*) $S_3 = \{p(x) \in P_3(-2,2) : p(0) = p(1), p(-1) = 0\}$.

10. Find the values of α, β, and γ for which the following system of equations has (i) a unique solution, (ii) infinitely many solutions, or (iii) no solution:

$$\begin{cases} 2x+3y-z=2, \\ x-4y+\alpha z=\beta, \\ 3x+10y-4z=\gamma. \end{cases}$$

11. (i) Let A be an $n\times n$ matrix. If $A^{n+1}=\mathbb{O}$, the zero matrix, then show that $A^k=\mathbb{O}$ for some $k\leq n$.

    (ii) If A is a 4×4 matrix such that $(A-2I)^3=\mathbb{O}$ but $(A-2I)^2\neq\mathbb{O}$, show that $\lambda=2$ is an eigenvalue of A with algebraic multiplicity 4 and geometric multiplicity 2.

12. Let $A^*=\overline{A}'$ be the adjoint of an $m\times n$ matrix A. Show that

    (i) $\text{rank}(A)=\text{rank}(A^*)$.

    (ii) $\text{Range}(A^*)$ is orthogonal to the null space of A.

    (iii) A^*A is positive definite if and only if $\text{Null}(A)=0$.

13. Show that:

    (i) If A is a square matrix with an eigenvalue λ, then $\overline{\lambda}$ is an eigenvalue of A^*.

    (ii) If A is a skew-symmetric matrix of odd order, then $\det(A)=0$.

    (iii) If $x'Ax=0$ for all $x\in\mathbb{R}^n$, then A is either a skew-symmetric or the zero matrix of order n.

    (iv) If A and B are matrices of order n, then AB and BA have the same set of eigenvalues.

14. If H is a unitary matrix and H^*AH and H^*BH are both diagonal matrices, then $AB=BA$.

15. Let A be a square matrix of order n. Show that:

    (i) A and A' have the same eigenvalues.

    (ii) If A is unitary, then all its eigenvalues have modulus 1.

    (iii) If $A^2=A$, then its eigenvalues are either 0 or 1.

    (iv) If $A^k=\mathbb{O}$ and $A^{k-1}\neq\mathbb{O}$ for some integer $k\geq 1$, then all the eigenvalues of A are zero.

(v) Every real matrix A can be written as the sum of a symmetric matrix and a skew-symmetric matrix.

16. Using the Gram–Schmidt process, construct an orthonormal set from the set $\{(1,1,0),(1,0,1),(0,2,3)\}$.

17. For the matrix $A = \begin{bmatrix} 3 & 0 & 0 \\ 1 & 3 & 0 \\ 2 & 0 & 3 \end{bmatrix}$, find a nonsingular matrix P such that $P^{-1}AP = J(3)$.

18. If A is a real symmetric positive definite matrix of order n with eigenpairs (λ_i, v_i), $i = 1,2,\ldots,n$, show that there exists a constant $c > 0$ such that $Av \cdot v > c\|v\|^2$ for all $v \in \mathbb{R}^n$.

3

Normed Linear Spaces

3.1 Introduction

In this chapter, key definitions, theorems, and techniques from linear algebra and functional analysis are briefly reviewed through examples. The purpose is to introduce notations and concepts that will recur in subsequent chapters.

In Section 3.2, definitions of vector space, norm, and inner product are given along with examples. In Section 3.3, linear dependence and independence of vector valued functions and certain properties of time-varying matrices are discussed. Section 3.4 contains definitions and examples of bounded linear operators, Banach spaces, Hilbert spaces, and adjoint operator. In Section 3.5, the Banach fixed point theorem and some examples are given. In Section 3.6, compact set, compact operator, closed operator, and open mapping are discussed briefly.

Section 3.6 is included primarily to support Chapter 10 on infinite dimensional systems. Readers without a background in basic functional analysis may skip both Section 3.6 and Chapter 10.

3.2 Vector Space

Definition 3.2.1 A **field** $\mathbb{F}$ is a set of elements, called **scalars**, together with two operations, addition and multiplication, for which the following axioms hold:

(i) For $a, b \in \mathbb{F}$, $a+b \in \mathbb{F}$ and $ab \in \mathbb{F}$. Moreover, $a+b=b+a$ and $ab=ba$ (the addition and multiplication are **commutative**).

(*ii*) For $a,b,c \in \mathbb{F}$, $a+(b+c)=(a+b)+c$ and $a(bc)=(ab)c$ (the addition and multiplication are **associative**).

(*iii*) For $a,b,c \in \mathbb{F}, a(b+c)=ab+ac$ (**distributive property**).

(*iv*) $\mathbb{F}$ contains the zero element 0, called the **additive identity**, such that $a+0=a$ for all $a \in \mathbb{F}$.

(*v*) $\mathbb{F}$ contains the unit element 1, called the **multiplicative identity**, such that $1a=a$ for all $a \in \mathbb{F}$.

(*vi*) For every element $a \in \mathbb{F}$, there exists an element $b \in \mathbb{F}$, called the **additive inverse**, such that $a+b=0$.

(*vii*) For every nonzero element $a \in \mathbb{F}$, there exists an element $b \in \mathbb{F}$, called the **multiplicative inverse**, such that $ab=1$.

***Example* 3.2.2** The following sets are fields with usual addition and multiplication of numbers: (*a*) The set of all real numbers $\mathbb{R}$. (*b*) The set of all complex numbers $\mathbb{C}$. (*c*) The set of all rational numbers $\mathbb{Q}$.

***Example* 3.2.3** The following sets are not fields with usual addition and multiplication: (*a*) The set of all natural numbers $\mathbb{N}$. (The axiom (*iv*) is not satisfied.) (*b*) The set of all nonnegative real numbers $\mathbb{R}_0^+$. (The axiom (*vi*) is not satisfied.) (*c*) The set of all integers $\mathbb{Z}$. (The axiom (*vii*) is not satisfied.)

Vector Space

Definition 3.2.4 A **vector space** V **over a field** $\mathbb{F}$ is a set, whose elements are called vectors, together with two operations called **vector addition** $(+)$ and **scalar multiplication** $(\cdot)$ for which the following axioms hold:

(*i*) For $x,y \in V, x+y \in V$. Moreover, $x+y=y+x$. ($+$ **is commutative**)

(*ii*) For $x \in V$ and $\alpha \in \mathbb{F}, \alpha \cdot x \in V$.

(*iii*) For $x,y,z \in V, (x+y)+z=x+(y+z)$. ($+$ **is associative**)

(*iv*) For $x,y \in V$ and $\alpha \in \mathbb{F}, \alpha \cdot (x+y)=\alpha \cdot x+\alpha \cdot y$. ($+$ **is distributive**)

(*v*) For $x \in V$ and $\alpha,\beta \in \mathbb{F}, \alpha \cdot (\beta \cdot x)=\alpha\beta \cdot x=\beta \cdot (\alpha \cdot x)$ and $(\alpha+\beta) \cdot x=\alpha \cdot x+\beta \cdot x$.

(*vi*) V contains the **zero vector** Θ (identity element) such that $x+\Theta=x$ for all $x \in V$.

(*vii*) For every $x \in V$, there exists $-x \in V$, called the **additive inverse**, such that $x+(-x)=\Theta$.

$(viii)$ For $x \in V$, $1 \cdot x = x$, where 1 denotes the unit element of $\mathbb{F}$.

A vector space can also be called a **linear space**.

***Example* 3.2.5** (i) $\mathbb{R}^n$ over the field $\mathbb{R}$ is an important vector space consisting of n-tuples of real numbers. It is called n-dimensional real vector space. The vector addition and scalar multiplication are defined as follows.
Vector addition: For $x = (x_1, x_2, \ldots, x_n)$ and $y = (y_1, y_2, \ldots, y_n)$,

$$x + y = (x_1 + y_1, x_2 + y_2, \ldots, x_n + y_n).$$

Scalar multiplication: For $x = (x_1, x_2, \ldots, x_n)$ and $\alpha \in \mathbb{R}$,

$$\alpha \cdot x = (\alpha x_1, \alpha x_2, \ldots, \alpha x_n).$$

Analogously, $\mathbb{C}^n$ over the field $\mathbb{C}$ is an n-dimensional complex vector space.
(ii) $\mathbb{R}^{m\times n}$ over $\mathbb{R}$ is a real vector space, where $\mathbb{R}^{m\times n}$ denotes the set of all $m \times n$ real matrices, with the vector addition and scalar multiplication defined as follows.
Vector addition: For $A = \{a_{ij}\}_{i=1,2,\ldots,m;j=1,2,\ldots,n}$ and $B = \{b_{ij}\}_{i=1,2,\ldots,m;j=1,2,\ldots,n}$,

$$A + B = \{a_{ij} + b_{ij}\}_{i=1,2,\ldots,m;j=1,2,\ldots,n}.$$

Scalar multiplication: For $A = \{a_{ij}\}_{i=1,2,\ldots,m;j=1,2,\ldots,n}$ and $\alpha \in \mathbb{R}$,

$$\alpha \cdot A = \{\alpha a_{ij}\}_{i=1,2,\ldots,m;j=1,2,\ldots,n}.$$

***Example* 3.2.6** (i) For $n \in \mathbb{N}$, $P_n[a,b]$, the set of all polynomials of degree $\leq n$ with real coefficients, is a vector space over $\mathbb{R}$. On $P_n[a,b]$, the vector addition and scalar multiplication are defined as follows:
Vector addition: If $P(t) = \sum_{j=0}^{n} \alpha_j t^j$ and $Q(t) = \sum_{j=0}^{n} \beta_j t^j$ for $t \in [a,b]$, the coefficients $\alpha_j, \beta_j \in \mathbb{R}$, then

$$(P+Q)(t) = P(t) + Q(t) = \sum_{j=0}^{n} (\alpha_j + \beta_j) t^j.$$

Scalar multiplication: If $P(t) = \sum_{j=0}^{n} \alpha_j t^j$ and $\alpha \in \mathbb{R}$, then

$$(\alpha P)(t) = \alpha P(t) = \sum_{j=0}^{n} \alpha \alpha_j t^j.$$

Basis: The set $\{1, t, t^2, \ldots, t^n\}$ forms the standard basis of $P_n[a,b]$ and hence the dimension of the space $P_n[a,b]$ is $n+1$.

Similarly, $P[a,b]$, the set of polynomials defined on $[a,b]$ with real coefficients, is a vector space over $\mathbb{R}$.

$P[a,b]$ is an infinite dimensional vector space since the standard basis of the space is $\{1, t^1, t^2, ..., t^n, ...\}$.
(ii) For a finite interval $[a,b] \subset \mathbb{R}$, $C[a,b]$ over $\mathbb{R}$ is a real vector space, where $C[a,b]$ denotes the set of all real valued continuous functions on $[a,b]$, with the vector addition and scalar multiplication defined as follows.
Vector addition: For $f, g \in C[a,b]$,

$$(f+g)(t) = f(t) + g(t) \text{ for all } t \in [a,b].$$

Scalar multiplication: For $f \in C[a,b]$ and $\alpha \in \mathbb{R}$,

$$(\alpha \cdot f)(t) = \alpha f(t) \text{ for all } t \in [a,b].$$

(iii) For $a, b \in \mathbb{R}$, $L_p[a,b]$, where $1 \leq p < \infty$, over $\mathbb{R}$ is a real vector space of all functions f defined on $[a,b]$ such that $\int_a^b |f(t)|^p dt < \infty$, i.e.,

$$L_p[a,b] = \{f : [a,b] \to \mathbb{R} : \int_a^b |f(t)|^p dt < \infty\}.$$

In $L_p[a,b]$, the vector addition and scalar multiplication are defined below.
Vector addition: For $f, g \in L_p[a,b]$,

$$(f+g)(t) = f(t) + g(t) \text{ for all } t \in [a,b].$$

Scalar multiplication: For $f \in L_p[a,b]$ and $\alpha \in \mathbb{R}$,

$$(\alpha \cdot f)(t) = \alpha f(t) \text{ for all } t \in [a,b].$$

In $L_p[a,b]$, two functions f and g are said to be **equivalent** if $\int_a^b |f(t) - g(t)|^p dt = 0$. In other words, we say that f and g are equal almost everywhere (a.e.) on $[a,b]$ and denote this notion by $f = g$ a.e. on $[a,b]$ or $f(t) = g(t)$ for almost every $t \in [a,b]$. For example, if $f(t) = 1$ for $t \in [-1,1]$, and

$$g(t) = \begin{cases} 1, & t \in [-1,1] \setminus \{0\}, \\ 0, & t = 0. \end{cases}$$

Then $f = g$ a.e. on $[-1,1]$.

Normed Linear Space

Norm

Definition 3.2.7 A **norm** on a vector space V over $\mathbb{F}$ is a real valued function on V whose value for $x \in V$ is denoted by $\|x\|$, and it satisfies the following axioms:

(*i*) $\|x\| > 0$ for all $x \in V$, $x \neq \theta$, (*Positivity*)
$\|x\| = 0 \Longleftrightarrow x = \theta$,

(*ii*) $\|\alpha x\| = |\alpha|\|x\|$ for all $x \in V$ and $\alpha \in \mathbb{F}$, (*Homogeneity*)

(*iii*) $\|x+y\| \leq \|x\| + \|y\|$ for all $x, y \in V$. (*Triangular inequality*)

Definition 3.2.8 A vector space provided with a norm is called a **normed linear space**.

***Example* 3.2.9** Let $V = \mathbb{R}^n$ over $\mathbb{R}$. Then for any $x = (x_1, x_2, \ldots, x_n) \in \mathbb{R}^n$, one can define several norms that satisfy Definition 3.2.7, such as

(*a*) $\|x\|_p = \left(|x_1|^p + |x_2|^p + \cdots + |x_n|^p\right)^{\frac{1}{p}}$ for $p \geq 1$,

(*b*) $\|x\|_{\max} = \max\{|x_1|, |x_2|, \ldots, |x_n|\}$.

The axioms of norm stated in Definition 3.2.7 can be verified for these examples.

The norm $\|\cdot\|_2$ is known as the Euclidean norm in $\mathbb{R}^n$.

***Example* 3.2.10** On the linear space $V = L_p[a,b]$, $p \geq 1$, the norm is defined by

$$\|f\|_{L_p} = \left(\int_a^b |f(x)|^p dx\right)^{\frac{1}{p}} \quad \text{for } f \in V. \tag{3.1}$$

***Example* 3.2.11** Let $V = L_p[[a,b] \times [a,b]]$, the space of all p-integrable functions on $[a,b] \times [a,b]$ of two variables. V is a vector space over $\mathbb{R}$, and the norm of an element $k \in V$ is defined by

$$\|k\|_{L_p} = \left(\int_a^b \int_a^b |k(t,s)|^p dt ds\right)^{\frac{1}{p}}.$$

***Example* 3.2.12** For the vector space $V = C[a,b]$ over $\mathbb{R}$, the following norms can be defined:

(*i*) The sup norm

$$\|f\|_\infty = \sup_{t \in [a,b]} |f(t)| \quad \text{for } f \in V.$$

(*ii*) For $p \geq 1$, the L_p norm on $C[a,b]$ is

$$\|f\|_{L_p} = \left(\int_a^b |f(t)|^p dt\right)^{\frac{1}{p}} \quad \text{for } f \in V.$$

Note that $\|f\|_{L_p} \leq \|f\|_\infty (b-a)^{\frac{1}{p}}$. Hence, $C[a,b] \subset L_p[a,b]$.

Example **3.2.13** Let $f(t) = t^2 - 2$ for $t \in [0,1]$. Then, as a function in $C[0,1]$

(*i*) the supremum norm of f is $\|f\|_\infty = \sup_{t\in[0,1]} |t^2 - 2| = 2$,

(*ii*) the L_1 norm of f is $\|f\|_{L_1} = \int_0^1 |t^2 - 2| dt = \int_0^1 (2 - t^2) dt = \frac{5}{3}$,

(*iii*) the L_2 norm of f is $\|f\|_{L_2} = \left(\int_0^1 |t^2 - 2|^2 dt\right)^{\frac{1}{2}} = \left(\int_0^1 [4 + t^4 - 4t^2] dt\right)^{\frac{1}{2}} = \sqrt{\frac{43}{15}}$.

Example **3.2.14** Let $V = \mathbb{R}^{m\times n}$, the set of all $m \times n$ real matrices. The following norms can be defined on V. If $A = \{a_{ij}\}_{i=1,2,\ldots,m;j=1,2,\ldots,n}$, then

(*i*) $\|A\|_p = \left(\sum_{i=1}^{m}\sum_{j=1}^{n} |a_{ij}|^p\right)^{\frac{1}{p}}, \quad p \geq 1,$

(*ii*) $\|A\|_R = \max_i \left[\sum_{j=1}^{n} |a_{ij}|\right]$ (row sum),

(*iii*) $\|A\|_C = \max_j \left[\sum_{i=1}^{m} |a_{ij}|\right]$ (column sum),

(*iv*) $\|A\|_\infty = \max_{i,j} |a_{ij}|$.

Lemma 3.2.15 Let $A = \{a_{ij}\}_{i=1,2,\ldots,m;j=1,2,\ldots,p}$ and $B = \{b_{jk}\}_{j=1,2,\ldots,p;k=1,2,\ldots,n}$ be two real matrices. Then

$$\|AB\|_2 \leq \|A\|_2 \|B\|_2.$$

Proof. Let $AB = \{c_{ij}\}_{i=1,2,\ldots,m;j=1,2,\ldots,n}$, where $c_{ij} = \sum_{k=1}^{p} a_{ik} b_{kj}$. By the Cauchy–Schwarz inequality (Theorem 2.9.5), we have

$$|c_{ij}|^2 = \left|\sum_{k=1}^{p} a_{ik} b_{kj}\right|^2 \leq \sum_{k=1}^{p} |a_{ik}|^2 \cdot \sum_{k=1}^{p} |b_{kj}|^2. \tag{3.2}$$

Summing over all i and j, we get

$$\begin{aligned}\|AB\|_2^2 &= \sum_{i=1}^{m}\sum_{j=1}^{n} |c_{ij}|^2 \\ &\leq \sum_{i=1}^{m}\sum_{j=1}^{n}\left(\sum_{k=1}^{p} |a_{ik}|^2 \cdot \sum_{k=1}^{p} |b_{kj}|^2\right) \quad \text{(by (3.2))}\end{aligned}$$

$$= \left(\sum_{i=1}^{m}\sum_{k=1}^{p}|a_{ik}|^2\right)\left(\sum_{j=1}^{n}\sum_{k=1}^{p}|b_{kj}|^2\right)$$
$$= \|A\|^2\|B\|^2.$$

Taking square roots on both sides yields the desired inequality. □

Lemma 3.2.16 Let A be an $n \times n$ matrix in $\mathbb{R}^{n\times n}$ with the norm $\|\cdot\|_2$. Let $\lambda_i : i = 1,2,\ldots,k(\leq n)$ be distinct eigenvalues of A. Then there exist constants $M > 0$ and $\alpha \in \mathbb{R}$ such that $\|e^{At}\|_2 \leq Me^{\alpha t}$ for $t > 0$.

Proof. We know that $e^{At} = Pe^{Jt}P^{-1}$, where J is the Jordan canonical form of A. Denoting $\|\cdot\|_2$ by $\|\cdot\|$ for simplicity,

$$\|e^{At}\| \leq \|P\|\|e^{Jt}\|\|P^{-1}\| \leq K\|e^{Jt}\|,$$

where $K \geq \|P\|\|P^{-1}\|$. Let J_i be the ith Jordan block in J. Then, from (2.36), it can be shown that

$$\|e^{J_i t}\| \leq M_i e^{\alpha_i t}$$

for some positive constants M_i and α_i. Then,

$$\|e^{At}\| \leq Me^{\alpha t},$$

where $M = \sum_{i=1}^{k} M_i$ and $\alpha = \max\{\alpha_1, \alpha_2, \ldots, \alpha_k\}$. □

***Example* 3.2.17** Find a finite upper bound for $\|e^{4A}\|$, where $A = \begin{bmatrix} 2 & 1 & 0 \\ 0 & 2 & 0 \\ 1 & 1 & 3 \end{bmatrix}$.

Solution In Example 2.7.5, it was shown that $P^{-1}AP = J = \begin{bmatrix} 2 & 1 & 0 \\ 0 & 2 & 0 \\ 0 & 0 & 3 \end{bmatrix}$, where $P = \begin{bmatrix} -1 & 1 & 0 \\ 0 & -1 & 0 \\ 1 & 1 & 1 \end{bmatrix}$ and $P^{-1} = \begin{bmatrix} -1 & -1 & 0 \\ 0 & -1 & 0 \\ 1 & 2 & 1 \end{bmatrix}$. Hence, $\|P\|_2 = \sqrt{6}$ and $\|P^{-1}\|_2 = 3$. From (2.32) and (2.35), $\|e^{4J_1}\|_2 = \sqrt{18}e^8$ and $\|e^{4J_2}\|_2 = e^{12}$. Hence, $\|e^{4J}\|_2 \leq \sqrt{19}e^{12}$. Thus, $\|e^{4A}\|_2 \leq \|P\|_2\|e^{4J}\|_2\|P^{-1}\|_2 \leq 3\sqrt{114}e^{12}$. By taking a different norm, the upper bound may be reduced. Thus, $\|e^{4A}\|_3 \leq 5226\frac{1}{3}e^{12}$.

Equivalence of norms

Theorem 3.2.18 On a finite dimensional vector space V, all the norms are equivalent, i.e., if $\|\cdot\|_A$ and $\|\cdot\|_B$ are two norms defined on V, then there exists constants $\alpha, \beta > 0$ such that

$$\alpha\|\cdot\|_A \leq \|\cdot\|_B \leq \beta\|\cdot\|_A.$$

However, in general, this results is not valid on infinite dimensional spaces.

Example **3.2.19** On $\mathbb{R}^2$, prove that

$$\frac{1}{2}\|x\|_1 \leq \|x\|_2 \leq \|x\|_1,$$

where $x = (x_1, x_2)$.

Solution Observe that $2|x_1||x_2| \leq |x_1|^2 + |x_2|^2$. Then

$$\begin{aligned}\frac{1}{4}\|x\|_1^2 = \frac{1}{4}(|x_1| + |x_2|)^2 &= \frac{1}{4}(|x_1|^2 + |x_2|^2 + 2|x_1||x_2|)\\ &\leq \frac{1}{2}(|x_1|^2 + |x_2|^2) \leq \|x\|_2^2 \leq |x_1|^2 + |x_2|^2 + 2|x_1||x_2| = \|x\|_1^2.\end{aligned}$$

Taking square roots gives the desired inequality.

Inner Product Space

Definition 3.2.20 Let V be a vector space over the field F ($\mathbb{R}$ or $\mathbb{C}$). An **inner product** on V is defined as a mapping from $V \times V$ into F, denoted by $\langle\cdot,\cdot\rangle$, that satisfies the following conditions:

(*i*) $\langle v_1, v_2\rangle = \overline{\langle v_2, v_1\rangle}$ for all $v_1, v_2 \in V$,

(*ii*) $\langle v, v\rangle \geq 0$ for all $v \in V$,
$\langle v, v\rangle = 0 \Longleftrightarrow v = 0$,

(*iii*) $\langle \alpha v_1, v_2\rangle = \alpha\langle v_1, v_2\rangle$ for all $v_1, v_2 \in V$ and $\alpha \in F$,

(*iv*) $\langle v_1 + v_2, v_3\rangle = \langle v_1, v_3\rangle + \langle v_2, v_3\rangle$ for all $v_1, v_2, v_3 \in V$.

Definition 3.2.21 A vector space provided with an inner product is called an **inner product space**.

Note that the inner product induces a norm on the vector space as $\|v\| = \sqrt{\langle v, v\rangle}$ for all $v \in V$. Hence, every inner product space is a normed linear space. However, the converse is generally not true (see Exercise 8).

Example **3.2.22** (i) Let $x, y \in \mathbb{R}^n$, where $x = (x_1, x_2, \ldots, x_n)$ and $y = (y_1, y_2, \ldots, y_n)$, with $x_i, y_i \in \mathbb{R}$ for $i = 1, 2, \ldots, n$. Then $\langle x, y \rangle = \sum_{i=1}^n x_i y_i$ defines an inner product on $\mathbb{R}^n$.

Let $x, y \in \mathbb{C}^n$, where $x = (x_1, x_2, \ldots, x_n) : x_j = a_j + ib_j$, and $y = (y_1, y_2, \ldots, y_n) : y_j = c_j + id_j$ with $a_j, b_j, c_j, d_j : j = 1, 2, \ldots, n$ are real numbers and $i = \sqrt{-1}$.
(ii) If $V = \mathbb{C}^n$ over $\mathbb{C}$, then

$$\langle x, y \rangle = \sum_{j=1}^n x_j \bar{y}_j = \sum_{j=1}^n ((a_j c_j + b_j d_j) + i(b_j c_j - a_j b_j))$$

is an inner product. Note that this inner product is nothing but the dot product of two vectors x and y in $\mathbb{C}^n$.
(iii) If $V = \mathbb{C}^n$ over $\mathbb{R}$, then

$$\langle x, y \rangle = \sum_{j=1}^n (a_j c_j + b_j d_j)$$

is an inner product.

Example **3.2.23** Let $V = L_2[a, b]$ over $\mathbb{R}$, which is the set of all functions f such that $\int_a^b |f(t)|^2 dt < \infty$. For $f, g \in V$, the inner product is defined as

$$\langle f, g \rangle = \int_a^b f(x) \overline{g(x)} dx. \tag{3.3}$$

Note that $\langle f, f \rangle = \|f\|_2^2$, as in equation (3.1) with $p = 2$.

The space $L_2([a, b], \mathbb{R}^n)$

Example **3.2.24** Let $V = L_2([a, b], \mathbb{R}^n)$ be the set of all vector functions x defined on $[a, b]$, where

$$x(t) = \begin{bmatrix} x_1(t) \\ x_2(t) \\ \vdots \\ x_n(t) \end{bmatrix} \in \mathbb{R}^n \text{ for each } t \in [a, b]$$

and

$$\int_a^b |x_i(t)|^2 dt < \infty, \quad i = 1, 2, \ldots, n.$$

For $x \in V$, the norm is defined as

$$\|x\|_V = \left(\int_a^b \|x(t)\|_{\mathbb{R}^n}^2 dt \right)^{\frac{1}{2}} = \left(\int_a^b \left[x_1^2(t) + x_2^2(t) + \cdots + x_n^2(t) \right] dt \right)^{\frac{1}{2}}. \tag{3.4}$$

Example **3.2.25** Let $V = L_2([a,b],\mathbb{R}^n)$. For $x,y \in V$, the inner product in V is defined as

$$\langle x,y\rangle_V = \int_0^1 \langle x(t),y(t)\rangle_{\mathbb{R}^n} dt = \int_0^1 [x_1(t)y_1(t)+x_2(t)y_2(t)+\cdots+x_n(t)y_n(t)]\,dt,$$

where

$$x(t) = \begin{bmatrix} x_1(t) \\ x_2(t) \\ \vdots \\ x_n(t) \end{bmatrix}, y(t) = \begin{bmatrix} y_1(t) \\ y_2(t) \\ \vdots \\ y_n(t) \end{bmatrix} \in \mathbb{R}^n.$$

Example **3.2.26** If $V = L_2([0,1],\mathbb{R}^2)$ and $x(t) = \begin{bmatrix} 1 \\ t \end{bmatrix} \in V$, then find $\|x\|_V$.

Solution From (3.4), we have

$$\|x\|_V = \left(\int_0^1 (1+t^2)dt\right)^{\frac{1}{2}} = \frac{2}{\sqrt{3}}.$$

Example **3.2.27** Let $V = L_2([0,1],\mathbb{R}^2)$, $x(t) = \begin{bmatrix} 1 \\ t \end{bmatrix}$, and $y(t) = \begin{bmatrix} t \\ t^2 \end{bmatrix}$, then

$$\langle x,y\rangle_V = \int_0^1 (t+t^3)dt = \frac{3}{4}.$$

Parallelogram Identity

Let $\langle\cdot,\cdot\rangle$ be an inner product defined on a vector space V over $\mathbb{R}$. Then for $x \in V$, $\sqrt{\langle x,x\rangle}$ induces a norm on V, denoted by $\|x\|$. For $x,y \in V$, we have

$$\|x \pm y\|^2 = \langle x\pm y, x\pm y\rangle = \|x\|^2 + \|y\|^2 \pm 2\langle x,y\rangle.$$

From these identities, we obtain the **parallelogram identity**:

$$\|x+y\|^2 + \|x-y\|^2 = 2(\|x\|^2 + \|y\|^2). \tag{3.5}$$

We conclude that if a norm is induced by an inner product, then the norm must satisfy the parallelogram identity. In this case, the inner product can be obtained from the norm as

$$\langle x,y\rangle = \frac{1}{4}(\|x+y\|^2 - \|x-y\|^2). \tag{3.6}$$

Also see Exercise 8 for analogous result on complex vector spaces.

Exercise 3.2.1 Find $M > 0$ and $\alpha \in \mathbb{R}$ such that $\|e^{At}\| \leq Me^{\alpha t}$ for $t \geq 0$. Here, $A = \begin{bmatrix} A_1 & 0 \\ 0 & A_2 \end{bmatrix}$ with blocks $A_1 = \begin{bmatrix} 3 & 2 \\ 2 & 3 \end{bmatrix}$ and $A_2 = \begin{bmatrix} 5 & 0 \\ 1 & 2 \end{bmatrix}$.

Exercise 3.2.2 Show that the sup norm on $C[0,1]$ does not satisfy the parallelogram identity.

Exercise 3.2.3 Let $f(t) = \begin{bmatrix} \sin t \\ \cos 2t \end{bmatrix}$ and $g(t) = \begin{bmatrix} \cos t \\ \sin 4t \end{bmatrix}$. Prove that f and g are orthogonal vector functions in $L_2([0,\pi],\mathbb{R}^2)$.

3.3 Vector Valued Functions

Let $J = [t_0, \tau] \subset \mathbb{R}$. A function v is called a vector valued function on J if $v(t) \in \mathbb{R}^n$ for $t \in J$, and we denote it as $v : J \to \mathbb{R}^n$. We can represent it as

$$v(t) = \begin{bmatrix} x_1(t) \\ x_2(t) \\ \vdots \\ x_n(t) \end{bmatrix},$$

where $x_i(t) : i = 1,2,\ldots,n$ are real valued functions of t. We can also refer to $v(t)$ as an $\mathbb{R}^n$ valued function. If $x_i(t) : i = 1,2,\ldots,n$ are continuous functions, then $v \in C(J,\mathbb{R}^n)$. Let $v_1, v_2, \ldots, v_m$ be $\mathbb{R}^n$ valued functions on J, where

$$v_i(t) = \begin{bmatrix} x_{i1}(t) \\ x_{i2}(t) \\ \vdots \\ x_{in}(t) \end{bmatrix}, \quad i = 1,2,\ldots,m.$$

They are called linearly dependent on J if there exist scalars, not all zero, such that

$$\alpha_1 v_1(t) + \alpha_2 v_2(t) + \cdots + \alpha_m v_m(t) = 0 \quad \text{for all } t \in J. \tag{3.7}$$

The functions $v_i : i = 1, 2, \ldots, m$ are called linearly independent if equation (3.7) holds only when $\alpha_i = 0$ for all i.

Note that in the above definitions, linearly independence and dependence are considered simultaneously for all $t \in [t_0, \tau]$. For example, consider the vectors $\begin{bmatrix} 2 \\ t \end{bmatrix}$ and $\begin{bmatrix} 2t \\ t^2 \end{bmatrix}$ for $t > 0$. They are linearly independent since $\alpha_1 \begin{bmatrix} 2 \\ t \end{bmatrix} + \alpha_2 \begin{bmatrix} 2t \\ t^2 \end{bmatrix} = \begin{bmatrix} 0 \\ 0 \end{bmatrix}$ for all $t > 0$ implies $2\alpha_1 + 2t\alpha_2 = 0$ and $t\alpha_1 + t^2\alpha_2 = 0$ for $t > 0$. At $t = 3$ (say), we have $\alpha_1 + 3\alpha_2 = 0$, and at $t = 1$, we have $\alpha_1 + \alpha_2 = 0$. From these equations, it follows that $\alpha_1 = \alpha_2 = 0$. Note that for each fixed t, the vectors $\begin{bmatrix} 2 \\ t \end{bmatrix}$ and $\begin{bmatrix} 2t \\ t^2 \end{bmatrix}$ are linearly dependent. However, they are linearly independent as functions of t.

The vector functions $\begin{bmatrix} 2 \\ 3t \end{bmatrix}$ and $\begin{bmatrix} 2t \\ 3 \end{bmatrix}$ are linearly independent as functions of t but are linearly dependent at $t = 1$ as vectors in $\mathbb{R}^2$. Thus, for a fixed t, if vectors are linearly dependent, it does not imply anything about the linear independence or dependence of the set of vector functions of t. Another important difference between constant vectors and vector functions is the following. If M is a constant $n \times n$ matrix, then $\det(M) = 0$ implies that the column vectors are linearly dependent. However, if $M(t)$ is a matrix whose elements are functions of t, and if $\det(M(t)) = 0$ for all t, we cannot arrive at any conclusion about the linear dependence or independence of the column vector functions. For example, consider the matrix $\begin{bmatrix} 1 & t \\ t & t^2 \end{bmatrix}$. Here, the determinant is 0 for all t, but the vectors $\begin{bmatrix} 1 \\ t \end{bmatrix}$ and $\begin{bmatrix} t \\ t^2 \end{bmatrix}$ are linearly independent. The columns of $\begin{bmatrix} 1 & t \\ 0 & 0 \end{bmatrix}$ are linearly independent, but the rows are linearly dependent. However, in the case of constant matrices, the number of linearly independent rows = the number of linearly independent columns.

In case $v_1(t), v_2(t), \ldots, v_n(t)$ are solutions of a system of differential equations of the form

$$\dot{x} = Ax, \quad t \in J,$$

where A is an $n \times n$ matrix, they have the following property.

$$\text{If } M(t) = \begin{bmatrix} v_1(t) & v_2(t) & \cdots & v_n(t) \\ \downarrow & \downarrow & \cdots & \downarrow \end{bmatrix}_{n\times n},$$

then $\det(M(t_0)) = 0$ implies that the vectors $v_i(t)$ are linearly dependent on J, and $\det(M(t_0)) \neq 0$ implies that the vectors are linearly independent on J. This important aspect on the solutions of a system of differentials equations will be discussed in Chapter 4 in the section on the fundamental matrix.

Another important result on matrix functions is given in the following lemma, which shows that the number of linearly independent rows of an $n \times m$ matrix function is same as the number of linearly independent rows of the corresponding $n \times n$ **Gramian matrix** W. This result will be useful in Chapter 5 while studying the controllability concept. Let

$$P(s) = \begin{bmatrix} r_1(s) \\ r_2(s) \\ \vdots \\ r_n(s) \end{bmatrix}_{n\times m}$$

be an $n \times m$ matrix function with $r_i(s) = [r_{i1}(s), r_{i2}(s), \ldots, r_{im}(s)]$ as the ith row such that the entries $r_{ij}(s) : j = 1, 2, \ldots, m$ are continuous functions of $s \in J$. Let W be the $n \times n$ Gramian matrix defined as

$$W = \int_{t_0}^{\tau} P(s)P'(s)ds. \tag{3.8}$$

Lemma 3.3.1 The rows $r_i(s) : i = 1, 2, \ldots, n$ are linearly independent as functions of s if and only if the $n \times n$ constant matrix W is nonsingular.

Proof. Suppose W is nonsingular, but the rows of $P(s)$ are linearly dependent as functions of $s \in J$. Then, there exist some scalars $\alpha_1, \alpha_2, \ldots, \alpha_n$ (not all zeros) such that

$$\alpha_1 r_1(s) + \alpha_2 r_2(s) + \cdots + \alpha_n r_n(s) = \Theta. \tag{3.9}$$

Equation (3.9) can be written as

$$\begin{bmatrix} \alpha_1 & \alpha_2 & \cdots & \alpha_n \end{bmatrix} \begin{bmatrix} r_{11}(s) & r_{12}(s) & \cdots & r_{1m}(s) \\ r_{21}(s) & r_{22}(s) & \cdots & r_{2m}(s) \\ \vdots & & & \\ r_{n1}(s) & r_{n2}(s) & \cdots & r_{nm}(s) \end{bmatrix} = \begin{bmatrix} 0 & 0 & \cdots & 0 \end{bmatrix}, \tag{3.10}$$

which can be written as

$$\alpha' P(s) = \Theta,$$

where

$$\alpha = \begin{bmatrix} \alpha_1 \\ \alpha_2 \\ \vdots \\ \alpha_n \end{bmatrix} \in \mathbb{R}^n.$$

This implies $\alpha' P(s) P'(s) \alpha = 0$ for all $s \in [t_0, \tau]$. Integrating both sides from t_0 to τ, we obtain

$$\alpha' W \alpha = 0. \tag{3.11}$$

Since α is a nonzero vector and W is symmetric, equation (3.11) implies that W is singular, a contradiction. Therefore, the rows of $P(s)$ are linearly independent.

Conversely, assume that the rows of $P(s)$ are linearly independent, but W is singular. Then, there exists a nonzero vector $\alpha \in \mathbb{R}^n$ such that $\alpha' W \alpha = 0$, which implies $\alpha' \int_{t_0}^{\tau} P(s) P'(s) ds \alpha = 0$. This gives $\int_{t_0}^{\tau} \|\alpha' P(s)\|^2 ds = 0$, leading to $\alpha' P(s) = \Theta$. Since $\alpha \neq \Theta$, the rows of $P(s)$ are linearly dependent, which is a contradiction. Hence, W is nonsingular. □

Corollary 3.3.2 Let A and B be $n \times n$ and $n \times m$ constant matrices, respectively. Define

$$W = \int_{t_0}^{\tau} e^{As} B B' e^{A's} ds.$$

Then W is nonsingular if and only if the rows of $e^{At} B$ are linearly independent as functions of t. To prove this, set $P(t) = e^{At} B$ in Lemma 3.3.1.

Exercise 3.3.1 (i) Find the intervals in $\mathbb{R}$ on which the columns of the matrix $\begin{bmatrix} 4 & 1 \\ t & t^2 \end{bmatrix}$ are linearly independent vector functions.

(ii) Determine the intervals on which the inverse of the matrix exists.

Exercise 3.3.2 Evaluate $W = \int_0^1 e^{At}BB'e^{A't}dt$ for the following cases: (i) $A = \begin{bmatrix} 0 & 1 \\ 0 & 0 \end{bmatrix}$ and $B = \begin{bmatrix} 0 \\ 1 \end{bmatrix}$; (ii) $A = \begin{bmatrix} 0 & 1 \\ 0 & 0 \end{bmatrix}$ and $B = \begin{bmatrix} 1 \\ 0 \end{bmatrix}$.

3.4 Linear Transformation

Definition 3.4.1 Let U and V be two vector spaces over a field F(either $\mathbb{R}$ or $\mathbb{C}$). A mapping L from U into V, denoted by $L : U \to V$, is called a **linear transformation** if

(i) $L(u + \tilde{u}) = Lu + L\tilde{u}$

(ii) $L(\alpha u) = \alpha L(u)$

for all $u, \tilde{u} \in U$ and $\alpha \in F$.

Example **3.4.2** Let $U = \mathbb{R}^n, V = \mathbb{R}^m$, and M be an $m \times n$ matrix with real entries. Then M is a linear transformation from U into V, as it satisfies Definition 3.4.1.

Example **3.4.3** Suppose that $L : \mathbb{R}^3 \to \mathbb{R}^2$ is defined as $L(x_1,x_2,x_3) = (x_2 + x_3, x_1)$. Then L is linear transformation by Definition 3.4.1, as shown below: For all $(x_1,x_2,x_3), (y_1,y_2,y_3) \in \mathbb{R}^3$ and $\alpha \in \mathbb{R}$, we have

$$\begin{aligned} L((x_1,x_2,x_3) + (y_1,y_2,y_3)) &= (x_2 + y_2 + x_3 + y_3, x_1 + y_1) \\ &= (x_2 + x_3, x_1) + (y_2 + y_3, y_1) \\ &= L(x_1,x_2,x_3) + L(y_1,y_2,y_3) \end{aligned}$$

and

$$L(\alpha(x_1,x_2,x_3)) = (\alpha x_2 + \alpha x_3, \alpha x_1) = \alpha(x_2 + x_3, x_1) = \alpha L(x_1,x_2,x_3).$$

Example **3.4.4** The transformation $L : \mathbb{R}^2 \to \mathbb{R}$ defined as $L(x,y) = x + y + 1$ is not a linear transformation, because

$$L(\alpha(x,y)) = \alpha x + \alpha y + 1 \neq \alpha x + \alpha y + \alpha = \alpha L(x,y) \quad \text{for all } \alpha \neq 1.$$

Example **3.4.5** Let $V = C[0,1]$ and $T : V \to V$. For any $f \in V$, define T as

$$(Tf)(t) = \int_0^t k(t,s)f(s)ds,$$

where $k(t,s)$ is a continuous function of two variables t and s. It is easy to verify that T is linear since

$$(T(f+\tilde{f}))(t) = \int_0^t k(t,s)(f(s)+\tilde{f}(s))ds = \int_0^t k(t,s)f(s)ds + \int_0^t k(t,s)\tilde{f}(s)ds$$
$$= (Tf)(t) + (T\tilde{f})(t) \quad \text{for } f, \tilde{f} \in V$$

and

$$(T(\alpha f))(t) = \int_0^t k(t,s)(\alpha f(s))ds = \alpha (Tf)(t) \quad \text{for } f \in V \text{ and } \alpha \in \mathbb{R}.$$

Example **3.4.6** Let $U = C^1[0,1]$ and $V = C[0,1]$. Define $T : U \to V$ as $Tf = \frac{df}{dt}$ for $f \in U$. Then T is a linear transformation since

(*i*) $(T(f+\tilde{f}))(t) = \frac{d}{dt}(f+\tilde{f}) = \frac{df}{dt} + \frac{d\tilde{f}}{dt} = (Tf)(t) + (T\tilde{f})(t) \quad$ for $f, \tilde{f} \in V$

(*ii*) $(T(\alpha f))(t) = \frac{d}{dt}(\alpha f) = \alpha\frac{df}{dt} = \alpha(Tf)(t) \quad$ for $f \in V$ and $\alpha \in \mathbb{R}$.

Bounded Linear Operator

Definition 3.4.7 Let U and V be normed linear spaces and $T : U \to V$ a linear transformation. T is said to be a **bounded linear operator** if there exists a constant $c > 0$ such that

$$\|Tu\|_V \leq c\|u\|_U \quad \text{for all } u \in U. \tag{3.12}$$

Note that the name 'bounded linear operator' does not imply that the range of T is bounded in V. It only states that the bounded sets in U are mapped into bounded sets in V. For example, if $T : \mathbb{R} \to \mathbb{R}$ is defined as $Tx = 2x$, then T is linear. The range of T is $\mathbb{R}$, which is not bounded. However, T is called a bounded linear operator since $|Tx| \leq 2|x|$ for all $x \in \mathbb{R}$.

A nonlinear operator $f : U \to V$ is said to be bounded if the range of f is a bounded set in V. For example, $f : \mathbb{R} \to \mathbb{R}$ defined by $f(x) = \sin x$ for all $x \in \mathbb{R}$ is not linear but is bounded.

Example **3.4.8** As in Example 3.4.3, let $T : \mathbb{R}^3 \to \mathbb{R}^2$ be defined as $T(x_1,x_2,x_3) = (x_2+x_3,x_1)$. Then T is linear and $\|T(x_1,x_2,x_3)\|^2 = (x_2+x_3)^2 + x_1^2$ (using the Euclidean norm in $\mathbb{R}^2$ and $\mathbb{R}^3$). Now

$$(x_2+x_3)^2 + x_1^2 = x_1^2 + x_2^2 + x_3^2 + 2x_2x_3$$

$$\leq x_1^2 + x_2^2 + x_3^2 + x_2^2 + x_3^2$$
$$\leq 2\left(x_1^2 + x_2^2 + x_3\right).$$

Hence $\|T(x_1,x_2,x_3)\| \leq \sqrt{2}\|(x_1,x_2,x_3)\|$, implying T is a bounded linear operator.

If we consider the $\|\cdot\|_1$ norm on $\mathbb{R}^2$ and $\mathbb{R}^3$ we get

$$\|T(x_1,x_2,x_3)\|_{\mathbb{R}^2} = \|(x_2+x_3,x_1)\|_{\mathbb{R}^2}$$
$$= |x_2+x_3| + |x_1| \leq |x_1| + |x_2| + |x_3| = \|(x_1x_2,x_3)\|_{\mathbb{R}^3}.$$

Hence, T is a bounded linear operator.

An important result on finite dimensional normed linear spaces is given below.

***Example* 3.4.9** Let $K(t,s) = t - 2s$ in the operator T defined in Example 3.4.5. Show that it is bounded.

Proof. If we consider sup norm on $C[0,1]$, then

$$\|Tf\| = \sup_{t\in[0,1]} \left| \int_0^t K(t,s)f(s)ds \right| \leq \sup_{t\in[0,1]} \int_0^t |K(t,s)|\,|f(s)|ds$$
$$\leq \sup_{t\in[0,1]} \int_0^1 |K(t,s)|\,|f(s)|ds \leq \sup_{t,s\in[0,1]} \{|K(t,s)|\} \sup_{s\in[0,1]} \{|f(s)|\} = c\|f\|,$$

where $c = \sup_{t,s\in[0,1]}\{|K(t,s)|\} = \sup_{t,s\in[0,1]}\{|t-2s|\} = 2.$ □

Space of Bounded Linear Operators

Let U and V be normed linear spaces. The set of all bounded linear operators from U into V is denoted by $BL(U,V)$. It can be easily verified that $BL(U,V)$ is a vector space over $\mathbb{R}$ under the addition and scalar multiplication of linear operators.

Definition 3.4.10 (Norm of a linear operator) Let $T \in BL(U,V)$. The norm of T is defined as

$$\|T\|_{BL(U,V)} = \sup_{\|x\|_U=1} \|Tx\|_V.$$

***Example* 3.4.11** For the operator T in Example 3.4.8, $\|T\| \leq \sqrt{2}$ under the Euclidean norm on both $\mathbb{R}^2$ and $\mathbb{R}^3$, and $\|T\| \leq 1$ under the $\|\cdot\|_1$ norm on both the spaces.

Unbounded Linear Operator

The differential operator defined in Example 3.4.6 is unbounded since the sequence $f_n(t) = t^n$ in $C^1[0,1]$ has norm 1 for all $n = 1, 2, \ldots$, under the sup norm, (since $\|t^n\| = \sup_{t\in[0,1]} |t^n| = 1$) and

$$\|Tf_n\| = \left\| \frac{df_n}{dt} \right\| = \|nt^{n-1}\| = n.$$

Hence, $\|Tf_n\| \to \infty$ as $n \to \infty$, which implies that T is not a bounded linear operator according to Definition 3.4.7.

Banach Space and Hilbert Space

Definition 3.4.12 Let X be a normed linear space with a norm denoted by $\|\cdot\|$. A sequence $\{x_n\}_{n=1}^{\infty}$ in X is said to **converge** to $x \in X$ if for any $\varepsilon > 0$, there exists an integer N such that

$$\|x_n - x\| < \varepsilon \quad \text{whenever } n > N. \tag{3.13}$$

This is also represented as $\lim_{n\to\infty} x_n = x$ or $\|x_n - x\| \to 0$ as $n \to \infty$.

Definition 3.4.13 (Cauchy sequence) A sequence $\{x_n\}_{n=1}^{\infty}$ in a normed linear space X is called a Cauchy sequence if for any $\varepsilon > 0$, there exists an integer N such that

$$\|x_m - x_k\| < \varepsilon \quad \text{whenever } m, k > N. \tag{3.14}$$

From the above definitions, it can be seen that every convergent sequence in a normed linear space is a Cauchy sequence, since

$$\begin{aligned} \|x_m - x_k\| = \|x_m - x + x - x_k\| &\le \|x_m - x\| + \|x_k - x\| \\ &< 2\varepsilon \quad \text{whenever } m, k > N. \end{aligned} \tag{3.15}$$

However, every Cauchy sequence need not be convergent.

***Example* 3.4.14** Let $P[0,1]$ be the vector space of all polynomials with the sup norm. The sequence $\{f_n(x)\}$, where $f_n(x) = \sum_{k=0}^{n} \frac{x^k}{k!}$, is a Cauchy sequence in $P[0,1]$. However, it does not converge in $P[0,1]$ because $\lim_{n\to\infty} f_n(x) = e^x$, which is not a polynomial.

Definition 3.4.15 A normed linear space V is said to be **complete** if every Cauchy sequence in V converges.

A complete normed linear space is called a **Banach space**, and a complete inner product space is called a **Hilbert space**. Note that every Hilbert space is a Banach space under the norm induced by the inner product.

***Example* 3.4.16** Some examples of Banach spaces and Hilbert spaces, which often arise in our discussions, are listed below without proof. To establish each statement, one needs to verify that the axioms of Banach space and Hilbert space are satisfied using various tools from real analysis and functional analysis.

(*i*) $C[a,b]$ with the sup norm is a Banach space.

(*ii*) $L_p[a,b], p \geq 1$, are Banach spaces.

(*iii*) $L_2[a,b]$ with the inner product defined as in Example 3.2.23 is a Hilbert space.

(*iv*) $\mathbb{R}^n$ over $\mathbb{R}$ is a Banach space under any norm.

(*v*) $\mathbb{R}^n$ over $\mathbb{R}$ is a Hilbert space under the Euclidean norm.

(*vi*) $C[a,b]$ is **not** a Banach space under L_p norm. (**Hint:** Show that the sequence $(\frac{t-a}{b-a})^n$ is a Cauchy sequence in $C[a,b]$ under the L_2 norm and converges to the discontinuous function

$$f(t) = \begin{cases} 0, & a \leq t < b, \\ 1, & t = b. \end{cases}$$

Hence, $C[a,b]$ is not complete.)

(*vii*) $C[a,b]$ is **not** a Hilbert space under sup norm. (**Hint:** Show that the sup norm does not satisfy the parallelogram identity. Hence, $C[a,b]$ is not an inner product space under the sup norm.)

Theorem 3.4.17 Let U and V be normed linear spaces. If V is a Banach space, then $BL(U,V)$ is also a Banach space.

Lemma 3.4.18 Let X be a Banach space and $T \in BL(X)$. If $\|I-T\| < 1$, then T is invertible.

Proof. Since $\|I-T\| < 1$, the series $I + (I-T) + (I-T)^2 + \cdots$ converges in $BL(X)$ to the limit $(I-(I-T))^{-1} = T^{-1}$. □

Theorem 3.4.19 ([10], Uniform boundedness theorem) Let $\{T_n\}$ be a sequence of bounded linear operators from a Banach space X into a normed linear space Y such that for every $x \in X$, $\|T_n x\| \leq c_x$ for $n = 1, 2, \ldots$, where $c_x > 0$ is a real number that depends on x. Then the sequence $\{\|T_n\|\}$ is bounded, i.e., there is a constant $c > 0$ such that $\|T_n\| \leq c$ for $n = 1, 2, \ldots$.

Adjoint Operator

Definition 3.4.20 Let U and V be Hilbert spaces over the same field F, and let $T: U \to V$ be a bounded linear map. Then the **adjoint operator** T^* of T is a linear map from V into U defined by the following relation

$$\langle Tu, v\rangle_V = \langle u, T^*v\rangle_U, \tag{3.16}$$

where $u \in U, v \in V$, and $\langle \cdot, \cdot \rangle$ denotes the inner product.

Example **3.4.21** Let T be an $m \times n$ matrix. Then T can be considered as a linear transformation from $\mathbb{R}^n$ to $\mathbb{R}^m$. Hence, $T: \mathbb{R}^n \to \mathbb{R}^m$ and $T^*: \mathbb{R}^m \to \mathbb{R}^n$. Since the inner product in $\mathbb{R}^n$ can be taken as the dot product of vectors, we can write (3.16) as

$$Tu \cdot v = u \cdot T^*v, \tag{3.17}$$

where $u = (u_1, u_2, \ldots, u_n)' \in \mathbb{R}^n$ and $v = (v_1, v_2, \ldots, v_m)' \in \mathbb{R}^m$.

In (3.17), the LHS is

$$v'(Tu) = (v'T)u = (T'v)'u, \tag{3.18}$$

and the RHS of (3.17) is

$$u \cdot T^*v = (T^*v)'u, \tag{3.19}$$

Comparing (3.18) and (3.19), we get $T^* = T'$.

Thus for any real matrix the adjoint is its transpose and similarly, the adjoint of a complex matrix is its conjugate transpose.

Example **3.4.22** Find the adjoint of the integral operator $T: L_2[a,b] \to L_2[a,b]$ defined as follows. Let $f \in L_2[a,b]$ and $k \in L_2[[a,b] \times [a,b]]$ over the field $\mathbb{R}$. Define the linear transformation T as

$$(Tf)(t) = \int_a^b k(t,s)f(s)ds. \tag{3.20}$$

The function $k(t,s)$ is called the **kernel** of the integral operator T. From the relation (3.16), the adjoint T^* satisfies the following equation

$$\langle Tf, g\rangle_V = \langle f, T^*g\rangle_U \quad \text{for all } f, g \in L_2[a,b]. \tag{3.21}$$

Now the LHS of (3.21) is

$$\langle Tf, g\rangle_V = \int_a^b (Tf)(t)g(t)dt = \int_a^b \left(\int_a^b k(t,s)f(s)ds\right) g(t)dt$$

$$= \int_a^b \int_a^b k(t,s)f(s)g(t)dsdt.$$

Interchanging the order of integration, we get

$$\langle Tf, g\rangle_V = \int_a^b \left(\int_a^b k(t,s)g(t)dt \right) f(s)ds. \tag{3.22}$$

The RHS of (3.21) is

$$\langle f, T^*g\rangle_U = \int_a^b f(s)(T^*g)(s)ds. \tag{3.23}$$

Comparing the RHSs of (3.22) and (3.23), we get

$$(T^*g)(s) = \int_a^b k(t,s)g(t)dt.$$

Thus, if $k(t,s)$ is the kernel of T, then $k(s,t)$ is the kernel of T^*. In case $L_2[a,b]$ is a complex vector space over $\mathbb{C}$, then $\overline{k(s,t)}$ would be the kernel of T^*.

Lemma 3.4.23 Let $J = [t_0, \tau]$ and $T(t)$ be an $n \times m$ matrix function of t. Let $Z = L_2(J, \mathbb{R}^m)$ and $L : Z \to \mathbb{R}^n$ be a linear transformation defined by

$$Lu = \int_{t_0}^{\tau} T(t)u(t)dt. \tag{3.24}$$

Then $L^* = T'(t)$.

Proof. From (3.16), $L^* : \mathbb{R}^n \to Z$ is defined by the relation

$$\langle Lu, v\rangle_{\mathbb{R}^n} = \langle u, L^*v\rangle_Z. \tag{3.25}$$

The LHS of (3.25) is

$$Lu \cdot v = v'(Lu) = v' \int_{t_0}^{\tau} T(t)u(t)dt = \int_{t_0}^{\tau} v'T(t)u(t)dt, \tag{3.26}$$

and the RHS of (3.25) is

$$\int_{t_0}^{\tau} \langle u(t), (L^*v)\rangle_{\mathbb{R}^m} dt = \int_{t_0}^{\tau} (L^*v)' u(t)dt. \tag{3.27}$$

Comparing the RHSs of (3.26) and (3.27), we obtain $(L^*v)' = v'T(t) \implies L^*v = T'(t)v \implies L^* = T'(t)$, the transpose of $T(t)$. □

Orthogonal Sets

Let S_1 and S_2 be two subsets of a Hilbert space U over $\mathbb{C}$. S_1 and S_2 are orthogonal to each other if for all $x \in S_1$ and $y \in S_2$, we have $\langle x, y \rangle = 0$. If S_1 is orthogonal to S_2, it is denoted by $S_1 \perp S_2$. Let U and V be Hilbert spaces over $\mathbb{C}$ and let $T : U \to V$ be a linear operator and T^* be the adjoint of T. Then the following theorem holds.

Theorem 3.4.24 (i) $\text{Range}(T) \perp \text{Null}(T^*)$ and (ii) $\text{Range}(T^*) \perp \text{Null}(T)$.

Proof. (i) Let $y \in \text{Range}(T)$ and $z \in \text{Null}(T^*)$. Since $y \in \text{Range}(T)$, there exists some $x \in U$ such that $y = Tx$. Hence,

$$\langle y, z \rangle_V = \langle Tx, z \rangle_V = \langle x, T^* z \rangle_U = 0 \text{ as } z \in \text{Null}(T^*).$$

Conversely, let $y \in \text{Range}(T)$ and $z \in V$ satisfying $(y, z) = 0$. Then $\langle Tx, z \rangle = 0$ that implies $\langle x, T^* z \rangle = 0$ for all $x \in U$. Hence $T^* z = \theta$ implying $z \in \text{Null}(T^*)$.

(ii) The proof is similar to that of (i). □

Exercise 3.4.1 Let $T : C[0,2] \to C[0,2]$ be defined by $(Tx)(t) = \int_0^t s^2 x(s) ds$. Show that

(*i*) $\|T\| \leq \frac{8}{3}$ under the sup norm.

(*ii*) $\|T\| \leq \sqrt{\frac{32}{5}}$ under the L_2 norm.

Exercise 3.4.2 Let $A \in \mathbb{R}^{n \times n}, B \in \mathbb{R}^{n \times m}$, and $Z = L_2([0, \tau], \mathbb{R}^m)$. Let $L : Z \to \mathbb{R}^n$ be defined by $Lz = \int_0^\tau e^{At} Bz(t) dt$. Show that the adjoint L^* of L is given by $L^* v = B^* e^{A^* t} v$ for $v \in \mathbb{R}^n$. If $A = \begin{bmatrix} 0 & 1 \\ -1 & 0 \end{bmatrix}$ and $B = \begin{bmatrix} 0 \\ 2 \end{bmatrix}$, show that $\|L^*\| \leq 2\sqrt{\tau}$.

Exercise 3.4.3 Find the range and the null space of $T : L_2([0,1], L_2[0,1])$ defined by $(Tf)(t) = \int_0^1 t(s+1) f(s) ds$ and verify Theorem 3.4.24.

3.5 Contraction Mapping

Definition 3.5.1 Let X be a normed linear space with norm denoted by $\|\cdot\|$. A mapping $T : X \to X$ is called a **contraction mapping** if

$$\|Tx_1 - Tx_2\| \leq \alpha \|x_1 - x_2\| \quad \text{for all } x_1, x_2 \in X,$$

where α satisfies $0 < \alpha < 1$.

Definition 3.5.2 Let $f : U \to U$ be a mapping. A point $u \in U$ is said to be a **fixed point** of f if $fu = u$.

Example 3.5.3 Let $X = \mathbb{R}$ and $T : \mathbb{R} \to \mathbb{R}$ be defined as $Tx = \frac{x}{2}$. Since $|Tx_1 - Tx_2| \leq \frac{1}{2}|x_1 - x_2|$, T is a contraction mapping and $x = 0$ is the unique fixed point.

Example 3.5.4 Let $T : \mathbb{R}^2 \to \mathbb{R}^2$ be defined as $T(x,y) = \left(\frac{\sin x}{4}, \frac{y}{3} + 2\right)$ for $x, y \in \mathbb{R}$. Here, $\mathbb{R}^2$ is a Banach space under the norm

$$\|(x,y)\| = |x| + |y|.$$

Since

$$\begin{aligned}
\|T(x_1,y_1) - T(x_2,y_2)\| &= \frac{1}{4}|\sin x_1 - \sin x_2| + \frac{1}{3}|y_1 - y_2| \\
&\leq \frac{1}{4}|x_1 - x_2| + \frac{1}{3}|y_1 - y_2| \\
&\leq \frac{1}{3}\|(x_1,y_1) - (x_2,y_2)\|,
\end{aligned}$$

T is a contraction mapping. It can be seen that T has a unique fixed point $(0,3)$.

Example 3.5.5 Let $X = C[0,1]$, the set of all continuous functions f defined on the interval $[0,1]$ with the norm

$$\|f\| = \sup_{t \in [0,1]} |f(t)|.$$

Define the map $T : X \to X$ as

$$(Tf)(t) = 1 + \int_0^t sf(s)ds.$$

It can be shown that T is a contraction since

$$\begin{aligned}
\|Tf_1 - Tf_2\| &= \sup_{t \in [0,1]} \left| \int_0^t s\,(f_1(s) - f_2(s))\,ds \right| \leq \sup_{t \in [0,1]} \int_0^t s\,|f_1(s) - f_2(s)|\,ds \\
&\leq \sup_{t \in [0,1]} \{|f_1(t) - f_2(t)| \int_0^1 sds\} \leq \frac{1}{2}\|f_1 - f_2\|.
\end{aligned}$$

Hence, T is a contraction mapping.

Banach Fixed Point Theorem

Theorem 3.5.6 (Banach fixed point theorem) Let X be a Banach space and $T : X \to X$ be a contraction map. Then T has a **unique** fixed point in X.

Proof. We can complete the proof in three steps.
Step 1: Construct a sequence $\{a_n\}$ in X and show that the sequence converges to an element $a_f \in X$.
Step 2: Show that a_f is unique.
Step 3: Show that a_f is the fixed point of T.
Proof of Step 1: Let $a_0 \in X$ be arbitrary, and let $a_n = Ta_{n-1}, n = 1,2,\ldots$. Let T be a contraction mapping as given in Definition 3.5.1. Then

$$\begin{aligned}||a_n - a_{n-1}|| = ||Ta_{n-1} - Ta_{n-2}|| \leq \alpha ||a_{n-1} - a_{n-2}|| &= \alpha ||Ta_{n-2} - Ta_{n-3}|| \\ &\leq \alpha^2 ||a_{n-2} - a_{n-3}|| \\ &\vdots \\ &\leq \alpha^{n-1} ||a_1 - a_0||. \end{aligned} \tag{3.28}$$

Let N be a positive integer and $n > m > N$. Then

$$\begin{aligned}||a_n - a_m|| &\leq ||a_n - a_{n-1}|| + ||a_{n-1} - a_{n-2}|| + \cdots + ||a_{m+1} - a_m|| \\ &\leq \left(\alpha^{n-1} + \alpha^n + \cdots + \alpha^m\right) ||a_1 - a_0|| \\ &= \alpha^m \left(1 + \alpha + \cdots + \alpha^{n-m-1}\right) ||a_1 - a_0|| \quad \text{(using (3.28))} \\ &= \alpha^m \left(\frac{1-\alpha^{n-m}}{1-\alpha}\right) ||a_1 - a_0|| \\ &\leq \frac{\alpha^N}{1-\alpha} ||a_1 - a_0||.\end{aligned}$$

Now, since $\alpha < 1$, for any given $\varepsilon > 0$, one can find an N such that

$$\frac{\alpha^N}{1-\alpha} ||a_1 - a_0|| < \varepsilon.$$

Hence, $\{a_n\}$ is a Cauchy sequence in X. Since X is complete, the sequence $\{a_n\}$ converges to a limit $a_f \in X$.
Proof of Step 2: Now we show that the limit a_f is independent of the initial selection $a_0 \in X$. For this, let $b_0 \in X$ with $b_0 \neq a_0$. Let $\{b_n\}$ be the sequence satisfying $b_n = Tb_{n-1}, n = 1,2,\ldots$. As in **Step 1**, it can be shown that $\{b_n\}$ is convergent in X and $\lim_{n\to\infty} b_n = b_f$ (say). To complete the proof, we need to show that $a_f = b_f$. Observe that

$$||a_f - b_f|| = ||a_f - a_n + a_n - b_n + b_n - b_f||.$$

Then

$$\|a_f - b_f\| \le \|a_f - a_n\| + \|a_n - b_n\| + \|b_n - b_f\|. \tag{3.29}$$

Now

$$\begin{aligned}\|a_n - b_n\| = \|Ta_{n-1} - Tb_{n-1}\| &\le \alpha\|a_{n-1} - b_{n-1}\| = \alpha\|Ta_{n-2} - Tb_{n-2}\| \\ &\le \alpha^2\|a_{n-2} - b_{n-2}\| \\ &\vdots \\ &\le \alpha^n\|a_0 - b_0\|.\end{aligned}$$

Since $a_n \to a_f, b_n \to b_f$, and $\alpha^n \to 0$, for any $\varepsilon > 0$, there exists an N such that

$$\|a_f - a_n\| < \frac{\varepsilon}{3}, \|b_f - b_n\| < \frac{\varepsilon}{3}, \text{ and } \|a_n - b_n\| < \frac{\varepsilon}{3} \quad \text{for } n > N.$$

Hence, from (3.29), $\|a_f - b_f\| < \varepsilon$ for any $\varepsilon > 0$, which implies that $a_f = b_f$.
Proof of Step 3: Finally, we show that a_f is the fixed point of T, i.e., $Ta_f = a_f$. Now

$$\begin{aligned}\|Ta_f - a_f\| &= \|Ta_f - Ta_n + Ta_n - a_n + a_n - a_f\| \\ &\le \alpha\|a_f - a_n\| + \alpha\|a_n - a_{n-1}\| + \|a_n - a_f\|,\end{aligned}$$

which, from **Step 2**, can be made arbitrarily small for sufficiently large n. Hence $Ta_f = a_f$. □

Corollary 3.5.7 Let X be a Banach space and $T : X \to X$ be a mapping such that T^k is a contraction for some positive integer k. Then T has a unique fixed point in X.

Proof. Since T^k is a contraction mapping, there exists a **unique** fixed point $a_f \in X$ such that $T^k a_f = a_f$. Hence $T^{k+1}a_f = T^k(Ta_f) = Ta_f$, which implies Ta_f is also a fixed point of T^k. Due to the uniqueness property of the fixed point, $Ta_f = a_f$. □

The contraction mapping theorem provides only a sufficient condition for the existence of a unique fixed point. The following examples illustrate this point.

***Example* 3.5.8** Consider the matrix $A = \begin{bmatrix} 2 & 0 \\ 0 & 3 \end{bmatrix}$. Then $A : \mathbb{R}^2 \to \mathbb{R}^2$ is a mapping and $\|AX_1 - AX_2\| \le 3\|X_1 - X_2\|$. A^k, for any integer $k \ge 0$, is not a contraction, but A has a unique fixed point $X = \begin{bmatrix} 0 \\ 0 \end{bmatrix}$.

Example **3.5.9** Let $X = C[0,1]$, the set of all continuous functions f defined on the interval $[0,1]$ with the norm

$$\|f\| = \sup_{t\in[0,1]} |f(t)|.$$

Define the map $T : X \to X$ as

$$(Tf)(t) = 1 + \int_0^t sf(s)ds.$$

It can be shown that T has a unique fixed point. From Example 3.5.5, T is a contraction mapping on X. Hence, according to Theorem 3.5.6, T has a unique fixed point which can be obtained by solving the equation $Tf = f$, i.e.,

$$1 + \int_0^t sf(s)ds = f(t).$$

The above equation is a Volterra integral equation.

We can obtain the solution as $f(t) = e^{\frac{t^2}{2}}$, which is the fixed point of T.

Example **3.5.10** Let X be as in the previous example, and $T : X \to X$ be defined as

$$(Tf)(t) = 2t \int_0^t f(s)ds + 1.$$

In this case,

$$\|Tf_1 - Tf_2\| \leq \sup_{t\in[0,1]} 2t \int_0^t |f_1(s) - f_2(s)|ds \leq 2\|f_1 - f_2\| \quad \text{for } f_1, f_2 \in X.$$

Moreover, if $f_1(s) = 1$ and $f_2(s) = 0$, then $\|Tf_1 - Tf_2\| = 2\|f_1 - f_2\|$. So, T is not a contraction mapping. Now

$$\begin{aligned}
\|T^2 f_1 - T^2 f_2\| &= \sup_{t\in[0,1]} \left|(T^2 f_1)(t) - (T^2 f_2)(t)\right| \\
&= \sup_{t\in[0,1]} \left|2t \int_0^t [(Tf_1)(s) - (Tf_2)(s)]\, ds\right| \\
&\leq 2 \sup_{t\in[0,1]} \int_0^t 2s \left|\int_0^s [f_1(s_1) - f_2(s_1)]\, ds_1\right| ds \\
&\leq 4 \sup_{t\in[0,1]} |f_1(t) - f_2(t)| \cdot \sup_{t\in[0,1]} \int_0^t s \left(\int_0^s ds_1\right) ds
\end{aligned}$$

$$\leq 4\|f_1 - f_2\| \int_0^1 s^2 ds$$
$$= \frac{4}{3}\|f_1 - f_2\|.$$

Similarly,

$$\|T^3 f_1 - T^3 f_2\| \leq 8\|f_1 - f_2\| \int_0^t s \int_0^{s_1} s_1 \int_0^{s_2} ds_2 ds_1 ds \leq \frac{8}{15}\|f_1 - f_2\|.$$

Hence, T^3 is a contraction mapping (whereas T and T^2 are not), and by Corollary 3.5.7, T has a unique fixed point $f \in X$ satisfying $Tf = f$, i.e.,

$$f(t) = 1 + 2t \int_0^t f(s) ds.$$

By solving this Volterra integral equation we can obtain the fixed point.

***Example* 3.5.11** The identity matrix $I : \mathbb{R}^n \to \mathbb{R}^n$ is not a contraction mapping. It has every point in $\mathbb{R}^n$ as its fixed point.

***Example* 3.5.12** Let $T : \mathbb{R}^2 \to \mathbb{R}^2$ be defined by $Tx = Ax + y$, where $A = \begin{bmatrix} 1 & 1 \\ 2 & 2 \end{bmatrix}$ and $y = \begin{bmatrix} 1 \\ -1 \end{bmatrix}$. It can be verified that T is not a contraction mapping, but it has a unique fixed point $x = y$.

***Example* 3.5.13** Let $L : C^1[0,1] \to C[0,1]$ be defined by

$$(Lf)(t) = f'(t) - f(t).$$

It is easy to see that $f(t) \equiv 0$ and $f(t) = ce^{2t}$, where $c \in \mathbb{R}$, are fixed points of the operator L. Hence, $L^k, k \geq 1$, cannot be a contraction mapping.

Exercise 3.5.1 If T_1 and T_2, defined from X into itself, are contraction mappings, show that $T_1 T_2 : X \to X$ is also a contraction. Furthermore, if x is a fixed point of $T_1 T_2$, then $T_2 x$ is a fixed point of $T_2 T_1$.

3.6 Compact Set, Compact Operator, and Closed Operator

Compact operators and closed operators play important roles in the analysis of infinite dimensional systems, which are discussed in Chapter 10. This portion may be skipped by those readers who do not deal with systems involving partial differential equations.

Definition 3.6.1 (Compact set) Let X be a normed linear space. A subset S of X is said to be **compact** if every sequence in S has a convergent subsequence, i.e., if $\{x_n\} \subset S$, then there exists a subsequence $\{x_{n_k}\} \subset \{x_n\}$ such that $x_{n_k} \to x$ and $x \in S$.

Definition 3.6.2 (Relatively compact set) A subset S of a normed linear space X is called **relatively compact** if the closure of S, $\overline{S}$, is compact.

Lemma 3.6.3 If $X = \mathbb{R}^n$ with a norm defined on it, then every closed and bounded subset of X is compact.

Lemma 3.6.4 Let S be a compact subset of a normed linear space. Then S is closed and bounded.

However, the converse need not be true in general. In a finite dimensional normed linear space, a subset S is compact if and only if S is closed and bounded. In infinite dimensional spaces, this is not necessarily the case.

Example **3.6.5** Let $X = \ell_1 = \{x = (x_1, x_2, \ldots) : \sum_{i=1}^{\infty} |x_i| < \infty\}$ with the norm defined as $\|x\| = \sum_{i=1}^{\infty} |x_i|$. Let $S = \{x \in \ell_1 : \|x\| \leq 1\}$. Then clearly S is a closed and bounded set in ℓ_1. However, S is not compact since the sequence $\{e_n = (0, 0, \ldots, 1_{n\text{th}}, 0, \ldots)\}_{n=1}^{\infty}$, which forms the standard basis of X, has no convergent subsequence. This is because $\|e_n - e_m\| = 2$ for any $n, m \in \mathbb{N}$, which implies that any subsequence of $\{e_n\}$ is not even a Cauchy sequence.

Definition 3.6.6 Let X and Y be normed linear spaces. A linear operator $T : X \to Y$ is said to be **compact** if T maps bounded sets to relatively compact sets. In other words, if S is a bounded subset of X, then $\overline{T(S)}$ is a compact set in Y.

Example **3.6.7** Let A be an $m \times n$ matrix. Then $A : \mathbb{R}^n \to \mathbb{R}^m$ is a compact linear operator. Linearity of A can be easily verified. For every bounded set S in $\mathbb{R}^n$, $A(S)$ is bounded in $\mathbb{R}^m$, and $\overline{A(S)}$ is compact in $\mathbb{R}^m$ according to Lemma 3.6.3. Hence, A is compact.

Definition 3.6.8 Let $A : D(A) \subset X \to Y$ be a linear operator, where $D(A)$ denotes the domain of A. Then the graph $G(A)$ of A is defined as

$$G(A) = \{(x, Ax) \in X \times Y : x \in D(A)\}.$$

Definition 3.6.9 A linear operator A is called **closed** if its graph $G(A)$ is a closed subset of $X \times Y$. In other words, A is closed if $\{x_n\} \subset D(A)$, $\lim_{n\to\infty} x_n = x$, and $Ax_n \to y$ imply $Ax = y$.

Example **3.6.10** Any bounded linear operator defined on a closed domain is closed. Let $D(A)$ be the domain of A and $\{x_n\} \subset D(A)$. If $x_n \to x$, then $Ax_n \to Ax$ (since A is continuous). Hence, $x \in D(A)$, and A is closed.

Example **3.6.11** Let X be a normed linear space, $V \subset X$ be a subspace of X, and V not be closed. Let $I : V \to X$ be the identity map. Then I is bounded but not closed.

Definition 3.6.12 (Open mapping) Let X and Y be normed linear spaces. Then $T : D(T) \subset X \to Y$ is called an **open mapping** if, for every open set in $D(T)$, its image is an open set in Y.

Theorem 3.6.13 ([10], Open mapping theorem) Let X and Y be Banach spaces, and let $T : X \to Y$ be a bounded linear and onto operator. Then T is an open mapping. Moreover, if T is one-to-one, then T^{-1} is continuous and hence bounded.

Some important properties of compact operators, which will be useful in Chapter 10, are listed below without proofs.

Property 3.6.14

(*i*) If S is a compact subset of a normed linear space, then S is closed and bounded.

(*ii*) If X is a finite dimensional normed linear space, then every closed and bounded subset of X is compact.

(*iii*) If X is a normed linear space in which every closed and bounded set is compact, then X is a finite dimensional space.

(*iv*) Let $T : X \to Y$ be a compact operator, where X and Y are Banach spaces. If Range(T) of T is closed in Y, then Range(T) is a finite dimensional subspace of Y.

(*v*) Let X and Y be normed linear spaces. If $T : X \to Y$ is compact, then it is bounded. The converse is not true in general. However, if X is a finite dimensional space, then boundedness of T implies its compactness.

(*vi*) Let $T : X \to Y$ be a bounded linear operator. If T maps X into a finite dimensional subspace of Y, then T is compact.

(*vii*) Let $T_n : X \to Y$, $n = 1, 2, \ldots$, be a sequence of compact operators on a normed linear space X into a Banach space Y. If $T_n : n = 1, 2, \ldots$ is compact and $\lim_{n\to\infty} \|T_n - T\| = 0$, then T is compact.

(*viii*) Let X be a normed linear space. If $T : X \to X$ is compact and $S : X \to X$ is bounded, then $TS : X \to X$ and $ST : X \to X$ are also compact.

(*ix*) The identity operator I on a normed linear space is not compact in general. I defined on a normed linear space X is compact if and only if X is finite dimensional.

3.7 Summary

- Vector spaces $\mathbb{R}^n, \mathbb{R}^{m\times n}, P[a,b], C[a,b]$, and $L_p[a,b]$ over the field $\mathbb{R}$ (or $\mathbb{C}$) are defined in Examples 3.2.5 and 3.2.6.
- Different norms are defined on these vector spaces in Examples 3.2.9–3.2.14.
- Inner products are defined on the vector spaces $\mathbb{R}^n$, $\mathbb{C}^n$, $L_2[a,b]$, and $L_2[[a,b],\mathbb{R}^n]$ in Examples 3.2.22–3.2.27.
- Parallelogram identity is given in equation (3.5). If any norm on a real vector space satisfies the parallelogram identity, then it defines an inner product as given in equation (3.6). Thus, the parallelogram identity is the necessary and sufficient condition for a normed linear space to be an inner product space. Exercise 8 provides the expression of inner product in terms of the norm defined on a complex vector space, provided the norm satisfies the parallelogram identity.
- Vector valued and matrix valued functions of t are considered in Section 3.3. It is shown that even if the columns of a square matrix $M(t)$ are linearly independent, the determinant of $M(t)$ may be zero, contrary to constant matrix theory.
- If the columns of an $n \times n$ matrix $M(t)$ are n linearly independent solutions of a linear system $\dot{x} = A(t)x, t \in J$, where $A(t)$ is an $n \times n$ matrix, then $(M(t))^{-1}$ exists for all $t \in J$.
- Contraction mapping on a normed linear space along with examples are given in Section 3.5. Banach fixed point theorem ensures the existence of a unique fixed point for a contraction mapping defined on a Banach space. Example 3.5.12 illustrates that the contraction property is a sufficient condition for the existence of a unique fixed point but not necessary. In Chapter 4, existence and uniqueness of solution to IVP (4.18) will be proved using the Banach fixed point theorem.
- Banach space, Hilbert space, Adjoint operator and space of bounded linear operators are briefly discussed in Section 3.4. Properties of compact

operators and closed operators are stated in Section 3.6. Extensive use of these properties can be seen in Chapter 10, where infinite dimensional control systems are considered. Readers without a background from basic functional analysis may skip these topics from Section 3.6 and Chapter 10.

3.8 Exercise

1. Verify that the norms given in Example 3.2.9 satisfy the axioms provided in Definition 3.2.7.

2. Construct a set of three vectors in $P_3[0,1]$, denoted as $f_i(t)$ for $i = 1,2,3$, such that any two of them are linearly independent, but all three together are linearly dependent.

3. Let $\mathbb{R}^+$ denote the set of positive real numbers. Show that $\mathbb{R}^+$ is a vector space over $\mathbb{R}$ with respect to vector addition '+' defined by

   $$x + y = xy$$

   and scalar multiplication '$\cdot$' defined by

   $$\alpha x = x^{\alpha}.$$

4. (i) Show that $V = \mathbb{R} \times \mathbb{R}^+$ is a vector space with the following operations of vector addition and scalar multiplication:

   $$(x_1, y_1) + (x_2, y_2) = (x_1 + x_2, y_1 y_2) \text{ and } \alpha(x_1, y_1) = (\alpha x_1, y_1^{\alpha}).$$

   (ii) Show that the set $S = \{(x, a^x) : x \in R, a > 0 \text{ is a constant}\}$ is a subspace of V.

   (iii) Show that the set $\{(1,3),(2,9)\}$ is linearly dependent in V.

   (iv) Show that the set $\{(1,2),(4,8)\}$ is linearly independent in V.

5. Let (x_1, y_1) and (x_2, y_2) be two vectors in $\mathbb{R}^2$ over $\mathbb{R}$. Show that the function $\langle \cdot, \cdot \rangle$ defined by

   $$\langle (x_1, y_1), (x_2, y_2) \rangle = 2x_1x_2 + 5y_1y_2,$$

   is an inner product on $\mathbb{R}^2$.

6. Let V be an n-dimensional vector space and $B = \{v_1, v_2, \ldots, v_n\}$ be a basis of V. For arbitrary vectors $x, y \in V$, define $\langle x, y \rangle$ as

   $$\langle x, y \rangle = \langle [x]_B, [y]_B \rangle_{\mathbb{C}^n},$$

where $[x]_B$ denotes the coordinate vector of x with respect to the basis B, and the inner product on the RHS is the standard inner product in $\mathbb{C}^n$. Show that $\langle \cdot, \cdot \rangle$ is an inner product on V. Furthermore, show that B is orthonormal in the inner product space V.

7. Show that the function

$$\langle (x_1, y_1), (x_2, y_2) \rangle = (y_1 - x_1)(y_2 - x_2) + (2x_1 - y_1)(2x_2 - y_2)$$

defines an inner product on $\mathbb{R}^2$ with respect to the basis $\{(1,2), (1,1)\}$.

8. Let V be a real (or complex) vector space equipped with a norm $\|\cdot\|$.

   (*i*) Show that V is an inner product space if and only if the norm satisfies the **parallelogram identity**, that is,

$$\|x+y\|^2 + \|x-y\|^2 = 2(\|x\|^2 + \|y\|^2) \quad \text{for all } x, y \in V.$$

   (*ii*) If V is a real vector space, then the inner product can be obtained from the norm via the relation

$$\langle x, y \rangle = \frac{1}{4}\left[\|x+y\|^2 - \|x-y\|^2\right].$$

   (*iii*) If V is a complex vector space, then

$$\text{Re}\langle x, y \rangle = \frac{1}{4}\left[\|x+y\|^2 - \|x-y\|^2\right]$$

   and

$$\text{Im}\langle x, y \rangle = \frac{1}{4}\left[\|x+iy\|^2 - \|x-iy\|^2\right].$$

9. Let $X = C[0,1]$, the space of all continuous functions defined on $[0,1]$. Let $f_n : [0,1] \to \mathbb{R}$ be a sequence in X defined by $f_n(t) = t^n$. If the norm on X is given by

$$\|f\| = \max_{t \in [0,1]} |f(t)|,$$

show that $\{f_n\}$ is not a Cauchy sequence in X.

10. Let $X = C[0,1]$ with the norm defined by

$$\|f\|^2 = \int_0^1 |f(t)|^2 dt.$$

Let $f_n(t) = t^n$ be a sequence of functions in X. Show that $\{f_n\}$ is a Cauchy sequence in X but does not converge in X.

11. Let $T : L_2[0,1] \to L_2[0,1]$ be defined by $(Tx)(t) = \int_0^1 s(t-s)^2 x(s)ds$. Show that T is linear and continuous. Also, find T^*.

12. Let $T : L_2[a,b] \to L_2[a,b]$ be defined by

$$Tx = \sum_{n=1}^{\infty} e^{\lambda_n} \langle x, \phi_n \rangle \phi_n,$$

where $\{\lambda_n\}$ is a sequence with $\sup_n \lambda_n < \infty$, $\{\phi_n\}$ is an orthogonal basis of $L_2[a,b]$, and $\langle \cdot, \cdot \rangle$ denotes the inner product in $L_2[a,b]$. Show that T is a bounded linear operator.

13. Let $\{\phi_n\}$ be an orthogonal basis of $L_2[a,b]$, and let $T_k : L_2[a,b] \to L_2[a,b]$ for $k \in \mathbb{N}$ be defined by

$$T_k x = \sum_{n=1}^{k} e^{-n} \langle x, \phi_n \rangle \phi_n.$$

Prove that T_k is a compact linear operator for each $k \in \mathbb{N}$. Hence, show that the operator $T : L_2[a,b] \to L_2[a,b]$ defined by

$$Tx = \sum_{n=1}^{\infty} e^{-n} \langle x, \phi_n \rangle \phi_n$$

is compact.

14. If $T : \mathbb{C}^n \to \mathbb{C}^m$ is an $m \times n$ complex matrix, show that the adjoint T^* is the complex conjugate of T', i.e., $(\overline{T})'$.

15. Show that the matrix $A = \begin{bmatrix} 1 & 0 \\ 0 & 3 \end{bmatrix}$ is not a contraction on $\mathbb{R}^2$ but has infinitely many fixed points.

16. Show that the operator $T : C[0,1] \to C[0,1]$ defined by

$$(Tf)(t) = \frac{t}{2} \int_0^1 f(s)ds$$

with the sup norm on $C[0,1]$ has the unique fixed point $f \equiv 0$.

17. Let $T : C[0,1] \to C[0,1]$ be defined by

(i)

$$(Tf)(t) = 2t\int_0^1 f(s)ds$$

with the L_2 norm. Show that T is not a contraction but has infinitely many fixed points in $C[0,1]$.

(ii)

$$(Tf)(t) = \frac{\alpha t}{a^2}\int_0^a f(s)ds$$

with the sup norm on $C[0,a]$. Find the value of α for which T is a contraction.

4

Ordinary Differential Equations

4.1 Introduction

Systems of first order ordinary differential equations (ODEs) play a vital role in control systems theory. In this chapter, results from the theory of ODEs , limited to those necessary for developing subsequent chapters, are presented.

In Section 4.2, the definition and classification of ODEs of order n are given. In Section 4.3, solutions to first order ODEs and initial value problems (IVPs) are discussed. Moreover, in Theorem 4.3.8, a sufficient condition for the existence of solutions to first order ODEs is given. The existence and uniqueness of solutions to IVPs is proved in Section 4.4 using the Banach fixed point theorem. In Sections 4.5 and 4.6, system of ODEs and its solution in series form are discussed. In Section 4.7, solutions of time-invariant linear systems in terms of eigenpairs of the coefficient matrix is presented. Section 4.8 contains some results on fundamental matrices and the state transition matrix. In Section 4.9, solution to nonhomogeneous linear system is discussed. Gronwall's inequality is briefly discussed in Section 4.10. Theorem 4.5.1 on the existence and uniqueness of solutions of IVPs, Section 4.6 on systems of linear ODEs, Section 4.8 on fundamental matrices and the state transition matrix, and Theorem 4.10.1 on Gronwall's inequality are important focal points for a better understanding of the contents of Chapters 5–8.

4.2 n^{th} order ODE

Definition 4.2.1 An ODE is a mathematical relation involving an independent variable, a dependent variable, and its derivatives.

Throughout this book, the independent variable is assumed to be real valued, and the dependent variable belongs to a real vector space.

The **order** of an ODE is the order of the highest derivative present in the equation.

An ODE of order n can be represented as

$$F(t, x(t), \dot{x}(t), \ldots, x^{(n)}(t)) = 0. \tag{4.1}$$

In (4.1), $t \in \mathbb{R}$ is the independent variable, and x is the dependent variable. $\dot{x}(t)$ and $\ddot{x}(t)$ denote the first and second order derivatives of x with respect to t, respectively, and $x^{(k)}(t) : k = 3, \ldots, n$ denotes the kth derivative of x with respect to t. The function F relates the $n+2$ variables $t, x, \dot{x}, \ldots, x^{(n)}$. In explicit form, an nth order ODE can be expressed as

$$x^{(n)} = g(t, x, \dot{x}, \ldots, x^{(n-1)}). \tag{4.2}$$

Definition 4.2.2 Consider the differential equation (4.2). If the function g is linear in its arguments $x, \dot{x}, \ldots, x^{(n-1)}$, then the differential equation (4.2) is called a **linear** ODE. Otherwise, it is called a **nonlinear** ODE.

Thus, an nth order linear ODE is of the form

$$a_0(t)x^{(n)}(t) + a_1(t)x^{(n-1)}(t) + \cdots + a_n(t)x(t) = f(t), \quad t \in [t_0, \tau], \tag{4.3}$$

where $a_i(t) : i = 0, 1, 2, \ldots, n$ and $f(t)$ are continuous functions of t on $[t_0, \tau]$, and $a_0(t) \neq 0$. If $f(t) \equiv 0$ on $[t_0, \tau]$, then (4.3) is called a **homogeneous** equation.

***Example* 4.2.3** The equation
(i) $\dot{x} + tx = 0$ for $t \in \mathbb{R}$ is a homogeneous first order linear ODE.
(ii) $\ddot{x} + t^2\dot{x} + 2x + 3t = 0$ for $t \in \mathbb{R}$ is a nonhomogeneous second order linear ODE.
(iii) $t\ddot{x} = \sin x$ is a second order nonlinear ODE.

Exercise 4.2.1 Classify the following differential equations as linear or nonlinear, and homogeneous or nonhomogeneous

(i) $x\dot{x} + tx = e^t$.

(ii) $t^2\ddot{x} + 2x - 1 = 0$.

(*iii*) $(\sin t)\ddot{x} + t\dot{x} + x = x^2$.

(*iv*) $\ddot{x} + (1-t^2)\dot{x} = e^t x$.

4.3 First Order ODE

Let D be a domain in $\mathbb{R}^2$ and $f : D \to \mathbb{R}$ be a continuous function. Consider the first order ODE

$$\dot{x} = f(t,x). \tag{4.4}$$

Definition 4.3.1 A real valued function x of t, defined on an interval $[t_0, \tau] =: J \subset \mathbb{R}$, is called a **solution** of the differential equation (4.4) if (*i*) for all $t \in J$, $(t, x(t)) \in D$; (*ii*) x is differentiable on J; and (*iii*) $\dot{x}(t) = f(t, x(t))$ for all $t \in J$.

Let (t_0, x_0) be an interior point in D.

The first order ODE (4.4) with the initial condition as given below

$$\begin{cases} \dot{x} = f(t,x), \\ x(t_0) = x_0 \end{cases} \tag{4.5}$$

is known as an **initial value problem**.

A function $x : J \to \mathbb{R}$ is said to be a **solution** of the IVP (4.5) if (*i*) for all $t \in J$, $(t, x(t)) \in D$; (*ii*) x is differentiable on J; (*iii*) $\dot{x}(t) = f(t, x(t))$ for all $t \in J$; and (*iv*) $x(t_0) = x_0$.

There are differential equations that may not have any solution. For example, the equation $|\dot{x}| + 1 = 0$ has no solution. The equation $\dot{x}^2 + 1 = 0$ has no real solution; however, it has a solution of the form $x(t) = it + c$, where $i = \sqrt{-1}$ and c is a constant.

Physical systems are expected to have real solutions. Therefore, to construct a useful mathematical model, it is essential to know whether a solution exists. Thus, identifying the class of differential equation that admits solutions is a very important problem in the theory of differential equations, and a theorem that provides conditions for the existence of solutions is referred to as an **existence theorem**.

There are differential equations that have many solutions for a given initial condition.

***Example* 4.3.2** Consider the IVP

$$\begin{cases} \dot{x} = x^{\frac{3}{4}}, \quad t > 0, \\ x(0) = 0. \end{cases}$$

Writing the equation as $x^{-\frac{3}{4}}dx = dt$ and integrating from 0 to t, we obtain a solution $x(t) = \frac{t^4}{256}$ for $t > 0$. However, the IVP has infinitely many solutions given by

$$x(t) = \begin{cases} 0, & 0 \leq t \leq c, \\ \dfrac{(t-c)^4}{256}, & t > c, \end{cases}$$

for some $c > 0$.

Since differential equations originate from physical systems, they are generally expected to have a unique solution for a given set of conditions. Hence, a desirable mathematical model is one that has a unique solution. Thus, we refer to the problem of finding conditions such that a given differential equation has exactly one solution as the **uniqueness problem**.

Example **4.3.3** Consider the differential equation

$$\begin{cases} \dot{x} = x, \\ x(0) = \alpha \in \mathbb{R}. \end{cases} \tag{4.6}$$

Let $x_\alpha(t)$ be the solution of (4.6). Then $x_\alpha(t) = \alpha e^t$ for $t \geq 0$, and $|x_\alpha(t) - x_\beta(t)| = |\alpha - \beta|e^t$ for $t \geq 0$. Note that $|x_\alpha(t) - x_\beta(t)| \to \infty$ as $t \to \infty$, irrespective of how small $|\alpha - \beta|$ is. This shows that closeness of the initial conditions does not necessarily imply the closeness of the solutions for all t. In other words, the solutions do not change continuously with respect to the initial conditions, and a small change in initial condition causes a large deviation in the solution.

In general, a mathematical model of a physical system is considered useful if it has the following important properties:

(i) There exists a unique solution that satisfies the initial condition.

(ii) The solution is continuous with respect to the initial condition.

A mathematical problem having these properties is called a **well-posed problem** in the sense of **Hadamand**. If any of the above conditions are violated, the problem is called **ill-posed**.

In many practical phenomena, the corresponding mathematical models are often ill-posed. Even for well-posed problems, finding an analytical solution can be challenging. There are several equations of significant importance in various areas of science and technology for which explicit solutions cannot be obtained. For such equations, numerical methods can be applied to obtain the approximate solution.

***Example* 4.3.4** The differential equation $\dot{x} + e^{t^2}x = 0$ is a simple example that cannot be solved explicitly, although a solution of the form $x(t) = c\exp(-\int e^{t^2}dt)$ exists, where c is a constant. Here, $\int e^{t^2}dt$ cannot be evaluated analytically.

In the following, we shall consider simple first order differential equations of the form (4.4) to illustrate the various conditions that influence the existence of a solution.

Linear Homogeneous Equations

Let us consider the first order homogeneous linear differential equation

$$\dot{x} - p(t)x = 0, \tag{4.7}$$

where p is a continuous function on an interval $J = [t_0, \tau]$. The solution x of (4.7) can be obtained by separating the variables as

$$\frac{dx}{x} = p(t)dt. \tag{4.8}$$

Integrating both sides from t_0 to t, we get

$$\log x(t) = \int_{t_0}^{t} p(s)ds + k, \tag{4.9}$$

where k is an arbitrary constant. Setting $k = \log\alpha$ in (4.9), we obtain

$$x(t) = \alpha e^{\int_{t_0}^{t} p(s)ds}. \tag{4.10}$$

Equation (4.10) provides the general solution of (4.7).

Let $x(t_0) = x_0$ be the initial condition for the equation (4.7). Then, from (4.10), we have $\alpha = x_0$, so the solution of the IVP is

$$x(t) = x_0 e^{\int_{t_0}^{t} p(s)ds}. \tag{4.11}$$

If $x_0 = 0$, then $x(t) = 0$ for all t, which is known as the **trivial** solution.

Linear Nonhomogeneous Equations

Consider the linear nonhomogeneous first order equation

$$\dot{x} - p(t)x = q(t), \tag{4.12}$$

where p and q are continuous functions on J. To find a solution to (4.12), let us assume that $x(t)$ is of the form

$$x(t) = c(t)e^{\int_{t_0}^{t} p(s)ds}, \tag{4.13}$$

where $c(t)$ is a differentiable function. Substituting (4.13) into (4.12), we obtain

$$(\dot{c}(t)+c(t)p(t))e^{\int_{t_0}^{t} p(s)ds} - p(t)c(t)e^{\int_{t_0}^{t} p(s)ds} = q(t),$$

which simplifies to

$$\dot{c}(t) = e^{-\int_{t_0}^{t} p(s)ds} q(t).$$

Hence,

$$c(t) = \int_{t_0}^{t} e^{-\int_{t_0}^{s} p(r)dr} q(s)ds + c_1, \text{where } c_1 \text{ is a constant.}$$

Substituting $c(t)$ back into (4.13), we obtain the general solution to (4.12) as

$$x(t) = c_1 e^{\int_{t_0}^{t} p(s)ds} + \int_{t_0}^{t} e^{\int_{s}^{t} p(r)dr} q(s)ds. \tag{4.14}$$

Note that the first term is the general solution of the homogeneous equation and the second term is a particular solution of the nonhomogeneous equation. If $x(t_0) = x_0$ is the initial condition, then $c_1 = x(t_0)$ and

$$x(t) = x_0 e^{\int_{t_0}^{t} p(s)ds} + \int_{t_0}^{t} e^{\int_{s}^{t} p(r)dr} q(s)ds.$$

If $p(t) = p$, a constant, then

$$x(t) = x_0 e^{p(t-t_0)} + \int_{t_0}^{t} e^{p(t-s)} q(s)ds.$$

***Example* 4.3.5** Find the general solution of the differential equation

$$\dot{x} + x\sin t + 2 \sin t = 0.$$

Solution Comparing the differential equation with (4.12), we have $p(t) = -\sin t$ and $q(t) = -2t\cos t$. Then, from (4.14), the general solution is

$$\begin{aligned} x(t) &= c_1 e^{-\int^{t} \sin t dt} + \int^{t} e^{-\int_{s}^{t} \sin r dr} (-2\sin s)\, ds \\ &= c_1 e^{\cos t} - 2e^{\cos t}\left[\int^{t} e^{-\cos s} \sin s ds\right] \\ &= c_1 e^{\cos t} - 2. \end{aligned}$$

Note: To derive the formulas (4.11) and (4.14), the continuity of the functions p and q is assumed. However, these are only sufficient conditions. It can be observed that the integrability of $p(t)$ and $q(t)$ on a suitable interval is a necessary and sufficient condition for obtaining both (4.11) and (4.14).

Example **4.3.6** The general solution of

$$\dot{x} = x\cot t + 2\cos t \tag{4.15}$$

is

$$x(t) = Ce^{\int^t \cot s ds} + 2\int^t e^{\int_s^t \cot r dr}(\cos s)\, ds = C\sin t + 2\int^t \frac{\sin t}{\sin s}\cos s ds,$$

where C is a constant. Hence,

$$x(t) = \begin{cases} C\sin t + 2\sin t\log(\sin t), & \text{if } \sin t > 0, \\ C\sin t + 2\sin t\log(-\sin t), & \text{if } \sin t < 0. \end{cases} \tag{4.16}$$

The solution $x(t)$ in (4.16) is differentiable in those intervals where $\sin t$ is non zero.

Example **4.3.7** Consider the differential equation

$$\begin{cases} x\dot{x} + (1+x^2)\sin t = 0, \\ x(0) = 1. \end{cases} \tag{4.17}$$

The above equation can be written as $d(x^2)/dt + 2x^2\sin t + 2\sin t = 0$ with initial condition $x^2(0) = 1$. From (4.14) the solution $x^2(t)$ of the above equation is given by $x^2(t) = (2e^{2(\cos t - 1)} - 1)$. Hence $x(t) = \pm(2e^{-4\sin^2(\frac{t}{2})} - 1)^{\frac{1}{2}}$. To satisfy the initial condition, we select the expression with the positive sign. Therefore, $x(t) = (2e^{-4\sin^2(\frac{t}{2})} - 1)^{\frac{1}{2}}$ is a real solution when $2e^{-4\sin^2(\frac{t}{2})} \geq 1$, i.e., $0 \leq t \leq \frac{1}{2}\sqrt{\log 2}$. This example illustrates that a real solution to a differential equation may be available only on a restricted domain.

Existence Theorem

Consider the following first order nonlinear IVP

$$\begin{cases} \dot{x} = f(t,x), \\ x(t_0) = x_0. \end{cases} \tag{4.18}$$

Here, f is a given continuous function on some domain D in $\mathbb{R}^2$ containing the point (t_0, x_0). Our aim is to obtain a function x satisfying the differential equation in some neighborhood of t_0 such that the graph of x, defined by $\{(t, x(t)) : t_0 \leq t \leq t_0 + h\}$, contains the point (t_0, x_0) and lies in D.

We have seen that there are IVPs which may have one solution, more than one solution, or no solution. The following theorem provides sufficient conditions that guarantees the existence of a solution to the IVP (4.18). The proof is beyond the scope of this book.

Theorem 4.3.8 Suppose that f is continuous on a domain $D \subset \mathbb{R}^2$ containing the point (t_0, x_0) and that $|f(t,x)| \leq M$ for all $(t,x) \in D$. Then there exists an interval $J_h = (t_0, t_0 + h), h > 0$, such that the IVP (4.18) has a solution x on J_h.

Example **4.3.9** Consider the following differential equation

$$\begin{cases} \dot{x} + \dfrac{2x}{\sqrt{|t-1|}} = 0, \quad 0 \leq t \leq 2, \\ x(0) = 1. \end{cases} \tag{4.19}$$

From (4.14), the IVP (4.19) has a solution given by

$$x(t) = ce^{-\int_0^t \frac{2}{\sqrt{|t-1|}} dt} = e^{1-\sqrt{|t-1|}}.$$

At this point, we may conclude that the conditions of Theorem 4.3.8 are not necessary for ensuring the existence of a solution. Here, $f(t,x) = -\frac{2x}{\sqrt{|t-1|}}$ is not continuous along $t = 1$.

Example **4.3.10** Consider the differential equation $\dot{x} = \frac{1}{x}$. Since the function $f(t,x) = \frac{1}{x}$ is not continuous at $x = 0$, the solution is valid only for nonzero initial conditions. (Since $x\dot{x} = 1 \implies d(x^2) = 2dt$, this yields the solution $x(t) = \sqrt{2t + c}$ for $t > -\frac{c}{2}$, where $c \in \mathbb{R}$.) If t_0 is the initial value of t and $x(t_0) = x_0$ is the initial condition on x, then $c = x_0^2 - 2t_0$.

Example **4.3.11** The differential equation $\dot{x} = \frac{x}{t}$ has a solution even though $f(t,x) = \frac{x}{t}$ is not continuous at $t = 0$. By separating the variables, the equation becomes $\frac{dx}{x} = \frac{dt}{t}$. Integrating both sides, all possible solutions can be obtained as $x(t) = ct$, where c is a constant. This example shows that even if $f(t,x)$ is not continuous, the equation (4.18) may have a solution. Note that $x(0) = 0$ for all possible solutions. Hence, the equation has no solution for any nonzero initial condition at $t = 0$.

Suppose that f is continuous in D and that (t_0, x_0) is an interior point of D. The first step towards the existence result is to show that the IVP (4.18) is equivalent to the following integral equation:

$$x(t) = x_0 + \int_{t_0}^{t} f(s, x(s)) ds, \quad t \in J. \tag{4.20}$$

The precise equivalence is given as follows.

Lemma 4.3.12 Consider the IVP (4.18) in which f is continuous in a neighborhood of (t_0, x_0). A function $x(t)$ is a solution of (4.18) on a domain D containing (t_0, x_0) if and only if $x(t)$ is a solution of the integral equation (4.20) on J.

Proof. If x is a solution of (4.18) on J, we have

$$\dot{x}(t) = f(t, x(t)), \quad t \in J. \tag{4.21}$$

Integrating from t_0 to t, we obtain

$$x(t) - x(t_0) = \int_{t_0}^{t} f(s, x(s))ds, \quad t \in J. \tag{4.22}$$

Imposing the given initial condition, we see that x satisfies (4.20).

Conversely, if x is a solution of the integral equation (4.20), then by the continuity of the function $f(t, x(t))$, the RHS of (4.20) is differentiable for all $t \in J$. Differentiating both sides of (4.20), we have

$$\dot{x} = f(t, x), \tag{4.23}$$

and by substituting $t = t_0$ in (4.20), we obtain $x(t_0) = x_0$. □

With the help of this lemma we will establish the existence of a solution of (4.18) by proving the existence of a solution to (4.20). To begin, let us first define the Lipschitz continuity of a function.

Definition 4.3.13 A function f of a single variable x is said to be **Lipschitz continuous** on an interval $(a, b) \subset \mathbb{R}$ if for $x_1, x_2 \in (a, b)$, $|f(x_1) - f(x_2)| \leq \alpha |x_1 - x_2|$ for some constant $\alpha > 0$, which is called the **Lipschitz constant**.

***Example* 4.3.14** For the function $f(x) = x + x^2, 0 \leq x \leq 2$,

$$|f(x_1) - f(x_2)| = |x_1 - x_2 + x_1^2 - x_2^2| \leq |x_1 - x_2| + |x_1^2 - x_2^2| \leq 5|x_1 - x_2|$$

for all $x_1, x_2 \in (0, 2)$. Therefore, f is Lipschitz continuous on $[0, 2]$ with the Lipschitz constant $\alpha = 5$.

Definition 4.3.15 Let f be a function of two variables t and x, defined on a domain $D \subset \mathbb{R}^2$. Then f is said to be Lipschitz continuous with respect to x if it satisfies the inequality

$$|f(t, x_1) - f(t, x_2)| \leq \alpha |x_1 - x_2| \quad \text{for some constant } \alpha > 0 \tag{4.24}$$

for all (t, x_1) and (t, x_2) in the domain D.

***Example* 4.3.16** Show that the function $f(t,x) = tx^r$, where $r \in (0,1)$ is a constant, does not satisfy the Lipschitz condition in the rectangle $S = \{(t,x) : |t| \leq 1, 0 \leq x \leq 2\}$.

Solution We need only to produce a pair of points for which (4.24) fails to hold with any constant $\alpha > 0$. Consider the pair of points (t,x_1) and $(t,0)$ in S. Then, we have

$$\left|\frac{f(t,x_1) - f(t,0)}{x_1 - 0}\right| = |x_1^{r-1}t| \leq |x_1^{r-1}|.$$

Now, by choosing $x_1 > 0$ sufficiently small, we can show that x_1^{r-1} can be made larger than any preassigned constant α. Therefore, (4.24) cannot hold for any α. Hence, $f(t,x)$ does not satisfy the Lipschitz condition.

In Example 4.3.16, the function $f(t,x)$ does not satisfy the Lipschitz continuity in the region S. However, it may satisfy Lipschitz condition in some other region; for example, in $S_1 := \{(t,x) : |t| \leq 1, |x-2| < 1\}$, the function f satisfies the Lipschitz condition. In addition, we may observe that S_1 does not contain the point $(t,0)$, and both f and $\frac{\partial f}{\partial x}$ are continuous and bounded on S_1.

The following lemma provides sufficient conditions on f such that f satisfies the Lipschitz continuity.

Lemma 4.3.17 Let $f : D \subset \mathbb{R}^2 \to \mathbb{R}$ be such that (i) f is continuous on D with $|f(t,x)| \leq k_1$, a constant, and (ii) $\frac{\partial f}{\partial x}$ is continuous on D with $\left|\frac{\partial f}{\partial x}\right| \leq k$, a constant. Then f is Lipschitz continuous with respect to x.

Proof. From (ii), for any given $\varepsilon > 0$, there exists a $\delta > 0$ such that

$$\left|\frac{f(t,x_1) - f(t,x_2)}{x_1 - x_2} - \frac{\partial f}{\partial x}\right| < \varepsilon \quad \text{whenever } |x_1 - x_2| < \delta,$$

implying that

$$|f(t,x_1) - f(t,x_2)| \leq (k+\varepsilon)|x_1 - x_2|.$$

Now if $|x_1 - x_2| \geq \delta$, then

$$\left|\frac{f(t,x_1) - f(t,x_2)}{x_1 - x_2}\right| \leq \frac{2k_1}{\delta} = k_2 \text{ (say)}.$$

Hence,

$$|f(t,x_1) - f(t,x_2)| \leq \begin{cases} k_2|x_1 - x_2|, & \text{if } |x_1 - x_2| \geq \delta, \\ (k+\varepsilon)|x_1 - x_2|, & \text{if } |x_1 - x_2| < \delta. \end{cases}$$

Combining the above two inequalities, we obtain

$$|f(t,x_1) - f(t,x_2)| \le L|x_1 - x_2|, \quad \text{where } L = \max\{k_2, k+\varepsilon\}.$$

□

Lemma 4.3.17 provides a sufficient condition for a function to be Lipschitz continuous. The following example illustrates that it is not a necessary condition.

***Example* 4.3.18** Consider the function $f(t,x) = t|x|$ in the rectangle $S = \{(t,x) : |t| \le a, |x| \le b\}$, where a and b are two positive real constants. Show that f satisfies the Lipschitz condition with respect to x even though $\frac{\partial f}{\partial x}$ is not continuous on S.

Solution Observe that $\frac{\partial f}{\partial x}$ does not exist at $x = 0$. However, we have

$$|f(t,x_1) - f(t,x_2)| \le |t|\,||x_1| - |x_2|| \le a|x_1 - x_2|.$$

Exercise 4.3.1 Show that the solution of the differential equation $\dot{x} = x^n$ with the initial condition $x(t_0) = x_0$ is valid in the interval $(-\infty, b)$, where (i) $b = t_0 + \frac{1}{(n-1)x_0^{n-1}}$ if $n \ge 2$ and $x_0 > 0$; (ii) $b = \infty$ if $n = 2k$, where $k = 1, 2, \ldots$, and $x_0 < 0$. Also, show that for $n \ge 1, x(t) > 0$ if $x_0 > 0$ and $x(t) < 0$ if $x_0 < 0$.

Exercise 4.3.2 Show that the functions (i) $t \sin t$ and (ii) $|t| \sin t$ have bounded derivatives on every bounded interval in $\mathbb{R}$. Hence, show that they are Lipschitz continuous on any finite interval in $\mathbb{R}$.

Exercise 4.3.3 Show that $f(t) = |t| \cos t$ is not differentiable at $t = 0$ but is Lipschitz continuous on $[-\frac{\pi}{2}, \frac{\pi}{2}]$.

Exercise 4.3.4 Find the solution of

$$\dot{x} - x = \begin{cases} e^t, & 0 \le t \le 1, \\ 1, & 1 < t \le 2 \end{cases}$$

that satisfies the initial condition $x(0) = 2$, assuming that x is continuous on $[0, 2]$.

4.4 Existence and Uniqueness Theorems

Consider the IVP (4.5) in the rectangular region S. If $f(t,x(t))$ is integrable as a function of t, then integrating both sides of the differential equation in (4.18) from t_0 to t and using the initial condition, we get

$$x(t) = x_0 + \int_{t_0}^{t} f(s,x(s))ds. \tag{4.25}$$

Thus, if $x(t)$ is a solution of (4.18), then it satisfies (4.25). In general, the converse is not true. However, if $f(t,x(t))$ is continuous as a function of t and $x(t)$ satisfies (4.25), then x is a solution of (4.18) (Lemma 4.3.12).

Hence, to establish the existence and uniqueness of a solution to (4.18), it is enough to prove the existence and uniqueness of a solution to the integral equation (4.25). We shall prove this under the following condition on the function $f(t,x)$.

(**A**) (**Condition on** f) Let f be a real valued function defined on a domain D in $\mathbb{R}^2$ satisfying the following:

- (i) f is continuous on the domain D.
- (ii) f satisfies the Lipschitz continuity with respect to the second argument x, i.e., if (t,x_1) and (t,x_2) are in D then

  $$|f(t,x_1) - f(t,x_2)| \leq \alpha |x_1 - x_2|$$

  for some positive real number α.

Theorem 4.4.1 Let f be a real valued function defined on a domain D in $\mathbb{R}^2$ satisfying the condition (**A**). Let $(t_0,x_0) \in D$. Then the IVP (4.18) has a unique solution in D.

Proof. In order to prove the theorem, we shall show that the integral equation (4.25) has a unique solution. Let S be the rectangular region in D defined as $S = \{(t,x) \in D : t_0 - \tau \leq t \leq t_0 + \tau, x_0 - b \leq x \leq x_0 + b\}$ for some $\tau, b > 0$. Let $C(J)$ be the set of all real valued continuous functions defined on $J = [t_0, \tau]$.

Let $T : C(J) \to C(J)$ be the operator defined as

$$(Tx)(t) = x_0 + \int_{t_0}^{t} f(s,x(s))ds. \tag{4.26}$$

Since $C(J)$ is a Banach space under the supremum norm, if T^k is a contraction mapping on $C(J)$ for some positive integer k, then, from Corollary 3.5.7, there exists a unique $x^* \in C(J)$ such that

$$(Tx^*)(t) = x^*(t). \tag{4.27}$$

From (4.26) and (4.27), it is clear that x^* is the unique solution of (4.25), which in turn is the required unique solution of the IVP (4.18).

Now let us show that the operator $T^k : C(J) \to C(J)$ is a contraction map for some positive integer k. Let $x, \tilde{x} \in C(J)$. Then

$$\|T^k x - T^k \tilde{x}\| = \sup_{t \in J} \left|(T^k x)(t) - (T^k \tilde{x})(t)\right|.$$

From (4.26),

$$\begin{aligned}
|Tx(t) - T\tilde{x}(t)| &= \left|\int_{t_0}^{t} (f(s,x(s)) - f(s,\tilde{x}(s)))\, ds\right| \\
&\leq \int_{t_0}^{t} |f(s,x(s)) - f(s,\tilde{x}(s))|\, ds \\
&\leq \alpha \int_{t_0}^{t} |x(s) - \tilde{x}(s)|\, ds \ \text{(using the Lipschitz continuity of } f) \quad (4.28) \\
&\leq \alpha \sup_{s \in J} |x(s) - \tilde{x}(s)| \int_{t_0}^{t} ds \\
&= \alpha \|x - \tilde{x}\|(t - t_0). \qquad (4.29)
\end{aligned}$$

Now

$$\begin{aligned}
|(T^2 x)(t) - (T^2 \tilde{x})(t)| &= |(T(Tx))(t) - (T(T\tilde{x}))(t)| \\
&\leq \alpha \int_{t_0}^{t} |(Tx)(s) - (T\tilde{x})(s)|\, ds \quad \text{(from (4.28))} \\
&\leq \alpha^2 \|x - \tilde{x}\| \cdot \int_{t_0}^{t} (s - t_0) ds \quad \text{(using (4.29))} \\
&= \alpha^2 \|x - \tilde{x}\| \frac{(t - t_0)^2}{2}.
\end{aligned}$$

Proceeding in a similar manner, it can be shown that

$$|(T^k x)(t) - (T^k \tilde{x})(t)| \leq \alpha^k \frac{(t - t_0)^k}{k!} \|x - \tilde{x}\|. \qquad (4.30)$$

Hence,

$$\|T^k x - T^k \tilde{x}\| = \sup_{t \in J} \left|(T^k x)(t) - (T^k \tilde{x})(t)\right| \leq \alpha^k \frac{(\tau - t_0)^k}{k!} \|x - \tilde{x}\|. \qquad (4.31)$$

For fixed α and τ, we known that $\lim_{k \to \infty} \frac{\alpha^k (\tau - t_0)^k}{k!} = 0$. Hence, for a sufficiently large value of k, we can obtain

$$\sigma := \frac{\alpha^k (\tau - t_0)^k}{k!} < 1. \qquad (4.32)$$

From (4.31) and (4.32), it follows that there exists an integer k such that $T^k : C(J) \to C(J)$ is a contraction map, i.e.,

$$\|T^k x - T^k \tilde{x}\| \leq \sigma \|x - \tilde{x}\|, \quad 0 < \sigma < 1.$$

Hence, from Corollary 3.5.7, T has a unique fixed point $x^* \in C(J)$, which is the unique solution of the integral equation (4.25) and hence the unique solution of the IVP (4.18). □

In the proof of the Banach fixed point theorem (Theorem 3.5.6), it was shown that for any arbitrary element a_0 in the Banach space X, the sequence $\{a_n\} \subset X$ converges to the fixed point a^*, where

$$a_n = Ta_{n-1}, \quad n = 1, 2, \ldots. \tag{4.33}$$

Hence, for any given $\varepsilon > 0$, we can find an integer N such that $\|a_n - a^*\| < \varepsilon$ whenever $n > N$. Thus, for sufficiently large n, a_n is an approximation to a^*.

The above procedure can be used to compute an approximation solution for the IVP (4.18) by constructing a sequence as given below.

For this, consider the operator defined in (4.26)

$$(Tx)(t) = x_0 + \int_{t_0}^{t} f(s, x(s))ds.$$

Let $x_0(t) = x_0$ be a constant function in $C(J)$, and define $x_n(t) = (Tx_{n-1})(t)$ for $t \in J$. Note that $x_n(t_0) = x_0$ for all $n \geq 1$. In Theorem 4.4.1, it was shown that $\lim_{n\to\infty} \|x_n - x^*\| = 0$, implying $\lim_{n\to\infty} x_n(t) = x^*(t)$ uniformly for all $t \in J$. Hence, for a sufficiently large n, $x_n(t)$ can be considered an approximation to the unique solution $x^*(t)$ of the IVP (4.18).

***Example* 4.4.2** Consider the IVP

$$\begin{cases} \dot{x} = t + x, & 0 \leq t \leq 1, \\ x(0) = 2. \end{cases} \tag{4.34}$$

Starting with $x_0(t) = 2$, find the first three elements of the sequence $\{x_n\}_{n=1}^{\infty}$, which converges to the solution of (4.34).

Solution Given that the initial approximation is $x_0(t) = 2$, $0 \leq t \leq 1$. Then, from (4.33), we obtain

$$x_1(t) = (Tx_0)(t) = 2 + \int_0^t (s+2)ds = 2 + \frac{t^2}{2} + 2t,$$

$$x_2(t) = (Tx_1)(t) = 2 + \int_0^t \left(s + \left(2 + 2s + \frac{s^2}{2}\right)\right) ds = 2 + 2t + \frac{3t^2}{2} + \frac{t^3}{6},$$
$$x_3(t) = (Tx_2)(t) = 2 + \int_0^t \left(s + \left(2 + 2s + \frac{3s^2}{2} + \frac{s^3}{6}\right)\right) ds = 2 + 2t + \frac{3}{2}t^2 + \frac{t^3}{2} + \frac{t^4}{24}.$$

It can be easily seen that the exact solution to the given equation is $x(t) = 3e^t - t - 1$.

If we take $x_3(t)$ as approximate solution, then the error is given by

$$\begin{aligned} \|x - x_3\| &= \sup_{t\in[0,1]} |x(t) - x_3(t)| = \sup_{t\in[0,1]} \left| \frac{t^3}{3!} + \frac{t^4}{12} + \frac{t^5}{5!} + \cdots \right| \\ &\leq \sup_{t\in[0,1]} \left| e^t - 1 - t - \frac{t^2}{2} - \frac{t^4}{12} \right|. \end{aligned}$$

Since e^t increases more rapidly than $\left(1 + t + \frac{t^2}{2} + \frac{t^4}{12}\right)$ in the interval $[0,1]$, the maximum occurs at $t = 1$. Hence,

$$\|x - x_3\| \leq e - \frac{31}{12} \approx e - 2.5833 \approx 0.1349.$$

The error in the approximation can be reduced further if the approximate solution is taken as $x_n(t)$ for large n.

4.5 System of ODEs

Consider the system of n first order differential equations of the form

$$\dot{x}_i = f_i(t, x_1, x_2, \ldots, x_n) \quad \text{for } t \in [t_0, \tau] =: J, i = 1, 2, \ldots, n, \tag{4.35}$$

where $x_i(t) \in \mathbb{R}$ for each t and $f_i : J \times \mathbb{R}^n \to \mathbb{R}$.

Let $x : J \to \mathbb{R}^n$ and $f : J \times \mathbb{R}^n \to \mathbb{R}^n$ be represented by

$$x(t) = \begin{bmatrix} x_1(t) \\ x_2(t) \\ \vdots \\ x_n(t) \end{bmatrix} \text{ and } f(t,x) = \begin{bmatrix} f_1(t, x_1, x_2, \ldots, x_n) \\ f_2(t, x_1, x_2, \ldots, x_n) \\ \vdots \\ f_n(t, x_1, x_2, \ldots, x_n) \end{bmatrix}.$$

Let

$$X = C(J, \mathbb{R}^n)$$

$$= \{x = \begin{bmatrix} x_1 \\ x_2 \\ \vdots \\ x_n \end{bmatrix} : x(t) \in \mathbb{R}^n \text{ for each } t \text{ and every } x_i \text{ is continuous on } J\}.$$

The norm on the space X is denoted by $\|\cdot\|_X$. For $x \in C(J, \mathbb{R}^n)$, the norm is defined by

$$\|x\|_X = \sup_{t \in J} \|x(t)\|_1 = \sup_{t \in J} \sum_{i=1}^{n} |x_i(t)|.$$

Now consider the IVP involving the system of first order differential equations (4.35) with initial conditions $x_i(t_0) = x_{0i}$ for $i = 1, 2, \ldots, n$. In vector form, the IVP can be written as

$$\begin{cases} \dot{x} = f(t, x), & t \in J, \\ x(t_0) = x_0. \end{cases} \tag{4.36}$$

The following condition ensures the existence and uniqueness of solutions to the system (4.36).

(B) **(Condition on f)** Let $f : J \times \mathbb{R}^n \to \mathbb{R}^n$ be such that

(*i*) f is continuous on $J \times \mathbb{R}^n$.

(*ii*) f satisfies the Lipschitz continuity with respect to the second argument x, i.e., there exists a constant $\alpha > 0$ such that

$$\|f(t, x_1) - f(t, x_2)\| \leq \alpha \|x_1 - x_2\|$$

for all $t \in J$ and $x_1, x_2 \in \mathbb{R}^n$. The norm in the above inequality condition can be any norm defined on $\mathbb{R}^n$.

Theorem 4.5.1 Under the condition (**B**), the IVP (4.36) has a unique solution in $C(J, \mathbb{R}^n)$.

Proof. As in Theorem 4.4.1, define the operator $T : C(J, \mathbb{R}^n) \to C(J, \mathbb{R}^n)$ as

$$(Tx)(t) = x_0 + \int_{t_0}^{t} f(s, x(s))ds = \begin{bmatrix} x_{01} \\ x_{02} \\ \vdots \\ x_{0n} \end{bmatrix} + \begin{bmatrix} \int_0^t f_1(s, x(s))ds \\ \int_0^t f_2(s, x(s))ds \\ \vdots \\ \int_0^t f_n(s, x(s))ds \end{bmatrix} \tag{4.37}$$

for $t \in J$. It can be shown that the operator T^k is a contraction mapping on $C(J,\mathbb{R}^n)$ for some positive integer k, i.e., for any $x, \tilde{x} \in C(J,\mathbb{R}^n)$

$$\|T^k x - T^k \tilde{x}\|_X \leq \sigma \|x - \tilde{x}\|_X, \quad 0 < \sigma < 1.$$

The above inequality can be proved using the expressions (4.28)–(4.32), by replacing the modulus sign with $\|\cdot\|_1$.

Hence, T has a unique fixed point in $C(J,\mathbb{R}^n)$, which is the unique solution of the IVP (4.36). □

Notice that (**B**) is only a sufficient condition.

4.6 System of Linear ODEs

Consider the linear time-varying system

$$\begin{cases} \dot{x} = A(t)x, & t \in J, \\ x(t_0) = x_0 \in \mathbb{R}^n, \end{cases} \tag{4.38}$$

where $A(t) = \{a_{ij}(t)\}_{i,j=1,2,\ldots,n}$ and every a_{ij} is continuous on J. Comparing (4.38) with (4.36), we observe that $f(t,x) = A(t)x$. It is easy to verify that the condition **(B)** is satisfied by $A(t)x$ as a_{ij} are conditions on J and

$$\|A(t)x_1 - A(t)x_2\| \leq K\|x_1 - x_2\|,$$

where $K = \max_{t\in J; i,j=1,2,\ldots,n} |a_{ij}(t)|$. Then, from Theorem 4.5.1, it can be concluded that for every initial condition $x(t_0) = x_0$, equation (4.38) has a unique solution $x \in C(J,\mathbb{R}^n)$.

Integrating both sides of (4.38) from t_0 to t, we obtain the following integral equation

$$x(t) = x(t_0) + \int_{t_0}^{t} A(s_1)x(s_1)ds_1. \tag{4.39}$$

From Lemma 4.3.12, it is known that the solution of (4.38) and (4.39) are the same. Now, let us solve (4.39) using the successive substitution method. Substituting $x(s_1)$ (obtained by replacing t with s_1 and s_1 with s_2 in (4.39)) into the RHS of (4.39), we get

$$\begin{aligned} x(t) &= x_0 + \int_{t_0}^{t} A(s_1)\left[x_0 + \int_{t_0}^{s_1} A(s_2)x(s_2)ds_2\right] ds_1 \\ &= x_0 + \int_{t_0}^{t} A(s_1)ds_1 x_0 + \int_{t_0}^{t} A(s_1)\left(\int_{t_0}^{s_1} A(s_2)x(s_2)ds_2\right) ds_1. \end{aligned} \tag{4.40}$$

Repeating this process successively, we obtain

$$x(t) = \left[I + \int_{t_0}^{t} A(s_1)ds_1 + \int_{t_0}^{t}\int_{t_0}^{s_1} A(s_1)A(s_2)ds_2ds_1 + \cdots \right.$$
$$\left. + \int_{t_0}^{t}\int_{t_0}^{s_1}\cdots\int_{t_0}^{s_{r-1}} A(s_1)A(s_2)\ldots A(s_r)ds_r\ldots ds_2ds_1 + \cdots \right] x_0. \tag{4.41}$$

Equation (4.41) can be expressed as

$$x(t) = \phi(t,t_0)x_0, \tag{4.42}$$

where the $n\times n$ matrix $\phi(t,t_0)$ is given by

$$\phi(t,t_0) = I + \int_{t_0}^{t} A(s_1)ds_1 + \int_{t_0}^{t}\int_{t_0}^{s_1} A(s_1)A(s_2)ds_2ds_1 + \cdots$$
$$+ \int_{t_0}^{t}\int_{t_0}^{s_1}\cdots\int_{t_0}^{s_{r-1}} A(s_1)A(s_2)\ldots A(s_r)ds_r\ldots ds_2ds_1 + \cdots. \tag{4.43}$$

The infinite series on the RHS of (4.43) is called the **Peano–Baker** series.

Proposition 4.6.1 The Peano–Baker series converges on any finite interval $[t_0,\tau]$ for any matrix $A(t)$ of order n whose elements are continuous functions of t on J.

Proof. Since the elements of $A(t)$ are bounded on J, $\|A(t)\| \le K$, a constant, for all t.

Let $\phi_m(t,t_0)$ be the $n\times n$ matrix representing the partial sum of the first m terms of the series (4.43). Then

$$\begin{aligned}\|\phi_m(t,t_0)\| &= \left\| I + \int_{t_0}^{t} A(s_1)ds_1 + \int_{t_0}^{t}\int_{t_0}^{s_1} A(s_1)A(s_2)ds_2ds_1 + \cdots \right.\\ &\quad \left. + \int_{t_0}^{t}\int_{t_0}^{s_1}\cdots\int_{t_0}^{s_{m-1}} A(s_1)A(s_2)\cdots A(s_m)ds_m\cdots ds_2ds_1 \right\|\\ &\le 1 + K(t-t_0) + K^2\frac{(t-t_0)^2}{2!} + \cdots + K^m\frac{(t-t_0)^m}{m!}\\ &\le e^{K(t-t_0)}.\end{aligned}$$

The RHS is independent of m. Hence,

$$\lim_{m\to\infty} \|\phi_m(t,t_0)\| \le e^{K(t-t_0)} \le e^{K(\tau-t_0)},$$

which implies that $\lim_{m\to\infty}\|\phi_m(t,t_0)\|$ is finite.

Now let us prove that $\{\phi_m(t,t_0)\}$ is a Cauchy sequence in $\mathbb{R}^{n\times n}$ (the space of real $n\times n$ matrices). Let l be a sufficiently large integer and $r > l$. Denote $K(\tau-t_0)$ by γ. Since γ is a constant, choose l such that $\frac{\gamma}{l} < \frac{1}{2}$. Then

$$\|\phi_r(t,t_0) - \phi_l(t,t_0)\| \le \frac{\gamma^l}{l!} + \frac{\gamma^{l+1}}{(l+1)!} + \cdots + \frac{\gamma^{r-l}}{(r-l)!}$$

$$\leq \frac{\gamma^l}{l!}\left[1+\frac{1}{2}+\frac{1}{2^2}+\cdots+\frac{1}{2^{r-l}}\right]$$
$$\leq 2\frac{\gamma^l}{l!} \to 0 \text{ as } l \to \infty.$$

Hence, $\{\phi_m(t,t_0)\}$ is a uniform Cauchy sequence in $\mathbb{R}^{n\times n}$. Since $\mathbb{R}^{n\times n}$ is a Banach space under the matrix norm, the sequence converges uniformly. Hence, $\lim_{m\to\infty}\phi_m(t,t_0)$ exists, and we denote this limit by $\phi(t,t_0)$, which is called the **state transition matrix**. □

***Example* 4.6.2** Find the first four terms of the Peano–Baker series for the matrix $A(t) = \begin{bmatrix} t & 1 \\ 0 & 2 \end{bmatrix}$ for $0 \leq t \leq 5$.

Solution From (4.43), the first four terms of $\phi(t,t_0)$ are

$$\begin{aligned}\phi(t,t_0) &\approx I + \int_{t_0}^{t} A(s_1)ds_1 + \int_{t_0}^{t} A(s_1)\left(\int_{t_0}^{s_1} A(s_2)ds_2\right)ds_1 \\ &\quad + \int_{t_0}^{t} A(s_1)\left(\int_{t_0}^{s_1} A(s_2)\left(\int_{t_0}^{s_{r-1}} A(s_3)ds_3\right)ds_2\right)ds_1 \\ &= \sum_{m=0}^{4} S_m(t),\end{aligned} \tag{4.44}$$

where

$$S_0(t) = I \text{ and } S_m(t) = \int_{t_0}^{t} A(r)S_{m-1}(r)dr \quad \text{for } m = 1,2,3.$$

Now for $t_0 = 0$, we have

$$S_1(t) = \int_0^t \begin{bmatrix} r & 1 \\ 0 & 2 \end{bmatrix} dr = \begin{bmatrix} \frac{t^2}{2} & t \\ 0 & 2t \end{bmatrix},$$

$$S_2(t) = \int_0^t \begin{bmatrix} \frac{r^3}{2} & r^2+2r \\ 0 & 4r \end{bmatrix} dr = \begin{bmatrix} \frac{t^4}{2^3} & \frac{t^3}{3}+t^2 \\ 0 & 2t^2 \end{bmatrix},$$

$$S_3(t) = \int_0^t \begin{bmatrix} \frac{r^5}{2^3} & \frac{r^4}{3}+r^3+2r^2 \\ 0 & 4r^2 \end{bmatrix} dr = \begin{bmatrix} \frac{t^6}{3\cdot 2^4} & \frac{t^5}{5\cdot 3}+\frac{t^4}{4}+\frac{2t^3}{3} \\ 0 & \frac{4t^3}{3} \end{bmatrix}.$$

Then from (4.44), we have

$$\phi(t,0) \approx \begin{bmatrix} \phi_{11}(t) & \phi_{12}(t) \\ \phi_{21}(t) & \phi_{22}(t) \end{bmatrix}, \quad 0 \le t \le 5,$$

where

$$\phi_{11}(t) = 1 + \frac{t^2}{2} + \frac{t^4}{8} + \frac{t^6}{48}, \; \phi_{12}(t) = t + t^2 + \frac{t^3}{3} + \frac{t^4}{4} + \frac{t^5}{15}, \; \phi_{21}(t) = 0, \text{ and}$$
$$\phi_{22}(t) = 1 + 2t + 2t^2 + \frac{4t^3}{3}.$$

Remark 4.6.3 From the above example, it is clear that the exact values of the elements of $\phi(t,t_0)$ generally cannot be obtained for any given $A(t)$. However, due to the convergence of the Peano–Baker series, approximate values can be obtained by truncating the series. Exact values of $\phi(t,t_0)$ can be obtained in special cases, as demonstrated in the following lemma.

Lemma 4.6.4 If the matrices $A(t)$ and $A(s)$ are commutative for all $t,s \in J$, i.e., if

$$A(t)A(s) = A(s)A(t) \quad \text{for all } t,s \in J,$$

then

$$\phi(t,t_0) = e^{\int_{t_0}^{t} A(r)dr}. \tag{4.45}$$

Proof. From (4.43), we have

$$\begin{aligned} \phi(t,t_0) &= I + \int_{t_0}^{t} A(s_1)ds_1 + \int_{t_0}^{t} A(s_1)\left(\int_{t_0}^{s_1} A(s_2)ds_2\right)ds_1 + \cdots \\ &\quad + \int_{t_0}^{t} A(s_1)\int_{t_0}^{s_1} A(s_2)\ldots\int_{t_0}^{s_{r-1}} A(s_r)ds_r\ldots ds_2 ds_1 + \cdots \\ &= \sum_{m=0}^{\infty} S_m(t), \end{aligned} \tag{4.46}$$

where

$$S_0(t) = I \text{ and } S_m(t) = \int_{t_0}^{t} A(r)S_{m-1}(r)dr \quad \text{for } m = 1,2,\ldots.$$

If $A(t)$ and $A(r)$ commute for all t and r, then we can show that

$$\int_{t_0}^{t}\int_{t_0}^{s} A(s)A(s_1)ds_1 ds = \frac{1}{2!}\left(\int_{t_0}^{t} A(s)ds\right)^2. \tag{4.47}$$

For this, consider the identity

$$\int_{t_0}^{t} \frac{d}{ds}\left(\int_{t_0}^{s} A(s_1)ds_1\right)^2 ds = \left(\int_{t_0}^{t} A(s_1)ds_1\right)^2. \tag{4.48}$$

If $A(s_1)A(s) = A(s)A(s_1)$, the LHS of (4.48) can be written as

$$\begin{aligned}\int_{t_0}^{t} \frac{d}{ds}\left(\int_{t_0}^{s} A(s_1)ds_1\right)^2 ds &= \int_{t_0}^{t} 2\left(\int_{t_0}^{s} A(s_1)ds_1\right)A(s)ds \\ &= 2\int_{t_0}^{t} A(s)\int_{t_0}^{s} A(s_1)ds_1 ds. \end{aligned} \tag{4.49}$$

Hence, from (4.48) and (4.49), we get

$$\left(\int_{t_0}^{t} A(s_1)ds_1\right)^2 = 2\int_{t_0}^{t} A(s)\int_{t_0}^{s} A(s_1)ds_1 ds, \tag{4.50}$$

which is the same as (4.47). Now we prove that

$$S_j(t) = \frac{1}{j!}\left(\int_{t_0}^{t} A(s_1)ds_1\right)^j \quad \text{for } j = 1, 2, \ldots. \tag{4.51}$$

This can be proved by induction. If (4.51) is true for $j = k$, then

$$S_k(t) = \int_{t_0}^{t} A(s_1)S_{k-1}(s_1)ds_1 = \frac{1}{k!}\left(\int_{t_0}^{t} A(s_1)ds_1\right)^k.$$

Now for $j = k+1$, we have

$$\begin{aligned} S_{k+1}(t) = \int_{t_0}^{t} A(s_1)S_k(s_1)ds_1 &= \frac{1}{k!}\int_{t_0}^{t} A(s_1)\left(\int_{t_0}^{s_1} A(s_2)ds_2\right)^k ds_1 \\ &= \frac{1}{k!}\int_{t_0}^{t} \frac{1}{k+1}\frac{d}{ds_1}\left(\int_{t_0}^{s_1} A(s_2)ds_2\right)^{k+1} ds_1 \\ &= \frac{1}{(k+1)!}\left(\int_{t_0}^{t} A(s_2)ds_2\right)^{k+1}. \end{aligned} \tag{4.52}$$

As the result is true for $j = 2$, by induction it is true for all integers $j \geq 2$. Hence, from (4.46), we get

$$\phi(t, t_0) = I + \int_{t_0}^{t} A(s_1)ds_1 + \cdots + \frac{1}{j!}\left(\int_{t_0}^{t} A(s_1)ds_1\right)^j + \cdots = e^{\int_{t_0}^{t} A(s)ds}.$$

□

Example **4.6.5** Let $A(t) = \begin{bmatrix} 1 & t \\ 0 & 1 \end{bmatrix}$ for $t \in J$. Note that $A(t)A(s) = A(s)A(t)$ for all $t, s \in J$. Hence, from Lemma 4.6.4, the state transition matrix is

$$\phi(t, t_0) = e^{\int_{t_0}^{t} A(s)ds}.$$

Now

$$\int_{t_0}^{t} A(s)ds = \begin{bmatrix} t - t_0 & \frac{(t-t_0)^2}{2} \\ 0 & t - t_0 \end{bmatrix} = (t - t_0)I + \frac{(t - t_0)^2}{2} \begin{bmatrix} 0 & 1 \\ 0 & 0 \end{bmatrix}.$$

Therefore,

$$\phi(t, t_0) = \begin{bmatrix} e^{t-t_0} & 0 \\ 0 & e^{t-t_0} \end{bmatrix} \begin{bmatrix} 1 & \frac{(t-t_0)^2}{2} \\ 0 & 1 \end{bmatrix} = e^{t-t_0} \begin{bmatrix} 1 & \frac{(t-t_0)^2}{2} \\ 0 & 1 \end{bmatrix} \quad \text{for } t \in J.$$

4.7 Autonomous System

For time-invariant systems, the state transition matrix has a simple expression as given below. Consider the autonomous system

$$\begin{cases} \dot{x}(t) = Ax(t), & t \in J = [t_0, \tau], \\ x(t_0) = x_0. \end{cases} \tag{4.53}$$

Since A is a constant matrix, from (4.43), we have

$$\phi(t, t_0) = I + A(t - t_0) + \frac{A^2(t - t_0)^2}{2!} + \cdots + \frac{A^r(t - t_0)^r}{r!} + \cdots = e^{(t-t_0)A}.$$

Thus, the solution of the IVP (4.53) is

$$x(t) = e^{A(t-t_0)}x_0. \tag{4.54}$$

Solution of Linear System in Terms of Eigenpairs

Consider the autonomous linear system

$$\dot{x} = Ax, \quad t \geq t_0, \tag{4.55}$$

where A is an $n \times n$ matrix.

Case I. *A is diagonalizable.* If A is diagonalizable with eigenpairs $(\lambda_i, X_i), i = 1,2,\ldots,n$, then it can be shown that the solution of (4.55) is

$$x(t) = c_1 e^{\lambda_1(t-t_0)} X_1 + c_2 e^{\lambda_2(t-t_0)} X_2 + \cdots + c_n e^{\lambda_n(t-t_0)} X_n, \tag{4.56}$$

where c_i are arbitrary constants. To prove this, note that the eigenvectors $\{X_1, X_2, \ldots, X_n\}$ form a basis of $\mathbb{R}^n$, and for each t, $x(t) \in \mathbb{R}^n$ has a unique expression as

$$x(t) = \alpha_1(t) X_1 + \alpha_2(t) X_2 + \cdots + \alpha_n(t) X_n. \tag{4.57}$$

Substituting (4.57) into (4.55) and equating the coefficients of X_i, we get

$$\dot{\alpha}_i(t) = \lambda_i \alpha_i(t), \quad i = 1,2,\ldots,n.$$

Hence,

$$\alpha_i(t) = e^{\lambda_i(t-t_0)} c_i, \tag{4.58}$$

where $c_i : i = 1,2,\ldots,n$ are arbitrary constants. Substituting $\alpha_i(t)$ from (4.58) into (4.57), we obtain the required identity (4.56). Note that each term in (4.56), $c_i e^{\lambda_i(t-t_0)} X_i$ for $i = 1,2,\ldots,n$ is a solution of (4.55) with initial condition $x(t_0) = c_i X_i$ and these solutions are linearly independent.

Case II. *A is non diagonalizable.* If A is not diagonalizable, let $\lambda_i : i = 1,2,\ldots,k$ be distinct eigenvalues with m_i and l_i as algebraic and geometric multiplicities, respectively. Then $l_i \leq m_i$ and $m_1 + m_2 + \cdots + m_k = n$. Let $X_1^i, X_2^i, \ldots, X_{l_i}^i$ be linearly independent eigenvectors corresponding to λ_i, and let $Y_{j1}^i, Y_{j2}^i, \ldots, Y_{jr_{ij}}^i$ be generalized eigenvectors (see Section 2.7) corresponding to $X_j^i : j = 1,2,\ldots,l_i$. Hence,

$$(r_{i1}+1) + (r_{i2}+1) + \cdots + (r_{il_i}+1) = m_i, \quad i = 1,2,\ldots,k.$$

The n linearly independent vectors given below

$$\bigcup_{i=1}^{k} \{X_1^i, Y_{11}^i, Y_{12}^i, \ldots, Y_{1r_{i1}}^i, X_2^i, Y_{21}^i, Y_{22}^i, \ldots, Y_{2r_{i2}}^i, \ldots, X_{l_i}^i, Y_{l_i1}^i, Y_{l_i2}^i, \ldots, Y_{l_ir_{il_i}}^i\}$$

form a basis of $\mathbb{R}^n$. Hence, the solution $x(t)$ of (4.55) can be written as

$$\begin{aligned} x(t) = {} & \alpha_1(t) X_1^1 + \alpha_2(t) Y_{11}^1 + \alpha_3(t) Y_{12}^1 + \cdots \\ & + \alpha_{r_{11}+1}(t) Y_{1r_{11}}^1 + \cdots + \alpha_n(t) Y_{l_kr_{kl_k}}^k, \quad t \geq t_0. \end{aligned} \tag{4.59}$$

Substituting $x(t)$ from (4.59) into (4.55) using (2.27) and comparing the coefficients of the basis vectors, we get

$$\dot{\alpha}_1(t)X_1^1 + \dot{\alpha}_2(t)Y_{11}^1 + \cdots + \dot{\alpha}_{r_{11}+1}(t)Y_{1r_{11}}^1 + \cdots = \alpha_1(t)\lambda_1 X_1^1 + \alpha_2(t)\left(\lambda_1 Y_{11}^1 + X_1^1\right)$$
$$+\alpha_3(t)\left(\lambda_1 Y_{12}^1 + Y_{11}^1\right) + \cdots + \alpha_{r_{11}+1}(t)\left(\lambda_1 Y_{1r_{11}}^1 + Y_{1(r_{11}-1)}^1\right) + \cdots. \quad (4.60)$$

Comparing coefficients on both sides, we get

$$\begin{cases} \dot{\alpha}_1(t) = \lambda_1\alpha_1(t) + \alpha_2(t), \\ \dot{\alpha}_2(t) = \lambda_2\alpha_1(t) + \alpha_3(t), \\ \vdots \\ \dot{\alpha}_{r_{11}}(t) = \lambda_1\alpha_{r_{11}}(t) + \alpha_{r_{11}+1}(t), \\ \dot{\alpha}_{r_{11}+1}(t) = \lambda_1\alpha_{r_{11}+1}(t) \end{cases} \quad (4.61)$$

for $t \geq t_0$. System (4.61) provides the coefficients $\alpha_1(t), \alpha_2(t), \ldots, \alpha_{r_{11}+1}(t)$ corresponding to X_1^1 and its generalized eigenvectors. The system (4.61) can be written as

$$\begin{bmatrix} \dot{\alpha}_1 \\ \dot{\alpha}_2 \\ \vdots \\ \dot{\alpha}_{r_{11}+1} \end{bmatrix} = \begin{bmatrix} \lambda_1 & 1 & 0 & \cdots & 0 & 0 \\ 0 & \lambda_1 & 1 & \cdots & 0 & 0 \\ \vdots & \vdots & & & & \\ 0 & 0 & 0 & \cdots & \lambda_1 & 1 \\ 0 & 0 & 0 & \cdots & 0 & \lambda_1 \end{bmatrix} \begin{bmatrix} \alpha_1 \\ \alpha_2 \\ \vdots \\ \alpha_{r_{11}+1} \end{bmatrix}, \quad t \geq t_0, \quad (4.62)$$

which gives (using (4.54) and (2.38))

$$\begin{bmatrix} \alpha_1(t) \\ \alpha_2(t) \\ \vdots \\ \alpha_{r_{11}+1}(t) \end{bmatrix} = e^{\lambda_1(t-t_0)} \begin{bmatrix} 1 & t-t_0 & \frac{(t-t_0)^2}{2!} & \cdots & \frac{(t-t_0)^{r_{11}}}{r_{11}!} \\ 0 & 1 & t-t_0 & \cdots & \frac{(t-t_0)^{r_{11}-1}}{(r_{11}-1)!} \\ \vdots & & & & \\ 0 & 0 & 0 & \cdots & 1 \end{bmatrix} \begin{bmatrix} c_1 \\ c_2 \\ \vdots \\ c_{r_{11}+1} \end{bmatrix}, \quad (4.63)$$

where $c_1, c_2, \ldots, c_{r_{11}+1}$ are arbitrary constants. Similarly, coefficients corresponding to other eigenvectors and their generalized eigenvectors can be obtained. Thus (4.59) represents the general solution of (4.55).

Remark 4.7.1 In (4.63), if $\mathrm{Re}(\lambda_1) < 0$, then $(t-t_0)^k e^{\lambda_1(t-t_0)} \to 0$ as $t \to \infty$ for any t_0 and k. Hence $\alpha_1(t), \alpha_2(t), \ldots, \alpha_{r_{11}+1}(t)$ all tend to 0 as $t \to \infty$. Thus, if all eigenvalues λ_i lie in the left half of the complex plane, then all the coefficients $\alpha_i(t) : i = 1, 2, \ldots, n$ in (4.59) tend to 0 as $t \to \infty$. Hence, $x(t) \to 0$ as $t \to \infty$.

***Example* 4.7.2** (For Case I) Find the general solution of the differential equation
$$\dot{x} = \begin{bmatrix} 1 & -1 & 4 \\ 3 & 2 & -1 \\ 2 & 1 & -1 \end{bmatrix} x.$$

Solution The characteristic equation of the coefficient matrix A is $det(A - \lambda I) = 0$, which gives the eigenvalues $\lambda_1 = 1$, $\lambda_2 = 3$, and $\lambda_3 = -2$. The corresponding eigenvectors are $\begin{bmatrix} -1 \\ 4 \\ 1 \end{bmatrix}$, $\begin{bmatrix} 1 \\ 2 \\ 1 \end{bmatrix}$, and $\begin{bmatrix} -1 \\ 1 \\ 1 \end{bmatrix}$. From (4.56), the solution $x(t)$ is of the form

$$x(t) = c_1 e^t \begin{bmatrix} -1 \\ 4 \\ 1 \end{bmatrix} + c_2 e^{3t} \begin{bmatrix} 1 \\ 2 \\ 1 \end{bmatrix} + c_3 e^{-2t} \begin{bmatrix} -1 \\ 1 \\ 1 \end{bmatrix} = \begin{bmatrix} -c_1 e^t + c_2 e^{3t} - c_3 e^{-2t} \\ 4c_1 e^t + 2c_2 e^{3t} + c_3 e^{-2t} \\ c_1 e^t + c_2 e^{3t} + c_3 e^{-2t} \end{bmatrix},$$

where $c_i : i = 1, 2, 3$ are arbitrary constants.

***Example* 4.7.3** (For Case II) Find the solution of $\dot{x} = Ax$, where $A = \begin{bmatrix} 2 & 1 & 0 \\ 0 & 2 & 0 \\ 1 & 1 & 3 \end{bmatrix}$ with the initial condition $x(5) = \begin{bmatrix} 0 \\ 1 \\ 4 \end{bmatrix}$.

Solution The eigenvalues of A are $\lambda_1 = 2$ and $\lambda_2 = 3$. The eigenvector corresponding to $\lambda_1 = 2$ is $\begin{bmatrix} 1 \\ 0 \\ -1 \end{bmatrix}$ with a generalized eigenvector $\begin{bmatrix} 1 \\ 1 \\ -3 \end{bmatrix}$. The eigenvector corresponding to $\lambda_2 = 3$ is $\begin{bmatrix} 0 \\ 0 \\ 1 \end{bmatrix}$. From (4.59), the general solution is

$$x(t) = \alpha_1(t) \begin{bmatrix} 1 \\ 0 \\ -1 \end{bmatrix} + \alpha_2(t) \begin{bmatrix} 1 \\ 1 \\ -3 \end{bmatrix} + \alpha_3(t) \begin{bmatrix} 0 \\ 0 \\ 1 \end{bmatrix}, \tag{4.64}$$

where $\alpha_i(t) : i = 1, 2, 3$ are obtained from (4.63) as follows

$$\begin{bmatrix} \alpha_1(t) \\ \alpha_2(t) \end{bmatrix} = e^{2(t-5)} \begin{bmatrix} 1 & t-5 \\ 0 & 1 \end{bmatrix} \begin{bmatrix} c_1 \\ c_2 \end{bmatrix} \text{ and } \alpha_3(t) = e^{3(t-5)} c_3.$$

Hence, $\alpha_1(t) = (c_1 + c_2(t-5))e^{2(t-5)}$, $\alpha_2(t) = c_2 e^{2(t-5)}$, and $\alpha_3(t) = c_3 e^{3(t-5)}$. The constants $c_i : i = 1, 2, 3$ can be obtained from (4.64) by using the initial condition $x(5)$. Then substituting $\alpha_i(t) : i = 1, 2, 3$ in (4.64) we obtain the solution as

$$x(t) = e^{2(t-5)} \begin{bmatrix} 1 & t-5 & 0 \\ 0 & 1 & 0 \\ e^{t-5} - 1 & 2e^{t-5} - t + 3 & e^{t-5} \end{bmatrix} \begin{bmatrix} 0 \\ 1 \\ 4 \end{bmatrix} = e^{2(t-5)} \begin{bmatrix} t-5 \\ 1 \\ 6e^{t-5} - t + 3 \end{bmatrix}.$$

***Example* 4.7.4** (For Case II) Solve the IVP $\dot{x} = \begin{bmatrix} 2 & 1 & 3 \\ 0 & 2 & -1 \\ 0 & 0 & 2 \end{bmatrix} x$ with the initial condition $x(0) = \begin{bmatrix} 1 \\ 2 \\ 1 \end{bmatrix}$.

Solution The coefficient matrix A has the eigenvalue $\lambda = 2$ with algebraic multiplicity 3, and the corresponding eigenvector is $\begin{bmatrix} 1 \\ 0 \\ 0 \end{bmatrix}$. The two linearly independent generalized eigenvectors are $\begin{bmatrix} 0 \\ 1 \\ 0 \end{bmatrix}$ and $\begin{bmatrix} 0 \\ 0 \\ 1 \end{bmatrix}$. Hence, from (4.63), the general solution is

$$x(t) = c_1 e^{2t} \begin{bmatrix} 1 \\ 0 \\ 0 \end{bmatrix} + c_2 t e^{2t} \begin{bmatrix} 0 \\ 1 \\ 0 \end{bmatrix} + c_3 t^2 e^{2t} \begin{bmatrix} 0 \\ 0 \\ 1 \end{bmatrix},$$

where $c_i : i = 1,2,3$ are arbitrary constants. Using the given initial condition, the unique solution of the IVP is obtained as

$$x(t) = e^{2t} \begin{bmatrix} 1 + 5t - \frac{1}{2}t^2 \\ 2 - t \\ 1 \end{bmatrix}.$$

4.8 Fundamental Matrix

Consider the system

$$\dot{x} = A(t)x, \quad t \in J, \tag{4.65}$$

where the elements of the coefficient matrix $A(t)$ of order n are continuous functions of $t \in J$. From Theorem 4.5.1, it is guaranteed that equation (4.65) has a unique solution for a given initial condition $x(t_0) = x_0$. The following theorem provides the number of linearly independent solutions of equation (4.65).

Theorem 4.8.1 Let $A(t) = \{a_{ij}(t)\}_{i,j=1,2,\ldots,n}$ be such that every $a_{ij}(t)$ is continuous on J. Then the system $\dot{x} = A(t)x$ has n linearly independent solutions.

Proof. Let $v_i : i = 1,2,\ldots,n$ be n linearly independent vectors in $\mathbb{R}^n$. Then for each i, the system (4.65) with the initial condition $x(t_0) = v_i$ has a unique solution, which can be denoted by $z_i(t) : i = 1,2,\ldots,n$. We will show that

(*i*) $z_i : i = 1,2,\ldots,n$ are linearly independent vector functions, and

(*ii*) any other solution of (4.65) can be expressed as a linear combination of $z_i : i = 1,2,\ldots,n$.

To prove (i), let

$$a_1z_1(t)+a_2z_2(t)+\cdots+a_nz_n(t)=\Theta \quad \text{for all } t, \tag{4.66}$$

where $a_j : j=1,2,\ldots,n$ are real numbers. Substituting $z_i(t_0)=v_i$, $i=1,2,\ldots,n$, into (4.66), we get

$$a_1v_1+a_2v_2+\cdots+a_nv_n=\Theta, \tag{4.67}$$

which implies that $a_i=0$ for $i=1,2,\ldots,n$ since v_i are linearly independent.

Hence, $z_i(t) : i=1,2,\ldots,n$ are linearly independent vector functions in the space $C(J,\mathbb{R}^n)$.

To prove (ii), let x be any other solution of (4.65) such that

$$x(t_0)=v, \tag{4.68}$$

where v is an arbitrary vector in $\mathbb{R}^n$. We can show that x is linearly dependent on $z_i : i=1,2,\ldots,n$ in $C(J,\mathbb{R}^n)$. Since $\{v_1,v_2,\ldots,v_n\}$ forms a basis of $\mathbb{R}^n$, v has a unique representation as

$$v=b_1v_1+b_2v_2+\cdots+b_nv_n. \tag{4.69}$$

Using these scalars $b_i : i=1,2,\ldots,n$, let $z(t)$ be such that

$$z(t)=b_1z_1(t)+b_2z_2(t)+\cdots+b_nz_n(t). \tag{4.70}$$

Then

$$z(t_0)=b_1z_1(t_0)+b_2z_2(t_0)+\cdots+b_nz_n(t_0)=b_1v_1+b_2v_2+\cdots+b_nv_n=v. \tag{4.71}$$

From (4.68) and (4.71), it follows that the solutions $x(t)$ and $z(t)$ have the same value at $t=t_0$. Now, the Uniqueness Theorem (4.5.1) implies that $x(t)\equiv z(t)$. Hence, the result follows. □

Definition 4.8.2 Let $z_i(t) : i=1,2,\ldots,n$ be linearly independent solutions of the system (4.65) and $M(t)$ be the matrix with $z_i(t)$ as its ith column. Then $M(t)$ is called a **fundamental matrix** of (4.65).

As shown in Theorem 4.8.1 (i), for each basis of R^n, one can find n linearly independent solutions of (4.65) with initial conditions given by the basis vectors. Hence, the fundamental matrix is **NOT** unique. Let $z_i(t) : i=1,2,\ldots,n$ be the solutions of the system (4.65) with the initial conditions $z_i(t_0)=v_i\in\mathbb{R}^n$ for $i=1,2,\ldots,n$. Define

$$M(t)=\begin{bmatrix} z_1(t) & z_2(t) & \cdots & z_n(t) \\ \downarrow & \downarrow & & \downarrow \end{bmatrix}_{n\times n}. \tag{4.72}$$

Then it is easy to see that

$$\frac{dM(t)}{dt} = A(t)M(t)$$

with the initial condition

$$M(t_0) = \begin{bmatrix} v_1 & v_2 & \cdots & v_n \\ \downarrow & \downarrow & & \downarrow \end{bmatrix}_{n\times n} = M_0. \tag{4.73}$$

Now we prove the following theorem.

Theorem 4.8.3 Let J be an open interval containing the point t_0, and let $M(t)$ be a fundamental matrix of the system (4.65) with $M(t_0) = M_0$. Then $M(t)$ is non-singular for all $t \in J$ if and only if M_0 is non-singular.

Proof. Let M_0 be non-singular. Consider the matrix differential equation

$$\begin{cases} \dfrac{dN}{dt} = -A'N, \\ N(t_0) = \left(M_0'\right)^{-1}, \end{cases} \tag{4.74}$$

where A' denotes the transpose of A. We can show that $M(t)N'(t) = I$, where I is the identity matrix of order n. This will imply that $M(t)$ is non-singular for each $t \in J$. For this, consider

$$\frac{d}{dt}\left(N'(t)M(t)\right) = \frac{dN'}{dt}M + N'\frac{dM}{dt} = -N'AM + N'AM = \mathbb{O}\ (\text{zero matrix}). \tag{4.75}$$

Hence, $N'(t)M(t)$ is a constant matrix. Since $N'(t_0)M(t_0) = M_0^{-1}M_0 = I$, $N'(t)M(t) = I$ for all t. Thus, $N'(t) = (M(t))^{-1}$ for all $t \in J$.

Conversely, if $M(t)$ is non-singular for all $t \in J$, then obviously M_0 must be non-singular. □

From the above theorem, it can be concluded that if $M(t)$ is singular for any $t_1 \in J$, then it is singular for all t in that interval. To prove this, let $z_i(t) : i = 1, 2, \ldots, n$ be the solutions of (4.65) on J with the initial conditions $z_i(t_0) = v_i \in \mathbb{R}^n$. Define $M(t)$ and M_0 as in (4.72) and (4.73). If M_0 is non-singular, then by Theorem 4.8.3, $M(t)$ is nonsingular for all $t \in J$, implying that $M(t_1)$ is also non-singular. This is a contradiction. Hence, if $M(t_1)$ is singular for some $t_1 \in J$, then M_0 must be singular.

In the above discussion, the interval J is any closed interval in which the system (4.65) satisfies the conditions specified in Theorem 4.8.1.

Example **4.8.4** For the system (4.65), consider $A(t) = \begin{bmatrix} \frac{t}{1+t} & \frac{1}{t} \\ t^2 & 1 \end{bmatrix}$. If the closed interval J is such that $J \subset (-\infty,-1)$ or $(-1,0)$ or $(0,\infty)$, then the elements of $A(t)$ are continuous functions of $t \in J$.

Example **4.8.5** If A is a constant matrix in (4.65), then the interval J can be taken as $(-\infty,\infty)$.

Example **4.8.6** A matrix function $\phi(t)$ may have a zero determinant for all t in some interval, even though its columns are linearly independent vector functions of t. For example, let

$$\phi(t) = \begin{bmatrix} 1 & t & t^2 \\ 0 & 2 & t \\ 0 & 0 & 0 \end{bmatrix}.$$

Clearly, $\det(\phi(t)) = 0$ for $-\infty < t < \infty$, yet the columns of $\phi(t)$ are linearly independent vector functions of t. Thus, in view of Theorem 4.8.3, we conclude that the columns of the matrix $\phi(t)$ together cannot form a solution matrix (fundamental matrix) for any system of the form (4.65).

From Theorem 4.8.3, we can conclude that fundamental matrices are nonsingular. Since the columns of $M(t)$ are the n linearly independent solutions of (4.65), with the columns of M_0 as their respective initial conditions, we have

$$\begin{cases} \dfrac{dM}{dt} = A(t)M(t), \\ M(t_0) = M_0. \end{cases} \tag{4.76}$$

Let $x(t_0) = x_0 \in \mathbb{R}^n$ be the initial condition for the system (4.65). Then, from (4.42), $x(t) = \phi(t,t_0)x_0$, where $\phi(t,t_0)$ is the state transition matrix of $A(t)$. Define

$$y(t) = M(t)M^{-1}(t_0)x_0. \tag{4.77}$$

Differentiating both sides of (4.77) with respect to t, we obtain

$$\frac{dy}{dt} = \frac{dM}{dt}M^{-1}(t_0)x_0 = A(t)M(t)M^{-1}(t_0)x_0 = A(t)y \quad \text{(by (4.77))}.$$

Moreover, $y(t_0) = x_0$. Therefore, y is a solution of (4.65) with the initial condition x_0.

Since the solution of (4.65) is unique for a given initial condition, $y(t)$ and $x(t)$ must be identical. Hence, from (4.42) and (4.77), we obtain the following:

Theorem 4.8.7 If $\phi(t,t_0)$ is the state transition matrix and $M(t)$ is a fundamental matrix of (4.65), then

$$\phi(t,t_0) = M(t)M^{-1}(t_0). \tag{4.78}$$

Properties of the State Transition Matrix

Theorem 4.8.8 Let $\phi(t,s)$ be the state transition matrix corresponding to the equation (4.65). Then the following properties hold:

(1) $\frac{d}{dt}\phi(t,s) = A(t)\phi(t,s)$,

(2) $\phi(t,t) = I$,

(3) $\phi(t,r)\phi(r,s) = \phi(t,s)$,

(4) $(\phi(t,s))^{-1} = \phi(s,t)$

for all $t,s,r \geq 0$.

Proof. (1) Replacing t_0 by s in equation (4.78), we obtain

$$\phi(t,s) = M(t)M^{-1}(s). \tag{4.79}$$

Then,

$$\begin{aligned}\frac{d}{dt}\phi(t,s) = \frac{d}{dt}M(t)M^{-1}(s) &= A(t)M(t)M^{-1}(s) \quad (\text{from}(4.76))\\ &= A(t)\phi(t,s).\end{aligned}$$

The other properties can be derived directly from (4.79). □

Now consider the system (4.65) and its **dual system**

$$\dot{x} = -A'(t)x. \tag{4.80}$$

If $N(t)$ is a fundamental matrix for system (4.80), then $N(t)$ must satisfy the matrix differential equation

$$\frac{dN(t)}{dt} = -A'(t)N(t),$$

which is the same as (4.74). Hence from Theorem 4.8.3, it follows that

$$N'(t) = M^{-1}(t), \tag{4.81}$$

where $M(t)$ is a fundamental matrix of (4.65) with $M(t_0) = (N'(t_0))^{-1}$.

If $\phi(t,s)$ and $\psi(t,s)$ are the state transition matrices of (4.65) and (4.80), respectively, then from Theorem 4.8.7, we have $\phi(t,s) = M(t)M^{-1}(s)$ and

$$\begin{aligned}\psi(t,s) = N(t)N^{-1}(s) &= \left(M'(t)\right)^{-1} M'(s) \quad \text{(from (4.81))}\\ &= \left(M(s)M^{-1}(t)\right)'\\ &= \phi'(s,t).\end{aligned}$$

Thus we conclude the following.

Theorem 4.8.9 If $\phi(t,s)$ and $\psi(t,s)$ are the state transition matrices of (4.65) and its dual system (4.80), respectively, then $\psi(t,s) = \phi'(s,t)$.

***Example* 4.8.10** Show that $M(t) = \begin{bmatrix} t^2 & t \\ 2t & 1 \end{bmatrix}$ is a fundamental matrix for the system $\dot{x} = A(t)x$, where $A(t) = \begin{bmatrix} 0 & 1 \\ \frac{-2}{t^2} & \frac{2}{t} \end{bmatrix}$ on any interval J **not containing** zero.

Solution Let $x = \begin{bmatrix} x_1 \\ x_2 \end{bmatrix}$. Then, we have the following system

$$\begin{cases} \dot{x}_1 = x_2, \\ \dot{x}_2 = \dfrac{-2}{t^2}x_1 + \dfrac{2}{t}x_2, \end{cases}$$

which implies

$$t^2\ddot{x}_1 - 2t\dot{x}_1 + 2x_1 = 0. \tag{4.82}$$

Let $x_1 = t^r$. Then, from (4.82), we get

$$r(r-1) - 2r + 2 = 0 \implies (r-1)(r-2) = 0.$$

Thus, we get $x_1(t) = t^2$ or t. Then $\dot{x}_1(t) = x_2(t) = 2t$ or 1. Hence $\begin{bmatrix} t^2 \\ 2t \end{bmatrix}$ and $\begin{bmatrix} t \\ 1 \end{bmatrix}$ are the two linearly independent solutions provided that $t \neq 0$. Hence $M(t) = \begin{bmatrix} t^2 & t \\ 2t & 1 \end{bmatrix}$ for $t \neq 0$ is a fundamental matrix.

Example **4.8.11** If $M(t) = \begin{bmatrix} 2 & t^2 \\ t & 2t \end{bmatrix}$, $t \in [0,3]$, find the intervals in which $M(t)$ is a fundamental matrix of a system $\dot{x} = A(t)x$. Also, find $A(t)$.

Solution Since $\det(M(t)) = 0$ at $t = 0$ and $t = 2$, $M(t)$ is invertible in the subintervals $(0,2)$ and $(2,3]$. Moreover, if $M(t)$ is a fundamental matrix, then $\frac{dM(t)}{dt} = A(t)M(t)$. Hence, we obtain $A(t) = \dot{M}(t)M^{-1}(t) = \frac{1}{4t-t^3}\begin{bmatrix} 0 & 2t \\ 1 & 2 \end{bmatrix}\begin{bmatrix} 2t & -t^2 \\ -t & 2 \end{bmatrix} = \frac{1}{4t-t^3}\begin{bmatrix} -2t^2 & 4t \\ 0 & 4-t^2 \end{bmatrix}$ for $t \in (0,2)$ or $(2,3]$.

Example **4.8.12** Find a fundamental matrix of the system of differential equations

$$\dot{x} = \begin{bmatrix} 1 & -1 & 4 \\ 3 & 2 & -1 \\ 2 & 1 & -1 \end{bmatrix} x. \tag{4.83}$$

Solution We can see that 1, 2, and -2 are the eigenvalues of the coefficient matrix, and $\begin{bmatrix} -1 \\ 4 \\ 1 \end{bmatrix}$, $\begin{bmatrix} 1 \\ 2 \\ 1 \end{bmatrix}$, and $\begin{bmatrix} -1 \\ 1 \\ 1 \end{bmatrix}$, respectively, are the corresponding eigenvectors. Hence, $e^t\begin{bmatrix} -1 \\ 4 \\ 1 \end{bmatrix}$, $e^{3t}\begin{bmatrix} 1 \\ 2 \\ 1 \end{bmatrix}$, and $e^{-2t}\begin{bmatrix} -1 \\ 1 \\ 1 \end{bmatrix}$ are three linearly independent solutions of (4.83). By Definition 4.8.2,

$$M(t) = \begin{bmatrix} -e^t & e^{3t} & -e^{-2t} \\ 4e^t & 2e^{3t} & e^{-2t} \\ e^t & e^{3t} & e^{-2t} \end{bmatrix}$$

is a fundamental matrix solution of (4.83).

Example **4.8.13** Find e^{At} for the matrix $A = \begin{bmatrix} 1 & 1 & 1 \\ 0 & 3 & 2 \\ 0 & 0 & 5 \end{bmatrix}$.

Solution To find e^{At}, we use the result of Theorem 4.8.7. For this, we need to find a fundamental matrix solution. Since A is an upper triangular matrix, the eigenvalues of A are $1, 3$, and 5. The corresponding eigenvectors are $\begin{bmatrix} 1 \\ 0 \\ 0 \end{bmatrix}$, $\begin{bmatrix} 1 \\ 2 \\ 0 \end{bmatrix}$, and $\begin{bmatrix} 1 \\ 2 \\ 2 \end{bmatrix}$. Hence, a fundamental matrix of $\dot{x} = Ax$ is given by

$$M(t) = \begin{bmatrix} e^t & e^{3t} & e^{5t} \\ 0 & 2e^{3t} & 2e^{5t} \\ 0 & 0 & 2e^{5t} \end{bmatrix}.$$

From Theorem 4.8.7, $e^{At} = \phi(t, 0) = M(t)M^{-1}(0)$. Since

$$M^{-1}(0) = \begin{bmatrix} 1 & -\frac{1}{2} & 0 \\ 0 & \frac{1}{2} & -\frac{1}{2} \\ 0 & 0 & \frac{1}{2} \end{bmatrix},$$

we obtain

$$e^{At} = \begin{bmatrix} e^t & \frac{-1}{2}e^t + \frac{1}{2}e^{3t} & \frac{-1}{2}e^{3t} + \frac{1}{2}e^{5t} \\ 0 & e^{3t} & -e^{3t} + e^{5t} \\ 0 & 0 & e^{5t} \end{bmatrix}.$$

Exercise 4.8.1 Consider the system $\dot{x} = A(t)x$, where $A(t) = \begin{bmatrix} 1 & t \\ 0 & 1 \end{bmatrix}$. Find the following: (i) the state transition matrix and (ii) the fundamental matrix $M(t)$ satisfying $M(0) = \begin{bmatrix} 1 & 1 \\ 2 & 1 \end{bmatrix}$.

Exercise 4.8.2 Show that $M(t) = \begin{bmatrix} 1 & 1 \\ t & e^t \end{bmatrix}$ is a fundamental matrix for a system $\dot{x} = A(t)x$ on any interval in $\mathbb{R}$. Additionally, determine $A(t)$.

4.9 Solution of Nonhomogeneous System

Consider the nonhomogeneous system

$$\begin{cases} \dot{x} = A(t)x + g(t), & t \in J, \\ x(t_0) = x_0. \end{cases} \tag{4.84}$$

If we denote the RHS of (4.84) by $f(t,x)$, then f satisfies Lipschitz continuity with respect to x, since

$$\|f(t,x_1) - f(t,x_2)\| = \|A(t)x_1 - A(t)x_2\| \leq \|A(t)\|\|x_1 - x_2\| \leq \alpha\|x_1 - x_2\|,$$

where $\alpha = \max_{t\in J}\|A(t)\|$. So, by Theorem 4.5.1, the IVP (4.84) has a unique solution. Let us assume that the solution can be written in the form

$$x(t) = \phi(t,t_0)v(t) \tag{4.85}$$

for some suitable function v of t. If such v exists, then by differentiating (4.85) with respect to t, we get

$$\begin{aligned} \dot{x}(t) &= \frac{d}{dt}(\phi(t,t_0)v) \\ &= \frac{d\phi(t,t_0)}{dt}v + \phi(t,t_0)\dot{v} \\ &= A(t)\phi(t,t_0)v + \phi(t,t_0)\dot{v} \quad \text{(from Theorem 4.8.8}(i)) \\ &= A(t)x(t) + \phi(t,t_0)\dot{v}. \end{aligned} \tag{4.86}$$

Comparing the RHSs of (4.84) and (4.86), we get

$$\phi(t,t_0)\dot{v} = g(t).$$

Multiplying both sides by $\phi(t_0,t)$, we obtain $\dot{v} = \phi(t_0,t)g(t)$. By integrating from t_0 to t, we have

$$v(t) = v(t_0) + \int_{t_0}^{t} \phi(t_0,s)g(s)ds. \tag{4.87}$$

From (4.85), $v(t_0) = x(t_0)$. Substituting $v(t)$ from (4.87) into (4.85), the solution to (4.84) is expressed as

$$x(t) = \phi(t,t_0)\left[x(t_0) + \int_{t_0}^{t} \phi(t_0,s)g(s)ds\right]. \tag{4.88}$$

***Example* 4.9.1** Evaluate the solution of $\dot{x} = Ax + b(t)$ for $t \geq 5$, where $A = \begin{bmatrix} 2 & 1 & 0 \\ 0 & 2 & 0 \\ 1 & 1 & 3 \end{bmatrix}$ and $b(t) = \begin{bmatrix} 1 \\ 0 \\ t \end{bmatrix}$, with $x(5) = \begin{bmatrix} 0 \\ 1 \\ 4 \end{bmatrix}$.

Solution In Example 4.7.3, $e^{A(t-5)}$ has been evaluated as

$$\phi(t,5) = e^{A(t-5)} = e^{2(t-5)}\begin{bmatrix} 1 & t-5 & 0 \\ 0 & 1 & 0 \\ e^{t-5}-1 & e^{t-5}-t+3 & e^{t-5} \end{bmatrix}.$$

Hence, using the expression (4.88), the solution is given by

$$\begin{aligned} x(t) &= e^{2(t-5)}\begin{bmatrix} t-5 \\ 1 \\ 6e^{t-5}-t+3 \end{bmatrix} + e^{2t}\int_5^t e^{-2s}\begin{bmatrix} 1 \\ 0 \\ (s+1)e^{t-s}-1 \end{bmatrix} ds \\ &= e^{2(t-5)}\begin{bmatrix} t-5 \\ 1 \\ 6e^{t-5}-t+3 \end{bmatrix} + e^{2t}\begin{bmatrix} \frac{1}{10}e^{-10} - \frac{1}{2t}e^{-2t} \\ 0 \\ e^{t}\left(\frac{19}{9}e^{-15} - \frac{4}{3}e^{-3t} - \frac{1}{3}te^{-3t}\right) + \frac{e^{-2t}}{2} - \frac{e^{-10}}{2} \end{bmatrix}. \end{aligned}$$

4.10 Gronwall's Inequality

Theorem 4.10.1 Let x, p, and q be nonnegative continuous functions on an interval J, and let

$$x(t) \leq p(t) + \int_{t_0}^{t} q(s)x(s)ds. \tag{4.89}$$

Then

$$x(t) \leq p(t) + \int_{t_0}^{t} p(s)q(s)e^{\int_s^t q(r)dr}ds. \tag{4.90}$$

Proof. Define

$$r(t) = \int_{t_0}^{t} q(s)x(s)ds \quad \text{for } t \in J. \tag{4.91}$$

Then

$$r(t_0) = 0, \dot{r}(t) = q(t)x(t), \text{ and } x(t) \leq p(t) + r(t). \tag{4.92}$$

From (4.89), (4.91), and (4.92), we obtain $\dot{r}(t) = q(t)x(t) \leq q(t)p(t) + q(t)r(t)$. Multiplying both sides by the positive function $e^{-\int_{t_0}^t q(s_1)ds_1}$, we obtain

$$e^{-\int_{t_0}^t q(s_1)ds_1}\dot{r}(t) \leq e^{-\int_{t_0}^t q(s_1)ds_1}\left(q(t)p(t) + q(t)r(t)\right).$$

Hence,

$$\frac{d}{dt}\left(e^{-\int_{t_0}^t q(s_1)ds_1}r(t)\right) \leq e^{-\int_{t_0}^t q(s_1)ds_1}q(t)p(t). \tag{4.93}$$

Now integrating both sides from t_0 to s, we get

$$e^{-\int_{t_0}^s q(s_1)ds_1}r(s) - r(t_0) \leq \int_{t_0}^{s} q(t)p(t)e^{-\int_{t_0}^t q(s_1)ds_1}dt.$$

Since $r(t_0) = 0$, it follows that

$$r(s) \leq \int_{t_0}^{s} q(t)p(t)e^{\int_t^s q(s_1)ds_1}dt. \tag{4.94}$$

Using (4.94) in (4.89), the required inequality (4.90) is established. □

Corollary 4.10.2 Under the assumptions of Theorem 4.10.1, if $p(t) =$ a nonnegative constant a for $t \in J$, then (4.90) becomes

$$x(t) \leq a + a\int_{t_0}^{t} q(s)e^{\int_s^t q(r)dr}ds = a\left[1 + \int_{t_0}^{t} \frac{d}{ds}\left(-e^{\int_s^t q(r)dr}\right)ds\right].$$

Hence,

$$x(t) \le ae^{\int_{t_0}^{t} q(r)dr}. \tag{4.95}$$

Corollary 4.10.3 If $x(t) \ge 0$ for $t \ge t_0$ such that

$$x(t) \le a + b \int_{t_0}^{t} x(s)ds,$$

where a and b are nonnegative constants, then from (4.95),

$$x(t) \le ae^{b(t-t_0)}, \quad t \ge t_0. \tag{4.96}$$

Corollary 4.10.4 If x and q are nonnegative functions of $t, t \ge t_0$, and satisfy

$$x(t) \le \int_{t_0}^{t} q(s)x(s)ds,$$

then $x(t) \equiv 0$ for $t \ge t_0$.

Proof. Set $a = 0$ in inequality (4.95). □

Corollary 4.10.5 A result similar to Theorem 4.10.1 holds if we replace the inequality '$\le$' with '$\ge$' in (4.89) and (4.90).

Proof. In the proof of Theorem 4.10.1, replace '$\le$' with '$\ge$'. □

4.11 Summary

- Solutions to first order homogeneous and nonhomogeneous linear ODEs are given in equations (4.11) and (4.14).
- Examples of IVPs having a unique solution, infinitely many solutions, or no solution are given in Section 4.3.
- **Existence theorem** (Theorem 4.3.8): The IVP

  $$\begin{cases} \dot{x} = f(t,x), \\ x(t_0) = x_0 \end{cases}$$

  has at least one solution if f is continuous and bounded on a domain D containing the initial point (t_0, x_0).
- **Existence and uniqueness theorem** (Theorem 4.4.1): If $f(t,x)$ is continuous and bounded in $D \subset \mathbb{R}^2$, and $f(t,x)$ is Lipschitz continuous with respect to x on D, then the IVP with initial point $(t_0, x_0) \in D$ has a unique solution. The

proof of this theorem is based on the Banach fixed point theorem (Theorem 3.5.6) and Corollary 3.5.7.

- Theorem 4.5.1 is on the existence and uniqueness of solutions to n-dimensional IVPs. It is an extension of Theorem 4.4.1 to n-dimensional systems.
- The solution to time-varying linear system $\dot{x} = A(t)x$ with $x(t_0) = x_0$ can be written in the form of the Peano–Baker series (4.43).
- For the time-invariant system $\dot{x} = Ax$, the solution can be written in terms of eigenvalues and generalized eigenvectors of A (see equations (4.56), (4.59), and (4.63)).
- The relation between fundamental matrix and the state transition matrix of a linear system is given in equation (4.78).
- In Theorem 4.8.8, four important properties of the state transition matrix have been proved using equation (4.78).
- Theorem 4.8.9 provides the relation between the state transition matrices of a system and its dual.
- In Theorem 4.10.1, Gronwall's inequality is proved. This inequality plays an important role in Chapter 8 on stability.

4.12 Exercise

1. Convert the equation $\dddot{y} - 4\ddot{y} + 3\dot{y} + y = 0$ into an equivalent system of first order equations.

2. Convert the pair of second order equations

$$\begin{cases} \ddot{y} + 4\dot{y} + 3z = 0, \\ \ddot{z} - 3\dot{z} - y = 0 \end{cases}$$

into an equivalent system of four first order equations. Express the resulting system in matrix form.

3. Let $x(t) = \begin{bmatrix} x_1(t) \\ x_2(t) \end{bmatrix}$ be a solution of the system of equations $\dot{x} = \begin{bmatrix} 0 & 1 \\ -1 & -1 \end{bmatrix} x$. Show that $y = x_1(t)$ is a solution of the equation $\ddot{y} + \dot{y} + y = 0$.

4. Solve the system $\dot{x} = 2x + y + 1, \dot{y} = 2y + x + 1$, and $\dot{z} = 3z$, with initial conditions $x(0) = 0$, $y(0) = 1$, and $z(0) = 2$, by finding $\exp(At)$, where A is the coefficient matrix.

5. (*i*) If $A(t) = \begin{bmatrix} 1 & t \\ 0 & 0 \end{bmatrix}$, show that $\frac{d}{dt}A^2(t) \neq 2A(t)\frac{d}{dt}A(t)$.

   (*ii*) Show that $\frac{d}{dt}A^2(t) = 2A(t)\frac{d}{dt}A(t)$ if $A(t)$ and $A(s)$ commute for all $t, s \in \mathbb{R}$.

6. Solve the IVP:

   (*i*)
$$\dot{x} = \begin{bmatrix} 2 & 1 & 3 \\ 0 & 2 & -1 \\ 0 & 0 & 3 \end{bmatrix} x \text{ with the initial condition } x(0) = \begin{bmatrix} 1 \\ 2 \\ 1 \end{bmatrix}.$$

   (*ii*)
$$\dot{x} = \begin{bmatrix} 1 & 0 & 0 \\ 0 & 1 & -1 \\ 0 & 1 & 1 \end{bmatrix} x \text{ with the initial condition } x(0) = \begin{bmatrix} 1 \\ 1 \\ 1 \end{bmatrix}.$$

7. Three solutions of the equation $\dot{x} = Ax$ are given below:
$$x_1(t) = (e^t, e^{2t}, 0)',\ x_2(t) = (e^t + e^{3t}, 2e^{3t}, e^{3t})',\ x_3(t) = (e^t - e^{3t}, -2e^{3t}, -e^{3t})'.$$
   Find the matrix A and its eigenpairs.

8. Let $A = \begin{bmatrix} \lambda & \alpha & 0 & \cdots & 0 \\ 0 & \lambda & \alpha & 0 & \vdots \\ \vdots & & & & \alpha \\ 0 & 0 & 0 & \cdots & \lambda \end{bmatrix}$. Find $e^{\beta A}$ for real constants α, β, and λ.

9. Let $M(t)$ be a fundamental matrix solution of $\dot{x} = Ax$. Prove that the solution $x(t)$ of the IVP $\dot{x} = Ax$ with $x(t_0) = x_0$ is $x(t) = M(t)M^{-1}(t_0)x_0$.

10. Are the matrices $M(t)$ given below fundamental matrices of $\dot{x} = A(t)x, t \in [0,3]$, for some $A(t)$? If yes, find $A(t)$. If no, find the subintervals of $[1,3]$ in which $M(t)$ is a fundamental matrix.

$$(i) \begin{bmatrix} 1 & t \\ t & 0 \end{bmatrix}; (ii) \begin{bmatrix} 1 & t & t^2 \\ t & 2t^2 & t^3 \\ t^3 & t^4 & t^5 \end{bmatrix}; (iii) \begin{bmatrix} 1 & t-2 \\ t & -1 \end{bmatrix}.$$

11. Solve the IVP:

(i)

$$x = \begin{bmatrix} 1 & 0 & 0 \\ 2 & 1 & -2 \\ 0 & 0 & 1 \end{bmatrix} x + \begin{bmatrix} 0 \\ 0 \\ e^t \end{bmatrix}, \quad x(0) = \begin{bmatrix} 1 \\ 0 \\ 1 \end{bmatrix}.$$

(ii)

$$\dot{x} = \begin{bmatrix} 3 & 0 \\ 0 & 2 \end{bmatrix} x + \begin{bmatrix} t^2 \\ 1+t \end{bmatrix}, \quad x(0) = \begin{bmatrix} 0 \\ 1 \end{bmatrix}.$$

12. Prove the existence and uniqueness of a solution to the following IVP:

$$\begin{cases} \dot{x}(t) = t^2 \cos(x(t)), & t \in [0,2], \\ x(0) = \dfrac{\pi}{4}. \end{cases}$$

Additionally, determine an upper bound for the solution on the interval $[0,2]$ using Gronwall's inequality.

13. Let f and g be continuous nonnegative functions on $[0,\tau]$ that satisfy the differential inequality

$$f'(t) \leq af(t) + bg(t),$$

where a and b are nonnegative constants. Show that

$$f(t) \leq e^{at}\left[f(0) + b\int_0^t e^{-as} g(s)ds\right] \quad \text{for all } t \in [0,\tau].$$

14. If $u(t)$ is a nonnegative continuous function satisfying the inequality

$$u(t) \leq K_1 + \varepsilon(t-\alpha) + K_2 \int_{\alpha}^{t} u(s)ds$$

on an interval $\alpha \leq t \leq \beta$, where ε, K_1, and K_2 are given positive constants, show that

$$u(t) \leq K_1 e^{K_2(t-\alpha)} + \frac{\varepsilon}{K_2}\left[e^{K_2(t-\alpha)} - 1\right].$$

15. Find $C > 1$ and $\alpha > 0$ such that $\|e^{At}\| \leq Ce^{\alpha t}$, where $A = \begin{bmatrix} 2 & 1 \\ 0 & 2 \end{bmatrix}$.

5

Controllability and Observability

5.1 Introduction

Steering a system to a desired state by applying suitable control is an important problem in the study of control systems. In this context, the **controllability** of a control system plays a vital role. Another key property is the ability to obtain the trajectory of a dynamical system using specific observations or measurements over a time interval. If the trajectory of the system can be uniquely determined from the observation data, the system is called **observable**. This chapter discusses both of these concepts for linear control systems.

In Sections 5.2 and 5.3, controllability of a linear control system, necessary and sufficient conditions for controllability, and computation of control for some control systems are discussed. In Section 5.4, the method of converting a controllable system into its companion form is presented. Computation of feedback control is shown in Section 5.5. Observability and the duality theorem for linear systems are explained in Section 5.6. For uncontrollable and unobservable linear systems, controllable and observable subsystems, respectively, are derived in Sections 5.7 and 5.8. Applications such as "state observer using observations" and "trajectory tracking using feedback control" are discussed in Section 5.9.

5.2 Controllability

In this chapter, we consider **linear** control systems of the form

$$\dot{x} = Ax + Bu, \quad t \in [t_0, \tau] =: J, \tag{5.1}$$

where A and B are $n \times n$ and $n \times m$ matrices, respectively, and $x(t) \in \mathbb{R}^n$ and $u(t) \in \mathbb{R}^m$ for each $t \in J$.

Many real-life problems can be expressed mathematically as a linear control system in the form of (5.1). In chapter 1, mathematical models representing various phenomena were considered. Let us now recall the equation of motion of a mass under the influence of a force.

Example **5.2.1** Consider the control system representing the straight-line motion of a particle of mass m under a force $f(t)$, as described in Section 1.3. The system equation is

$$\begin{bmatrix} \dot{x}_1 \\ \dot{x}_2 \end{bmatrix} = \begin{bmatrix} 0 & 1 \\ 0 & 0 \end{bmatrix} \begin{bmatrix} x_1 \\ x_2 \end{bmatrix} + \begin{bmatrix} 0 \\ \frac{1}{m} \end{bmatrix} f, \tag{5.2}$$

where $x_1(t)$ and $x_2(t)$ denote the displacement and velocity, respectively, of the particle at time $t \in J$. Equation (5.2) is in the form of (5.1).

If $m = 1, f(t) = t(2-t), x_1(t_0) = 1, x_2(t_0) = 0, t_0 = 0$, and $\tau = 2$, then from (4.86) the solution of the system is

$$\begin{aligned}
\begin{bmatrix} x_1(t) \\ x_2(t) \end{bmatrix} &= e^{A(t-t_0)} \left(\begin{bmatrix} x_1(0) \\ x_2(0) \end{bmatrix} + \int_{t_0}^{t} e^{A(t_0-s)} \begin{bmatrix} 0 \\ 1 \end{bmatrix} f(s)ds \right) \\
&= \begin{bmatrix} 1 & t \\ 0 & 1 \end{bmatrix} \left(\begin{bmatrix} 1 \\ 0 \end{bmatrix} + \int_0^t \begin{bmatrix} 1 & -s \\ 0 & 1 \end{bmatrix} \begin{bmatrix} 0 \\ 1 \end{bmatrix} f(s)ds \right) \\
&= \begin{bmatrix} 1 & t \\ 0 & 1 \end{bmatrix} \left(\begin{bmatrix} 1 \\ 0 \end{bmatrix} + \int_0^t \begin{bmatrix} -s \\ 1 \end{bmatrix} (2s - s^2)ds \right) \\
&= \begin{bmatrix} 1 + \frac{t^3}{3} + \frac{7}{12}t^4 \\ t^2 - \frac{t^3}{3} \end{bmatrix}.
\end{aligned}$$

At the final time $\tau = 2$, the displacement is 13 and the velocity is $\frac{4}{3}$.

In the above example, the control function f (force) is specified to find the solution x. However, in many practical problems, it becomes necessary to determine a control function such that the solution attains a preassigned value at the final time τ. This situation leads us to the concept of **controllability** of a system which guarantees the ability of the system to move from any initial state to an arbitrary final state using a suitable control within a prescribed time interval.

Consider the control system

$$\dot{x} = f(t, x(t), u(t)). \tag{5.3}$$

The controllability of the system (5.3) is defined as follows:

Definition 5.2.2 A control system $\dot{x} = f(t, x(t), u(t))$ is said to be **controllable** in the time interval J if, for arbitrary vectors x_0 and x_1 in $\mathbb{R}^n$, there exists a control function $u(t)$ such that the solution $x(t)$ of (5.3) satisfies $x(t_0) = x_0$ and $x(\tau) = x_\tau$.

***Example* 5.2.3** Consider the simple first order linear differential equation

$$\dot{x} = ax + bu, \quad t \in J, \tag{5.4}$$

where a and b are real constants. Let $x(t)$ and $u(t)$ denote the state and control variables, respectively, of the control system at time t. For any given initial condition $x(t_0) = x_0$, the solution is

$$x(t) = e^{a(t-t_0)} \left[x_0 + b \int_{t_0}^{t} e^{a(t_0-s)} u(s) ds \right]. \tag{5.5}$$

At final time $t = \tau$, the state of the system is

$$x(\tau) = e^{a(\tau-t_0)} \left[x_0 + b \int_{t_0}^{\tau} e^{a(t_0-s)} u(s) ds \right]. \tag{5.6}$$

Let the final state be $x(\tau) = x_1$. Rearranging terms in the above equation, we can write

$$\frac{e^{a(t_0-\tau)} x_1 - x_0}{b} = \int_{t_0}^{\tau} e^{a(t_0-s)} u(s) ds.$$

Denoting the LHS of the above equation by K, we get

$$K = \int_{t_0}^{\tau} e^{a(t_0-s)} u(s) ds. \tag{5.7}$$

Now, it is easy to see that the control function

$$u(s) = \frac{K}{\tau - t_0} e^{-a(t_0-s)} \tag{5.8}$$

satisfies the equation (5.7), which means that given arbitrary initial and final conditions x_0 and x_1, one can find a control function u such that the solution x satisfies $x(t_0) = x_0$ and $x(\tau) = x_1$. In other words, the system (5.4) is controllable according to Definition 5.2.2. In fact, there are many control functions u that satisfy (5.7). For instance,

$$u(s) = \frac{Ka}{1 - e^{a(t_0 - \tau)}}$$

also satisfies (5.7).

***Example* 5.2.4** Consider the mixing problem as described in Section 1.4. Recall the equation (1.16)

$$\begin{cases} \dot{x} = -\dfrac{x}{25} + 4u, \\ x(0) = 2. \end{cases}$$

Here, $x(t)$ represents the amount of salt present in the tank at time t, and $u(t)$ is the control representing the amount of salt in the inflow. Since the one dimensional system is controllable, we can find $u(t)$ for a given final condition $x(\tau) = x_1$ as in (5.8). Note that in (5.8) (with $t_0 = 0$), K is negative if $e^{-a\tau}x_1 < x_0$. In that case, $u(t)$ would be negative. In this example, $u(t)$ represents the amount of salt, which cannot be negative. Thus, we conclude the following: Even though controllability guarantees a mathematical control function for any x_0 and x_1, in practice, certain final conditions cannot be reached at a given final time $t = \tau$.

Note that (5.4) is a one dimensional system since the state x is a real valued function. In this case, proving controllability and finding a suitable control for a given pair of initial and final conditions are straightforward. However, in an n-dimensional problem, the concept of controllability requires more analysis.

***Example* 5.2.5** Under ideal conditions, the equation of motion of a mass m moving under the influence of a force $u(t)$ during a time interval J is given by

$$m\ddot{x} = u, \tag{5.9}$$

where x, $\dot{x}$, and $\ddot{x}$ denote the position, velocity, and acceleration of the object, respectively. Taking $\dot{x}$ as v, equation (5.9) can be written as

$$\dot{v} = \frac{u}{m}, \quad t \in J. \tag{5.10}$$

Here, $v(t)$ and $u(t)$ are the state and control variables. Integrating both sides of (5.10) from t_0 to t, we get

$$v(t) = \frac{1}{m}\int_{t_0}^{t} u(s)ds + v(t_0), \quad t \in J. \tag{5.11}$$

For any two values v_0 and v_τ, we can find a control $u(s)$ such that $v(t_0) = v_0$ and $v(\tau) = v_\tau$. For example, if $u(s) = k$, a constant, then from (5.11),

$$v(\tau) = \frac{k}{m}(\tau - t_0) + v_0, \quad \text{where } k = \frac{m(v_\tau - v_0)}{\tau - t_0}.$$

Hence, the system (5.10) is controllable. Another control $u(s)$ for this problem is $u(s) = cs$, where $c = \frac{2m(v_1 - v_0)}{\tau^2 - f_0^2}$. One can find infinitely many control functions $u(s)$ for this problem.

Examples 5.2.1 and 5.2.5 represent the same equation of motion. In Example 5.2.1, displacement and velocity are considered as state variables, whereas in Example 5.2.5, only velocity is considered as the state variable of the system.

Now consider the n-dimensional linear control system

$$\begin{cases} \dot{x} = Ax + Bu, & t \in J, \\ x(t_0) = x_0. \end{cases} \tag{5.12}$$

Using (4.88), the solution of (5.12) can be written as

$$x(t) = \phi(t, t_0)\left[x(t_0) + \int_{t_0}^{t} \phi(t_0, s)B(s)u(s)ds\right], \tag{5.13}$$

where $\phi(t, s)$ is the state transition matrix corresponding to A.

In the next section, necessary and sufficient conditions for the controllability of linear autonomous and nonautonomous systems will be considered.

5.3 Controllability Conditions

Consider the linear control system

$$\dot{x} = A(t)x + B(t)u, \quad t \in J, \tag{5.14}$$

where A and B are $n \times n$ and $n \times m$ matrices, respectively, whose entries are continuous functions of $t \in J$.

Definition 5.3.1 Let $\phi(t, s)$ be the state transition matrix of $A(t)$ as in equation (4.43). The **controllability Gramian matrix** of order n corresponding to the system (5.14) is defined as

$$W[t_0, \tau] = \int_{t_0}^{\tau} \phi(t_0, s)B(s)B'(s)\phi'(t_0, s)ds. \tag{5.15}$$

Note that $W[t_0, \tau]$ is an $n \times n$ constant matrix for a given pair $[t_0, \tau]$. Hereafter, we shall denote $W[t_0, \tau]$ simply by W.

Example **5.3.2** For the system considered in Example 5.2.1,

$$\phi(t,s) = e^{A(t-s)} = \begin{bmatrix} 1 & t-s \\ 0 & 1 \end{bmatrix}.$$

Hence, from (5.15), we obtain

$$W = \int_0^2 \begin{bmatrix} 1 & -s \\ 0 & 1 \end{bmatrix} \begin{bmatrix} 0 \\ 1 \end{bmatrix} \begin{bmatrix} 0 & 1 \end{bmatrix} \begin{bmatrix} 1 & 0 \\ -s & 1 \end{bmatrix} ds = \begin{bmatrix} \frac{8}{3} & -2 \\ -2 & 2 \end{bmatrix}.$$

Theorem 5.3.3 The system (5.14) is controllable on J if and only if the Gramian matrix W is nonsingular.

Proof. **Necessity.** Let the system (5.14) be controllable. We need to show that W is a nonsingular matrix. If not, then there exists a nonzero vector $\alpha \in \mathbb{R}^n$ such that $W\alpha = 0$ (see Remark 2.6.5). Consequently, $\alpha' W \alpha = 0$, i.e.,

$$\begin{aligned}
&\alpha' \left[\int_{t_0}^{\tau} \phi(t_0,s)B(s)B'(s)\phi'(t_0,s)ds \right] \alpha = 0 \\
\Longrightarrow\ & \int_{t_0}^{\tau} \left(B'(s)\phi'(t_0,s)\alpha\right)' \left(B'(s)\phi'(t_0,s)\alpha\right) ds = 0 \\
\Longrightarrow\ & \int_{t_0}^{\tau} \|B'(s)\phi'(t_0,s)\alpha\|^2 ds = 0 \\
\Longrightarrow\ & B'(s)\phi'(t_0,s)\alpha = 0 \quad \text{for all } s \in J \\
\Longrightarrow\ & \alpha'\phi(t_0,s)B(s) = 0 \quad \text{for all } s \in J.
\end{aligned} \tag{5.16}$$

Since the system (5.14) is controllable, for arbitrary vectors x_0 and x_1 in $\mathbb{R}^n$ as initial and final conditions, respectively, we can find a control function u such that the solution x satisfies $x(t_0) = x_0$ and $x(\tau) = x_1$. The solution of (5.14) is

$$x(t) = \phi(t,t_0) \left[x(t_0) + \int_{t_0}^{t} \phi(t_0,s)B(s)u(s)ds \right].$$

At $t = \tau$, we have

$$x_1 = x(\tau) = \phi(\tau,t_0) \left[x_0 + \int_{t_0}^{\tau} \phi(t_0,s)B(s)u(s)ds \right]. \tag{5.17}$$

Since x_0 and x_1 can be chosen arbitrarily in $\mathbb{R}^n$, let $x_0 = -\alpha$ and $x_1 = 0$. As $\phi(\tau, t_0)$ is nonsingular, it follows from (5.17) that

$$\alpha = \int_{t_0}^{\tau} \phi(t_0, s)B(s)u(s)ds.$$

Multiplying both sides by α', we get

$$\|\alpha\|^2 = \alpha'\alpha = \int_{t_0}^{\tau} \left(\alpha'\phi(t_0, s)B(s)\right) u(s)ds = 0 \quad \text{(from (5.16))}.$$

Therefore, $\alpha = 0$, which contradicts the assumption that $\alpha \neq 0$. Hence, W is nonsingular.

Sufficiency. Let W be nonsingular. To show that the system (5.14) is controllable, let x_0 and x_1 be arbitrary vectors in $\mathbb{R}^n$. Multiplying both sides of (5.17) by $\phi(t_0, \tau)$, and using Theorem 4.8.8 (iv), we obtain

$$\phi(t_0, \tau)x_1 - x_0 = \int_{t_0}^{\tau} \phi(t_0, s)B(s)u(s)ds. \tag{5.18}$$

Let $\phi(t_0, \tau)x_1 - x_0 = P$. Since W^{-1} exists, we can show that the control function u given by

$$u(s) = B'(s)\phi'(t_0, s)W^{-1}(P) \tag{5.19}$$

satisfies equation (5.18). This can be proved directly by substituting $u(s)$ given in (5.19) into the RHS of (5.18).

$$\begin{aligned}
\text{The RHS of (5.18)} &= \int_{t_0}^{\tau} \phi(t_0, s)B(s)\left[B'(s)\phi'(t_0, s)W^{-1}(P)\right] ds \\
&= \left(\int_{t_0}^{\tau} \phi(t_0, s)B(s)B'(s)\phi'(t_0, s)ds\right) W^{-1}P \\
&= P = \text{the LHS of (5.18)}.
\end{aligned}$$

Hence, system (5.14) is controllable. □

Equation (5.19) provides a formula to compute one control function u that steers the system from an initial state x_0 to a final state x_1. In fact, we can find infinitely many such control functions which are given by $u + v$, where v is any function that maps the RHS of (5.18) (with v replacing u) to the zero vector.

Alternative proof of Theorem 5.3.3 *(using operator notion).* The solution x of the system (5.14) can be considered as an element of $L_2(J, \mathbb{R}^n)$, and the control u can be considered as an element of $L_2(J,\mathbb{R}^m)$. Here, $u(t) \in \mathbb{R}^m$ for each t, and $B : \mathbb{R}^m \to \mathbb{R}^n$. Now, define the linear operator M from $L_2(J,\mathbb{R}^m)$ into $\mathbb{R}^n$ by

$$Mu = \int_{t_0}^{\tau} \phi(\tau,t)Bu(t)dt. \tag{5.20}$$

Using (5.20), equation (5.17) can be written as

$$\phi(t_0,\tau)x_1 - x_0 = Mu \tag{5.21}$$

for any two vectors x_0 and x_1 in $\mathbb{R}^n$. Let $\phi(t_0,\tau)x_1 - x_0 = P$. Then, equation (5.21) becomes

$$Mu = P. \tag{5.22}$$

From Definition 5.3.1, the system (5.14) is controllable if, for any $P \in \mathbb{R}^n$, we can find a $u \in L_2(J,\mathbb{R}^m)$ satisfying (5.22). In other words, the system (5.14) is controllable if and only if $\text{Range}(M) = \mathbb{R}^n$. Since $\text{Range}(M)$ is orthogonal to $\text{Null}(M^*)$ (Theorem 3.4.25), to prove controllability, we need to show that

$$\text{Null}(M^*) = \{0\}, \tag{5.23}$$

so that $\text{Range}(M) = \mathbb{R}^n$. From Lemma 3.4.23, we obtain

$$M^* = (\phi(t_0,t)B(t))' = B'(t)\phi'(t_0,t), \tag{5.24}$$

which is an $m \times n$ matrix. Now from (5.20) and (5.24), we have

$$\begin{aligned} MM^*v = M(M^*v) &= \int_{t_0}^{\tau} \phi(t_0,t)B(t)(M^*v)(t)dt \\ &= \int_{t_0}^{\tau} \phi(t_0,t)B(t)B'(t)\phi'(t_0,t)vdt. \end{aligned}$$

As $v \in \mathbb{R}^n$ is independent of t, it can be factored out of the above integral. Hence, we get

$$MM^* = \int_{t_0}^{\tau} \phi(t_0,t)B(t)B'(t)\phi'(t_0,t)dt,$$

which is the same as the controllability Gramian matrix given in (5.15). □

Hence the Theorem 5.3.3 can be restated as follows:

Theorem 5.3.4 The system (5.14) is controllable if and only if the matrix MM^* is nonsingular.

Proof. **Necessity.** Let the system be controllable. Then, Range$(M) = \mathbb{R}^n$, and from (5.23), Null$(M^*) = \{0\}$. If MM^* is singular, then there exists a **nonzero** vector $\alpha \in \mathbb{R}^n$ such that

$$MM^*\alpha = 0 \implies \langle MM^*\alpha, \alpha\rangle_{\mathbb{R}^n} = 0.$$

Hence, by Definition 3.4.20, we have

$$\langle M^*\alpha, M^*\alpha\rangle_{\mathbb{R}^n} = 0 \implies \|M^*\alpha\|^2 = 0 \implies M^*\alpha = 0 \implies \alpha \in \text{Null}(M^*),$$

which contradicts (5.23). Hence, MM^* is nonsingular.

Sufficiency. Let MM^* be nonsingular. To show that the system is controllable, we need to find a control u satisfying the equation $Mu = P$, as in equation (5.22). If we choose

$$u = M^*\left(MM^*\right)^{-1}P, \tag{5.25}$$

then (5.22) is satisfied. It can be seen that u given in (5.25) is the same as the RHS of (5.19). □

The above theorem is also true when A and/or B are time-varying matrices. If A and B are both constant matrices, then one can obtain simpler condition for controllability which is known as the **Kalman rank condition**.

Remark 5.3.5 If the system $\dot{x}(t) = A(t)x(t) + B(t)u(t) : t_0 \le t \le \tau$ is controllable in R^n then we can obtain a control function $u(t) = B'(t)\varphi'(t_0, t)W^{-1}(P)$ (as given in (5.18)) that steers the system from $x(t_0) = x_0$ to $x(\tau) = x_1$. Now consider the system $\dot{y}(t) = A(t)y(t) + B(t)u(t) + f(t)$ where $f \in C[t_0, \tau : R^n]$. It can be shown that this system is also controllable and the control function v that steers the system from $y(0) = x_0$ to $y(\tau) = x_1$ is given by $v(t) = B'(t)\varphi'(t_0, t)W^{-1}(Q)$ where $Q = P - \int_{t_0}^{\tau}\varphi(t_0, , s)f(s)ds$.

Theorem 5.3.6 (Kalman rank condition) If A and B are constant matrices, then the system (5.14) is controllable if and only if

$$\text{rank}\begin{bmatrix} B & AB & A^2B & \cdots & A^{n-1}B \end{bmatrix}_{n\times nm} = n. \tag{5.26}$$

Proof. Since A is a constant matrix, $\phi(t, s) = e^{A(t-s)}$. From Theorem 5.3.3, the system (5.14) is controllable if and only if

$$W = \int_{t_0}^{\tau} e^{A(t_0-s)}BB'e^{A'(t_0-s)}ds$$

is nonsingular. From Lemma 3.3.1, it follows that the rows of the $n \times m$ matrix $e^{A(t_0-s)}B$ must be linearly independent. Let $r_1(s), r_2(s), \ldots, r_n(s)$ be the rows such that

$$e^{A(t_0-s)}B = \begin{bmatrix} r_1(s) \rightarrow \\ r_2(s) \rightarrow \\ \vdots \\ r_n(s) \rightarrow \end{bmatrix}_{n\times m}. \tag{5.27}$$

By the Cayley–Hamilton theorem, $e^{A(t_0-s)}$ can be expressed as

$$e^{A(t_0-s)} = \alpha_0(s)I + \alpha_1(s)A + \cdots + \alpha_{n-1}(s)A^{n-1},$$

where $\alpha_i(s) : i = 0, 1, \ldots, n-1$ are suitable functions of s. Now since

$$e^{A(t_0-s)}B = \alpha_0(s)B + \alpha_1(s)AB + \cdots + \alpha_{n-1}(s)A^{n-1}B,$$

(5.27) can be written as

$$\begin{bmatrix} B & AB & A^2B & \cdots & A^{n-1}B \end{bmatrix}_{n\times nm} \begin{bmatrix} \alpha_0(s) & 0 & \cdots & 0 \\ 0 & \alpha_0(s) & \cdots & 0 \\ \vdots & & \ddots & \\ 0 & 0 & \cdots & \alpha_0(s) \\ \alpha_1(s) & 0 & \cdots & 0 \\ 0 & \alpha_1(s) & \cdots & 0 \\ \vdots & & \ddots & \\ 0 & 0 & \cdots & \alpha_1(s) \\ \vdots & \vdots & \vdots & \vdots \\ \alpha_{n-1}(s) & 0 & \cdots & 0 \\ 0 & \alpha_{n-1}(s) & \cdots & 0 \\ \vdots & & \ddots & \\ 0 & 0 & \cdots & \alpha_{n-1}(s) \end{bmatrix}_{nm\times m}$$

$$= \begin{bmatrix} r_1(s) \rightarrow \\ r_2(s) \rightarrow \\ \vdots \\ r_n(s) \rightarrow \end{bmatrix}_{n \times m} \tag{5.28}$$

which can be written as $KG(s) = r(s)$, where $K = \begin{bmatrix} B & AB & A^2B & \cdots & A^{n-1}B \end{bmatrix}$, $G(s)$ is the $nm \times m$ matrix and $r(s)$ is the $n \times m$ matrix in the RHS. Let $R_1, R_2, \ldots, R_n$ be rows of K. If $\text{rank}(K) < n$, then at least one row is a linear combination of the other rows. For example, let the jth row be

$$R_j = a_1R_1 + a_2R_2 + \cdots + a_{j-1}R_{j-1} + a_{j+1}R_{j+1} + \cdots + a_nR_n, \tag{5.29}$$

for some scalars a_i, not all are zeros. Substituting (5.29) in the matrix K on the LHS of (5.28), we can show that

$$r_j(s) = a_1r_1(s) + a_2r_2(s) + \cdots + a_{j-1}r_{j-1}(s) + a_{j+1}r_{j+1}(s) + \cdots + a_nr_n(s),$$

which implies that the rows of $e^{A(t_0-s)}B$ are linearly dependent. From Lemma 3.3.1, it follows that W is singular, which is a contradiction. Hence $\text{rank}(K) = n$.

Conversely, if the system is not controllable, then W is singular, and hence from Lemma 3.3.1, $r_1(s), r_2(s), \ldots, r_n(s)$ are linearly dependent. Now, we can show that $\text{rank}(K) < n$. For this, if $r_j(s)$ is a linear combination of the remaining rows, then it can be shown from (5.28) that R_j can be written as a linear combination of the other rows. Hence, the result follows. □

***Example* 5.3.7** Consider the control system

$$\begin{cases} \dot{x_1} = 2x_1 + x_2 - 2x_3, \\ \dot{x_2} = x_1 + x_3 + 3u, \\ \dot{x_3} = x_2. \end{cases}$$

To test the controllability of the system, find the rank of $K = \begin{bmatrix} B & AB & A^2B \end{bmatrix}$, where $A = \begin{bmatrix} 2 & 1 & -2 \\ 1 & 0 & 1 \\ 0 & 1 & 0 \end{bmatrix}$ and $B = \begin{bmatrix} 0 \\ 3 \\ 0 \end{bmatrix}$. We obtain $K = \begin{bmatrix} 0 & 3 & 0 \\ 3 & 0 & 6 \\ 0 & 3 & 0 \end{bmatrix}$ and $\text{rank}(K) = 2$. Hence, the system is not controllable.

Example **5.3.8** The control system describing the motion of a particle of unit mass without friction under the influence of a force $f(t)$ (control) is given by

$$\begin{bmatrix} \dot{x}_1 \\ \dot{x}_2 \end{bmatrix} = \begin{bmatrix} 0 & 1 \\ 0 & 0 \end{bmatrix} \begin{bmatrix} x_1 \\ x_2 \end{bmatrix} + \begin{bmatrix} 0 \\ 1 \end{bmatrix} f(t).$$

For details, see Section 1.3. Determine a control function f such that $x_1(0) = 0, x_2(0) = 0, x_1(10) = 25$, and $x_2(10) = 0$.

Solution Here $A = \begin{bmatrix} 0 & 1 \\ 0 & 0 \end{bmatrix}$ and $B = \begin{bmatrix} 0 \\ 1 \end{bmatrix}$. The system is controllable since $\text{rank} \begin{bmatrix} B & AB \end{bmatrix} = \text{rank} \begin{bmatrix} 0 & 1 \\ 1 & 0 \end{bmatrix} = 2$. We now determine a control function $f(t)$ using the formula (5.19). Thus,

$$f(t) = B'\phi'(t_0, t)W^{-1}\left[\phi(t_0, \tau)x_\tau - x_0\right],$$

where $t_0 = 0, \tau = 10, x_0 = \begin{bmatrix} 0 \\ 0 \end{bmatrix}, x_\tau = \begin{bmatrix} 25 \\ 0 \end{bmatrix}$, and $\phi(t_0, t) = e^{A(t_0 - t)}$. Now, $e^{A(t_0-t)} = e^{-At} = \begin{bmatrix} 1 & -t \\ 0 & 1 \end{bmatrix}$. Therefore,

$$f(t) = \begin{bmatrix} 0 & 1 \end{bmatrix} \begin{bmatrix} 1 & 0 \\ -t & 1 \end{bmatrix} W^{-1} \begin{bmatrix} 1 & -10 \\ 0 & 1 \end{bmatrix} \begin{bmatrix} 25 \\ 0 \end{bmatrix}, \tag{5.30}$$

where

$$\begin{aligned} W &= \int_0^{10} e^{-As}BB'e^{-A's}ds \\ &= \int_0^{10} \begin{bmatrix} 1 & -s \\ 0 & 1 \end{bmatrix} \begin{bmatrix} 0 \\ 1 \end{bmatrix} \begin{bmatrix} 0 & 1 \end{bmatrix} \begin{bmatrix} 1 & 0 \\ -s & 1 \end{bmatrix} ds \\ &= \int_0^{10} \begin{bmatrix} -s \\ 1 \end{bmatrix} \begin{bmatrix} -s & 1 \end{bmatrix} ds \end{aligned}$$

$$= \int_0^{10} \begin{bmatrix} s^2 & -s \\ -s & 1 \end{bmatrix} ds = \frac{1}{3} \begin{bmatrix} 1000 & -150 \\ -150 & 30 \end{bmatrix}.$$

Consequently,

$$W^{-1} = \frac{1}{250} \begin{bmatrix} 3 & 15 \\ 15 & 100 \end{bmatrix}.$$

Substituting into (5.30), we obtain

$$f(t) = \begin{bmatrix} -t & 1 \end{bmatrix} W^{-1} \begin{bmatrix} 25 \\ 0 \end{bmatrix} = -\frac{3t}{10} + \frac{3}{2}, \quad 0 \le t \le 10,$$

which is the desired control function.

The corresponding trajectory $x(t) = \begin{bmatrix} x_1(t) \\ x_2(t) \end{bmatrix}$ on $[0, 10]$ is given (see Example 5.2.1) by

$$x(t) = \begin{bmatrix} 1 & t \\ 0 & 1 \end{bmatrix} \int_0^t \begin{bmatrix} 1 & -s \\ 0 & 1 \end{bmatrix} \begin{bmatrix} 0 \\ 1 \end{bmatrix} f(s)ds = \frac{3}{2} \begin{bmatrix} \frac{t^2}{2} - \frac{t^3}{30} \\ t - \frac{t^2}{10} \end{bmatrix},$$

which satisfies the initial and final conditions.

Example 5.3.9 (Inverted pendulum) Using the rank condition, it can be shown that the control system given in equation (1.27) that represents the motion of an inverted pendulum is controllable.

Controllable Subspace

If a system is not controllable, i.e., if the controllability Gramian matrix W is singular, it does not mean that we cannot find a control u for any pair x_0 and x_1. In what follows, we shall show that one can always find a control for some x_0 and x_1. For the system $\dot{x} = Ax + Bu, t \in J$, the solution is

$$x(t) = \phi(t, t_0)x(t_0) + \int_{t_0}^{t} \phi(t, s)B(s)u(s)ds.$$

At $t = \tau$, we have

$$\phi(t_0, \tau)x(\tau) - x(t_0) = \int_{t_0}^{\tau} \phi(t_0, s)B(s)u(s)ds.$$

Let the LHS be denoted by P. Then, from (5.20)

$$P = \int_{t_0}^{\tau} \phi(t_0, s)B(s)u(s)ds = Mu. \tag{5.31}$$

Setting

$$u(s) = B'(s)\phi'(t_0, s)\alpha = M^*\alpha, \tag{5.32}$$

the equation (5.31) becomes

$$P = W\alpha = MM^*\alpha. \tag{5.33}$$

Equation (5.33) is solvable for α if and only if $P \in \text{Range}(W)$. So, if $x(t_0)$ and $x(\tau)$ are chosen such that the equation (5.33) is solvable for α, then the required control is given by (5.32).

For the special case when A and B are constant matrices, using Lemma 3.3.1 and equation (5.28), it can be shown that the Range(W) and Range(K) are the same, implying that equation (5.33) is solvable if $P \in \text{Range}(K) =$ the span of the columns of K. Thus, if P is a linear combination of the columns of K then the control $u = M^*\alpha$ steers the system from x_0 to x_τ. In Section 5.9, for a system which is not controllable, the concept of controllable subsystem is discussed.

***Example* 5.3.10** For the given matrices $A = \begin{bmatrix} 1 & 3 \\ 2 & 0 \end{bmatrix}$ and $B = \begin{bmatrix} 1 \\ -1 \end{bmatrix}$, can we find a control u that steers the system $\dot{x} = Ax + Bu$ from $x(0) = \begin{bmatrix} 1 \\ 2 \end{bmatrix}$ to $x(5) = \begin{bmatrix} 0 \\ 1 \end{bmatrix}$?

Solution The state transition matrix of the system is

$$\phi(t, t_0) = e^{At} = \frac{3}{5}\begin{bmatrix} \frac{2}{3}e^{-2t} + e^{3t} & -e^{-2t} + e^{3t} \\ -\frac{2}{3}e^{-2t} + \frac{2}{3}e^{3t} & e^{-2t} + \frac{2}{3}e^{3t} \end{bmatrix} \text{ and } K = \begin{bmatrix} 1 & -2 \\ -1 & 2 \end{bmatrix}$$

Since rank of K is 1, the given system is not controllable.

From (5.33), we know that

$$P = \phi(t_0, \tau)x_1 - x_0 = e^{-5A}\begin{bmatrix} 0 \\ 1 \end{bmatrix} - \begin{bmatrix} 1 \\ 2 \end{bmatrix}$$

$$= \frac{3}{5}\begin{bmatrix} \frac{2}{3}e^{10}+e^{-15} & -e^{10}+e^{-15} \\ -\frac{2}{3}e^{10}+\frac{2}{3}e^{-15} & e^{10}+\frac{2}{3}e^{-15} \end{bmatrix}\begin{bmatrix} 0 \\ 1 \end{bmatrix} - \begin{bmatrix} 1 \\ 2 \end{bmatrix}$$

$$= \frac{3}{5}\begin{bmatrix} -e^{10}+e^{-15} \\ e^{10}+\frac{2}{3}e^{-15} \end{bmatrix} - \begin{bmatrix} 1 \\ 2 \end{bmatrix}.$$

Note that, $P \notin \text{Range}(K) =$ the span of $\begin{bmatrix} 1 \\ -1 \end{bmatrix}$. Hence, we cannot find a control u.

***Example* 5.3.11** Let A and B be as in the above example. Find the set of all possible final states $x(5) = \begin{bmatrix} a \\ b \end{bmatrix}$ such that $x(0) = \begin{bmatrix} 1 \\ 0 \end{bmatrix}$ is the initial state.

Solution We know that

$$P = e^{-5A}\begin{bmatrix} a \\ b \end{bmatrix} - \begin{bmatrix} 1 \\ 0 \end{bmatrix} = \frac{3}{5}\begin{bmatrix} e^{10}\left(\frac{2}{3}a-b\right)+e^{-15}(a+b) \\ e^{10}\left(-\frac{2}{3}a+b\right)+e^{-15}\left(\frac{2}{3}(a+b)\right) \end{bmatrix} - \begin{bmatrix} 1 \\ 0 \end{bmatrix}$$

$$= \begin{bmatrix} \frac{3}{5}e^{10}\left(\frac{2}{3}a-b\right)+\frac{3}{5}e^{-15}(a+b)-1 \\ \frac{3}{5}e^{10}\left(-\frac{2}{3}a+b\right)+\frac{2}{5}e^{-15}(a+b) \end{bmatrix} = \begin{bmatrix} \alpha \\ \beta \end{bmatrix} \text{ (say).}$$

To find a control u, P must be in $\text{Range}(K) =$ the span of $\begin{bmatrix} 1 \\ -1 \end{bmatrix}$. Hence, $\alpha = -\beta$, i.e.,

$$\frac{3}{5}e^{10}\left(\frac{2}{3}a-b\right)+\frac{3}{5}e^{-15}(a+b)-1 = -\frac{3}{5}e^{10}\left(-\frac{2}{3}a+b\right)-\frac{2}{5}e^{-15}(a+b)$$

$$\implies ae^{-15} = -be^{-15}-1$$

$$\implies b = -e^{15}-a.$$

Therefore, the possible final positions are $x(5) = \begin{bmatrix} a \\ -e^{15}-a \end{bmatrix}$ for any constant a.

Optimal Control

Recall that if the system

$$\dot{x} = Ax+Bu, \quad t \in J, \tag{5.34}$$

is controllable, then the control

$$u(t) = B'(t)\phi'(t_0,t)W^{-1}\left[\phi(t_0,\tau)x_1 - x_0\right] \tag{5.35}$$

steers the solution x from $x(t_0) = x_0$ to $x(\tau) = x_1$. We can now show that the control given in (5.35) is optimal in the following sense. Let u_1 be any other control function that steers the corresponding solution z_1 from $z_1(t_0) = x_0$ to $z_1(\tau) = x_1$. Then

$$\int_{t_0}^{\tau} \|u(t)\|^2 dt \leq \int_{t_0}^{\tau} \|u_1(t)\|^2 dt. \tag{5.36}$$

The norm inside the integral is the Euclidean norm in $\mathbb{R}^m$. Here, $u(t)$ and $u_1(t)$ are in $\mathbb{R}^m$ for each t, and the functions u and u_1 belong to the space $Z = L_2(J, \mathbb{R}^m)$ with the norm

$$\|u\|_Z = \left(\int_{t_0}^{\tau} \|u(t)\|^2 dt\right)^{\frac{1}{2}}. \tag{5.37}$$

The inner product in Z is defined by

$$\langle u, u_1\rangle_Z = \int_{t_0}^{\tau} \langle u(t), u_1(t)\rangle dt, \tag{5.38}$$

where $\langle u(t), u_1(t)\rangle$ denotes the dot product of the vectors $u(t)$ and $u_1(t)$ in $\mathbb{R}^m$. Using (5.37), the inequality (5.36) can be written as

$$\|u\|_Z^2 \leq \|u_1\|_Z^2. \tag{5.39}$$

Now let us prove the inequality (5.39). We know that

$$\begin{aligned} \|u - u_1\|_Z^2 \geq 0 \implies 0 &\leq \langle u - u_1, u - u_1\rangle_Z \\ &= \langle u, u\rangle_Z - \langle u, u_1\rangle_Z - \langle u_1, u\rangle_Z + \langle u_1, u_1\rangle_Z. \end{aligned} \tag{5.40}$$

As u and u_1 are vectors with real components, $\langle u, u_1\rangle_Z = \langle u_1, u\rangle_Z$. From (5.40), we get

$$0 \leq \|u\|_Z^2 + \|u_1\|_Z^2 - 2\langle u_1, u\rangle_Z. \tag{5.41}$$

Since the solutions x and z_1 of (5.34) corresponding to u and u_1 have the same initial and final conditions, we have

$$x(\tau) = x_1 = \phi(\tau, t_0)\left[x_0 + \int_{t_0}^{\tau} \phi(t_0, s)Bu(s)ds\right]$$

and

$$z_1(\tau) = x_1 = \phi(\tau, t_0)\left[x_0 + \int_{t_0}^{\tau} \phi(t_0, s)Bu_1(s)ds\right].$$

From the above two equations, we obtain

$$\int_{t_0}^{\tau} \phi(t_0,s)Bu(s)ds = \int_{t_0}^{\tau} \phi(t_0,s)Bu_1(s)ds,$$

which implies (see (5.20))

$$Mu = Mu_1.$$

Hence, $u - u_1 \in \text{Null}(M)$. Since $\text{Null}(M)$ is orthogonal to $\text{Range}(M^*)$, $u - u_1$ is orthogonal to $\text{Range}(M^*)$. We know from (5.25) and (5.35) that

$$u = M^* \left(MM^*\right)^{-1} \left(\phi(t_0,\tau)x_1 - x_0\right),$$

which belongs to $\text{Range}(M^*)$. Hence, u is orthogonal to $(u - u_1)$, implying $\langle u, u - u_1\rangle_Z = 0$, i.e., $\langle u,u\rangle_Z - \langle u,u_1\rangle_Z = 0$. Thus,

$$\langle u,u_1\rangle_Z = \|u\|_Z^2. \tag{5.42}$$

Using (5.42) in the inequality (5.41), we get

$$0 \le \|u\|_Z^2 + \|u_1\|_Z^2 - 2\|u\|_Z^2 \implies \|u\|_Z^2 \le \|u_1\|_Z^2,$$

which is the same as (5.36). Thus, we obtained the following:

Theorem 5.3.12 For the system (5.34), the control function $u(t)$ given in (5.35) has the minimum norm. In other words, if $u_1(t)$ is any other control that steers the system from x_0 at t_0 to x_1 at τ, then

$$\int_{t_0}^{\tau} |u(t)|^2 dt \le \int_{t_0}^{\tau} |u_1(t)|^2 dt.$$

***Example* 5.3.13** Find the optimal control u that steers the solution of the system $\dot{x} = Ax + Bu$ from $x(0) = x_0 = \begin{bmatrix} 1 \\ 1 \end{bmatrix}$ to $x(2) = x_1 = \begin{bmatrix} 3 \\ 0 \end{bmatrix}$ such that $\int_0^2 \|u(t)\|^2 dt$ is minimum among all such control functions, given that $A = \begin{bmatrix} 2 & 1 \\ 0 & 2 \end{bmatrix}$ and $B = \begin{bmatrix} 0 \\ 1 \end{bmatrix}$.

Solution Since $\text{rank}\begin{bmatrix} B & AB \end{bmatrix} = 2$, the given system is controllable. Hence, from (5.35), the optimal control u is given by

$$u(t) = B'(t)\phi'(t_0,t)W^{-1}\left[\phi(t_0,\tau)x_1 - x_0\right]. \tag{5.43}$$

Here,

$$t_0 = 0, \tau = 2, \phi(t,s) = e^{A(t-s)}, \text{ and } W = \int_0^2 \phi(0,s)BB'\phi'(0,s)ds.$$

Since $\phi(t,s) = eA(t-s)$,

$$\phi(t,s) = \begin{bmatrix} e^{2(t-s)} & (t-s)e^{2(t-s)} \\ 0 & e^{2(t-s)} \end{bmatrix}$$

and

$$W = \int_0^2 \begin{bmatrix} e^{-2s} & -se^{-2s} \\ 0 & e^{-2s} \end{bmatrix} \begin{bmatrix} 0 \\ 1 \end{bmatrix} \begin{bmatrix} 0 & 1 \end{bmatrix} \begin{bmatrix} e^{-2s} & 0 \\ -se^{-2s} & e^{-2s} \end{bmatrix} ds$$
$$= \begin{bmatrix} \frac{1}{4} - \frac{13}{4}e^{-4} & -\frac{1}{4} + \frac{5}{4}e^{-4} \\ -\frac{1}{4} + \frac{5}{4}e^{-4} & \frac{1}{2} - \frac{1}{2}e^{-4} \end{bmatrix}.$$

The optimal control u can now be obtained by substituting the corresponding values into (5.43).

Exercise 5.3.1 Let $A = \begin{bmatrix} 1 & 3 \\ 3 & 1 \end{bmatrix}$ and $B = \begin{bmatrix} 1 \\ 0 \end{bmatrix}$. Show that the system $\dot{x} = Ax + Bu$ is controllable. Using the Gramian matrix, find a control that steers the system from $x(0) = \begin{bmatrix} 1 \\ 2 \end{bmatrix}$ to $x(5) = \begin{bmatrix} 2 \\ 1 \end{bmatrix}$.

Exercise 5.3.2 For the system $\dot{x} = Ax + Bu$, where $A = \begin{bmatrix} 3 & 0 \\ 0 & 2 \end{bmatrix}$ and $B = \begin{bmatrix} 0 \\ 1 \end{bmatrix}$, with the initial condition $x(0) = \begin{bmatrix} 1 \\ 1 \end{bmatrix}$, find all possible final conditions $x(1) =$

$\begin{bmatrix} a \\ b \end{bmatrix}$ such that a control function $u(t)$ can be obtained using the controllability Gramian matrix.

5.4 Companion Form and Feedback Control

A linear control system

$$\dot{z} = Cz + dv \tag{5.44}$$

is said to be in **companion form** if the matrices C and d have the following form:

$$C = \begin{bmatrix} 0 & 1 & 0 & \cdots & 0 \\ 0 & 0 & 1 & \cdots & 0 \\ \vdots & \vdots & & \ddots & \\ 0 & 0 & & \cdots & 1 \\ -\gamma_n & -\gamma_{n-1} & & \cdots & -\gamma_1 \end{bmatrix}_{n\times n},$$

where $\gamma_i : i = 1, 2, \ldots, n$ are real numbers, and

$$d = \begin{bmatrix} 0 \\ 0 \\ \vdots \\ 0 \\ 1 \end{bmatrix}_{n\times 1}.$$

It can easily be verified that rank $\begin{bmatrix} d & Cd & C^2d & \cdots & C^{n-1}d \end{bmatrix} = n$. Hence, any system in companion form (5.44) is always controllable. Note that the characteristic equation of the matrix C is $\lambda^n + \gamma_1\lambda^{n-1} + \gamma_2\lambda^{n-2} + \ldots. + \gamma_n = 0$.

***Example* 5.4.1** Consider the nth order differential equation

$$\frac{d^n y}{dt^n} + r_1 \frac{d^{n-1} y}{dt^{n-1}} + \cdots + r_n y = v(t) : r_i \in R \text{ for } i := 1, 2, \ldots, n. \tag{5.45}$$

Let

$$z_1 = y, z_2 = \frac{dy}{dt}, \ldots, z_n = \frac{d^{n-1}y}{dt^{n-1}}.$$

Then

$$\dot{z} = \begin{bmatrix} \dot{z}_1 \\ \dot{z}_2 \\ \vdots \\ \dot{z}_n \end{bmatrix} = \begin{bmatrix} 0 & 1 & 0 & \cdots & 0 \\ 0 & 0 & 1 & \cdots & 0 \\ \vdots & \vdots & & \ddots & \\ 0 & 0 & & \cdots & 1 \\ -\gamma_n & -\gamma_{n-1} & & \cdots & -\gamma_1 \end{bmatrix} \begin{bmatrix} z_1 \\ z_2 \\ \vdots \\ z_n \end{bmatrix} + \begin{bmatrix} 0 \\ 0 \\ \vdots \\ 0 \\ 1 \end{bmatrix} v$$

is in the companion form (5.44).

Now one can ask whether any control system

$$\dot{x} = Ax + bu \tag{5.46}$$

can be converted into a companion form (5.44), where A and b are matrices of order $n \times n$ and $n \times 1$, respectively. The following theorem shows that, in general, this is not true.

Theorem 5.4.2 A control system of the form (5.46) can be converted into its companion form (5.44) uniquely if and only if it is controllable. In that case, there exists a nonsingular matrix T such that

$$C = TAT^{-1} \tag{5.47}$$

and

$$d = Tb. \tag{5.48}$$

Proof. **Sufficiency.** Let the system be controllable. Then by Theorem 5.3.4,

$$\text{rank}(K) = n,$$

where $K = \begin{bmatrix} b & Ab & A^2b & \cdots & A^{n-1}b \end{bmatrix}_{n\times n}$. Let $\alpha = \begin{bmatrix} \alpha_1 \\ \alpha_2 \\ \vdots \\ \alpha_n \end{bmatrix} \in \mathbb{R}^n$ be a vector such that

$$K'\alpha = d = \begin{bmatrix} 0 \\ 0 \\ \vdots \\ 1 \end{bmatrix}. \tag{5.49}$$

As $\text{rank}(K) = n$, $(K')^{-1}$ exists. Hence

$$\alpha = (K')^{-1}d, \tag{5.50}$$

and $\alpha' = \begin{bmatrix} 0 & 0 & \cdots & 1 \end{bmatrix} K^{-1}$.

Now let T be an $n \times n$ matrix defined by

$$T = \begin{bmatrix} \alpha' & \rightarrow \\ \alpha' A & \rightarrow \\ \vdots & \\ \alpha' A^{n-1} & \rightarrow \end{bmatrix}. \tag{5.51}$$

If T^{-1} exists, then let

$$T^{-1} = \begin{bmatrix} S_1 & S_2 & \cdots & S_n \\ \downarrow & \downarrow & & \downarrow \end{bmatrix}_{n\times n}$$

with $S_i : i = 1, 2, \ldots, n$ as columns. Now

$$TT^{-1} = I \Longrightarrow \begin{bmatrix} \alpha' & \rightarrow \\ \alpha' A & \rightarrow \\ \vdots & \\ \alpha' A^{n-1} & \rightarrow \end{bmatrix} \begin{bmatrix} S_1 & S_2 & \cdots & S_n \\ \downarrow & \downarrow & & \downarrow \end{bmatrix} = \begin{bmatrix} 1 & 0 & \cdots & 0 \\ 0 & 1 & \cdots & 0 \\ & & \ddots & \\ 0 & 0 & \cdots & 1 \end{bmatrix}.$$

Therefore,

$$\alpha' A^{i-1} S_j = \begin{cases} 1, & \text{if } i = j, \\ 0, & \text{if } i \neq j \end{cases} \tag{5.52}$$

for $i, j = 1, 2, \ldots, n$. Moreover,

$$TAT^{-1} = \begin{bmatrix} \alpha' A & \rightarrow \\ \alpha' A^2 & \rightarrow \\ \vdots & \\ \alpha' A^n & \rightarrow \end{bmatrix} \begin{bmatrix} S_1 & S_2 & \cdots & S_n \\ \downarrow & \downarrow & & \downarrow \end{bmatrix}$$

$$= \begin{bmatrix} \alpha' A S_1 & \alpha' A S_2 & \cdots & \alpha' A S_n \\ \alpha' A^2 S_1 & \alpha' A^2 S_2 & \cdots & \alpha' A^2 S_n \\ \vdots & & & \\ \alpha' A^{n-1} S_1 & \alpha' A^{n-1} S_2 & \cdots & \alpha' A^{n-1} S_n \\ \alpha' A^n S_1 & \alpha' A^n S_2 & \cdots & \alpha' A^n S_n \end{bmatrix}.$$

Now by using (5.52), we get

$$TAT^{-1} = \begin{bmatrix} 0 & 1 & 0 & \cdots & 0 \\ 0 & 0 & 1 & \cdots & 0 \\ \vdots & & & & \\ 0 & 0 & 0 & \cdots & 1 \\ \alpha' A^n S_1 & \alpha' A^n S_2 & \alpha' A^n S_3 & \cdots & \alpha' A^n S_n \end{bmatrix}, \tag{5.53}$$

which is in the form of C. Furthermore, using (5.50)

$$Tb = \begin{bmatrix} \alpha' & \rightarrow \\ \alpha' A & \rightarrow \\ \vdots & \\ \alpha' A^{n-1} & \rightarrow \end{bmatrix} \begin{bmatrix} b \\ \downarrow \end{bmatrix} = \begin{bmatrix} \alpha' b & \rightarrow \\ \alpha' A b & \rightarrow \\ \vdots & \\ \alpha' A^{n-1} b & \rightarrow \end{bmatrix} = \begin{bmatrix} d' K^{-1} b & \rightarrow \\ d' K^{-1} A b & \rightarrow \\ \vdots & \\ d' K^{-1} A^{n-1} b & \rightarrow \end{bmatrix} = \begin{bmatrix} 0 \\ 0 \\ \vdots \\ 1 \end{bmatrix}, \tag{5.54}$$

since $K^{-1}K = I$

$$K^{-1}(A^{i-1}b) = e_i \quad \text{for } i = 1,2,\ldots,n-1, \tag{5.55}$$

where e_i is the ith column of the identity matrix of order n.

Finally, it can be easily seen that the matrix T given in (5.51) is nonsingular by showing that its rows are linearly independent. For this, let $\beta_i : i = 1,2,\ldots,n$ be scalars such that

$$\beta_1\alpha' + \beta_2\alpha'A + \cdots + \beta_n\alpha'A^{n-1} = 0. \tag{5.56}$$

By multiplying both the sides by b on the right, using $\alpha' = \begin{bmatrix} 0 & 0 & \cdots 1 \end{bmatrix} K^{-1}$, and applying (5.55), we get

$$\beta_n = 0.$$

Now the equation (5.56) becomes

$$\beta_1\alpha' + \beta_2\alpha'A + \cdots + \beta_{n-1}\alpha'A^{n-2} = 0.$$

By multiplying both the sides by Ab on the right, and applying (5.55), we obtain

$$\beta_{n-1} = 0.$$

Continuing this process, it can be seen that all $\beta_i = 0$, and hence T is nonsingular. Thus it has been proved that if the system (5.46) is controllable, it can be converted into its unique companion form (5.44) using a nonsingular matrix T given by (5.51).

Necessity. Assume that the system (5.46) can be converted into the companion form $\dot{z} = Cz + dv$, where $C = TAT^{-1}$ and $d = Tb$, with T being a nonsingular matrix. We know that rank $\begin{bmatrix} d & Cd & C^2d & \cdots & C^{n-1}d \end{bmatrix} = n$. Hence,

$$\text{rank}\begin{bmatrix} Tb & TAT^{-1}Tb & TA^2T^{-1}Tb & \cdots & TA^{n-1}T^{-1}Tb \end{bmatrix} = n$$

$$\Longrightarrow \text{rank}\left(T\begin{bmatrix} b & Ab & A^2b & \cdots & A^{n-1}b \end{bmatrix}\right) = n.$$

Since T is nonsingular, rank $\begin{bmatrix} b & Ab & A^2b & \cdots & A^{n-1}b \end{bmatrix} = n$. Hence, the system is controllable. □

To convert the equation $\dot{x} = Ax + bu$ into the companion form, multiply both sides by T and let $y = Tx$. Then

$$T\dot{x} = TAx + Tbu \Longrightarrow \dot{y} = TAT^{-1}y + Tbu \Longrightarrow \dot{y} = Cy + du.$$

***Example* 5.4.3** Convert the system $\dot{x} = Ax + bu$, where $A = \begin{bmatrix} 2 & 1 \\ 1 & 2 \end{bmatrix}$ and $b = \begin{bmatrix} 1 \\ 0 \end{bmatrix}$, into its companion form.

Solution Let $K = \begin{bmatrix} b & Ab \end{bmatrix} = \begin{bmatrix} 1 & 2 \\ 0 & 1 \end{bmatrix}$. Since $\text{rank}(K) = 2$, the given system is controllable by Theorem 5.3.4. From Theorem 5.4.2, we have $TAT^{-1} = C$ and $Tb = d = \begin{bmatrix} 0 \\ 1 \end{bmatrix}$, where $T = \begin{bmatrix} \alpha' \\ \alpha' A \end{bmatrix}$ and $\alpha' = \begin{bmatrix} 0 & 1 \end{bmatrix} K^{-1}$ (from (5.49)). Hence $\alpha' = \begin{bmatrix} 0 & 1 \end{bmatrix} \begin{bmatrix} 1 & -2 \\ 0 & 1 \end{bmatrix} = \begin{bmatrix} 0 & 1 \end{bmatrix}$, $\alpha' A = \begin{bmatrix} 1 & 2 \end{bmatrix}$, and $T = \begin{bmatrix} 0 & 1 \\ 1 & 2 \end{bmatrix}$. From (5.47), we obtain $C = \begin{bmatrix} 0 & 1 \\ -3 & 4 \end{bmatrix}$. Finally, the companion form can be obtained by substituting C and d in the equation

$$\dot{y} = Cy + du, \quad \text{where } y = Tx \text{ and } d = \begin{bmatrix} 0 \\ 1 \end{bmatrix}.$$

5.5 Feedback Control

Consider the autonomous linear control system:

$$\dot{x} = Ax + bu, \tag{5.57}$$

where A and b are $n \times n$ and $n \times 1$ matrices, respectively. The control u is said to be a **linear feedback control** if it can be expressed as a linear function of the state vector x, i.e.,

$$u(t) = \kappa x(t), \tag{5.58}$$

where κ is a $1 \times n$ matrix. Substituting $u(t)$ from (5.58) into (5.57), we get

$$\dot{x} = (A + b\kappa)x \tag{5.59}$$

In some practical problems, it is required to find a feedback control u such that the matrix $A + b\kappa$ has preassigned eigenvalues so that the solution x of (5.59) is in a desired form. In that case, the following question arises:

Given matrices A and b, is it possible to find a matrix κ such that $A + b\kappa$ has a preassigned set $\{\omega_1, \omega_2, \ldots, \omega_n\}$ as eigenvalues?

The answer is generally negative. Under suitable assumptions on A and b, we have the following:

Theorem 5.5.1 Let A and b be $n \times n$ and $n \times 1$ matrices, respectively, such that $\operatorname{rank}\begin{bmatrix} b & Ab & A^2b & \cdots & A^{n-1}b \end{bmatrix} = n$, and let $S = \{\omega_1, \omega_2, \ldots, \omega_n\}$ be a set of scalars (real or complex). Then, there exists a $1 \times n$ matrix κ such that $A + b\kappa$ has S as its set of eigenvalues.

Proof. From Theorem 5.4.2, we can find a nonsingular matrix T such that $TAT^{-1} = C$ and $Tb = d$, where

$$C = \begin{bmatrix} 0 & 1 & 0 & \cdots & 0 \\ 0 & 0 & 1 & \cdots & 0 \\ \vdots & & & \ddots & \\ 0 & 0 & 0 & \cdots & 1 \\ -\alpha_n & -\alpha_{n-1} & -\alpha_{n-2} & \cdots & -\alpha_1 \end{bmatrix} \text{ and } d = \begin{bmatrix} 0 \\ 0 \\ \vdots \\ 1 \end{bmatrix}. \tag{5.60}$$

Note that $\alpha_i : i = 1, 2, \ldots, n$ are the coefficients of the characteristic polynomial of C (see Exercise 7). As A is similar to C, the characteristic equation of A is

$$\lambda^n + \alpha_1 \lambda^{n-1} + \alpha_2 \lambda^{n-2} + \cdots + \alpha_n = 0. \tag{5.61}$$

We need to find $\kappa = \begin{bmatrix} \kappa_n & \kappa_{n-1} & \cdots & \kappa_1 \end{bmatrix}$ such that $A + b\kappa$ has eigenvalues $\omega_i : i = 1, 2, \ldots, n$. Hence, the characteristic equation of $A + b\kappa$ must be $(\lambda - \omega_1)(\lambda - \omega_2) \cdots (\lambda - \omega_n) = 0$, which can be rewritten as

$$\lambda^n + \beta_1 \lambda^{n-1} + \beta_2 \lambda^{n-2} + \cdots + \beta_n = 0. \tag{5.62}$$

Therefore, κ is such that $A + b\kappa$ is similar to

$$D = \begin{bmatrix} 0 & 1 & 0 & \cdots & 0 \\ 0 & 0 & 1 & \cdots & 0 \\ \cdots & \cdots & \cdots & \cdots & \cdots \\ 0 & 0 & 0 & \cdots & 1 \\ -\beta_n & -\beta_{n-1} & -\beta_{n-2} & \cdots & -\beta_1 \end{bmatrix}. \tag{5.63}$$

Now,

$$T(A+b\kappa)T^{-1} = TAT^{-1} + Tb\kappa T^{-1} = C + d\kappa T^{-1} \tag{5.64}$$

$$\text{Let } \kappa T^{-1} = h = \begin{bmatrix} h_n & h_{n-1} & \cdots & h_1 \end{bmatrix}. \tag{5.65}$$

$$\text{Then } Tb\kappa T^{-1} = d\kappa T^{-1} = \begin{bmatrix} 0 \\ 0 \\ \vdots \\ 1 \end{bmatrix} h = \begin{bmatrix} 0 & 0 & \cdots & 0 \\ 0 & 0 & \cdots & 0 \\ \vdots & & & \\ h_n & h_{n-1} & \cdots & h_1 \end{bmatrix} = H \ (say),$$

Now, the RHS of (5.64) can be written as $C+H$ which is in the companion form D for the matrix $A+B\kappa$. By comparing the elements of $C+H$ and D we obtain

$$-\alpha_i + h_i = -\beta_i \implies h_i = \alpha_i - \beta_i, \quad i = 1, 2, \ldots, n.$$

Now since h and T are known κ can be obtained from (5.65) as $\kappa = hT$. □

***Example* 5.5.2** For $A = \begin{bmatrix} 2 & 1 \\ 1 & 2 \end{bmatrix}$ and $b = \begin{bmatrix} 2 \\ 3 \end{bmatrix}$, find $\kappa = \begin{bmatrix} \kappa_2 & \kappa_1 \end{bmatrix}$ such that $A + b\kappa$ has $\{1,4\}$ as its eigenvalues.

Solution Since rank $\begin{bmatrix} b & Ab \end{bmatrix} = 2$, we can find T such that $TAT^{-1} = C$ and $Tb = d$, where $C = \begin{bmatrix} 0 & 1 \\ -3 & 4 \end{bmatrix}$ and $d = \begin{bmatrix} 0 \\ 1 \end{bmatrix}$. (Note that $\lambda^2 - 4\lambda + 3$ is the characteristic polynomial of both A and C.) Compared with (5.61), we have $\alpha_1 = -4$ and $\alpha_2 = 3$. Since 1 and 4 are the desired eigenvalues of $A + b\kappa$, the characteristic equation of $A + b\kappa$ is $(\lambda - 1)(\lambda - 4) = 0$, i.e., $\lambda^2 - 5\lambda + 4 = 0$. So $\beta_1 = -5$ and $\beta_2 = 4$. From (5.65), we find $h_1 = 1$ and $h_2 = -1$. Using (5.49)–(5.51), we obtain $T = \begin{bmatrix} \frac{3}{5} & -\frac{2}{5} \\ \frac{4}{5} & -\frac{1}{5} \end{bmatrix}$. Hence,

$$\kappa = \begin{bmatrix} -1 & 1 \end{bmatrix} T = \begin{bmatrix} \frac{1}{5} & \frac{1}{5} \end{bmatrix}.$$

General Case

Theorem 5.5.1 can be extended to the general case when the $n \times 1$ matrix b is replaced by an $n \times m$ matrix B such that

$$K = \begin{bmatrix} B & AB & \cdots & A^{n-1}B \end{bmatrix}_{n\times mn} \quad \text{and } \operatorname{rank}(K) = n.$$

Let $b_1, b_2, \ldots, b_m$ be columns of B. We choose n independent columns of K as per the following procedure:

The columns $b_1, Ab_1, \ldots, A^{\gamma_1-1}b_1$ $(\gamma_1 \leq n)$, respectively, are the first columns of each block in K such that they are linearly independent and $A^{\gamma_1}b_1$ is linearly dependent on them. If $\gamma_1 < n$, then let $b_2, Ab_2, \ldots, A^{\gamma_2-1}b_2$ along with $b_1, Ab_1, \ldots, A^{\gamma_1-1}b_1$ be linearly independent and $A^{\gamma_2}b_2$ be dependent on them.

Proceeding this way, we can find a nonsingular $n \times n$ matrix G as given below

$$G = \left[\begin{array}{cccc|cccc|cccc|c} b_1 & Ab_1 & \cdots & A^{\gamma_1-1}b_1 & b_2 & Ab_2 & \cdots & A^{\gamma_2-1}b_2 & b_3 & Ab_3 & \cdots & A^{\gamma_3-1}b_3 & \cdots \\ \downarrow & \downarrow & & \downarrow & \downarrow & \downarrow & & \downarrow & \downarrow & \downarrow & & \downarrow & \cdots \end{array}\right]_{n\times n} \tag{5.66}$$

Let N be the $m \times n$ matrix defined as

$$N = \begin{bmatrix} 0 & 0 & \cdots & 0 & 0 & \cdots & 0 & \cdots & \cdots \\ \downarrow & \downarrow & & 1 & \downarrow & & 0 & & \\ & & & 0 & & & 1 & & \\ & & & \downarrow & & & 0 & & \\ & & & & & & \downarrow & & \end{bmatrix}_{m\times n}, \tag{5.67}$$

where the nonzero columns are at $\gamma_1, \gamma_1+\gamma_2, \gamma_1+\gamma_2+\gamma_3, \ldots$ places. Now, let

$$\kappa_1 = NG^{-1}, \tag{5.68}$$

and consider the matrices $A + B\kappa_1$ and b_1. If the matrices $A + B\kappa_1$ and b_1 (in place of A and b in Theorem 5.5.1) satisfy the rank condition of Theorem 5.5.1, then for a given set $S = \{\omega_1, \omega_2, \ldots, \omega_n\}$ of n numbers, we can find a $1 \times n$ matrix κ_2 such that $(A + B\kappa_1) + b_1\kappa_2$ has S as the set of its eigenvalues. To complete the proof of the Theorem 5.5.1 for the general case, first let us prove that

$$\operatorname{rank}\begin{bmatrix} b_1 & (A+B\kappa_1)b_1 & \cdots & (A+B\kappa_1)^{n-1}b_1 \\ \downarrow & \downarrow & & \downarrow \end{bmatrix} = n. \tag{5.69}$$

Note that $G^{-1}G = I$, the identity matrix, and hence

$$G^{-1}b_1 = e_1, G^{-1}Ab_1 = e_2, \ldots, G^{-1}A^{\gamma_1 - 1}b_1 = e_{\gamma_1}, \ldots,$$

where $e_i = (0, 0, \ldots, 1, 0, \ldots, 0)'$ with 1 at the ith place. So,

$$(A + B\kappa_1)b_1 = Ab_1 + B\kappa_1 b_1 = Ab_1 + BNG^{-1}b_1 = Ab_1 + BNe_1 = Ab_1.$$

Similarly,

$$\begin{aligned}(A + B\kappa_1)^2 b_1 &= (A + B\kappa_1)(A + B\kappa_1)b_1 = (A + B\kappa_1)Ab_1 \\ &= A^2 b_1 + BNG^{-1}Ab_1 = A^2 b_1 + BNe_2 = A^2 b_1.\end{aligned}$$

Proceeding further, we get

$$(A + B\kappa_1)^{\gamma_1 - 1} b_1 = A^{\gamma_1 - 1} b_1.$$

and

$$\begin{aligned}(A + B\kappa_1)^{\gamma_1} b_1 &= (A + B\kappa_1)A^{\gamma_1 - 1}b_1 = A^{\gamma_1} b_1 + B\kappa_1 A^{\gamma_1 - 1} b_1 \\ &= A^{\gamma_1} b_1 + BNG^{-1}A^{\gamma_1 - 1} b_1 = A^{\gamma_1} b_1 + BNe_{\gamma_1} \\ &= A^{\gamma_1} b_1 + B\begin{bmatrix}0\\1\\\vdots\\0\end{bmatrix} = A^{\gamma_1} b_1 + b_2.\end{aligned}$$

Similarly, we have

$$(A + B\kappa_1)^{\gamma_1 + \gamma_2} b_1 = A^{\gamma_1 + \gamma_2} b_1 + A^{\gamma_2} b_2 + b_3$$

and so on. It can be seen that the columns of the matrix in equation (5.69) are linearly independent by comparing them with the colums of G. Hence equation (5.69) holds.

Now, from Theorem 5.5.1 (by considering $A + B\kappa_1$ and b_1 in places of A and b, respectively, in Theorem 5.5.1), we can find a $1 \times n$ matrix κ_2 such that

$$(A + B\kappa_1) + b_1\kappa_2 \tag{5.70}$$

has the pre-assigned set S as the set of eigenvalues. Note that $b_1\kappa_2$ can be written as $b_1\kappa_2 = B\kappa_3$, where κ_3 is an $m \times n$ matrix with κ_2 as the first row and the remaining rows are zeros. Hence (5.70) can be written as $A + B(\kappa_1 + \kappa_3) = A + B\kappa$, where $\kappa = \kappa_1 + \kappa_3$.

Thus, if rank$(K) = n$, then for any given arbitrary set $S = \{\omega_1, \omega_2, \ldots, \omega_n\}$ of scalars, one can find an $m \times n$ matrix κ such that $A + B\kappa$ has eigenvalues $\omega_i : i = 1, 2, \ldots, n$.

Thus we obtain the following theorem.

Theorem 5.5.3 Theorem 5.5.1 holds with B, an $n \times m$ matrix, in place of b.

Applications of this theorem can be seen in Section 5.9 on state observer and in Chapter 8 on stability.

***Example* 5.5.4** Let $A = \begin{bmatrix} 1 & 0 & 1 \\ 0 & 1 & 1 \\ 2 & -1 & 0 \end{bmatrix}$ and $B = \begin{bmatrix} 1 & 0 \\ 1 & 1 \\ 0 & 1 \end{bmatrix}$. Convert the system $\dot{x} = Ax + Bu$ into an equivalent system $\dot{z} = (A + B\kappa_1)z + b_1 v$, where b_1 is the first column of B.

Solution Note that $b_1 = \begin{bmatrix} 1 \\ 1 \\ 0 \end{bmatrix}$, $Ab_1 = \begin{bmatrix} 1 \\ 1 \\ 1 \end{bmatrix}$, $A^2 b_1 = \begin{bmatrix} 2 \\ 2 \\ 1 \end{bmatrix}$, and $b_2 = \begin{bmatrix} 0 \\ 1 \\ 1 \end{bmatrix}$. Since $A^2 b_1$ is linearly dependent on b_1 and Ab_1, we have from (5.66) and (5.67) that $G = \begin{bmatrix} b_1 & Ab_1 & | & b_2 \\ \downarrow & \downarrow & | & \downarrow \end{bmatrix} = \left[\begin{array}{cc|c} 1 & 1 & 0 \\ 1 & 1 & 1 \\ 0 & 1 & 1 \end{array}\right]$ and $N = \begin{bmatrix} 0 & 0 & 0 \\ 0 & 1 & 0 \end{bmatrix}$. Then $\kappa_1 = NG^{-1} = \begin{bmatrix} 0 & 0 & 0 \\ 1 & -1 & 1 \end{bmatrix}$ and $B\kappa_1 = \begin{bmatrix} 0 & 0 & 0 \\ 1 & -1 & 1 \\ 1 & -1 & 1 \end{bmatrix}$. Now, consider the equivalent system $\dot{z} = (A + B\kappa_1)z + b_1 v = \begin{bmatrix} 1 & 0 & 1 \\ 1 & 0 & 2 \\ 3 & 2 & 1 \end{bmatrix} z + \begin{bmatrix} 1 \\ 1 \\ 0 \end{bmatrix} v$. From (5.69), we conclude that this system is controllable.

Exercise 5.5.1 Convert the system $\dot{x} = Ax + Bu$ into an equivalent form $\dot{y} = (A + Bk)y + b_1 v$, where $A = \begin{bmatrix} 3 & 3 & 0 \\ 0 & 0 & 2 \\ 1 & 1 & 0 \end{bmatrix}$, $B = \begin{bmatrix} 0 & 0 \\ 1 & 0 \\ 0 & 2 \end{bmatrix}$, and b_1 is the first column of B.

5.6 Observability

Consider the following system:

$$\begin{cases} \dot{x} = A(t)x + B(t)u, \\ y = C(t)x \end{cases} \tag{5.71}$$

for $t \in J$, where A, B, and C are $n \times n$, $n \times m$, and $p \times n$ time-varying matrices, respectively, and $x(t) \in \mathbb{R}^n$, $u(t) \in \mathbb{R}^m$, and $y(t) \in \mathbb{R}^p$ represent the state, control and observations of the system at time t, respectively. If $x(t_0) = x_0$ is the initial condition, then we know that

$$x(t) = \phi(t, t_0)x_0 + \int_{t_0}^{t} \phi(t,s)B(s)u(s)ds \tag{5.72}$$

and

$$y(t) = C(t)\left[\phi(t, t_0)x_0 + \int_{t_0}^{t} \phi(t,s)B(s)u(s)ds\right]. \tag{5.73}$$

Note that if $u(t)$, $t \in J$, and x_0 are known, then the state $x(t)$ can be uniquely determined using (5.72). However, in many practical situations, a measurement or observation $y(t)$ of the system (5.72) and the control function u are known for $t \in J$. Using this information, it is required to find the state $x(t)$ of the system (5.71) uniquely. From (5.72), it is clear that the initial condition x_0 must be uniquely determined to ensure that $x(t)$ is unique. Thus, the observability of the system (5.71) can be defined as follows:

Definition 5.6.1 The system (5.71) is said to be **observable** on the time interval J if the initial state x_0 can be uniquely determined from the knowledge of the control u and the observation y on J.

Example **5.6.2** Consider the system

$$\begin{bmatrix} \dot{x}_1 \\ \dot{x}_2 \end{bmatrix} = \begin{bmatrix} 2 & 4 \\ 0 & -2 \end{bmatrix} \begin{bmatrix} x_1 \\ x_2 \end{bmatrix}, \quad t \in [0, 1], \tag{5.74}$$

with the initial condition $x_1(0) = a_1$ and $x_2(0) = a_2$. It can be seen that the solution is

$$\begin{cases} x_1(t) = a_1 e^{2t} + a_2\left(e^{2t} - e^{-2t}\right), \\ x_2(t) = a_2 e^{-2t} \end{cases} \tag{5.75}$$

for $t \in [0, 1]$. If the observation of the system in the interval $[0, 1]$ is

$$z(t) = x_1(t) + x_2(t),$$

then

$$z(t) = e^{2t}(a_1 + a_2). \tag{5.76}$$

Note that the above observation (5.76) can also be obtained for different initial conditions $x_0 = \begin{bmatrix} a_1 + ka_2 \\ (1-k)a_2 \end{bmatrix}$ for $k \in \mathbb{R}$. This shows that the same observation (5.76) can be obtained from different solutions $\begin{bmatrix} x_1(t) \\ x_2(t) \end{bmatrix}$ with different initial conditions. Hence, the system (5.74) is **not** observable with the observation (5.76).

Example **5.6.3** For the system (5.74), let the observation be

$$z(t) = x_1(t), \quad t \in [0, 1]. \tag{5.77}$$

Then we can show that different initial conditions result in different observations. Hence, the system (5.74) with the observation (5.77) is observable.

Necessary and Sufficient Conditions for Observability

For the system (5.71), consider the observation equation (5.73) for $t \in J$, which can be written as

$$y(t) - \int_{t_0}^{t} C(t)\phi(t,s)B(s)u(s)ds = C(t)\phi(t,t_0)x_0, \quad t \in [t_0, \tau].$$

Let the LHS be denoted by $q(t)$. Since $y(t)$ and $u(t)$ are assumed to be known for all $t \in [\tau_0, \tau], q(t)$ is a known function and the above equation can be written as

$$C(t)\phi(t,t_0)x_0 = q(t), \quad t \in [t_0, \tau]. \tag{5.78}$$

To prove the observability of (5.71), we need to show that there exists a unique $x_0 \in \mathbb{R}^n$ satisfying (5.78). The following theorem provides a necessary and sufficient condition for the observability of (5.71).

Theorem 5.6.4 The system $\dot{x} = Ax + Bu$ with observation $y(t) = Cx(t)$ for $t \in J$ is observable if and only if the observability Gramian $M[t_0, \tau]$ is nonsingular, where

$M[t_0, \tau]$ is the $n \times n$ matrix defined as

$$M[t_0, \tau] = \int_{t_0}^{\tau} \phi'(t, t_0) C'(t) C(t) \phi(t, t_0) dt. \tag{5.79}$$

Proof. **Sufficiency.** Since t_0 and τ are fixed, hereafter we denote $M[t_0, \tau]$ simply by M. Let M be nonsingular. We need to show that equation (5.78) has a unique solution $x_0 \in \mathbb{R}^n$ for a given $q(t)$. Multiplying both sides of (5.78) by $\phi'(t, t_0) C'(t)$ and integrating from t_0 to τ, we obtain

$$Mx_0 = \int_{t_0}^{\tau} \phi'(t, t_0) C'(t) q(t) dt = \beta \text{ (say)} \in \mathbb{R}^n.$$

Since β is known and M is invertible, x_0 can be uniquely determined.

Necessity. Assuming that system (5.71) is observable. We need to show that M is nonsingular. If M were singular, there would exist a nonzero vector $\alpha \in \mathbb{R}^n$ such that $M\alpha = 0$. Then

$$\begin{aligned} 0 = \alpha' M \alpha &= \int_{t_0}^{\tau} \alpha' \phi'(t, t_0) C'(t) C(t) \phi(t, t_0) \alpha dt \\ &= \int_{t_0}^{\tau} (C(t) \phi(t, t_0) \alpha)' (C(t) \phi(t, t_0) \alpha) \, dt, \end{aligned}$$

which implies that

$$\int_{t_0}^{\tau} \|C(t) \phi(t, t_0) \alpha\|^2 dt = 0.$$

Hence,

$$C(t) \phi(t, t_0) \alpha = 0 \quad \text{for all } t \in [t_0, \tau]. \tag{5.80}$$

Adding (5.79) and (5.81) we obtain $C(t)\phi(t, \tau_0)(x_0 + \alpha) = q(t)$. Thus the equation (5.78) is satisfied by more than one vector namely, x_0 and $x_0 + \alpha$ which contradicts the observability of (5.71). Hence, α cannot be nonzero, implying that M is nonsingular. □

***Example* 5.6.5** Let $A(t) = \begin{bmatrix} 1 & t \\ 0 & 1 \end{bmatrix}$ for $t \in J = [0, 2]$ and $C = [0 \quad 1]$ a constant matrix. Test the observability of the system $\dot{x} = A(t)x$ with observation $y = Cx$.

Note that $A(t)A(s) = A(s)A(t)$ for all $t, s \in J$. Hence, from Lemma 4.6.4, the state transition matrix for $A(t)$ is

$$\phi(t, t_0) = e^{\int_{t_0}^{t} A(s) ds}.$$

Now

$$\int_{t_0}^{t} A(s)ds = \begin{bmatrix} t-t_0 & \frac{(t-t_0)^2}{2} \\ 0 & t-t_0 \end{bmatrix} = (t-t_0)I + \frac{(t-t_0)^2}{2}\begin{bmatrix} 0 & 1 \\ 0 & 0 \end{bmatrix}.$$

Therefore,

$$\phi(t,t_0) = \begin{bmatrix} e^{t-t_0} & 0 \\ 0 & e^{t-t_0} \end{bmatrix}\begin{bmatrix} 1 & \frac{(t-t_0)^2}{2} \\ 0 & 1 \end{bmatrix} = e^{t-t_0}\begin{bmatrix} 1 & \frac{(t-t_0)^2}{2} \\ 0 & 1 \end{bmatrix} \quad \text{for } t \in J.$$

By computing M as given in (5.79), we get

$$M = \begin{bmatrix} 0 & 0 \\ 0 & a \end{bmatrix},$$

where $a = (e^4 - 1)/2$. Since M is singular the system is not observable.

In the above example, if $C = [1 \quad 0]$ then M is nonsingular.

Dual System

The following two systems are said to be **dual** to each other

$$\begin{cases} \dot{x} = Ax + Bu, \\ y = Cx \end{cases} \tag{5.81}$$

and

$$\begin{cases} \dot{z} = -A'z + C'v, \\ w = B'z, \end{cases} \tag{5.82}$$

where A, B, and C are $n \times n$, $n \times m$, and $p \times n$ time-varying matrices, respectively. If $\phi(t,s)$ is the state transition matrix corresponding to the matrix $A(t)$, then from Theorem 4.8.9 we know that

$$\psi(t,s) = \phi'(s,t) \tag{5.83}$$

is the state transition matrix for $-A'(t)$. Now consider the system (5.82). It is observable in the interval J if and only if the observability Gramian matrix

$$M = \int_{t_0}^{\tau} \psi'(t,t_0)B(t)B'(t)\psi(t,t_0)dt \tag{5.84}$$

is nonsingular. Using (5.83), we obtain

$$M = \int_{t_0}^{\tau} \phi(t_0,t)B(t)B'(t)\phi'(t_0,t)dt. \tag{5.85}$$

Note that M given in (5.85) is the controllability Gramian of the system (5.81). Thus, it can be concluded that the observability of (5.82) implies the controllability of (5.81) and vice versa. Similarly, the controllability Gramian of (5.82) is

$$W = \int_{t_0}^{\tau} \psi(t_0,t)C'(t)C(t)\psi'(t_0,t)dt = \int_{t_0}^{\tau} \phi'(t,t_0)C'(t)C(t)\phi(t,t_0)dt,$$

which is the observability Gramian of (5.81). Hence, the controllability of (5.82) implies the observability of (5.81) and vice versa.

The above discussion can be summarized in the following theorem:

Theorem 5.6.6 (Duality Theorem)

(1) The system (5.81) is controllable if and only if the dual system (5.82) is observable.

(2) The system (5.81) is observable if and only if the dual system (5.82) is controllable.

Corollary 5.6.7 If A, B, and C in system (5.81) are constant matrices (time-invariant), then for observability, it is necessary and sufficient that

$$\operatorname{rank} \begin{bmatrix} C \\ CA \\ CA^2 \\ \vdots \\ CA^{n-1} \end{bmatrix}_{np\times n} = n. \tag{5.86}$$

Proof. As the observability of (5.81) is equivalent to the controllability of (5.82), by applying the controllability rank condition for (5.82) (Theorem 5.3.4),

$$\operatorname{rank}\begin{bmatrix} C' & -A'C' & (-A')^2C' & \cdots & (-A')^{n-1}C' \end{bmatrix} = n,$$

which is the same as (5.86). □

***Example* 5.6.8** Test the controllability and observability of the system

$$\begin{cases} \dot{x}_1 = 2x_1 + x_2, \\ \dot{x}_2 = x_1 + 2x_2 + u \end{cases}$$

with observation $y = x_1 + x_2$.

Solution Comparing the given system with (5.71), we get

$$A = \begin{bmatrix} 2 & 1 \\ 1 & 2 \end{bmatrix}, B = \begin{bmatrix} 0 \\ 1 \end{bmatrix}, \text{ and } C = \begin{bmatrix} 1 & 1 \end{bmatrix}.$$

Let

$$K = \begin{bmatrix} B & AB \end{bmatrix} = \begin{bmatrix} 0 & 1 \\ 1 & 2 \end{bmatrix}.$$

Since $\text{rank}(K) = 2$, from Theorem 5.3.4, the system is controllable. Now let

$$V = \begin{bmatrix} C \\ CA \end{bmatrix} = \begin{bmatrix} 1 & 1 \\ 3 & 3 \end{bmatrix}.$$

Since $\text{rank}(V) = 1$, from Corollary 5.6.7, the system is not observable.

***Example* 5.6.9** Consider the two tank problem discussed in Section 1.4 in Chapter 1, whose system equations are

$$\begin{cases} \dot{x}_1 = \frac{x_2}{50} - \frac{2x_1}{25} + 15(1 + \cos t), \\ \dot{x}_2 = \frac{2x_1}{25} - \frac{2x_2}{25}, \end{cases}$$

where $x_1(t)$ and $x_2(t)$ denote the amount of salt in tank-1 and tank-2, respectively, at time t. As the outflow is from tank-2, one can measure the salt content in tank-2. Hence, let $y(t) = x_2(t)$ for $t \in [0, \tau]$ be the observation. Now the system matrices are

$$A = \begin{bmatrix} -\frac{2}{25} & \frac{1}{50} \\ \frac{2}{25} & -\frac{2}{25} \end{bmatrix}, B = \begin{bmatrix} 1 \\ 0 \end{bmatrix}, C = \begin{bmatrix} 0 & 1 \end{bmatrix} \text{ and } u(t) = 15(1 + cost).$$

Let

$$V = \begin{bmatrix} C \\ CA \end{bmatrix} = \begin{bmatrix} 0 & 1 \\ \frac{2}{25} & -\frac{2}{25} \end{bmatrix}.$$

Since $\text{rank}(V) = 2$, the system is observable.

Example **5.6.10** Consider the inverted pendulum model given in (1.27). Here, $x_1(t)$ denotes the angle between the vertical line and the pendulum. In practice, this angle can be measured. Hence, we can consider $y(t) = x_1(t)$ to be the observation. Then $C = \begin{bmatrix} 1 & 0 & 0 & 0 \end{bmatrix}$ and

$$V = \begin{bmatrix} C \\ CA \\ CA^2 \\ CA^3 \end{bmatrix} = \begin{bmatrix} 1 & 0 & 0 & 0 \\ 0 & 1 & 0 & 0 \\ 100/19 & 0 & 1 & 0 \\ 0 & 100/19 & 0 & 1 \end{bmatrix}$$

Since $\text{rank}(V) = 4$ the system is observable.

Exercise 5.6.1 At time $t = 0$, each of the three tanks contains 500 liters of water, in which 15, 20, and 25 kgs of salt are dissolved in tanks-1, 2, and 3, respectively. For $t > 0$, 15 lt/sec of pure water flows into tank-1 from an overhead tank. Tank-2 receives an inflow of 10 lt/sec from another overhead tank, where the salt concentration is $u(t)$ kg/lt. Outflow rates are 20 lt/sec from tank-2 and 5 lt/sec from tank-3. The mixture in the three tanks is circulated as follows: 20 lt/sec from tank-1 to tank-2, 5 lt/sec from tank-2 to tank-1, 10 lt/sec from tank-2 to tank-3, and 5 lt/sec from tank-3 to tank-2. If $x_i(t), i = 1,2,3$, represent the amount of salt in tank-i at time t, write a control system in which the state variables are $x_i(t)$, $i = 1,2,3$, and $u(t)$ is the control function. Test the controllability of the system. If $Y(t) = \begin{bmatrix} x_2(t) \\ x_3(t) \end{bmatrix}$, $0 \le t \le \tau$, is the observation of the system, show that the system is observable.

5.7 Controllable Subsystem

Consider the time-invariant system

$$\dot{x} = Ax + Bu, \tag{5.87}$$

where A is an $n \times n$ matrix and B is an $n \times m$ matrix. Let

$$K = \begin{bmatrix} B & AB & A^2B & \cdots & A^{n-1}B \end{bmatrix}$$

be the $n \times mn$ Kalman matrix. From Theorem 5.3.4, it is known that the system (5.87) is controllable if and only if $\text{rank}(K) = n$. In the case where $\text{rank}(K) = r < n$, using a suitable transformation, we can find a subspace S of $\mathbb{R}^n$ and a subsystem of (5.87) which is controllable in S. To prove this, let $C_1, C_2, \ldots, C_r$ be linearly independent columns of K. Let $C_{r+1}, C_{r+2}, \ldots, C_n$ be $n-r$ linearly independent vectors in $\mathbb{R}^n$ such that the set $\{C_1, C_2, \ldots, C_r, C_{r+1}, \ldots, C_n\}$ forms a basis for $\mathbb{R}^n$. Let $P = \begin{bmatrix} P_1 & \vdots & P_2 \end{bmatrix}$ be the $n \times n$ matrix, where the block matrices P_1 and P_2 are given by

$$P_1 = \begin{bmatrix} C_1 & C_2 & \cdots & C_r \\ \downarrow & \downarrow & & \downarrow \end{bmatrix}_{n \times r}$$

and

$$P_2 = \begin{bmatrix} C_{r+1} & C_{r+2} & \cdots & C_n \\ \downarrow & \downarrow & & \downarrow \end{bmatrix}_{n \times (n-r)}.$$

Note that the columns of K can be generated by the columns of P_1

Let

$$P^{-1} = Q = \begin{bmatrix} Q_1 \\ \cdots \\ Q_2 \end{bmatrix},$$

where the blocks Q_1 and Q_2 are $r \times n$ and $(n-r) \times n$ matrices, respectively, such that

$$Q_1P_1 = I_r, Q_1P_2 = O_{r \times (n-r)}, Q_2P_1 = O_{(n-r) \times r}, \text{ and } Q_2P_2 = I_{(n-r) \times (n-r)}. \tag{5.88}$$

Note that $Q_2B = O_{(n-r) \times m}$ since $Q_2P_1 = O_{(n-r) \times r}$ and the columns of B are generated by the columns of P_1. Now, from $\dot{x} = Ax + Bu$, we get $P^{-1}\dot{x} = P^{-1}Ax + P^{-1}Bu$. Let

$$z = \begin{bmatrix} z_1 \\ \cdots \\ z_2 \end{bmatrix} \text{ and } x = Pz, \tag{5.89}$$

which imply $z = P^{-1}x = Qx$. Hence, the blocks $z_1 = Q_1 x$ and $z_2 = Q_2 x$. Writing (5.87) in terms of z we obtain

$$\dot{z} = P^{-1}APz + P^{-1}Bu,$$

which can be written as

$$\begin{bmatrix} \dot{z}_1 \\ \cdots \\ \dot{z}_2 \end{bmatrix} = \left[\begin{array}{c|c} Q_1AP_1 & Q_1AP_2 \\ \hline Q_2AP_1 & Q_2AP_2 \end{array}\right] \begin{bmatrix} z_1 \\ \cdots \\ z_2 \end{bmatrix} + \begin{bmatrix} Q_1Bu \\ \cdots \\ O \end{bmatrix}. \tag{5.90}$$

Since $AC_i : i = 1,2,\ldots,r$ are in the subspace of $\mathbb{R}^n$ generated by C_i, and since $Q_2C_i = 0$ for $i = 1,2,\ldots,r$ (as $Q_2P_1 = O$), we infer that $Q_2AP_1 = O_{(n-r)\times r}$. Hence, the equation (5.90) can be written as

$$\dot{z}_1 = (Q_1AP_1)z_1 + (Q_1AP_2)z_2 + Q_1Bu, \tag{5.91}$$

$$\dot{z}_2 = (Q_2AP_2)z_2. \tag{5.92}$$

Note that the subsystem (5.92) is not controllable. However, it is easy to see that, for any given z_2, the subsystem (5.91) is controllable in the r-dimensional subspace. This can be proved using the second part of the Remark 5.3.5, by replacing $f(t)$ with $(Q_1AP_2)z_2$. This is due to the fact

$$\begin{aligned} &\text{rank}\left[Q_1B \,\middle|\, (Q_1AP_1)(Q_1B) \,\middle|\, (Q_1AP_1)^2(Q_1B) \,\middle|\, \cdots \,\middle|\, (Q_1AP_1)^{r-1}(Q_1B) \right] \\ &= \text{rank}\begin{bmatrix} Q_1 \end{bmatrix}\begin{bmatrix} B & AB & A^2B & \cdots & A^{r-1}B \end{bmatrix} = \text{rank}(K) = r \quad (\text{since } Q_1 \text{ is nonsingular}). \end{aligned}$$

From (5.90), it can be concluded that the r-dimensional subspace given by $S = \{x \in \mathbb{R}^n : x = P_1z_1\}$ is the controllable subspace which is in the same Range (P_1). This subspace S is independent of the choice of the linearly independent vectors $C_1, C_2, \ldots, C_r$, as columns of P_1. This can be shown by replacing P_1 by P_1D where D is any nonsingular $r \times r$ matrix . The matrix P_1D provides a set of r linearly independent vectors generated by $C_1, C_2, \ldots, C_r$. Now replace P_1 by P_1D and Q_1 by $D^{-1}Q_1$ in the above discussion. From (5.89), $z_1 = D^{-1}Q_1x$ for any $x \in \mathbb{R}^n$ and hence the controllable subspace S is independent of the choice of the P_1 matrix, as $S = \{x \in \mathbb{R}^n : x = (P_1Dz_1)\} = \text{Range}(P_1)$.

Example **5.7.1** The system $\dot{x} = Ax + Bu$, where $A = \begin{bmatrix} 1 & 2 \\ 4 & -6 \end{bmatrix}$ and $B = \begin{bmatrix} 2 \\ 1 \end{bmatrix}$ is not controllable since $\text{rank}(K) = \text{rank}\begin{bmatrix} 2 & 4 \\ 1 & 2 \end{bmatrix} = 1$. To find the controllable subsystem, let $P_1 = \begin{bmatrix} 2 \\ 1 \end{bmatrix}$ and $P_2 = \begin{bmatrix} 1 \\ 0 \end{bmatrix}$. Then by following the steps in Section 5.7, $P = \begin{bmatrix} P_1 & \vdots & P_2 \end{bmatrix} = \begin{bmatrix} 2 & 1 \\ 1 & 0 \end{bmatrix}$ and $Q = P^{-1} = \begin{bmatrix} Q_1 \\ \cdots \\ Q_2 \end{bmatrix} = \begin{bmatrix} 0 & 1 \\ 1 & -2 \end{bmatrix}$. Since $Q_1AP_1 = 2$, $Q_1Bu = u$, $Q_2Bu = 0$, $Q_2AP_2 = -7$, $Q_1AP_2 = 4$, and $Q_2AP_1 = 0$ using the transformation $z = P^{-1}x$, we obtain the following subsystems as in (5.91)–(5.92)

$$\dot{z}_1 = 2z_1 + 4z_2 + u, \tag{5.93}$$

$$\dot{z}_2 = -7z_2. \tag{5.94}$$

Since

$$z = \begin{bmatrix} z_1 \\ \cdots \\ z_2 \end{bmatrix} = P^{-1}x = \begin{bmatrix} Q_1x = x_2 \\ \cdots \\ Q_2x = x_1 - 2x_2 \end{bmatrix}.$$

$$\begin{cases} z_1 = x_2, \\ z_2 = x_1 - 2x_2. \end{cases} \tag{5.95}$$

If the initial condition of the given system is $x(t_0) = x_0 = \begin{bmatrix} x_{01} \\ x_{02}, \end{bmatrix}$, then the corresponding initial conditions for (5.93) and (5.94) are $z_1(t_0) = x_{02}$, and $z_2(t_0) = x_{01} - 2x_{02}$. The solution of (5.94) for a given initial condition $z_2(t_0)$ is

$$z_2(t) = e^{-7(t-t_0)}z_2(t_0). \tag{5.96}$$

Note that $z_2(t)$ cannot be controlled, as (5.94) is independent of z_1 and u. From (5.95),

$$z_2(t) = x_1(t) - 2x_2(t) = e^{-7(t-t_0)}z_2(t_0), \tag{5.97}$$

i.e., for any t, the solution x must satisfy (5.97). Since the equation (5.93) is controllable, for any initial condition $z_1(t_0)(=x_2(t_0))$ and final condition $z_1(\tau)(=x_2(\tau))$, one can find a control $u(t)$. To illustrate this, let $t_0=0$ and $x(0)=\begin{bmatrix}3\\-4\end{bmatrix}$ be the initial condition. Then from (5.95) and (5.96), the solution $\begin{bmatrix}x_1(t)\\x_2(t)\end{bmatrix}$ is such that

$$x_1(t)-2x_2(t)=e^{-7t}(3-2(-4)) \Longrightarrow x_1(t)=2x_2(t)+11e^{-7t}. \tag{5.98}$$

Using (5.95) in the equation (5.93)we obtain

$$\begin{aligned}\dot{x}_2(t)&=2x_2+4(x_1-2x_2)+u\\&=-6x_2+4x_1+u\\&=-6x_2+4\left(2x_2+11e^{-7t}\right)+u\,(\text{using (5.98)}).\end{aligned}$$

Thus the controllable subsystem is $\dot{x}_2(t)=2x_2+44e^{-7t}+u;\quad x_2(0)=-4.$ (5.99)

Let τ be the final time. Since (5.99) is controllable, for an arbitrary final condition $x_2(\tau)$, one can find a control $u(t)$ so that $x_2(t)$ satisfies the initial and final conditions. Then from (5.98), $x_1(\tau)=2x_2(\tau)+11e^{-7\tau}$. Hence, $x_1(\tau)$ cannot be arbitrary but depends on $x_2(\tau)$, i.e, for the given initial condition $x(0)=(3,-4)$ the set of all reachable points in $\mathbb{R}^2$ are $x(\tau)=(2a+11e^{(-7\tau)},a)$, where $x_2(\tau)=a$ is any arbitrary real number.

5.8 Observable Subsystem

Consider the system

$$\begin{cases}\dot{x}=Ax+Bu,\\y=Cx,\end{cases} \tag{5.100}$$

where x, u, and y are state, control, and observation variables in the interval J, respectively. Here, we assume that A, B, and C are constant matrices of order $n\times n$, $n\times m$, and $p\times n$, respectively. In this section, we shall show that if (5.100) is not observable, then there exists an observable subsystem corresponding to (5.100). Since (5.100) is not observable,

$$\text{rank}(V) = \text{rank}\begin{bmatrix} C \\ CA \\ CA^2 \\ \vdots \\ CA^{n-1} \end{bmatrix} = k < n.$$

Let $\{r_1, r_2, \ldots, r_k\}$ be k linearly independent rows of the $np \times n$ matrix V. Let

$$Q_1 = \begin{bmatrix} r_1 \rightarrow \\ r_2 \rightarrow \\ \vdots \\ r_k \rightarrow \end{bmatrix}$$

be a $k \times n$ matrix with rank k, and

$$Q_2 = \begin{bmatrix} r_{k+1} \rightarrow \\ r_{k+2} \rightarrow \\ \vdots \\ r_n \rightarrow \end{bmatrix}$$

be an $(n-k) \times n$ matrix with rank $n-k$, whose rows are linearly independent of Q_1. Denote by Q the $n \times n$ matrix with Q_1 and Q_2 as blocks, i.e.,

$$Q = \begin{bmatrix} Q_1 \\ \cdots \\ Q_2 \end{bmatrix}.$$

As $\text{rank}(Q) = n$, Q^{-1} exists. Let

$$Q^{-1} = P = \begin{bmatrix} P_1 & \vdots & P_2 \end{bmatrix},$$

where P_1 and P_2 are $n \times k$ and $n \times (n-k)$ matrices, respectively. Hence

$$QP = I_n = \left[\begin{array}{c|c} Q_1P_1 & Q_1P_2 \\ \hline Q_2P_1 & Q_2P_2 \end{array}\right], \tag{5.101}$$

where $Q_1P_1 = I_k$, $Q_2P_2 = I_{(n-k)}$, $Q_1P_2 = \mathbb{O}_{k\times(n-k)}$, and $Q_2P_1 = \mathbb{O}_{(n-k)\times k}$. Note that $Q_1P_2 = \mathbb{O}_{k\times(n-k)}$ implies $CP_2 = \mathbb{O}_{p\times(n-k)}$ since rows of C are generated by rows of Q_1. Now, let us transform the variable x into z as $z = Qx$. Denoting

$$x = \begin{bmatrix} x_1 \\ \cdots \\ x_2 \end{bmatrix} \text{ and } z = \begin{bmatrix} z_1 \\ \cdots \\ z_2 \end{bmatrix},$$

and using $PQ = I$, the system (5.100) is equivalently written as

$$\begin{cases} Q\dot{x} = QAPQx + QBu, \\ y = CPQx. \end{cases}$$

Hence,

$$\begin{cases} \dot{z} = \left(QAQ^{-1}\right)z + QBu, \\ y = CPz, \end{cases} \tag{5.102}$$

which can be written as

$$\begin{bmatrix} \dot{z}_1 \\ \cdots \\ \dot{z}_2 \end{bmatrix} = \left[\begin{array}{c|c} Q_1AP_1 & Q_1AP_2 \\ \hline Q_2AP_1 & Q_2AP_2 \end{array}\right] \begin{bmatrix} z_1 \\ \cdots \\ z_2 \end{bmatrix} + \begin{bmatrix} Q_1B \\ \cdots \\ Q_2B \end{bmatrix} u$$

and

$$y = \begin{bmatrix} CP_1 & \vdots & CP_2 \end{bmatrix} z = \begin{bmatrix} CP_1 & \vdots & \mathbb{O} \end{bmatrix} \begin{bmatrix} z_1 \\ z_2 \end{bmatrix} = CP_1z_1.$$

Since $Q_1AP_2 = \mathbb{O}_{k\times(n-k)}$, the following subsystems are obtained:

$$\begin{cases} \dot{z}_1 = (Q_1AP_1)z_1 + Q_1Bu, \\ y = (CP_1)z_1, \end{cases} \tag{5.103}$$

and

$$\dot{z}_2 = (Q_2AP_1)z_1 + (Q_2AP_2)z_2 + Q_2Bu. \tag{5.104}$$

Note that the system (5.104) has no observations, while the system (5.103) is observable. Hence, the k-dimensional variable $z_1 = Q_1x$, $x \in \mathbb{R}^n$, is observable using the observation y.

***Example* 5.8.1** Consider the system (5.100), where

$$A = \begin{bmatrix} 0 & 1 \\ 0 & 0 \end{bmatrix}, B = \begin{bmatrix} 0 \\ 1 \end{bmatrix}, C = \begin{bmatrix} 0 & 1 \end{bmatrix}.$$

Then,

$$V = \begin{bmatrix} C \\ CA \end{bmatrix} = \begin{bmatrix} 0 & 1 \\ 0 & 0 \end{bmatrix}.$$

As $\text{rank}(V) = 1$, the system is not observable. Let $Q_1 = \begin{bmatrix} 0 & 1 \end{bmatrix}$. Since the rows of Q_2 must be linearly independent of the rows of Q_1, we choose $Q_2 = \begin{bmatrix} 1 & 0 \end{bmatrix}$. Let

$$Q = \begin{bmatrix} Q_1 \\ Q_2 \end{bmatrix} = \begin{bmatrix} 0 & 1 \\ 1 & 0 \end{bmatrix}.$$

Then,

$$\begin{bmatrix} P_1 & \vdots & P_2 \end{bmatrix} = P = Q^{-1} = \begin{bmatrix} 0 & 1 \\ 1 & 0 \end{bmatrix}.$$

Consider the transformation $z_1 = Q_1 x = x_2$ and $z_2 = Q_2 x = x_1$. To find the observable subsystem, we compute

$$Q_1 A P_1 = \begin{bmatrix} 0 & 1 \end{bmatrix} \begin{bmatrix} 0 & 1 \\ 0 & 0 \end{bmatrix} \begin{bmatrix} 0 \\ 1 \end{bmatrix} = \begin{bmatrix} 0 & 1 \end{bmatrix} \begin{bmatrix} 0 \\ 1 \end{bmatrix} = \mathbb{O}, Q_1 B = I, \text{ and } CP_1 = 1.$$

Hence, from (5.103), we get

$$\begin{cases} \dot{z}_1 = u, \\ y = z_1. \end{cases} \tag{5.105}$$

In (5.105) the observation z_1 itself is the state. Hence this subsystem is observable. From (5.104), we get

$$\dot{z}_2 = (Q_2 A P_1) z_1 + (Q_2 A P_2) z_2 + Q_2 B u = z_1. \tag{5.106}$$

It is clear that the subsystem (5.106) is not observable.

5.9 State Observer

In this section we consider the problem of estimating the state of a system in an interval J using an observation in J. Consider the time-invariant system

$$\begin{cases} \dot{x} = Ax + Bu, \\ y = Cx \end{cases} \tag{5.107}$$

for $t \in J = [t_0, \tau]$, where $x(t) \in \mathbb{R}^n$, $u(t) \in \mathbb{R}^m$, and $y(t) \in \mathbb{R}^p$ are the state, control, and observation at time t, respectively. Let the system be observable. Then,

$$\operatorname{rank} \begin{bmatrix} C \\ CA \\ CA^2 \\ \vdots \\ CA^{n-1} \end{bmatrix}_{np \times n} = n.$$

Equivalently,

$$\operatorname{rank} \left[C' \quad A'C' \quad (A')^2 C' \quad \cdots \quad (A')^{n-1} C' \right] = n. \tag{5.108}$$

From Theorems 5.5.3 and 5.6.6, it follows that for any given set $S = \{\mu_1, \mu_2, \ldots, \mu_n\}$, one can find a $p \times n$ matrix L' such that $A' + C'L'$ has $\mu_i : i = 1, 2, \ldots, n$ as eigenvalues. Hence, $A + LC$ has eigenvalues $\mu_i : i = 1, 2, \ldots, n$. Let $x(t)$ be the actual state and $x_E(t)$ denote the **estimate** of $x(t)$ at time t, and let $e(t) = x(t) - x_E(t)$ be the estimation error. Now consider the system

$$\dot{e} = (A + LC)e. \tag{5.109}$$

Let $\mu_i < 0$ for all $i = 1, 2, \ldots, n$. Since the μ_i are eigenvalues of $A + LC$, the solution $e(t)$ of (5.109) can be written as (see (4.55))

$$e(t) = \alpha_1 e^{\mu_1 t} X_1 + \alpha_2 e^{\mu_2 t} X_2 + \cdots + \alpha_n e^{\mu_n t} X_n, \tag{5.110}$$

where X_i are the eigenvectors corresponding to $\mu_i : i = 1, 2, \ldots, n$, and $\alpha_i : i = 1, 2, \ldots, n$ are constants. As $\mu_i < 0$ for $i = 1, 2, \ldots, n$, it follows from (5.110) that

$$\lim_{t \to \infty} e(t) = 0, \tag{5.111}$$

irrespective of the initial error $e(0)$. Hence, the estimate $x_E(t)$ will approach the actual state $x(t)$ as t increases. Equation (5.109) can be written as

$$\dot{x}(t) - \dot{x}_E(t) = (A + LC)(x(t) - x_E(t)).$$

Using (5.107), the above equation becomes

$$\dot{x}_E = (A + LC)x_E + Bu - Ly. \tag{5.112}$$

Equation (5.109) is known as a **state observer**, as it provides an estimate $x_E(t)$ of the actual state $x(t)$ based on the control $u(t)$ and the observation $y(t)$ over a large time interval since the estimation error $e(t)$ is small for large values of t.

***Example* 5.9.1** Consider the straight line motion of an object of mass m under the influence of a force $f(t)$. Taking into account the friction force, the motion can be mathematically modeled as

$$m\ddot{z} + k\dot{z} = f.$$

As in Section 1.3, the above equation can be written as

$$\dot{x} = Ax + Bu,$$

where

$$x = \begin{bmatrix} x_1 \\ x_2 \end{bmatrix}, A = \begin{bmatrix} 0 & 1 \\ 0 & -\frac{k}{m} \end{bmatrix}, B = \begin{bmatrix} 0 \\ \frac{1}{m} \end{bmatrix}, x_1 = z, x_2 = \dot{z}, \text{ and } u = f.$$

Let $y(t) = x_1(t)$ be the observation to the motion. Then

$$y = Cx,$$

where $C = \begin{bmatrix} 1 & 0 \end{bmatrix}$. As rank $\begin{bmatrix} C \\ CA \end{bmatrix} = 2$, the system is observable. Now, let us assume that the mass $m = 1$, the control $u = f(t) = 1$, and the constant $k = 1$. Then we obtain the actual state as

$$x(t) = \begin{bmatrix} x_1(t) \\ x_2(t) \end{bmatrix} = \begin{bmatrix} \alpha_1 + (1 - e^{-t})\,\alpha_2 + t - 1 + e^{-t} \\ e^{-t}\alpha_2 + 1 - e^{-t} \end{bmatrix} \tag{5.113}$$

and the observation as

$$y(t) = \alpha_1 + \left(1 - e^{-t}\right)\alpha_2 + t - 1 + e^{-t}, \tag{5.114}$$

where α_1 and α_2, respectively, are the initial displacement and velocity, i.e., $x_1(0) = \alpha_1$ and $x_2(0) = \alpha_2$. Now, let us show that we can obtain an estimate of the state vector $x(t)$ from the observer system (5.112). For this, we need to find a 2×1 matrix L such the $A+LC$ has negative eigenvalues. From Theorems 5.5.3 and 5.6.6, it follows that for any given set $S = \{-3, -1\}$ (say), one can find a 1×2 matrix L' such that $A' + C'L'$ (and hence $A+LC$) has eigenvalues -3 and -1. It can be verified that , $A+LC$ has eigenvalues -3 and -1 with

$$L = \begin{bmatrix} -3 \\ 0 \end{bmatrix}.$$

And, we obtain

$$A + LC = \begin{bmatrix} -3 & 1 \\ 0 & -1 \end{bmatrix}. \tag{5.115}$$

From (5.112),the state estimate x_E satisfies

$$\dot{x}_E = \begin{bmatrix} -3 & 1 \\ 0 & -1 \end{bmatrix} x_E + \begin{bmatrix} 0 \\ 1 \end{bmatrix} \cdot 1 - \begin{bmatrix} -3 \\ 0 \end{bmatrix} y, \tag{5.116}$$

where y is given by (5.114). On solving (5.116), we get

$$x_E = \begin{bmatrix} \beta_1 e^{-3t} + \frac{\beta_2}{2}\left(e^{-t} - e^{-3t}\right) + e^{-3t}\left(-3\alpha_1 + \frac{\alpha_2}{2}\right) + e^{-t}\left(1 - \frac{3\alpha_2}{2}\right) + \alpha_1 + \alpha_2 + t - 1 \\ \beta_2 e^{-t} + 1 - e^{-t} \end{bmatrix},$$

where β_1 and β_2 are the arbitrary initial conditions of x_E, i.e., $x_E(0) = \begin{bmatrix} \beta_1 \\ \beta_2 \end{bmatrix}$. The actual state $x(t)$ is given in (5.113).

From $x_E(t)$ and $x(t)$, one can see that for large values of t, both $x(t)$ and $x_E(t)$ are close to

$$\begin{bmatrix} \alpha_1 + \alpha_2 + t - 1 \\ 1 \end{bmatrix}$$

implying that x_E converges towards x as t increases.

***Example* 5.9.2** In the above example, let the control u be $f(t) = e^{-t}$. Then,

$$x(t) = \begin{bmatrix} \alpha_1 + (1-e^{-t})\,\alpha_2 + 1 - e^{-t} - te^{-t} \\ e^{-t}\alpha_2 + te^{-t} \end{bmatrix}$$

and

$$x_E(t) = \begin{bmatrix} \beta_1 e^{-3t} + \frac{\beta_2}{2}\left(e^{-t} - e^{-3t}\right) \\ \beta_2 e^{-t} \end{bmatrix}$$

$$+ \begin{bmatrix} \frac{1}{2}te^{-3t} - \frac{3}{2}te^{-t} - \left(\frac{5}{4} + \frac{3}{2}\alpha_2\right)\left(e^{-t} - e^{-3t}\right) + (\alpha_1 + \alpha_2 + 1)\left(1 - e^{-3t}\right) \\ 0 \end{bmatrix}.$$

As $t \to \infty$, both $x(t)$ and $x_E(t)$ tend to the same vector $\begin{bmatrix} \alpha_1 + \alpha_2 + 1 \\ 0 \end{bmatrix}$. Thus, the state observer system (5.116) is able to closely observe the actual state when t is large.

***Example* 5.9.3** For the control system

$$\begin{cases} \begin{bmatrix} \dot{x}_1 \\ \dot{x}_2 \end{bmatrix} = \begin{bmatrix} 0 & 1 \\ 0 & 0 \end{bmatrix} \begin{bmatrix} x_1 \\ x_2 \end{bmatrix} + \begin{bmatrix} 0 \\ 1 \end{bmatrix} u, \\ x_1(0) = 0, \quad x_2(0) = 4, \end{cases} \tag{5.117}$$

let the feedback control be $u = Kx$, where $K = \begin{bmatrix} k_2 & k_1 \end{bmatrix}$ so that $u = k_2x_1 + k_1x_2$. Assuming that $K = \begin{bmatrix} 3 & 2 \end{bmatrix}$, the system becomes

$$\begin{bmatrix} \dot{x}_1 \\ \dot{x}_2 \end{bmatrix} = \begin{bmatrix} 0 & 1 \\ 3 & 2 \end{bmatrix} \begin{bmatrix} x_1 \\ x_2 \end{bmatrix}, \quad \begin{bmatrix} x_1(0) \\ x_2(0) \end{bmatrix} = \begin{bmatrix} 0 \\ 4 \end{bmatrix},$$

whose solution is

$$\begin{bmatrix} x_1(t) \\ x_2(t) \end{bmatrix} = \begin{bmatrix} e^{3t} - e^{-t} \\ 3e^{3t} + e^{-t} \end{bmatrix}. \tag{5.118}$$

Now let us assume that the observation of the system (5.117) is

$$y(t) = x_1(t), \quad t > 0. \tag{5.119}$$

Hence, $C = \begin{bmatrix} 1 & 0 \end{bmatrix}$. As in equation (5.112), the observer for the system is defined as

$$\dot{x}_E = (A + LC)x_E + Bu - Ly, \tag{5.120}$$

where L is a 2×1 matrix chosen such that $A + LC$ has negative eigenvalues. (It can be shown that if $L = \begin{bmatrix} -4 \\ -3 \end{bmatrix}$, then $A + LC$ has eigenvalues -1 and -3.)

In state estimation problems, the control function $u(t)$ of the system under observation is not known a priori, in general. In such situations, in (5.120), instead of the actual control $u = Kx$, we can assume an estimate of the control in a feedback form as $\bar{u} = Kx_E$, then we get

$$\dot{x}_E = (A + LC + BK)x_E - Lx_1. \tag{5.121}$$

Substituting the matrices A, B, C, L and K in (5.121),

$$\dot{x}_E = \begin{bmatrix} -4 & 1 \\ 0 & 2 \end{bmatrix} x_E + \begin{bmatrix} 4 \\ 3 \end{bmatrix} x_1(t)$$

with some initial condition

$$\begin{bmatrix} x_{E1}(0) \\ x_{E2}(0) \end{bmatrix} = \begin{bmatrix} \bar{x}_{01} \\ \bar{x}_{02} \end{bmatrix}. \tag{5.122}$$

The solution of this initial value problem (IVP) is

$$\begin{bmatrix} x_{E1}(t) \\ x_{E2}(t) \end{bmatrix} = \begin{bmatrix} e^{-4t} & \frac{1}{6}\left(e^{2t} - e^{-4t}\right) \\ 0 & e^{2t} \end{bmatrix} \begin{bmatrix} \bar{x}_{01} \\ \bar{x}_{02} \end{bmatrix} + \begin{bmatrix} e^{3t} + \frac{2}{3}e^{-4t} - e^{-t} - \frac{2}{3}e^{2t} \\ 3e^{3t} + e^{-t} - 4e^{2t} \end{bmatrix}. \tag{5.123}$$

Note that the above estimate $x_E(t)$ coincides with the actual solution $x(t)$ of (5.117) if the initial condition (5.122) is the same as that of (5.117), i.e., $\begin{bmatrix} 0 \\ 4 \end{bmatrix}$. For any other initial condition in (5.122), $x_E(t) \to x(t)$ for large values of t. Therefore, we conclude that the system (5.117) can be observed using the observer (5.120) for large values of t (or $t - t_0$ if the initial time is t_0).

Exercise 5.9.1 Let $A = \begin{bmatrix} 3 & 0 & 1 \\ 1 & 1 & 0 \\ 0 & 1 & -1 \end{bmatrix}$, $B = \begin{bmatrix} 0 \\ 1 \\ 1 \end{bmatrix}$, and $C = \begin{bmatrix} 2 & 0 & 1 \end{bmatrix}'$. Construct a state observer system for the control system $\dot{x} = Ax + Bu$ with observation $y = cx$ in the following cases:
Case (i). The control function is $u(t) = 1 - e^{-t}$.
Case (ii). The control function $u(t)$ is in feedback form, $u(t) = Kx(t)$, where $K = \begin{bmatrix} 3 & 0 & 1 \end{bmatrix}'$.

In Case (ii), assuming that $x(0) = \begin{bmatrix} 1 & 1 & 0 \end{bmatrix}'$ as the actual initial condition, find $x(t)$. Assuming the initial condition of the state estimate as $x_E(0) = \begin{bmatrix} 1 & 0.5 & 0.1 \end{bmatrix}'$, find $x_E(t)$ and show that $|x(t) - x_E(t)| \to 0$ as $t \to \infty$.

Exercise 5.9.2 Let $A = \begin{bmatrix} 1 & 0 & 1 \\ 0 & 1 & 0 \\ 0 & 0 & 1 \end{bmatrix}$, $B = \begin{bmatrix} 1 \\ -1 \\ 1 \end{bmatrix}$, and $C = \begin{bmatrix} 0 & 1 & 0 \end{bmatrix}'$. Show that the system $\dot{x} = Ax + Bu$ with observation $y = Cx$ is neither controllable nor observable. Determine the controllable and observable subsystems.

5.10 Applications

Controlling an Object Moving on a Plane

Let $r(t) = (x_1(t), x_2(t))$ be the position vector of an object of mass m in $\mathbb{R}^2$. Let $v(t) = (v_1(t), v_2(t))$ and $a(t) = (a_1(t), a_2(t))$ be the velocity and acceleration vectors, respectively. Hence,

$$\dot{x}_i = v_i \text{ and } \dot{v}_i = a_i \quad \text{for } i = 1, 2. \tag{5.124}$$

If $f(t) = (f_1(t), f_2(t))$ is the force acting on the object, where $f_1(t)$ and $f_2(t)$ are the components of the force in the x_1 and x_2 directions, respectively, then from Newton's second law of motion, we obtain

$$m\ddot{x}_i = f_i \quad \text{for } i = 1, 2. \tag{5.125}$$

From (5.124) and (5.125), the control system representing the equation of motion of the object is

$$\dot{x}_i = v_i \text{ and } \dot{v}_i = \frac{f_i}{m} \quad \text{for } i = 1,2. \tag{5.126}$$

As the gravity is acting in $-x_2$ direction, the vertical force f_2 includes the natural force $-mg$, where g is the acceleration due to gravity. Hence, $f_2 = u_2 - mg$, where $u_2(t)$ is the vertical force (control) exerted on the object. Let $f_1 = u_1$ be the control force in the x_1 direction.

In matrix form, the system (5.126) can be written as

$$\begin{bmatrix} \dot{x}_1 \\ \dot{v}_1 \\ \dot{x}_2 \\ \dot{v}_2 \end{bmatrix} = \begin{bmatrix} 0 & 1 & 0 & 0 \\ 0 & 0 & 0 & 0 \\ 0 & 0 & 0 & 1 \\ 0 & 0 & 0 & 0 \end{bmatrix} \begin{bmatrix} x_1 \\ v_1 \\ x_2 \\ v_2 \end{bmatrix} + \begin{bmatrix} 0 & 0 \\ 1 & 0 \\ 0 & 0 \\ 0 & 1 \end{bmatrix} \begin{bmatrix} f_1 \\ f_2 \end{bmatrix}. \tag{5.127}$$

It can be easily verified that the system is controllable. Let us assume that a desired parabolic trajectory for the object is given by $x_2(t) = 1 - (1 - x_1(t))^2$ with the boundary conditions $x_1(0) = x_2(0) = \dot{x}_1(2) = \dot{x}_2(2) = 0$ and $x_1(2) = 2$. Let us assume that $x_1(t)$ is in the form of a polynomial $x_1(t) = a + bt + ct^2$ such that the desired boundary conditions are satisfied and $x_2(t)$ is such that $x_2 = 2x_1 - x_1^2$ .Thus the trajectory during the time interval $0 \leq t \leq 2$ is

$$\begin{cases} x_{1d}(t) = 2t - (1/2)t^2, \\ x_{2d}(t) = 2x_{1d}(t) - x_{1d}^2(t). \end{cases}$$

With this trajectory, the object will start from rest at $(0,0)$ and reach and stop at $(2,0)$ at $t = 2$.

Let $X(t) = \begin{bmatrix} x_1(t) \\ v_1(t) \\ x_2(t) \\ v_2(t) \end{bmatrix}$ be the actual state and $X_d(t) = \begin{bmatrix} x_{1d}(t) \\ v_{1d}(t) \\ x_{2d}(t) \\ v_{2d}(t) \end{bmatrix}$, the desired trajectory.

Let $e(t) = X(t) - X_d(t)$ be the error trajectory. Then we obtain the control system in terms of the error $e(t)$ as

$$\dot{e}(t) = Fe(t) + Gu(t), \tag{5.128}$$

where

$$F = \begin{bmatrix} 0 & 1 & 0 & 0 \\ 0 & 0 & 0 & 0 \\ 0 & 0 & 0 & 1 \\ 0 & 0 & 0 & 0 \end{bmatrix}, G = \begin{bmatrix} 0 & 0 \\ 1 & 0 \\ 0 & 0 \\ 0 & 1 \end{bmatrix}, \text{ and } u(t) = \begin{bmatrix} f_1 - a_{1d} \\ f_2 - a_{2d} \end{bmatrix}.$$

Now, let us define a feedback control $u = \kappa e$ such that $F + G\kappa$ has all negative eigenvalues. For example, let $\kappa = \begin{bmatrix} p & q & r & s \\ 0 & 0 & 0 & 0 \end{bmatrix}$, where the constants p, q, r, and s are chosen so that the characteristic equation of $F + G\kappa$ has negative eigenvalues. For example, let $\{-3, -4, -5, -6\}$ be eigenvalues of $F + G\kappa = \begin{bmatrix} 0 & 1 & 0 & 0 \\ p & q & r & s \\ 0 & 0 & 0 & 1 \\ 0 & 0 & 0 & 0 \end{bmatrix}$. By comparing the coefficients of the characteristic polynomial of $F + G\kappa$, i.e, $\det(F + G\kappa - \lambda I)$ with $(\lambda+3)(\lambda+4)(\lambda+5)(\lambda+6)$, we obtain $F + G\kappa = \begin{bmatrix} 0 & 1 & 0 & 0 \\ -119 & 18 & -360 & -342 \\ 0 & 0 & 0 & 1 \\ 0 & 0 & 0 & 0 \end{bmatrix}$. The error system (5.128) now becomes $\dot{e} = (F + Gk)e$, whose solution in terms of the eigenpairs of $F + G\kappa$ is $e(t) = \sum_{k=1}^{4} c_k e^{\lambda_k t} \begin{bmatrix} 1 \\ \lambda_k \\ \frac{1}{\lambda_k^2} \\ \frac{1}{\lambda_k} \end{bmatrix}$, where $c_k \in \mathbb{R}$ and $\lambda_k = -2 - k$ for $k = 1, 2, 3, 4$ are the eigenvalues. It can be easily seen that as t increases, $e(t)$ tends to zero, which guarantees tracking of the desired trajectory.

5.11 Summary

In this chapter, the concepts of controllability and observability of linear systems have been discussed. It was shown that the linear system (5.12) is controllable if and only if the controllability Gramian matrix is nonsingular. The linear system (5.71) is observable if and only if the observability Gramian matrix is nonsingular.
If the system (5.12) is not controllable, then a controllable subsystems (5.91) can be obtained using the procedure given in Section 5.7.
If a system is not observable, then Section 5.8 provides a method for obtaining an observable subsystem (5.103).
If a system is observable, then using the procedure discussed in Section 5.9, the state of the system can be observed by constructing a suitable state observer.
Trajectory tracking on a plane using a feedback control is shown in Section 5.9.

5.12 Exercise

1. Test the controllability and observability of the system

$$\begin{cases} \dot{x} = Ax + Bu, \\ Y = Cx, \end{cases}$$

where

$$A = \begin{bmatrix} 1 & 2 & 1 \\ 0 & 1 & 1 \\ 0 & 2 & 2 \end{bmatrix}, B = \begin{bmatrix} 1 \\ 0 \\ 0 \end{bmatrix}, \text{ and } C = \begin{bmatrix} 1 & 0 & 2 \end{bmatrix}.$$

2. For the system $\dot{x} = Ax + Bu$ for $0 \leq t \leq 5$, find the controllability Gramian matrix W if $A = \begin{bmatrix} 2 & 0 \\ 0 & 3 \end{bmatrix}$ and $B = \begin{bmatrix} 1 \\ -1 \end{bmatrix}$. Hence, find a control function u that steers the system from $x(0) = \begin{bmatrix} 1 \\ 0 \end{bmatrix}$ to $x(5) = \begin{bmatrix} 0 \\ 1 \end{bmatrix}$.

3. For the system $\dot{x} = Ax + Bu, 0 \leq t \leq 5$, where $A = \begin{bmatrix} 2 & 0 \\ 0 & 3 \end{bmatrix}$ and $B = \begin{bmatrix} 1 \\ 0 \end{bmatrix}$, find the

controllability Gramian matrix W and show that it is singular. Find the set of all vectors $\begin{bmatrix}\alpha\\ \beta\end{bmatrix}$ that can be reached using a control from the initial point $\begin{bmatrix}0\\ 0\end{bmatrix}$.

4. If $A = \begin{bmatrix}3 & 1 & 2\\ 0 & 1 & 0\\ 0 & 2 & 0\end{bmatrix}$ and $B = \begin{bmatrix}-1\\ 0\\ 1\end{bmatrix}$, can we find a control $u(t)$ in the interval $[0,5]$ such that the solution x satisfies the initial and final conditions $x(0) = \begin{bmatrix}1\\ 1\\ 3\end{bmatrix}$ and $x(5) = \begin{bmatrix}1\\ 0\\ -3\end{bmatrix}$, respectively? Find all vectors $\begin{bmatrix}a\\ b\\ c\end{bmatrix}$ such that the solution $x(t)$ satisfies $x(0) = \begin{bmatrix}1\\ 1\\ 3\end{bmatrix}$ and $x(5) = \begin{bmatrix}a\\ b\\ c\end{bmatrix}$.

5. Show that

$$\operatorname{rank}\begin{bmatrix}C' & -A'C' & (-A')^2C' & \cdots & (-A')^{n-1}C'\end{bmatrix} = \operatorname{rank}\begin{bmatrix}C\\ CA\\ CA^2\\ \vdots\\ CA^{n-1}\end{bmatrix},$$

where A and C are $n \times n$ and $p \times n$ matrices, respectively.

6. Show that the system

$$\begin{cases}\dot{x} = Ax + Bu,\\ Y = Cx,\end{cases}$$

where

$$A=\begin{bmatrix}0&1&0\\0&1&0\\1&2&1\end{bmatrix},B=\begin{bmatrix}1\\0\\2\end{bmatrix}, \text{ and } C=\begin{bmatrix}0&1&0\end{bmatrix}$$

is not observable, and find the observable subsystem.

7. If

$$C=\begin{bmatrix}0&1&0&\cdots&0\\0&0&1&\cdots&0\\\cdots&\cdots&\cdots&\cdots&\cdots\\0&0&0&\cdots&1\\-\alpha_n&-\alpha_{n-1}&-\alpha_{n-2}&\cdots&-\alpha_1\end{bmatrix},$$

then show that the characteristic equation of C is

$$\lambda^n+\alpha_1\lambda^{n-1}+\alpha_2\lambda^{n-2}+\cdots+\alpha_n=0.$$

8. For the system $\dot{x}=2x+y+u, \dot{y}=2y+u$, find a feedback control $u=Kx$ such that $A+BK$ has eigenvalues 1 and -3.

9. Find the controllability Gramian matrix for the control system $\dot{x}=Ax+Bu$ for $0\le t\le 2$, where $A=\begin{bmatrix}2&1\\0&3\end{bmatrix}$ and $B=\begin{bmatrix}0\\1\end{bmatrix}$. Moreover, find a matrix K such that the eigenvalues of $A+BK$ are -3 and -2.

10. For $A=\begin{bmatrix}0&1&0\\0&1&2\\1&-1&0\end{bmatrix}$ and $B=\begin{bmatrix}1&1\\1&1\\0&1\end{bmatrix}$, find a matrix $K_{2\times 3}$ such that $A+BK$ has eigenvalues $\{-1,-2,-3\}$.

11. Convert the following control system $\dot{x}=Ax+BU$ into an equivalent system of the form $\dot{Y}=(A+BK)Y+b_1V$, where $A=\begin{bmatrix}3&3&0\\0&0&2\\1&1&0\end{bmatrix}$, $B=\begin{bmatrix}0&0\\1&0\\0&2\end{bmatrix}$, and b_1 is the

first column of B.

12. Let $A = \begin{bmatrix} 1 & 0 & 1 \\ 0 & 1 & 2 \\ 0 & 0 & 1 \end{bmatrix}$, $B = \begin{bmatrix} 1 & 0 \\ 1 & 0 \\ 0 & 1 \end{bmatrix}$, and $C = \begin{bmatrix} 1 & 1 & -1 \end{bmatrix}$. For the system $\dot{x} = Ax + Bu$ with the observation $y = Cx$, write the dual system. Furthermore, write the observable subsystem.

13. Convert the control system $\dot{x} = Ax + Bu$ into an equivalent system of the form $\dot{z} = (A + BK_1)z + b_1 v$, where $A = \begin{bmatrix} 1 & 1 & 0 \\ 0 & 0 & 1 \\ 1 & 1 & 0 \end{bmatrix}$, $B = \begin{bmatrix} 0 & 0 \\ 1 & 0 \\ 0 & 1 \end{bmatrix}$, and b_1 is the first column of B. Hence, find a matrix K_2 and a feedback control $v = K_2 z$ such that $A + BK$ has eigenvalues $1, 2,$ and 3, where $K = K_1 + K_2$.

14. In Example 5.9.3, assume that the feedback matrix $K = \begin{bmatrix} 1 & 3 \end{bmatrix}$, and the initial conditions are $x_1(0) = 2$ and $x_2(0) = 1$. Let $y(t) = x_1(t)$ for $t > 0$ be the observation. Find the state observer $x_E(t)$ and show that $x_E(t) \to x(t)$ as $t \to \infty$.

6

Optimal Control

6.1 Introduction

In this chapter, we consider optimal control problems aimed at obtaining optimal solutions (both state and control) that minimize or maximize a given cost functional, also known as the performance index or objective functional, denoted by $\mathbb{J}$.

In Chapter 5, we obtained the optimal control $u(t)$ for a linear system $\dot{x} = Ax + Bu$, $t_0 \leq t \leq \tau$, with the initial condition $x(t_0) = x_0$ and the terminal condition $x(\tau) = x_1$, such that u minimizes the functional $\mathbb{J} = \int_{t_0}^{\tau} \|u(t)\|^2 dt$. The optimal control u was obtained using the concept of controllability and was expressed in equation (5.35). This is one type of optimal control problem. In general, there are various forms of objective functionals involving x, $\dot{x}$, and u that need to be optimized for both linear and nonlinear control systems. In the following sections, various conditions for obtaining optimal solutions are derived using methods from calculus of variations.

In Sections 6.2 and 6.3, definitions and examples of increments, differentials, derivatives, and variations of functionals and vector valued functions are provided. In Section 6.4, Euler's equations and boundary conditions for extremizing a functional $\mathbb{J}$ in the form of an integral that depends on a single variable x and its derivative with fixed and/or free boundary conditions, are derived. In Section 6.5, extremization of the functional $\mathbb{J}$ that depends on several variables and their derivatives is considered. Section 6.6 contains the Euler–Lagrange equations and boundary conditions for finding the extremum of $\mathbb{J}$ under constraints. If the constraints considered in Section 6.6 are replaced by a control system, then the extremization problem becomes an optimal control problem, which is dealt with in Section 6.7 using Lagrangian and Hamiltonian. As a particular case, linear quadratic (LQ) optimal control problem is presented in Section 6.8. Some special cases are considered in Section 6.9.

6.2 Increment and Differential

In this section, basic definitions, notations, and useful formulas are presented. The readers who are familiar with the concept of differentials, formulas involving derivatives, and differentials of vector functions may skip this section or quickly glance through it for notations.

Let $x : [t_0, \tau] \to \mathbb{R}$ be a smooth function of t. If $a \in [t_0, \tau]$ and Δa is an increment at $t = a$, then the **increment of the function** x at a due to the increment Δa is given by $(\Delta x)(a) = x(a+\Delta a) - x(a)$. Using Taylor's series expansion of x about a (if it exists),

$$(\Delta x)(a) = \dot{x}(a)\Delta a + \dot{x}^2(a)\frac{(\Delta a)^2}{2!} + \cdots.$$

Neglecting higher order terms and retaining the linear term, we obtain the **differential** d of x at a, denoted by $(dx)(a)$. Thus,

$$(dx)(a) = \dot{x}(a)\Delta a.$$

Similarly, if $x(t) = \begin{bmatrix} x_1(t) & x_2(t) & \cdots & x_n(t) \end{bmatrix}' \in \mathbb{R}^n$, then the increment of x at a is given by

$$(\Delta x)(a) = x(a+\Delta a) - x(a) = \begin{bmatrix} (\Delta x_1)(a) \\ (\Delta x_2)(a) \\ \vdots \\ (\Delta x_n)(a) \end{bmatrix}.$$

The differential at a is

$$(dx)(a) = \begin{bmatrix} \dot{x}_1(a) \\ \dot{x}_2(a) \\ \vdots \\ \dot{x}_n(a) \end{bmatrix} \Delta a = \dot{x}(a)\Delta a.$$

Thus, the differential dx is the linear approximation of the increment of x at a.

Variation of a Function

Let $t \in [t_0, \tau] \subset \mathbb{R}$ and x be a function of t. Let δx be another function of t defined on $[t_0, \tau]$, called the **variation** in x. Then $(x + \delta x)(t) = x(t) + \delta x(t), t \in [t_0, \tau]$, defines a new function on $[t_0, \tau]$.

Functional

Any function whose range is in $\mathbb{R}$ or $\mathbb{C}$ is called a **functional**. In this chapter, we mainly consider real valued functionals in the form of an integral.

***Example* 6.2.1** Let $x \in C([0,1], \mathbb{R}^2)$ and $\mathbb{J} : C([0,1], \mathbb{R}^2) \to \mathbb{R}$ be a functional defined as follows. For $x(t) = (x_1(t), x_2(t))', 0 \le t \le 1$,

$$\mathbb{J}(x) = \int_0^1 \|x(t)\|^2 dt = \int_0^1 (x_1^2(t) + x_2^2(t))dt.$$

Then the value of $\mathbb{J}(x)$ when $x(t) = (t^2, t)'$ is

$$\mathbb{J}(x) = \int_0^1 (t^4 + t^2)dt = \frac{8}{15}.$$

Increment of a Functional

Let $\mathbb{J} = \mathbb{J}(x)$ be a functional that depends on functions x of t defined on $[t_0, \tau]$. Let $x^*(t)$ be a given function and $(\delta x)(t)$ be a variation in $x^*(t)$ for $t \in [t_0, \tau]$. Then, the increment in $\mathbb{J}(x^*)$ due to δx is defined as

$$\Delta\mathbb{J} = \mathbb{J}(x^* + \delta x) - \mathbb{J}(x^*).$$

***Example* 6.2.2** Let

$$\mathbb{J}(x) = \int_0^1 (x^2(t) + 2x(t))dt.$$

Then the increment in $J(x)$ due to a variation δx is

$$\begin{aligned}\Delta\mathbb{J} &= \int_0^1 [(x(t) + \delta x(t))^2 + 2(x(t) + \delta x(t))]dt - \int_0^1 (x^2(t) + 2x(t))dt \\ &= \int_0^1 [2x(t)\delta x(t) + (\delta x(t))^2 + 2\delta x(t)]dt.\end{aligned}$$

***Example* 6.2.3** Let $\mathbb{J}(x) = \int_0^1 x^2(t)dt$. Find the increment $\Delta\mathbb{J}$ at x^*, when $x^*(t) = \sin(\frac{\pi}{2}t)$ and $(\delta x)(t) = \frac{1}{10}\sin(\pi t)$.

Solution

$$\begin{aligned}(\Delta\mathbb{J})(x^*) = \mathbb{J}(x^*+\delta x) - \mathbb{J}(x^*) &= \int_0^1 [2x^*(t)(\delta x)(t) + (\delta x)^2(t)]dt \\ &= \int_0^1 \left[\frac{1}{5}\sin(\frac{\pi}{2}t)\sin(\pi t) + \frac{1}{100}\sin^2(\pi t)\right] dt \\ &= \frac{4}{3\pi} + \frac{1}{200}.\end{aligned}$$

Variation of a Functional $\mathbb{J}$

Let $x^*(t)$ be a given function defined on $[t_0, \tau]$. The increment of $\mathbb{J}$ at x^* due to a variation δx in x^* is given by $(\Delta\mathbb{J})(x^*) = \mathbb{J}(x^*+\delta x) - \mathbb{J}(x^*)$. Using Taylor's series about $x^*(t)$ and neglecting second and higher order terms, we obtain the first variation in $\mathbb{J}$ at x^*, denoted by $(\delta\mathbb{J})(x^*)$.

Example **6.2.4** Let

$$\mathbb{J} = \int_{t_0}^{\tau} f(t, x(t))dt.$$

Then increment $\Delta\mathbb{J}$ at a given function x^* due to a variation δx in x^* is

$$\begin{aligned}(\Delta\mathbb{J})(x^*) &= \int_{t_0}^{\tau} f(t, x^*(t) + \delta x(t))dt - \int_{t_0}^{\tau} f(t, x^*(t))dt \\ &= \int_{t_0}^{\tau} [f(t, x^*(t) + \delta x(t)) - f(t, x^*(t))]dt.\end{aligned}$$

Using Taylor's expansion about $x^*(t)$ and considering only the linear term, we obtain the **first variation** $\delta\mathbb{J}$ at x^* as

$$(\delta\mathbb{J})(x^*) = \int_{t_0}^{\tau} \left(\frac{\partial f(t, x(t))}{\partial x}\right)_{x^*} \delta x(t)dt. \tag{6.1}$$

From Taylor's expansion, we obtain the **second variation** in $\mathbb{J}$ at x^* as

$$(\delta^2\mathbb{J})(x^*) = \frac{1}{2!}\int_{t_0}^{\tau} \left(\frac{\partial^2 f(t, x(t))}{\partial x^2}\right)_{x^*} (\delta x(t))^2 dt, \tag{6.2}$$

and so on.

Example **6.2.5** Let

$$\mathbb{J} = \int_0^1 x^3(t)dt.$$

Find the first and second variations in $\mathbb{J}$ at $x^*(t) = t^2$ corresponding to a variation δx of x^*.

Solution Here $f(t,x(t)) = x^3(t)$. From (6.1), we have

$$\delta\mathbb{J} = \int_0^1 3(x^*(t))^2 \delta x(t)dt = 3\int_0^1 t^4 \delta x(t)dt.$$

From (6.2), we have

$$\delta^2\mathbb{J} = \frac{6}{2!}\int_0^1 x^*(t)(\delta x(t))^2 dt = 3\int_0^1 t^2(\delta x(t))^2 dt.$$

Hereafter, the usage of "variation" refers to the "first variation."

Example **6.2.6** Let $x(t) = \begin{bmatrix} x_1(t) & x_2(t) & \cdots & x_n(t) \end{bmatrix}'$ be a vector valued function, and let

$$\mathbb{J}(x) = \int_{t_0}^{\tau} f(t,x(t))dt = \int_{t_0}^{\tau} f(t,x_1(t),x_2(t),\ldots,x_n(t))dt,$$

where f is a real valued function. Let $x^*(t) = \begin{bmatrix} x_1^*(t) & x_2^*(t) & \cdots & x_n^*(t) \end{bmatrix}'$ be a fixed function and $(\delta x)(t) = \begin{bmatrix} (\delta x_1)(t) & (\delta x_2)(t) & \cdots & (\delta x_n)(t) \end{bmatrix}'$ be a variation in x^*. Then the increment of $\mathbb{J}$ at x^* due to δx is

$$\begin{aligned}(\Delta\mathbb{J})(x^*) &= \int_{t_0}^{\tau} [f(t,(x^*+\delta x)(t) - f(t,x^*(t))]dt \\ &= \int_{t_0}^{\tau}\left[\left(\frac{\partial f}{\partial x_1}\right)_{x^*}\delta x_1(t) + \left(\frac{\partial f}{\partial x_2}\right)_{x^*}\delta x_2(t) + \cdots + \left(\frac{\partial f}{\partial x_n}\right)_{x^*}\delta x_n(t)\right]dt \\ &\quad + \text{higher order terms.}\end{aligned}$$

Hence, the variation in $\mathbb{J}$ at x^* due to δx is

$$(\delta\mathbb{J})(x^*) = \int_{t_0}^{\tau}\left[\left(\frac{\partial f}{\partial x_1}\right)_{x^*}\delta x_1(t) + \left(\frac{\partial f}{\partial x_2}\right)_{x^*}\delta x_2(t) + \cdots + \left(\frac{\partial f}{\partial x_n}\right)_{x^*}\delta x_n(t)\right]dt. \quad (6.3)$$

Exercise 6.2.1 If $\mathbb{J}(x) = x^2(1) + \int_0^1 x^2(t)dt$, find $\Delta\mathbb{J}$ and $\delta\mathbb{J}$ at $x^*(t) = t^3$ and $\delta x(t) = \frac{t^2}{10}$.

6.3 Derivative and Variation of Vector Functions

Derivative

(I) Let $x(t) = \begin{bmatrix} x_1(t) \\ x_2(t) \\ \vdots \\ x_n(t) \end{bmatrix} \in \mathbb{R}^n$ for each $t \in \mathbb{R}$. Then $\frac{dx}{dt} = \dot{x} = \begin{bmatrix} \dot{x}_1 \\ \dot{x}_2 \\ \vdots \\ \dot{x}_n \end{bmatrix}$.

(II) If $M(t) = \{m_{ij}(t)\}_{i=1,2,\ldots,m;j=1,2,\ldots,n}$, then $\frac{dM}{dt} = \left\{\frac{d}{dt}m_{ij}(t)\right\}_{i=1,2,\ldots,m;j=1,2,\ldots,n}$.

(III) If $x \in \mathbb{R}^n$ is a column vector written as $x = \begin{bmatrix} x_1 & x_2 & \cdots & x_n \end{bmatrix}'$ and $g(x) = g(x_1, x_2, \ldots, x_n)$ is a real valued function, then $\frac{dg}{dx}$, the gradient of g, is defined as

$$\frac{dg}{dx} = \begin{bmatrix} \frac{\partial g}{\partial x_1} \\ \frac{\partial g}{\partial x_2} \\ \vdots \\ \frac{\partial g}{\partial x_n} \end{bmatrix} = \begin{bmatrix} \frac{\partial g}{\partial x_1} & \frac{\partial g}{\partial x_2} & \cdots & \frac{\partial g}{\partial x_n} \end{bmatrix}'.$$

***Example* 6.3.1** Let $x = \begin{bmatrix} x_1 \\ x_2 \end{bmatrix}$ and $g(x) = x_1^2 + x_1x_2$. Then $\frac{dg}{dx} = \begin{bmatrix} \frac{\partial g}{\partial x_1} & \frac{\partial g}{\partial x_2} \end{bmatrix}' = \begin{bmatrix} 2x_1 + x_2 & x_1 \end{bmatrix}'$.

***Example* 6.3.2** If $x = \begin{bmatrix} x_1 & x_2 & \cdots & x_n \end{bmatrix}'$ and $y = \begin{bmatrix} y_1 & y_2 & \cdots & y_n \end{bmatrix}'$, then

For a fixed vector y, $\frac{d}{dx}(x'y) = \frac{d}{dx}(x_1y_1 + x_2y_2 + \cdots + x_ny_n) = \begin{bmatrix} y_1 & y_2 & \cdots & y_n \end{bmatrix}' = y$ (6.4)

and

For a fixed vector x, $\frac{d}{dy}(x'y) = \frac{d}{dy}(x_1y_1 + x_2y_2 + \cdots + x_ny_n) = \begin{bmatrix} x_1 & x_2 & \cdots & x_n \end{bmatrix}' = x.$ (6.5)

***Example* 6.3.3** Find $\frac{d}{dx}(x'Ax)$, where $x \in \mathbb{R}^n$ and A is an $n \times n$ constant matrix. From (6.4) and (6.5), we obtain

$$\frac{d}{dx}(x'Ax) = \frac{d}{dx}(x'(Ax)) + \frac{d}{dx}((x'A)x) = Ax + A'x = (A + A')x.$$

(IV) If $x = \begin{bmatrix} x_1 & x_2 & \cdots & x_n \end{bmatrix}' \in \mathbb{R}^n$ and $G(x)$ is a vector valued function given by

$$G(x) = \begin{bmatrix} g_1(x) \\ g_2(x) \\ \vdots \\ g_m(x) \end{bmatrix} \in \mathbb{R}^m,$$

then

$$\frac{dG}{dx} = \begin{bmatrix} \frac{dg_1}{dx} & \frac{dg_2}{dx} & \cdots & \frac{dg_m}{dx} \end{bmatrix}' = \begin{bmatrix} \frac{\partial g_1}{\partial x_1} & \frac{\partial g_2}{\partial x_1} & \cdots & \frac{\partial g_m}{\partial x_1} \\ \frac{\partial g_1}{\partial x_2} & \frac{\partial g_2}{\partial x_2} & \cdots & \frac{\partial g_m}{\partial x_2} \\ \vdots & \vdots & & \vdots \\ \frac{\partial g_1}{\partial x_n} & \frac{\partial g_2}{\partial x_n} & \cdots & \frac{\partial g_m}{\partial x_n} \end{bmatrix}'$$

$$= \begin{bmatrix} \frac{\partial g_1}{\partial x_1} & \frac{\partial g_1}{\partial x_2} & \cdots & \frac{\partial g_1}{\partial x_n} \\ \frac{\partial g_2}{\partial x_1} & \frac{\partial g_2}{\partial x_2} & \cdots & \frac{\partial g_2}{\partial x_n} \\ \vdots & & & \\ \frac{\partial g_m}{\partial x_1} & \frac{\partial g_m}{\partial x_2} & \cdots & \frac{\partial g_m}{\partial x_n} \end{bmatrix}_{m \times n}.$$

$\frac{dG}{dx}$ is also called the **Jacobian** of G with respect to x.

Example **6.3.4** Let $x = \begin{bmatrix} x_1 \\ x_2 \\ x_3 \end{bmatrix}$ and $G(x) = \begin{bmatrix} g_1(x) \\ g_2(x) \end{bmatrix} = \begin{bmatrix} x_1 x_2 \\ x_2 x_3^2 \end{bmatrix}$. Then $\frac{dG}{dx} = \begin{bmatrix} \frac{dg_1}{dx} & \frac{dg_2}{dx} \end{bmatrix}' =$ $\begin{bmatrix} x_2 & x_1 & 0 \\ 0 & x_3^2 & 2x_2x_3 \end{bmatrix}$.

Example **6.3.5** If $A = \{a_{ij}\}_{m \times n}$ and $x \in \mathbb{R}^n$, then from **(IV)**, $\frac{d}{dx}(Ax) = A$.

(V) If $G(x) = \begin{bmatrix} g_1(x) & g_2(x) & \cdots & g_n(x) \end{bmatrix}$, then $\frac{dG}{dx} = \begin{bmatrix} \frac{dg_1}{dx} & \frac{dg_2}{dx} & \cdots & \frac{dg_n}{dx} \end{bmatrix}$, which is the same as $\begin{bmatrix} \frac{dG'}{dx} \end{bmatrix}'$.

Example **6.3.6** Let $x = \begin{bmatrix} x_1 \\ x_2 \end{bmatrix}$ and $A = \begin{bmatrix} a_{11} & a_{12} \\ a_{21} & a_{22} \end{bmatrix}$. Then $\frac{d}{dx}(x'A) = A$. Since $x'A = \begin{bmatrix} a_{11}x_1 + a_{21}x_2 \\ a_{12}x_1 + a_{22}x_2 \end{bmatrix}'$, $\frac{d}{dx}(x'A) = \left[\frac{d}{dx}(a_{11}x_1 + a_{21}x_2) \quad \frac{d}{dx}(a_{12}x_1 + a_{22}x_2)\right] = \begin{bmatrix} a_{11} & a_{12} \\ a_{21} & a_{22} \end{bmatrix} = A.$

Variation

(VI) Let $x = \begin{bmatrix} x_1 & x_2 & \cdots & x_n \end{bmatrix}'$ and $g(x) = g(x_1, x_2, \ldots, x_n)$. Let $\delta x = \begin{bmatrix} \delta x_1 & \delta x_2 & \cdots & \delta x_n \end{bmatrix}'$ be a variation in x. Then, the increment and the variation in g corresponding to δx are $\Delta g = g(x + \delta x) - g(x) = g(x_1 + \delta x_1, x_2 + \delta x_2, \ldots, x_n + \delta x_n) - g(x_1, x_2, \ldots, x_n)$ and

$$\delta g = \frac{\partial g}{\partial x_1}\delta x_1 + \frac{\partial g}{\partial x_2}\delta x_2 + \cdots + \frac{\partial g}{\partial x_n}\delta x_n = \begin{bmatrix} \frac{\partial g}{\partial x_1} & \frac{\partial g}{\partial x_2} & \cdots & \frac{\partial g}{\partial x_n} \end{bmatrix} \begin{bmatrix} \delta x_1 \\ \delta x_2 \\ \vdots \\ \delta x_n \end{bmatrix} = \left[\frac{dg}{dx}\right]' \delta x.$$

(VII) If $x \in \mathbb{R}^n$ and $G(x) = \begin{bmatrix} g_1(x) \\ g_2(x) \\ \vdots \\ g_m(x) \end{bmatrix}$, then the variation in G is

$$\delta G = \begin{bmatrix} \delta g_1 \\ \delta g_2 \\ \vdots \\ \delta g_m \end{bmatrix} = \begin{bmatrix} \left[\frac{\partial g_1}{\partial x}\right]' \delta x \\ \vdots \\ \left[\frac{\partial g_m}{\partial x}\right]' \delta x \end{bmatrix} = \begin{bmatrix} \frac{\partial g_1}{\partial x_1} & \frac{\partial g_1}{\partial x_2} & \cdots & \frac{\partial g_1}{\partial x_n} \\ \frac{\partial g_2}{\partial x_1} & \frac{\partial g_2}{\partial x_2} & \cdots & \frac{\partial g_2}{\partial x_n} \\ \vdots & & & \\ \frac{\partial g_m}{\partial x_1} & \frac{\partial g_m}{\partial x_2} & \cdots & \frac{\partial g_m}{\partial x_n} \end{bmatrix} \begin{bmatrix} \delta x_1 \\ \delta x_2 \\ \vdots \\ \delta x_n \end{bmatrix} = \frac{dG}{dx}\delta x \quad \text{(from (IV))}.$$

Example **6.3.7** (*i*) $\delta(Ax) = \left[\frac{d}{dx}(Ax)\right] \delta x = A\, \delta x.$
(*ii*) $\delta(x'A) = \left[\frac{d}{dx}(x'A)\right] \delta x = A\, \delta x.$
(*iii*) $\delta(x'Ax) = \left[\frac{d}{dx}(x'Ax)\right]' \delta x = [Ax + A'x]' \delta x = x'(A' + A)\, \delta x$ (from **(VI)** and **(III)**).

(*iv*) Let $x(t) = \begin{bmatrix} x_1(t) & x_2(t) & \cdots & x_n(t) \end{bmatrix}'$ for $t \in [t_0, \tau]$ and

$$\mathbb{J} = \int_{t_0}^{\tau} (x'Px)dt,$$

where P is a constant real symmetric matrix. Let $\delta x(t) = \begin{bmatrix} \delta x_1(t) & \delta x_2(t) & \cdots & \delta x_n(t) \end{bmatrix}'$ be a variation in $x(t)$. Then, the corresponding variation in $\mathbb{J}$ is

$$\begin{aligned}
\delta \mathbb{J} &= \int_{t_0}^{\tau} \delta(x'(t)Px(t))dt \\
&= \int_{t_0}^{\tau} \left[\frac{d}{dx}(x'(Px)) + \frac{d}{dx}((x'P)x) \right]' \delta x(t)dt \quad \text{(from } \mathbf{(VI)}\text{)} \\
&= \int_{t_0}^{\tau} [Px + P'x]' \delta x(t)dt \quad \text{(from (6.4) and (6.5))} \\
&= 2\int_{t_0}^{\tau} (x'P\, \delta x)dt, \text{ since } P = P'.
\end{aligned}$$

Exercise 6.3.1 Let $F(x) = \begin{bmatrix} x_1^2 + x_3 & x_2 + x_3^2 \end{bmatrix}'$, where $x = \begin{bmatrix} x_1 & x_2 & x_3 \end{bmatrix}'$. Find: (*i*) $\frac{dF}{dx}$; (*ii*) ΔF and δF at $x^*(t) = \begin{bmatrix} t & t^2 & 1+t \end{bmatrix}'$ and $\delta x(t) = \begin{bmatrix} \frac{t}{10} & \frac{t^2}{5} & \frac{t}{20} \end{bmatrix}'$.

6.4 Euler's Equation and Boundary Conditions

In this section, the problem of maximizing or minimizing a functional $\mathbb{J}$ expressed in integral form is considered. Necessary and sufficient conditions for obtaining an optimal solution are derived using calculus of variation methods.

Let $\mathbb{J} : C[t_0, \tau] \to \mathbb{R}$ be a functional denoted by $\mathbb{J}(x)$. Let $S \subset C[t_0, \tau]$ be defined as $S = \{x \in C[t_0, \tau] : x(t_0) = x_0 \text{ and } x(\tau) = x_1\}$. Let $x^* \in S$ be a function such that $\mathbb{J}(x)$ attains its minimum (maximum) at x^* out of all $x \in S$ close to (in a neighborhood of) x^*, i.e, $\mathbb{J}(x^*) = \min_{x \in S, \|x - x^*\| < \varepsilon} \mathbb{J}(x)$ for a sufficiently small $\varepsilon > 0$.

The following lemma provides the necessary and sufficient conditions for $\mathbb{J}(x)$ to attain its minimum (maximum) at x^*.

Lemma 6.4.1 Necessary and sufficient conditions for the functional $\mathbb{J}(x)$ to attain a minimum (maximum) at $x = x^*$ are as follows:
Necessary condition. If

$$\mathbb{J}(x^*) = \underset{x \in S, \|x - x^*\| < \varepsilon}{\text{extremum}} \mathbb{J}(x)$$

for some $\varepsilon > 0$, then the first variation $\delta\mathbb{J} = 0$ at $x = x^*$.
Sufficient condition. If $\delta\mathbb{J} = 0$ and the second variation $\delta^2\mathbb{J} > 0\,(< 0)$ at $x = x^*$, then $\mathbb{J}(x^*)$ is a minimum (maximum) of $\mathbb{J}(x)$.

Proof. **Necessary condition.** Let $\eta : [t_0, \tau] \to \mathbb{R}$ be a continuous function such that $\eta(t_0) = \eta(\tau) = 0$. Let $\alpha \in \mathbb{R}$, and define $(\delta x)(t) = \alpha\eta(t)$ for sufficiently small value of α. Then

$$x(t) = x^*(t) + (\delta x)(t) = x^*(t) + \alpha\eta(t)$$

defines a class of functions close to x^*. By fixing η and varying α, we can express $\mathbb{J}(x)$ as a function of α, since

$$\mathbb{J}(x) = \mathbb{J}(x^* + \alpha\eta) = g(\alpha) \text{ (say)}. \tag{6.6}$$

Note that $\frac{dx}{d\alpha} = \eta$. Since the minimum (maximum) of $\mathbb{J}(x)$ occurs at $x = x^*$, i.e., at $\alpha = 0$, the necessary condition on g is

$$\left.\frac{dg}{d\alpha}\right|_{\alpha=0} = 0. \tag{6.7}$$

From (6.6), we have

$$\frac{dg}{d\alpha} = \frac{d\mathbb{J}}{dx}\frac{dx}{d\alpha} = \frac{d\mathbb{J}}{dx}\eta.$$

Hence,

$$\left.\frac{dg}{d\alpha}\right|_{\alpha=0} = \left.\frac{d\mathbb{J}}{dx}\right|_{x^*}\eta. \tag{6.8}$$

Since $\eta(t)$ is an arbitrary function with zero end conditions, it follows from (6.7) and (6.8) that

$$\left.\frac{d\mathbb{J}}{dx}\right|_{x^*} = 0.$$

Therefore,

$$\delta\mathbb{J}|_{x^*} = \left.\frac{d\mathbb{J}}{dx}\right|_{x^*}\delta x = 0$$

is the necessary condition for a minimum (maximum) of $\mathbb{J}$ at $x = x^*$.
Sufficient condition. A sufficient condition on g for attaining a minimum (maximum) at $\alpha = 0$ is $\left.\frac{d^2g}{d\alpha^2}\right|_{\alpha=0} > 0\,(< 0)$. Now,

$$\frac{d^2g}{d\alpha^2} = \frac{d^2\mathbb{J}}{dx^2}\left(\frac{dx}{d\alpha}\right)^2 + \frac{d\mathbb{J}}{dx}\frac{d^2x}{d\alpha^2}.$$

Therefore,

$$\left.\frac{d^2g}{d\alpha^2}\right|_{\alpha=0} = \left.\frac{d^2\mathbb{J}}{dx^2}\right|_{x^*} \eta^2 \quad \left(\text{since } \left.\frac{d\mathbb{J}}{dx}\right|_{x^*} = 0\right),$$

which implies that $\left.\frac{d^2g}{d\alpha^2}\right|_{\alpha=0}$ and $\left.\frac{d^2\mathbb{J}}{dx^2}\right|_{x^*}$ have the same sign. Thus, the sufficient condition for a minimum (maximum) at $x = x^*$ is

$$\left.\frac{d^2\mathbb{J}}{dx^2}\right|_{x^*} > 0\,(<0).$$

Hence, the second variation

$$\delta^2\mathbb{J}\big|_{x^*} = \left.\frac{d^2\mathbb{J}}{dx^2}\right|_{x^*} (\delta x)^2 > 0\,(<0)$$

is the sufficient condition for a minimum (maximum) of $\mathbb{J}$ at $x = x^*$. □

Problem I.
Case I. Fixed boundary conditions. First, let us consider the problem of extremizing the functional:

$$\mathbb{J}(x) = \int_{t_0}^{\tau} f(t, x(t), \dot{x}(t))dt \tag{6.9}$$

under the conditions $x(t_0) = x_0$ and $x(\tau) = x_1$. In other words, we want to find the optimal solution, called the optimal curve $x^*(t)$, which is chosen from the set of all differentiable curves x with end conditions $x(t_0) = x_0$ and $x(\tau) = x_1$ such that $\mathbb{J}$ attains the minimum or maximum at x^* (if it exists).

***Example* 6.4.2** It is well known that out of all curves $y = x(t)$ in $\mathbb{R}^2$ with end points $x(t_0) = x_0$ and $x(\tau) = x_1$, the one with minimum arc length is the segment of the straight line joining the points (t_0, x_0) and (τ, x_1). This can be expressed as the following minimization problem:

$$\text{minimize } \mathbb{J}(x) = \int_{t_0}^{\tau} ds = \int_{t_0}^{\tau} \sqrt{(dx)^2 + (dt)^2} = \int_{t_0}^{\tau} \sqrt{1+\dot{x}^2}dt$$

under the conditions $x(t_0) = x_0$ and $x(\tau) = x_1$.

Solution to this problem using variation method can be seen after introducing Euler's equation.

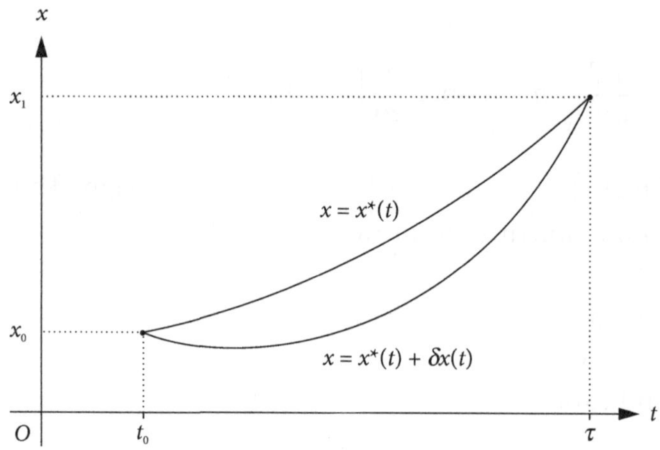

Figure 6.1 Fixed boundary

Euler's Equation

To obtain a **necessary condition** for the extremization problem (6.9), let $x^*(t)$ be the optimal curve and $x(t)$ be any other neighboring curve with the same boundary conditions. Let δx be a variation in x^* so that

$$x(t) = x^*(t) + (\delta x)(t). \tag{6.10}$$

Then

$$\frac{d}{dt}x(t) = \frac{d}{dt}x^*(t) + \frac{d}{dt}(\delta x)(t).$$

Hence

$$\dot{x}(t) = \dot{x}^*(t) + \frac{d}{dt}(\delta x)(t).$$

Hereafter, we shall denote

$$\frac{d}{dt}(\delta x)(t) = \delta \dot{x}(t). \tag{6.11}$$

(Note that in the above equation the RHS is used only as a notation for the LHS. In general, the derivative of a variation and a variation of the derivative are not the same.)

We can obtain a collection of curves $x(t)$ with the same boundary conditions as that of $x^*(t)$ if we take variation functions $\delta x(t)$ with $\delta x(a) = \delta x(\tau) = 0$. Now (6.9) can be written as

$$\mathbb{J}(x) = \int_{t_0}^{\tau} f(t, x^*(t) + \delta x(t), \dot{x}^*(t) + \delta \dot{x}(t))dt.$$

From Lemma 6.4.1, a necessary condition for obtaining an extremum of $\mathbb{J}(x)$ at x^* is that the first variation $\delta\mathbb{J} = 0$ at x^*. Hence using (6.3),

$$(\delta\mathbb{J})(x^*) = \int_{t_0}^{\tau} \left(\left(\frac{\partial f}{\partial x}\right)_{x^*} \delta x(t) + \left(\frac{\partial f}{\partial \dot{x}}\right)_{x^*} \delta \dot{x}(t) \right) dt = 0. \tag{6.12}$$

Now,

$$\begin{aligned}
\int_{t_0}^{\tau} \left(\frac{\partial f}{\partial \dot{x}}\right)_{x^*} \delta \dot{x}(t)dt &= \int_{t_0}^{\tau} \left(\frac{\partial f}{\partial \dot{x}}\right)_{x^*} \frac{d}{dt}(\delta x)(t)dt \quad \text{(from (6.11))} \\
&= \int_{t_0}^{\tau} \left(\frac{\partial f}{\partial \dot{x}}\right)_{x^*} d(\delta x)(t) \\
&= (\delta x)(t) \left(\frac{\partial f}{\partial \dot{x}}\right)_{x^*} \Bigg|_{t_0}^{\tau} - \int_{t_0}^{\tau} \left(\frac{d}{dt}\frac{\partial f}{\partial \dot{x}}\right)_{x^*} \delta x(t)dt. \qquad (6.13)
\end{aligned}$$

(using integration by parts)

Using (6.13) in equation (6.12), we obtain

$$\left(\left[\frac{\partial f}{\partial \dot{x}}\right]_{\tau} (\delta x)(\tau) - \left[\frac{\partial f}{\partial \dot{x}}\right]_{t_0} (\delta x)(t_0) \right) + \int_{t_0}^{\tau} \left(\frac{\partial f}{\partial x} - \frac{d}{dt}\left(\frac{\partial f}{\partial \dot{x}}\right) \right)_{x^*} \delta x(t)dt = 0. \tag{6.14}$$

Since $(\delta x)(\tau) = 0 = (\delta x)(t_0)$, we obtain the following equation:

$$\int_{t_0}^{\tau} \left(\frac{\partial f}{\partial x} - \frac{d}{dt}\left(\frac{\partial f}{\partial \dot{x}}\right) \right)_{x^*} \delta x(t)dt = 0. \tag{6.15}$$

Since $\delta x(t)$ is any arbitrary function (with zero boundary conditions), we get the following:

$$\frac{\partial f}{\partial x} - \frac{d}{dt}\left(\frac{\partial f}{\partial \dot{x}}\right) = 0 \text{ at } x^*. \tag{6.16}$$

Equation (6.16) is known as **Euler's equation**, which gives a necessary condition for minimizing or maximizing $\mathbb{J}$. The optimal function x^* (if it exists) can be obtained by solving Euler's equation (6.16) under the given boundary conditions.

Second variation test: To decide whether the solution $x^*(t)$ of Euler's equation (6.16) provides a maximum or minimum of the functional $\mathbb{J}$, the second variation of $\mathbb{J}$ at x^* can be used, similar to second derivative test for maxima/minima in differential calculus. (i) $\delta^2\mathbb{J}(x^*) > 0 \implies \mathbb{J}(x^*)$ is the minimum value of $\mathbb{J}(x)$. (ii) $\delta^2\mathbb{J}(x^*) < 0 \implies \mathbb{J}(x^*)$ is the maximum of $\mathbb{J}(x)$. From (6.9)–(6.11), and the second derivative terms in Taylor's series of f about x^* we get,

$$\delta^2\mathbb{J}(x^*) = \frac{1}{2}\int_{t_0}^{\tau} \begin{bmatrix} \delta x & \delta \dot{x} \end{bmatrix} \begin{bmatrix} f_{xx} & f_{x\dot{x}} \\ f_{x\dot{x}} & f_{\dot{x}\dot{x}} \end{bmatrix}_{x^*} \begin{bmatrix} \delta x \\ \delta \dot{x} \end{bmatrix} dt.$$

Now, since $\delta x(t)$ and $\delta \dot{x}(t)$ are arbitrary, $\delta^2\mathbb{J}(x^*) > 0$ (< 0) implies $\begin{bmatrix} f_{xx} & f_{x\dot{x}} \\ f_{x\dot{x}} & f_{\dot{x}\dot{x}} \end{bmatrix}_{x^*}$ is positive definite (negative definite), which provides a sufficient condition to ascertain that $\mathbb{J}(x^*)$ is a minimum (maximum) of the functional $\mathbb{J}(x)$. In the case where f is independent of x, then $f_{xx} = f_{x\dot{x}} = 0$ and $\delta^2\mathbb{J} = \frac{1}{2}\int_{t_0}^{\tau} f_{\dot{x}\dot{x}}(\delta\dot{x})^2 dt$. Hence, the sign of $\delta^2\mathbb{J}$ depends only on $(f_{\dot{x}\dot{x}})_{x^*}$. Thus, $(f_{\dot{x}\dot{x}})_{x^*} > 0 (< 0)$ implies $\mathbb{J}$ attains minimum (maximum) at x^*.

Solution (to Example 6.4.2) Given that $f(t,x,\dot{x}) = \sqrt{1+\dot{x}^2}$. Using Euler's equation (6.16), we obtain $\frac{d}{dt}\left(\frac{\dot{x}}{\sqrt{1+\dot{x}^2}}\right) = 0$, which implies that $\dot{x}$ is a constant, say c. Hence, $x(t) = ct + d$, where c and d can be obtained using the boundary conditions $x(t_0) = x_0$ and $x(\tau) = x_1$. Now since $f_{xx} = 0 = f_{x\dot{x}}$ and $f_{\dot{x}\dot{x}} > 0$, $\delta^2\mathbb{J} = \frac{1}{2}\int_{t_0}^{\tau} f_{\dot{x}\dot{x}}(\delta\dot{x})^2 dt > 0$. From the second variation test, we conclude that the solution obtained provides a minimum of the functional $\mathbb{J}$.

Case II. Free boundary conditions.

$$\text{Extremize } \mathbb{J} = \int_{t_0}^{\tau} f(t, x(t), \dot{x}(t))dt$$

with $x(t_0) = x_0$ but $x(\tau)$ is not specified.

Let x^* be the function that satisfies $x^*(t_0) = x_0$ and extremizes $\mathbb{J}$. To find a necessary condition to be satisfied by x^*, one can follow the same procedure as in Problem I and obtain equation (6.14)

$$(\delta\mathbb{J})(x^*) = \int_{t_0}^{\tau} \left[\frac{\partial f}{\partial x} - \frac{d}{dt}\left(\frac{\partial f}{\partial \dot{x}}\right)\right]_{x^*} \delta x(t)dt + \left(\delta x(t)\frac{\partial f}{\partial \dot{x}}\right)_{\tau} - \left(\delta x(t)\frac{\partial f}{\partial \dot{x}}\right)_{t_0} = 0. \tag{6.17}$$

Since $x(t_0)$ is fixed, we have $\delta x(t_0) = 0$, whereas $x(\tau)$ is unspecified, so $\delta x(\tau)$ need not be zero. However, to minimize $\mathbb{J}$, one must consider all variations $\delta x(t)$ with $\delta x(\tau) =$

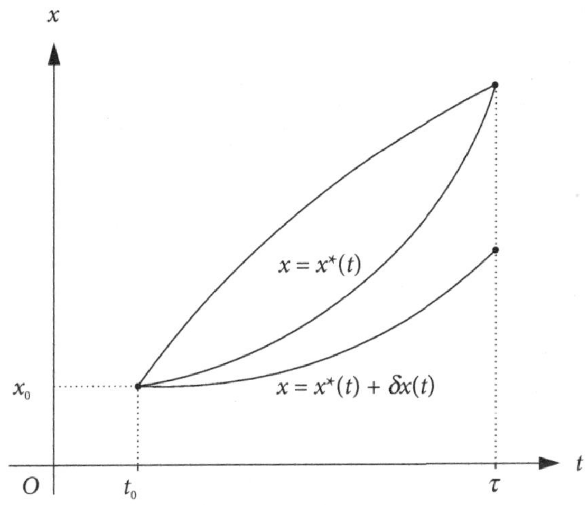

Figure 6.2 Free final condition

0 as well as $\delta x(\tau) \neq 0$. By considering all curves with $\delta x(\tau) = 0$ in equation (6.16), we obtain the Euler's equation (6.16) to be satisfied by x^* as in Case I. Substituting (6.16) into (6.17) for all possible functions $\delta x(t)$ (with $\delta x(t_0) = 0$ and $\delta x(\tau)$ arbitrary), we have

$$\delta x(\tau) \left(\frac{\partial f}{\partial \dot{x}} \right)_\tau = 0,$$

which implies

$$\left. \frac{\partial f}{\partial \dot{x}} \right|_{t=\tau} = 0. \tag{6.18}$$

Similarly, if $x(t_0)$ is unspecified, then the required condition at $t = t_0$ is

$$\left. \frac{\partial f}{\partial \dot{x}} \right|_{t=t_0} = 0.$$

Thus, $x^*(t)$ can be obtained by solving Euler's equation (6.16) under suitable boundary conditions.

***Example* 6.4.3** Find the extremum of

$$\mathbb{J}(x) = \int_0^1 (x^2(t) + \dot{x}^2(t))dt,$$

where $\mathbb{J}$ is defined for all differentiable functions $x(t)$ with boundary conditions:

(i) $x(0) = 1$ and $x(1) = 0$,

(ii) $x(0) = 1$ and $x(1)$ is free.

Solution We need to find the optimal function x^* such that $\mathbb{J}(x)$ attains the extremum value at x^*. Let

$$f(t,x,\dot{x}) = x^2 + \dot{x}^2.$$

Then Euler's equation (6.16) is

$$\frac{\partial f}{\partial x} - \frac{d}{dt}\left(\frac{\partial f}{\partial \dot{x}}\right) = 2x - \frac{d}{dt}(2\dot{x}) = 0 \implies \ddot{x} - x = 0.$$

Hence, $x(t) = c_1e^t + c_2e^{-t}$ where c_1 and c_2 are arbitrary constants. Now $x^*(t)$ can be obtained by finding c_1 and c_2 using suitable boundary conditions.

(i) As $x(0) = 1$ and $x(1) = 0$, $c_1 + c_2 = 1$ and $ec_1 + \frac{c_2}{e} = 0$. Hence, $x^*(t) = \frac{1}{1-e^2}(e^t - e^{2-t})$.

(ii) As $x(0) = 1$ and $\left(\frac{\partial f}{\partial \dot{x}}\right)_{t=1} = 0$, i.e., $\dot{x}(1) = 0$, we get $c_1 + c_2 = 1$ and $ec_1 - \frac{c_2}{e} = 0$. Hence, $x^*(t) = \frac{1}{1+e^2}(e^t + e^{2-t})$.

The following lemma will be useful for deriving boundary conditions when (i) final time τ is unspecified but $x(\tau)$ is specified (ii) both τ and $x(\tau)$ are not specified.

Lemma 6.4.4 Let $x^*(t)$ for $t_0 \leq t \leq \tau^*$ be the optimal curve AB (as shown in Figure 6.1) that provides the extremum value of functional $\mathbb{J}$ given in (6.9). Let δx be a variation in x^* such that $(x^* + \delta x)(t)$ represents a neighboring curve AC to AB. We consider the general case where both τ^* and $x^*(\tau^*) = x_{\tau^*}$ are unspecified. Let $\Delta\tau$ denote an increment in τ^* (note that $\Delta\tau$ can be positive, zero, or negative), and let $(x^* + \delta x)(t)$ be defined on the interval $t_0 \leq t \leq \tau^* + \Delta\tau$. Define δx_τ as the difference between the end points of x^* and $x^* + \delta x$, i.e.,

$$\delta x_\tau = (x^* + \delta x)(\tau^* + \Delta\tau) - x^*(\tau^*) = CF,$$

and let

$$(\delta x)(\tau^*) = (x^* + \delta x)(\tau^*) - x^*(\tau^*) = EF.$$

Then,

$$(\delta x)(\tau^*) = \delta x_\tau - \dot{x}(\tau^*)\Delta\tau.$$

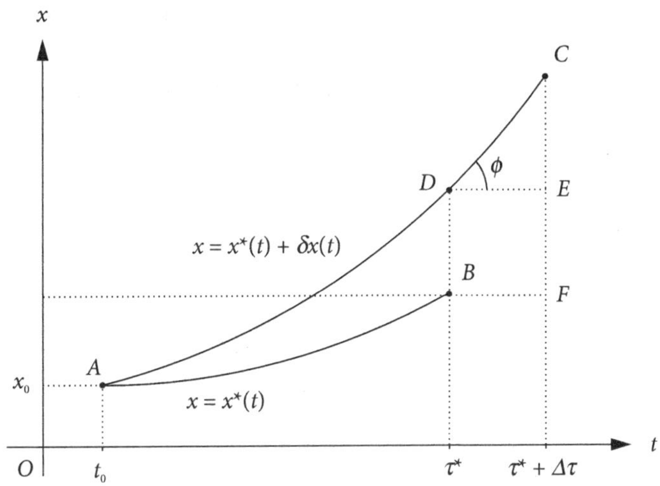

Figure 6.3 Relation between $(\delta x)(\tau^*)$, δx_τ, and $\Delta\tau$

Proof. The slope of the tangent to $x^* + \delta x$ at D is given by

$$\tan\phi = \frac{d}{dt}(x^* + \delta x)|_{t=\tau^*} \approx \frac{CE}{DE} = \frac{CF - EF}{DE} = \frac{\delta x_\tau - \delta x(\tau^*)}{\Delta\tau}.$$

Hence,

$$\dot{x}^*(\tau^*) + (\delta\dot{x})(\tau^*) = \frac{\delta x_\tau - \delta x(\tau^*)}{\Delta\tau}, \tag{6.19}$$

which implies

$\delta x(\tau^*) = \delta x_\tau - \dot{x}(\tau^*)\Delta\tau$ (by neglecting the second order variation $\Delta\tau(\delta\dot{x}(\tau^*))$). (6.20)

If the final time is already specified in a problem and x_τ is free (Case II), then $\Delta\tau = 0$, and from (6.20), we obtain

$$\delta x(\tau^*) = \delta x_\tau. \tag{6.21}$$

If τ is unspecified but x_τ is specified (Case III discussed below), then $\delta x_\tau = 0$, and from (6.20), we obtain

$$\delta x(\tau^*) = -\dot{x}(\tau^*)\Delta\tau. \tag{6.22}$$

If both τ and x_τ are unspecified, then equation (6.20) holds. □

Case III. Free final time.

$$\text{Extremize } \mathbb{J} = \int_{t_0}^{\tau} f(t, x(t), \dot{x}(t))dt \tag{6.23}$$

under the following conditions: t_0 is fixed, the final time τ is free (unspecified), and $x(t_0) = x_0$ and $x(\tau) = x_1$ are specified.

Let $x^*(t)$ be the optimal solution and τ^* the optimal final time, and let $x = x^* + \delta x$ and $\tau = \tau^* + \Delta\tau$, where δx is a variation in x^* and $\Delta\tau$ is an increment in τ. The increment in $\mathbb{J}$ at x^* is given by

$$\begin{aligned}\Delta\mathbb{J} &= \int_{t_0}^{\tau^*+\Delta\tau} f(t, x^* + \delta x, \dot{x}^* + \delta\dot{x})dt - \int_{t_0}^{\tau^*} f(t, x^*, \dot{x}^*)dt \\ &= \int_{t_0}^{\tau^*} (f(t, x^* + \delta x, \dot{x}^* + \delta\dot{x}) - f(t, x^*, \dot{x}^*))dt \\ &\quad + \int_{\tau^*}^{\tau^*+\Delta\tau} f(t, x^* + \delta x, \dot{x}^* + \delta\dot{x})dt. \end{aligned} \tag{6.24}$$

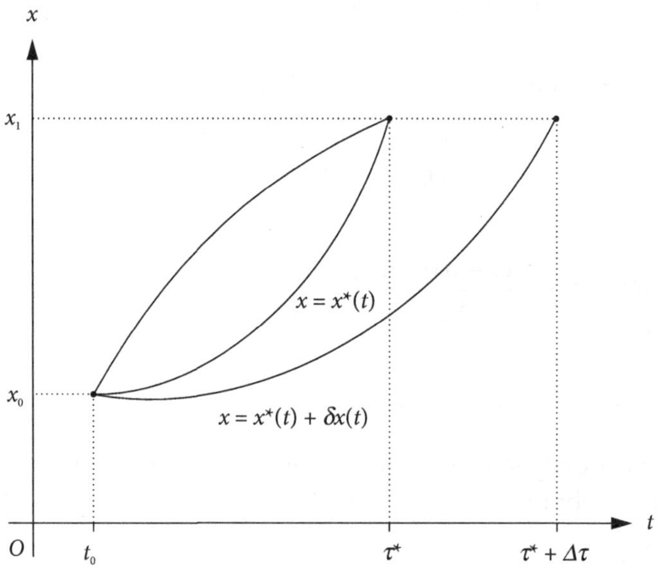

Figure 6.4 Free final time

Using Taylor's series expansion about x^* and $\dot{x}^*$, and neglecting higher order terms, we obtain the first variation in $\mathbb{J}$ as:

$$\delta\mathbb{J} = \int_{t_0}^{\tau^*} \left[\left(\frac{\partial f}{\partial x}\right)_{x^*} \delta x + \left(\frac{\partial f}{\partial \dot{x}}\right)_{x^*} \delta\dot{x} \right] dt + \int_{\tau^*}^{\tau^*+\Delta\tau} f(t, x^* + \delta x, \dot{x}^* + \delta\dot{x})dt.$$

For sufficiently small $\Delta\tau$ the last term is approximated by $(f(t,x(t),\dot{x}(t)))_{\tau^*}\Delta\tau$. Hence, we have

$$\delta\mathbb{J} \approx \int_{t_0}^{\tau^*}\left(\frac{\partial f}{\partial x}\right)_{x^*}\delta x\,dt + \int_{t_0}^{\tau^*}\left(\frac{\partial f}{\partial \dot{x}}\right)_{x^*}\delta\dot{x}\,dt + (f(t,x(t),\dot{x}(t)))_{\tau^*}\Delta\tau. \tag{6.25}$$

Using the notation $\delta\dot{x} = \frac{d}{dt}(\delta x)$ (see (6.11)) and applying integration by parts in the second term, we get

$$\begin{aligned}\int_{t_0}^{\tau^*}\left(\frac{\partial f}{\partial \dot{x}}\right)_{x^*}\delta\dot{x}\,dt &= \int_{t_0}^{\tau^*}\left(\frac{\partial f}{\partial \dot{x}}\right)_{x^*}\frac{d}{dt}\delta x\,dt = \int_{t_0}^{\tau}\left(\frac{\partial f}{\partial \dot{x}}\right)_{x^*}d(\delta x)\\ &= \left(\delta x(t)\,\frac{\partial f}{\partial \dot{x}}\right)\Bigg|_{t_0}^{\tau^*} - \int_0^{\tau^*}\left(\frac{d}{dt}\left(\frac{\partial f}{\partial \dot{x}}\right)\right)_{x^*}\delta x(t)\,dt.\end{aligned}$$

Since $\delta\mathbb{J} = 0$ is the necessary condition for extremizing $\mathbb{J}$, from (6.25), we obtain the following equation:

$$\begin{aligned}0 = \delta\mathbb{J} = &\int_0^{\tau^*}\left(\frac{\partial f}{\partial x} - \frac{d}{dt}\left(\frac{\partial f}{\partial \dot{x}}\right)\right)_{x^*}\delta x(t)\,dt\\ &+ \left(\delta x(t)\,\frac{\partial f}{\partial \dot{x}}\right)\Bigg|_{t_0}^{\tau^*} + f(t,x(t),\dot{x}(t))|_{\tau^*}\Delta\tau.\end{aligned}$$

Since $x(t_0)$ is fixed, $\delta x(t_0) = 0$. Hence,

$$\begin{aligned}0 = \delta\mathbb{J} = &\int_0^{\tau^*}\left(\frac{\partial f}{\partial x} - \frac{d}{dt}\left(\frac{\partial f}{\partial \dot{x}}\right)\right)_{x^*}\delta x(t)\cdot dt\\ &+ \delta x(\tau^*)\left(\frac{\partial f}{\partial \dot{x}}\right)_{\tau^*} + f(t,x(t),\dot{x}(t))|_{\tau^*}\Delta\tau.\end{aligned} \tag{6.26}$$

Note that the optimal curve x^* must remain optimal even when τ^* and $x(\tau^*)$ are specified, i.e., $\Delta\tau = 0$ and $\delta x(\tau^*) = 0$. Substituting into the above equation and using the arbitrariness of $\delta x(t)$, we obtain Euler's equation:

$$\frac{\partial f}{\partial x} - \frac{d}{dt}\left(\frac{\partial f}{\partial \dot{x}}\right) = 0 \text{ at } x = x^*. \tag{6.27}$$

Thus, **Euler's equation must be satisfied by the optimal function, irrespective of whether end conditions are specified or not.** Substituting (6.27) into (6.26), we get

$$0 = \delta\mathbb{J} = \delta x(\tau^*)\left(\frac{\partial f}{\partial \dot{x}}\right)_{\tau^*} + f(t,x(t),\dot{x}(t))|_{\tau^*}\Delta\tau. \tag{6.28}$$

Using the approximation $\delta x(\tau^*) \approx -\dot{x}^*(\tau^*)\Delta\tau$ (from Lemma 6.4.4, equation (6.22)), the above equation becomes

$$\left[f(t,x(t),\dot{x}(t)) - \dot{x}(t)\frac{\partial f}{\partial \dot{x}}\right]_{\tau^*} \Delta\tau = 0.$$

Since $\Delta\tau$ is arbitrary, we obtain

$$\left[f(t,x(t),\dot{x}(t)) - \dot{x}(t)\frac{\partial f}{\partial \dot{x}}\right]_{\tau^*} = 0. \tag{6.29}$$

Hence, (6.27) and (6.29), along with $x(t_0) = x_0$, are the necessary conditions for extremizing $\mathbb{J}$ in Case III.

Case IV. Free final time and free final condition. If τ and $x(\tau) = x_\tau$ are unspecified, then using (6.20) in (6.28), we get

$$\left(\frac{\partial f}{\partial \dot{x}}\right)_\tau \delta x_\tau + \left[f(t,x(t),\dot{x}(t)) - \dot{x}(t)\frac{\partial f}{\partial \dot{x}}\right]_\tau \Delta\tau = 0.$$

As both δx_τ and $\Delta\tau$ are arbitrary, we obtain the conditions $\left(\frac{\partial f}{\partial \dot{x}}\right)_\tau = 0$ and $\left[f(t,x(t),\dot{x}(t)) - \dot{x}(t)\frac{\partial f}{\partial \dot{x}}\right]_\tau = 0$, which imply $[f(t,x(t),\dot{x}(t))]_\tau = 0$. Using these two conditions, τ and $x(\tau)$ can be determined (**if these equations are consistent**).

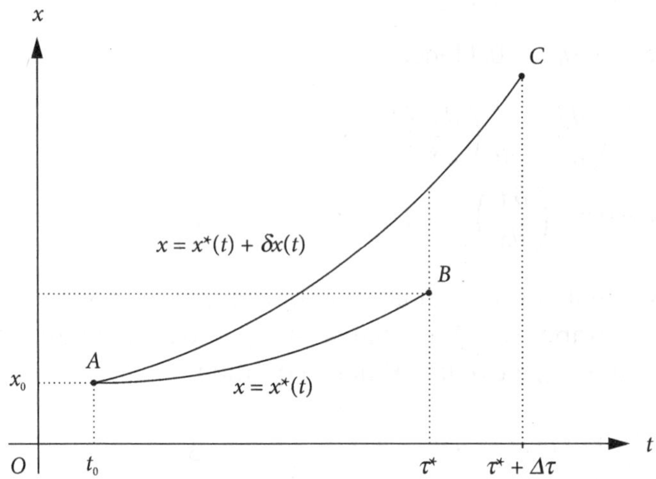

Figure 6.5 Free final time and final condition

Example **6.4.5**

Minimize $\mathbb{J}(x) = \int_0^\tau (x^2(t) + \dot{x}^2(t))dt$

over all functions x satisfying $x(0) = 1$ and $x(\tau) = 1$, where τ is free.

Solution Let $f(t,x,\dot{x}) = x^2 + \dot{x}^2$. Then, Euler's equation is

$$\frac{\partial f}{\partial x} - \frac{d}{dt}\left(\frac{\partial f}{\partial \dot{x}}\right) = 0 \implies \ddot{x} - x = 0 \implies x(t) = c_1 e^t + c_2 e^{-t}.$$

The final time τ can be obtained from (6.29).

$$\left(f(t,x(t),\dot{x}(t)) - \dot{x}\frac{\partial f}{\partial \dot{x}}\right)_\tau = 0 \implies \left(x^2 + \dot{x}^2 - \dot{x}(2\dot{x})\right)_\tau = 0 \implies x^2(\tau) - \dot{x}^2(\tau) = 0. \tag{6.30}$$

Now, since $x(0) = 1, c_1 + c_2 = 1$. Hence,

$$x(t) = c_1 e^t + (1 - c_1)e^{-t}. \tag{6.31}$$

The final condition $x(\tau) = 1$ implies

$$c_1 e^\tau + (1 - c_1)e^{-\tau} = 1. \tag{6.32}$$

Now from (6.30), $\left(x^2 - \dot{x}^2\right)_\tau = 0 \implies x(\tau) = \pm\dot{x}(\tau) \implies \dot{x}(\tau) = \pm 1$. If $\dot{x}(\tau) = 1$, then from (6.31),

$$c_1 e^\tau - (1 - c_1)e^{-\tau} = 1. \tag{6.33}$$

From (6.32) and (6.33), we get $c_2 e^{-\tau} = 0 \implies c_2 = 0$ or $\tau = \infty$. Hence, $c_1 = 1$ and $x(t) = e^t$. Now, $x(\tau) = 1 \implies \tau = 0$, which is a contradiction. Thus, we conclude that the given extremization problem has NO non-trivial solution.

6.5 Multi-variable Optimization Problem

Problem I. Consider the problem of finding the functions $x_i^*(t) : i = 1,2,\ldots,n$, $t_0 \leq t \leq \tau$, which extremize the performance index or cost functional

$$\mathbb{J}(x_1, x_2, \ldots, x_n) = \int_{t_0}^{\tau} f(t, x_1(t), x_2(t), \ldots, x_n(t), \dot{x}_1(t), \dot{x}_2(t), \ldots, \dot{x}_n(t))dt \tag{6.34}$$

subject to boundary conditions on $x_i(t) : x_i(t_0) = x_{i0}$ and $x_i(\tau) = x_{i\tau}$. Here, $x_i(t)$ and f are real valued differentiable functions. The optimal solutions $x_i^*(t)$ (if they exist) can be obtained by solving the Euler system of n equations as given below.

$$\frac{\partial f}{\partial x_i} - \frac{d}{dt}\left(\frac{\partial f}{\partial \dot{x}_i}\right) = 0, \quad i = 1,2,\ldots,n, \tag{6.35}$$

along with the boundary conditions at t_0 and τ, as in the following discussion.

The proof given below is similar to that of the one variable problem.

Let $x_i^*(t) : i = 1,2,\ldots,n$ form the optimal solution. Then the necessary condition is that the variation of $\mathbb{J}$ must be zero, i.e.,

$$\delta\mathbb{J} = 0 \text{ at } x_i^*, \quad i = 1,2,\ldots,n. \tag{6.36}$$

Now, (from (6.3)),

$$(\delta\mathbb{J})(x^*) = \int_{t_0}^{\tau} \sum_{i=1}^{n} \left[\left(\frac{\partial f}{\partial x_i}\right)_{x^*} \delta x_i(t) + \left(\frac{\partial f}{\partial \dot{x}_i}\right)_{x^*} \delta \dot{x}_i(t) \right] dt. \tag{6.37}$$

Since

$$\begin{aligned} \int_{t_0}^{\tau} \left(\frac{\partial f}{\partial \dot{x}_i}\right)_{x^*} \delta \dot{x}_i(t) dt &= \int_{t_0}^{\tau} \left(\frac{\partial f}{\partial \dot{x}_i}\right)_{x^*} \frac{d}{dt}(\delta x_i(t))\, dt \\ &= \left[\left(\frac{\partial f}{\partial \dot{x}_i}\right)_{x^*} \delta x_i(t) \right]_{t_0}^{\tau} - \int_{t_0}^{\tau} \frac{d}{dt}\left(\frac{\partial f}{\partial \dot{x}_i}\right)_{x^*} dt, \end{aligned} \tag{6.38}$$

(using integration by parts)

from (6.36)–(6.38), we get

$$(\delta\mathbb{J})(x^*) = \int_{t_0}^{\tau} \left(\sum_{i=1}^{n} \left(\frac{\partial f}{\partial x_i} - \frac{d}{dt}\left(\frac{\partial f}{\partial \dot{x}_i}\right)\right)_{x^*} \delta x_i(t) \right) dt + \sum_{i=1}^{n} \left[\left(\frac{\partial f}{\partial \dot{x}_i}\right)_{x^*} \delta x_i(t) \right]_{t_0}^{\tau} = 0. \tag{6.39}$$

As $x_i(t_0)$ and $x_i(\tau)$ are specified for all $i = 1,2,\ldots,n$, $\delta x_i(t_0) = 0 = \delta x_i(\tau)$. Hence, equation (6.39) becomes

$$\int_{t_0}^{\tau} \left(\sum_{i=1}^{n} \left(\frac{\partial f}{\partial x_i} - \frac{d}{dt}\left(\frac{\partial f}{\partial \dot{x}_i}\right)\right)_{x^*} \delta x_i(t) \right) dt = 0. \tag{6.40}$$

Since the variations $\delta x_i(t)$ are arbitrary for all $i = 1,2,\ldots,n$, equation (6.40) implies

$$\left(\frac{\partial f}{\partial x_i} - \frac{d}{dt}\left(\frac{\partial f}{\partial \dot{x}_i}\right)\right)_{x^*} = 0, \quad i = 1,2,\ldots,n. \tag{6.41}$$

In other words, the optimal solutions $x_i^* : i = 1,2,\ldots,n$ must satisfy the equations

$$\frac{\partial f}{\partial x_i} - \frac{d}{dt}\left(\frac{\partial f}{\partial \dot{x}_i}\right) = 0, \quad i = 1,2,\ldots,n, \tag{6.42}$$

along with the boundary conditions.

Euler's Equation in Vector Form

In vector form, the above problem can be written as given below.

$$\text{Extremize } \mathbb{J}(x) = \int_{t_0}^{\tau} f(t, x(t), \dot{x}(t))dt, \tag{6.43}$$

where $x(t) = \begin{bmatrix} x_1(t) & x_2(t) & \cdots & x_n(t) \end{bmatrix}' \in \mathbb{R}^n$ under boundary conditions $x(t_0) = x_0$ and $x(\tau) = x_\tau$.

Let $\delta x(t) = \begin{bmatrix} \delta x_1(t) & \delta x_2(t) & \cdots & \delta x_n(t) \end{bmatrix}'$ be a variation vector function. As $\frac{\partial f}{\partial x} = \begin{bmatrix} \frac{\partial f}{\partial x_1} & \frac{\partial f}{\partial x_2} & \cdots & \frac{\partial f}{\partial x_n} \end{bmatrix}'$ (see Section 6.3 (III)), the equation (6.39) can be written as

$$(\delta\mathbb{J})(x^*) = 0 \implies \int_{t_0}^{\tau} \left(\left(\frac{\partial f}{\partial x} \right)' - \frac{d}{dt} \left(\frac{\partial f}{\partial \dot{x}} \right)' \right)_{x^*} \delta x(t)dt + \left[\left(\frac{\partial f}{\partial \dot{x}} \right)'_{x^*} \delta x(t) \right]_{t_0}^{\tau} = 0. \tag{6.44}$$

Now, since $x(t_0)$ and $x(\tau)$ are specified, $\delta x(t_0) = 0 = \delta x(\tau)$ and since $\delta x(t)$ is arbitrary variation vector, from (6.44), we can infer that the optimal solution x^* satisfies

$$\frac{\partial f}{\partial x} - \frac{d}{dt} \left(\frac{\partial f}{\partial \dot{x}} \right) = 0, \tag{6.45}$$

which is the vector form of the system of Euler's equations given in (6.41).

Now, using arguments similar to the one variable case (with suitable modifications to suit vector notation), we obtain the following results for the four types of boundary conditions.

Case I. Fixed end conditions. $x(t_0) = x_0$ and $x(\tau) = x_\tau$ are specified in $\mathbb{R}^n$. The optimal vector function $x^*(t)$ (if it exists) is the solution of (6.45) (equivalently (6.41)) under the conditions $x(t_0) = x_0$ and $x(\tau) = x_1$.

Case II. $x(t_0) = x_0$ and τ are specified, but $x(\tau)$ is unspecified. In this case, x^* satisfies (6.45) under the conditions

$$x(t_0) = x_0 \text{ and } \left(\frac{\partial f}{\partial \dot{x}} \right)_\tau = 0. \tag{6.46}$$

Case III. $x(t_0) = x_0$ and $x(\tau) = x_\tau$ are specified, but τ is unspecified. To obtain τ and $x^*(t)$, solve (6.45) under the conditions

$$x(t_0) = x_0, x(\tau) = x_\tau \text{ and } \left[f(t, x(t), \dot{x}(t)) - \left(\frac{\partial f}{\partial \dot{x}} \right)' \dot{x}(t) \right]_\tau = 0. \tag{6.47}$$

Case IV. $x(t_0) = x_0$ is specified, but both τ and $x(\tau)$ are unspecified. In this case, solve (6.45) under the conditions

$$x(t_0) = x_0, \left(\frac{\partial f}{\partial \dot{x}}\right)_\tau = 0, \text{ and } (f(t,x(t),\dot{x}(t)))_\tau = 0. \tag{6.48}$$

***Example* 6.5.1** Find the extremum of

$$\mathbb{J}(x_1,x_2) = \int_1^2 (x_1^2 + \dot{x}_1\dot{x}_2 + x_2^2)dt,$$

where x_1 and x_2 satisfy the boundary conditions $x_1(1) = 1, x_1(2) = 0, x_2(1) = 2$, and $x_2(2) = 0$.

Solution Let $f(t,x_1,\dot{x}_1,x_2,\dot{x}_2) = x_1^2 + \dot{x}_1\dot{x}_2 + x_2^2$. Euler's equations to obtain the optimal functions $x_1^*(t)$ and $x_2^*(t)$ are

$$\frac{\partial f}{\partial x_1} - \frac{d}{dt}\left(\frac{\partial f}{\partial \dot{x}_1}\right) = 2x_1 - \frac{d}{dt}(\dot{x}_2) = 0$$

and

$$\frac{\partial f}{\partial x_2} - \frac{d}{dt}\left(\frac{\partial f}{\partial \dot{x}_2}\right) = 2x_2 - \frac{d}{dt}(\dot{x}_1) = 0,$$

which imply $x_1^{(4)} = 4x_1$ and $x_2 = \frac{1}{2}\ddot{x}_1$. Hence,

$$x_1^*(t) = c_1e^{\sqrt{2}t} + c_2e^{-\sqrt{2}t} + c_3e^{i\sqrt{2}t} + c_4e^{-i\sqrt{2}t} \tag{6.49}$$

and

$$x_2^*(t) = c_1e^{\sqrt{2}t} + c_2e^{-\sqrt{2}t} - c_3e^{i\sqrt{2}t} - c_4e^{-i\sqrt{2}t}, \tag{6.50}$$

where $c_i : i = 1,2,3,4$ are complex constants.

Since $x_1^*(t)$ and $x_2^*(t)$ are real valued, from (6.49) and (6.50), it follows that c_1 and c_2 are real constants and $c_3 = \overline{c_4}$. Let $c_3 = \alpha_3 + i\beta_3$. Then (6.49) and (6.50) can be written as

$$x_1^*(t) = c_1e^{\sqrt{2}t} + c_2e^{-\sqrt{2}t} + 2\alpha_3\cos(\sqrt{2}t) - 2\beta_3\sin(\sqrt{2}t)$$

and

$$x_2^*(t) = c_1e^{\sqrt{2}t} + c_2e^{-\sqrt{2}t} - 2\alpha_3\cos(\sqrt{2}t) + 2\beta_3\sin(\sqrt{2}t).$$

Now, using the boundary conditions, the constants c_1, c_2, α_3, and β_3 can be obtained.

Alternative procedure. Set $x_1 = y_1, \dot{x}_1 = y_2, x_2 = y_3$ and $\dot{x}_2 = y_4$ in the above Euler's equation. Then, $\dot{y}_1 = y_2, \dot{y}_2 = 2y_3, \dot{y}_3 = y_4$, and $\dot{y}_4 = 2y_1$, which can be expressed in matrix form as given below

$$\begin{bmatrix} \dot{y}_1(t) \\ \dot{y}_2(t) \\ \dot{y}_3(t) \\ \dot{y}_4(t) \end{bmatrix} = \begin{bmatrix} 0 & 1 & 0 & 0 \\ 0 & 0 & 2 & 0 \\ 0 & 0 & 0 & 1 \\ 2 & 0 & 0 & 0 \end{bmatrix} \begin{bmatrix} y_1 \\ y_2 \\ y_3 \\ y_4 \end{bmatrix}$$

with the boundary conditions $y_1(1) = 1$, $y_1(2) = 0$, $y_3(1) = 2$, and $y_3(2) = 0$. Using (4.56), the general solution of the system can be written as

$$\begin{bmatrix} y_1(t) \\ y_2(t) \\ y_3(t) \\ y_4(t) \end{bmatrix} = d_1 e^{\lambda_1 t} X_1 + d_2 e^{\lambda_2 t} X_2 + d_3 e^{\lambda_3 t} X_3 + d_4 e^{\lambda_4 t} X_4,$$

where λ_i and X_i $(i = 1,2,3,4)$ are eigenpairs of the 4×4 system matrix. The constants $d_i : i = 1,2,3,4$ can be determined using the given four boundary conditions. Finally, $y_1(t) = x_1^*(t)$ and $y_3(t) = x_2^*(t)$ are the required optimal functions.

***Example* 6.5.2** In the above example, if the boundary conditions are $x_1(1) = 1$ and $x_2(2) = 0$, while $x_1(2)$ and $x_2(1)$ are unspecified, find x_1^* and x_2^*.

Solution Since $x_1(2)$ is unspecified, from (6.16), the condition at $t = 2$ is $\left(\frac{\partial f}{\partial \dot{x}_1}\right)_{t=2} = 0$, i.e., $\dot{x}_2(2) = 0$. Similarly, since $x_2(1)$ is unspecified, $\left(\frac{\partial f}{\partial \dot{x}_2}\right)_{t=1} = 0$, implying $\dot{x}_1(1) = 0$. Now, the optimal solution (if it exists) can be obtained by solving (6.49) and (6.50), along with the conditions $x_1(1) = 1, x_2(2) = 0, \dot{x}_1(1) = 0$, and $\dot{x}_2(2) = 0$.

6.6 Constrained Optimization

In the previous section, it was shown that if the integrand f in the performance functional $\mathbb{J}$ is a function of n variables and their derivatives (x_i and $\dot{x}_i$ for $i = 1,2,\ldots,n$), then the optimal solution $x^*(t) = \begin{bmatrix} x_1^*(t) & x_2^*(t) & \cdots & x_n^*(t) \end{bmatrix}'$ can be obtained by solving the system (6.41) of n Euler's equations under suitable boundary conditions. In vector form, this system is written as in (6.45).

In this section we consider the problem of finding $x(t)$ that extremizes a cost functional $\mathbb{J}$ while satisfying a given constraint. We shall see that most of the optimization problems with performance functionals of the form (6.34), along with constraints, can be brought under the framework of system (6.45).

Problem II.

$$\text{Extremize}\, \mathbb{J}(x_1,x_2,\ldots,x_n) = \int_{t_0}^{\tau} f(t,x_1,x_2,\ldots,x_n,\dot{x}_1,\dot{x}_2,\ldots,\dot{x}_n)dt,$$

where the real valued functions x_i satisfy the constraints

$$g_j(t,x_1(t),x_2(t),\ldots,x_n(t)) = 0, \quad t_0 \le t \le \tau, j = 1,2,\ldots,m \tag{6.51}$$

with fixed boundary conditions $x_i(t_0) = x_{i0}$ and $x_i(\tau) = x_{i\tau}$ in $\mathbb{R}^n$. In vector form, the above problem can be stated as

$$\text{extremize}\, \mathbb{J}(x) = \int_{t_0}^{\tau} f(t,x(t),\dot{x}(t))dt \tag{6.52}$$

under the constraint

$$g(t,x(t)) = 0, \tag{6.53}$$

where $x(t) = \begin{bmatrix} x_1(t) & x_2(t) & \cdots & x_n(t) \end{bmatrix}'$, $x(t_0) = \begin{bmatrix} x_1(t_0) & x_2(t_0) & \cdots & x_n(t_0) \end{bmatrix}' = x_0$, $x(\tau) = \begin{bmatrix} x_1(\tau x_2(\tau) & \cdots & x_n(\tau) \end{bmatrix}' = x_\tau$, and $g(t,x) = \begin{bmatrix} g_1(t,x) & g_2(t,x) & \cdots & g_m(t,x) \end{bmatrix}'$. Note that $x(t_0)$ and $x(\tau)$ must also satisfy (6.53). Now let $x^*(t) = \begin{bmatrix} x_1^*(t) & x_2^*(t) & \cdots & x_n^*(t) \end{bmatrix}'$, $t \in [t_0,\tau]$, be the optimal function, which satisfies the constraint (6.53) and extremizes the functional $\mathbb{J}$. Then $g(t,x^*(t)) = 0$, $t_0 \le t \le \tau$, and $\min_{x \in S} \mathbb{J}(x) = \mathbb{J}(x^*)$, where S is the set of all functions $x : [t_0,\tau] \to \mathbb{R}^n$ such that $g(t,x(t)) = 0$ for all $t \in [t_0,\tau]$.

To find the optimal solution x^*, let us consider the following Lagrange multiplier method. Since there are m constraints, let us introduce m functions $\lambda_i(t) : i = 1,2,\ldots,m,\ t \in [t_0,\tau]$, called the **Lagrange multipliers**. Let $\lambda(t) = \begin{bmatrix} \lambda_1(t) & \lambda_2(t) & \cdots & \lambda_m(t) \end{bmatrix}'$ be the Lagrange multiplier vector. From (6.51), we can write

$$\sum_{i=1}^{m} \lambda_i(t) g_i(t,x_1(t),x_2(t),\ldots,x_n(t)) = 0,$$

which can be written in vector form as

$$\lambda'(t) g(t,x) = 0. \tag{6.54}$$

Integrating both sides from t_0 to τ, we get

$$\int_{t_0}^{\tau} \lambda'(t)g(t,x(t))dt = 0. \tag{6.55}$$

From (6.52), (6.53), and (6.55), the Problem II of extremizing $\mathbb{J}$ under constraints (6.51) along with boundary conditions can be redefined as the extremization of $\mathbb{J}_1(x,\lambda)$, where

$$\mathbb{J}_1(x,\lambda) = \int_{t_0}^{\tau} \left(f(t,x(t),\dot{x}(t)) + \lambda'(t)g(t,x(t))\right) dt \tag{6.56}$$

under the given boundary conditions at t_0 and τ. In $\mathbb{J}_1$, the total number of variables is $n+m$, consisting of $x_i(t) : i = 1,2,\ldots,n$ and $\lambda_j(t) : j = 1,2,\ldots,m$. Now let us denote the integrand in (6.56) as $L(t,x,\dot{x},\lambda)$. Hence,

$$L(t,x,\dot{x},\lambda) = f(t,x(t),\dot{x}(t) + \lambda'(t)g(t,x(t)) = f + \sum_{i=1}^{m} \lambda_i(t)g_i(t,x). \tag{6.57}$$

The function $L(t,x,\dot{x},\lambda)$ is called the **Lagrangian** of the extremization Problem II. Thus, Problem II can be stated as

$$\text{extremize}\ \mathbb{J}_1(x,\lambda) = \int_{t_0}^{\tau} L(t,x,\dot{x},\lambda)dt \tag{6.58}$$

under the boundary conditions $x(t_0) = x_0$ and $x(\tau) = x_\tau$. This is the same as Problem I (Section 6.5), except that the number of variables is increased, and the integrand is $L(t,x,\dot{x},\lambda)$ instead of $f(t,x,\dot{x})$.

Now, following arguments similar to those in Section 6.5, Euler's equations for the $n+m$ variables are

$$\frac{\partial L}{\partial x_i} - \frac{d}{dt}\left(\frac{\partial L}{\partial \dot{x}_i}\right) = 0, \quad i = 1,2,\ldots,n, \tag{6.59}$$

$$\frac{\partial L}{\partial \lambda_i} - \frac{d}{dt}\left(\frac{\partial L}{\partial \dot{\lambda}_i}\right) = 0, \quad i = 1,2,\ldots,m. \tag{6.60}$$

As the Lagrangian L is involved in this system of Euler's equation, it is called the system of **Euler–Lagrange equations**. However, since $\frac{\partial L}{\partial \dot{\lambda}_i} = 0$, the second equation becomes

$$\frac{\partial L}{\partial \lambda_i} = 0, \quad i = 1,2,\ldots,m. \tag{6.61}$$

Hence, from (6.57) and (6.61), $\frac{\partial L}{\partial \lambda_i} = 0 \implies g_i(t,x(t)) = 0$, which is the same as the given constraint (6.53).

In vector form, the equations (6.59) and (6.61) become

$$\begin{cases} \dfrac{\partial L}{\partial x} - \dfrac{d}{dt}\left(\dfrac{\partial L}{\partial \dot{x}}\right) = 0, \\ \dfrac{\partial L}{\partial \dot{\lambda}} = 0, \\ x(t_0) = x_0, \quad x(\tau) = x_\tau. \end{cases} \tag{6.62}$$

If $x^*(t)$ and $\lambda^*(t)$ are the solutions of the system (6.62), then they extremize $\mathbb{J}_1$ (in case $\mathbb{J}_1$ has an extremum), i.e.,

$$\mathbb{J}_1(x^*, \lambda^*) = \underset{x,\lambda}{\text{extremum}}\ \mathbb{J}_1(x, \lambda).$$

As $g(t, x^*(t)) = 0$ for $t_0 \leq t \leq \tau$, we obtain $\mathbb{J}_1(x^*, \lambda^*) = \mathbb{J}(x^*)$, which is the required extremum value of $\mathbb{J}(x)$.

Now, we can consider different cases of boundary conditions for the constrained Problem II. These are similar to conditions (6.46)–(6.48), in which f is replaced by L.

Case I. Fixed boundary conditions. Both $x(t_0) = x_0$ and $x(\tau) = x_\tau$ are specified. To obtain $x^*(t)$, solve (6.62) under the given boundary conditions.

Case II. $x(t_0) = x_0$ and τ are specified, but $x(\tau)$ is unspecified. For this case, solve (6.62) under the conditions:

$$x(t_0) = x_0 \text{ and } \left(\frac{\partial L}{\partial \dot{x}}\right)_\tau = 0. \tag{6.63}$$

Case III. $x(t_0) = x_0$ and $x(\tau) = x_\tau$ are specified, but the final time τ is not specified. For this case, solve (6.62) under the conditions:

$$x(t_0) = x_0,\ x(\tau) = x_\tau, \text{ and } \left(L - \left(\frac{\partial L}{\partial \dot{x}}\right)' \dot{x}\right)_\tau = 0. \tag{6.64}$$

Case IV. $x(t_0) = x_0$ is specified, while both τ and $x(\tau)$ are unspecified. In this case, $x^*(t)$ can be obtained by solving (6.62) under the conditions:

$$x(t_0) = x_0,\ \left(\frac{\partial L}{\partial \dot{x}}\right)_\tau = 0, \text{ and } (L)_\tau = 0. \tag{6.65}$$

Hereafter, for writing convenience, let us denote $\frac{\partial L}{\partial x}$ as L_x and $\frac{\partial L}{\partial \dot{x}}$ as $L_{\dot{x}}$.

***Example* 6.6.1**

$$\text{Extremize } \mathbb{J} = \int_0^\tau \left(x_1^2 + x_2^2 + \frac{3}{2}\dot{x}_1^2 - \dot{x}_2^2\right) dt$$

subject to the constraint $x_1 + x_2 = \frac{1}{2}$, with the following boundary conditions:

(*i*) $x_1(0) = 1, x_1(\tau) = \frac{3}{2}$, and $\tau = 2$.

(*ii*) $x_1(0) = 1, \tau = 2$, while $x_1(\tau)$ is unspecified.

(*iii*) $x_1(0) = 1, x_1(\tau) = \frac{5}{2}$, while τ is unspecified.

Solution The Lagrangian is given by

$$L(x_1, x_2, \dot{x}_1, \dot{x}_2, \lambda) = x_1^2 + x_2^2 + \frac{3}{2}\dot{x}_1^2 - \dot{x}_2^2 + \lambda\left(x_1 + x_2 - \frac{1}{2}\right),$$

and the corresponding Euler–Lagrange equations are

$$L_{x_1} - \frac{d}{dt}L_{\dot{x}_1} = 0 \implies 2x_1 + \lambda - 3\ddot{x}_1 = 0,$$

$$L_{x_2} - \frac{d}{dt}L_{\dot{x}_2} = 0 \implies 2x_2 + \lambda + 2\ddot{x}_2 = 0.$$

Eliminating λ from the above equations, we get

$$2(x_1 - x_2) - 3\ddot{x}_1 - 2\ddot{x}_2 = 0.$$

Substituting $x_2 = \frac{1}{2} - x_1$ (from the given constraint) into the above equation, we obtain

$$4x_1 - 1 - \ddot{x}_1 = 0.$$

The general solution of this second-order nonhomogeneous differential equation is

$$x_1(t) = \frac{1}{4} + c_1 e^{2t} + c_2 e^{-2t},$$

where c_1 and c_2 are arbitrary constants. These constants can be determined using appropriate boundary conditions for the three different cases.

(*i*) $x_1(0) = 1$ and $x_2(2) = \frac{3}{2}$.

(*ii*) $x_1(0) = 1$ and $(L_{\dot{x}_1})_{t=2} = 0$, i.e., $\dot{x}_1(2) = 0$.

(*iii*) $x_1(0) = 1$ and $x_1(\tau) = \frac{5}{2}$, where the final time τ is obtained from the relation $(L - \dot{x}L_{\dot{x}})_\tau = 0$, i.e., $(L - \dot{x}_1 L_{\dot{x}_1} - \dot{x}_2 L_{\dot{x}_2})_\tau = 0$. On simplifying the above, we get

$$\left(x_1^2 + x_2^2 - \frac{3}{2}\dot{x}_1^2 + \dot{x}_2^2\right)_\tau = 0.$$

Substituting $x_1(\tau) = \frac{5}{2}, x_2(\tau) = \frac{1}{2} - x_1(\tau)$, and $\dot{x}_2(\tau) = -\dot{x}_1(\tau)$, we deduce

$$\left(-\frac{1}{2}\dot{x}_1^2 + 5x_1^2 - x_1 + \frac{1}{4}\right)_\tau = 0.$$

Now,

$$x_1(0) = 1 \implies c_1 + c_2 = \frac{3}{4}$$

$$x_1(\tau) = \frac{10}{4} \implies c_1 e^{2\tau} + c_2 e^{-2\tau} = \frac{9}{4}.$$

Hence,

$$c_2 = \frac{3}{4} - c_1,$$

$$c_1 e^{2\tau} + \left(\frac{3}{4} - c_1\right) e^{-2\tau} = \frac{9}{4} \implies c_1 = \frac{\frac{9}{4} - \frac{3}{4}e^{-2\tau}}{e^{2\tau} - e^{-2\tau}}.$$

Thus, c_1 and c_2 are expressed in terms of τ. Since $x_1(t) = \frac{1}{4} + c_1 e^{2t} + c_2 e^{-2t}$, we have $\dot{x}_1(t) = 2c_1 e^{2t} - 2c_2 e^{-2t}$. Using the expressions of c_1 and c_2 in terms of τ, $\dot{x}_1(\tau)$ can be written in terms of τ. Now, substituting $\dot{x}_1(\tau)$ and $x_1(\tau) = \frac{5}{2}$ into the equation

$$\left(-\frac{1}{2}\dot{x}_1^2 + 5x_1^2 - x_1 + \frac{1}{4}\right)_\tau = 0,$$

we can solve for the value of τ.

Exercise 6.6.1

Extremize $\mathbb{J} = \int_0^{\frac{\pi}{2}} (x_1^2 + x_3^2 + \dot{x}_2^2 + \dot{x}_3^2)dt$

subject to the constraint $x_1^2 + x_2^2 = 1$ and the boundary conditions $x_2(0) = 1$, $x_2(\frac{\pi}{2}) = 0, x_3(0) = 2$, and $x_3(\frac{\pi}{2})$ unspecified.

6.7 Optimal Control Problem

Problem III. Consider the control system

$$\dot{x}(t) = g(t, x(t), u(t)), \quad t_0 \le t \le \tau, \tag{6.66}$$

where $x(t) = \begin{bmatrix} x_1(t) & x_2(t) & \cdots & x_n(t) \end{bmatrix}' \in \mathbb{R}^n$ and $u(t) = \begin{bmatrix} u_1(t) & u_2(t) & \cdots & u_m(t) \end{bmatrix}' \in \mathbb{R}^m$ are the state and control vector functions, respectively and the nonlinear function $g(t,x,u) = \begin{bmatrix} g_1(t,x,u) & g_2(t,x,u) & \cdots & g_n(t,x,u) \end{bmatrix}'$. The optimal control problem is to find the control function $u^*(t)$ and the corresponding state $x^*(t)$ that satisfy (6.66) and the given boundary conditions such that the performance functional

$$\mathbb{J}(x,u) = \int_{t_0}^{\tau} f(t,x(t),u(t))dt. \tag{6.67}$$

is extremized. In other words, $\mathbb{J}(x^*,u^*) = \min_{(x,u)\in S} \mathbb{J}(x,u)$, where $S = \{(x,u) : x$ and u satisfy (6.66) under appropriate boundary conditions$\}$. By treating the control system (6.66) as a constraint on x and u in the minimization of $\mathbb{J}(x,u)$, we can apply the Lagrange multiplier method as described in Problem II of the previous section.

Since the system (6.66) consists of n equations, let $\lambda(t) = \begin{bmatrix} \lambda_1(t) & \lambda_2(t) & \cdots & \lambda_n(t) \end{bmatrix}'$ be the Lagrange multiplier. Define the new objective functional $\mathbb{J}_2$ as

$$\mathbb{J}_2(x,u,\lambda) = \int_{t_0}^{\tau} \left(f(t,x,u) + \lambda'(g(t,x,u) - \dot{x})\right) dt. \tag{6.68}$$

Now, by defining the Lagrangian L as

$$L(t,x,\dot{x},u,\lambda) = f(t,x,u) + \lambda'(g(t,x,u) - \dot{x}), \tag{6.69}$$

we consider the problem of extremizing

$$\mathbb{J}_2(x,u,\lambda) = \int_{t_0}^{\tau} L(t,x,\dot{x},u,\lambda)dt \tag{6.70}$$

under suitable boundary conditions on $x(t)$ and $\lambda(t)$. Note that $\mathbb{J}_2(x,u,\lambda)$ in (6.70) is similar to $\mathbb{J}_1(x,\lambda)$ in (6.58), except that $\mathbb{J}_2$ includes the additional variable u. Therefore, following the same procedure as described in the previous section (with variables x, u, and λ), we obtain the **Euler–Lagrange equations** for Problem III as follows:

$$\begin{cases} L_x - \dfrac{d}{dt} L_{\dot{x}} = 0, \\ L_u = 0, \\ L_\lambda = 0. \end{cases} \tag{6.71}$$

The optimal pair x^* and u^* is obtained (if it exists) by solving the system (6.71) along with the boundary conditions for the following cases:

Case I. $x(t_0) = x_0$ **and** $x(\tau) = x_\tau$ **are specified.**
Case II. $x(t_0) = x_0$ **and** τ **are specified, but** $x(\tau)$ **is unspecified.** In this case, the condition at τ is

$$(L_{\dot{x}})_\tau = 0 \implies \lambda(\tau) = 0. \tag{6.72}$$

Case III. $x(t_0) = x_0$ **and** $x(\tau) = x_\tau$ **are specified, but** τ **is unspecified.** Here, the condition involving τ is

$$\left(L - \dot{x}' L_{\dot{x}}\right)_\tau = 0 \implies \left[f(t,x,u) + \lambda' g(t,x,u)\right]_\tau = 0. \tag{6.73}$$

Case IV. $x(t_0) = x_0$ **is given, but both** τ **and** $x(\tau)$ **are free.** In this case, the conditions for finding τ and $x(\tau)$ are

$$(L_{\dot{x}})_\tau = 0 \text{ and } \left(L - \dot{x}' L_{\dot{x}}\right)_\tau = 0, \text{ i.e., } (L)_\tau = 0. \tag{6.74}$$

Since $L = f + \lambda'(g - \dot{x})$ (in (6.70)), the system of Euler–Lagrange equations (6.71) is given by

$$\begin{cases} f_x + \lambda' g_x + \dot{\lambda} = 0, \\ f_u + \lambda' g_u = 0, \\ g - \dot{x} = 0. \end{cases} \tag{6.75}$$

Note that the third equation in (6.75) is the same as the given control system $\dot{x} = g(t,x,u)$. The second equation $f_u + \lambda' g_u = 0$ provides the expression for the optimal control $u^*(t)$ in terms of the Lagrange multiplier $\lambda(t)$. By eliminating $u(t)$ from the first equation in (6.75) and applying suitable boundary conditions, we can obtain the **optimal state** $x^*(t)$ and the **co-state** $\lambda^*(t)$, which, in turn, provides the expression for the optimal control function $u^*(t)$.

Hamiltonian

In this subsection, Hamiltonian function corresponding to (6.66) and (6.67) is defined. This will be useful in the study of constrained control problems. The Euler–Lagrange system (6.71) and the boundary conditions (6.72)–(6.74) take a simple form when we write these equations in terms of the Hamiltonian. The Hamiltonian is a real valued function defined as

$$H(t,x,u,\lambda) = \left[f(t,x,u) + \lambda' g(t,x,u)\right]_\tau. \tag{6.76}$$

Then, from (6.69), the Lagrangian is written as

$$L(t,x,\dot{x},u,\lambda) = H(t,x,u,\lambda) - \lambda' \dot{x}. \tag{6.77}$$

Hence $L_x = H_x, L_{\dot{x}} = -\lambda$, $L_u = H_u$ and $L_\lambda = H_\lambda - \dot{x}$.

In terms of the Hamiltonian, the Euler–Lagrange system (6.71) becomes

$$\begin{cases} H_x + \dot{\lambda} = 0, \\ H_u = 0, \\ H_\lambda - \dot{x} = 0. \end{cases}$$

The boundary conditions in terms of the Hamiltonian H for the various cases are as follows:

Case I. $x(t_0) = x_0$ **and** $x(\tau) = x_\tau$ **are specified.**

Case II. $x(t_0) = x_0$ **and** τ **are specified, but** $x(\tau)$ **is unspecified.** In this case, the condition at τ is

$$\lambda(\tau) = 0.$$

Case III. $x(t_0) = x_0$ **and** $x(\tau) = x_\tau$ **are specified, but** τ **is unspecified.** Here, the condition involving τ is

$$(H(t,x,u,\lambda))_\tau = 0.$$

Case IV. $x(t_0) = x_0$ **is given, but both** τ **and** $x(\tau)$ **are free.** In this case, the conditions for finding τ and $x(\tau)$ are

$$\lambda(\tau) = 0 \text{ and } (H(t,x,u,\lambda))_\tau = 0.$$

Exercise 6.7.1 Find the optimal pair x^* and u^* that extremizes the functional $\mathbb{J}(x,u) = \int_0^5 (2x^2 + u^2)dt$ subject to the constraint $\dot{x} = -x + 2u$ with the following boundary conditions: (i) $x(0) = 1, x(5) = 2$; (ii) $x(0) = 1, x(5)$ is unspecified.

Exercise 6.7.2 Write the Lagrangian and the Hamiltonian for the following optimal control problem.
Extremise $J(x,u) = \int_0^1 (4x^2 + u^2)\, dt$ where the state x and the control u satisfy $\dot{x} = -3x + u$ and find the optimal pair (x,u) under the boundary condition $x(0) = 1$ and $x(1)$ is unspecified.

6.8 Linear Quadratic Optimal Control Problems

If the integrand $f(t,x,u)$ in (6.67) takes a quadratic form in x and u, while $g(t,x,u)$ in (6.66) is linear in x and u, then the Euler–Lagrange equation (6.75) is obtained in a simpler form. In this section, the Euler–Lagrange equation for linear quadratic optimal control problem is derived.

Problem IV. Find the optimal control $u^*(t) \in \mathbb{R}^m$ and the corresponding optimal state $x^*(t) \in \mathbb{R}^n$ for the system $\dot{x} = Ax + Bu, x(t) \in \mathbb{R}^n$, such that the cost functional

$$\mathbb{J} = \int_{t_0}^{\tau} (x'Px + u'Qu)dt$$

is extremized, where $P_{n\times n}$ is a real symmetric positive semi-definite (RSPSD) matrix and $Q_{m\times m}$ is a real symmetric positive definite (RSPD) matrix. The boundary conditions for $x(t)$ are $x(t_0) = x_0$ and $x(\tau) = x_\tau$ in $\mathbb{R}^n$. To find the optimal control, we need to solve the Euler–Lagrange equations given below. From (6.69) the Lagrangian L is given by

$$L(x, \dot{x}, u, \lambda) = x'Px + u'Qu + \lambda'(Ax + Bu - \dot{x}).$$

The Euler–Lagrange equations are

$$\frac{\partial L}{\partial x} - \frac{d}{dt}\left(\frac{\partial L}{\partial \dot{x}}\right) = 0, \frac{\partial L}{\partial u} = 0, \text{ and } \frac{\partial L}{\partial \lambda} = 0,$$

which yield

$$2Px + A'\lambda + \frac{d}{dt}(\lambda) = 0, \tag{6.78}$$

$$2Qu + B'\lambda = 0, \tag{6.79}$$

and

$$\dot{x} - Ax - Bu = 0. \tag{6.80}$$

From these equations, we get

$$u(t) = -\frac{1}{2}Q^{-1}B'\lambda, \tag{6.81}$$

$$\dot{\lambda} = -2Px - A'\lambda, \tag{6.82}$$

and

$$\dot{x} = Ax + Bu. \tag{6.83}$$

The system (6.82) is called the **co-state system**. From (6.81), (6.82), and (6.83), we obtain

$$\begin{bmatrix} \dot{x} \\ \cdots \\ \dot{\lambda} \end{bmatrix} = \left[\begin{array}{c|c} A & -\frac{1}{2}BQ^{-1}B' \\ \hline -2P & -A' \end{array}\right] \begin{bmatrix} x \\ \lambda \end{bmatrix}. \tag{6.84}$$

The system (6.84) is known as the **canonical system** for state and co-state. We can find the optimal functions $x^*(t)$ and $\lambda^*(t)$ by solving the system (6.84) using suitable boundary conditions. Finally, the optimal control $u^*(t)$ is obtained using equation (6.81).

***Example* 6.8.1** Formulate the canonical system along with the boundary conditions for finding the optimal control $u(t)$ for the system $\dot{x} = Ax + Bu$, $0 \leq t \leq 1$, where $A = \begin{bmatrix} 1 & -1 \\ 0 & 2 \end{bmatrix}$ and $B = \begin{bmatrix} 2 \\ 1 \end{bmatrix}$, with the initial condition $x(0) = \begin{bmatrix} -1 \\ 2 \end{bmatrix}$, and the final condition $x(1) = \begin{bmatrix} 2 \\ 1 \end{bmatrix}'$ such that the performance index

$$\mathbb{J} = \int_0^1 \left[\|x(t)\|^2 + \|u(t)\|^2 \right] dt$$

is minimized.

Solution Since $\|x\|^2 = x'x$ and $\|u\|^2 = u^2$, we have $P = \begin{bmatrix} 1 & 0 \\ 0 & 1 \end{bmatrix}$ and $Q = 1$. From (6.84), the canonical system is given by

$$\begin{bmatrix} \dot{x} \\ \dot{\lambda} \end{bmatrix} = \begin{bmatrix} \dot{x}_1 \\ \dot{x}_2 \\ \dot{\lambda}_1 \\ \dot{\lambda}_2 \end{bmatrix} = \left[\begin{array}{cc|cc} 1 & -1 & -2 & -1 \\ 0 & 2 & -1 & -\frac{1}{2} \\ \hline -2 & 0 & -1 & 0 \\ 0 & -2 & 1 & 2 \end{array}\right] \begin{bmatrix} x_1 \\ x_2 \\ \lambda_1 \\ \lambda_2 \end{bmatrix}.$$

Now, using the boundary conditions $x_1(0) = -1, x_2(0) = 2$, and $x_1(1) = 2\,, x_2(1) = 1$ we can find the states and co-states $x_i(t)$ and $\lambda_i(t)$ for $i = 1, 2$. Finally, using $\lambda(t)$, the optimal control is obtained from (6.81) as

$$u(t) = \frac{1}{2} Q^{-1} B' \lambda = \begin{bmatrix} 1 & \frac{1}{2} \end{bmatrix} \begin{bmatrix} \lambda_1(t) \\ \lambda_2(t) \end{bmatrix}.$$

***Example* 6.8.2** Find the optimal control $u(t)$ that minimizes

$$\mathbb{J} = \int_1^2 (x^2(t) + u^2(t))dt$$

subject to the constraint $\dot{x} = 2x + 3u$ with the initial condition $x(1) = 2$, while the final condition $x(2)$ is unspecified.

Solution From (6.69) the Lagrange L is

$$L(t, x, \dot{x}, u, \lambda) = x^2 + u^2 + \lambda(t)(2x + 3u - \dot{x}).$$

and from (6.78)–(6.80) the Euler–Lagrange equations are

$$\begin{cases} 2x + 2\lambda + \dot{\lambda} = 0, & \text{(from (6.78))} \\ 2u + 3\lambda = 0, & \text{(from (6.79))} \\ \dot{x} - 2x - 3u = 0. & \text{(from (6.80))} \end{cases} \tag{6.85}$$

Since $x(2)$ is free, the condition at $t = 2$ is $\left(\frac{\partial L}{\partial \dot{x}}\right)_{t=2} = 0$, which implies $\lambda(2) = 0$. From (6.85), $u = -\frac{3}{2}\lambda$ and

$$\begin{cases} \dot{\lambda} = -2x - 2\lambda, \\ \dot{x} = 2x - \dfrac{9}{2}\lambda. \end{cases}$$

Hence, the canonical system is

$$\begin{bmatrix} \dot{x} \\ \dot{\lambda} \end{bmatrix} = \begin{bmatrix} 2 & -\frac{9}{2} \\ -2 & -2 \end{bmatrix} \begin{bmatrix} x \\ \lambda \end{bmatrix}$$

with the conditions $x(1) = 2$ and $\lambda(2) = 0$. By solving the above system (see (4.56)), we obtain

$$\begin{bmatrix} x(t) \\ \lambda(t) \end{bmatrix} = c_1 e^{\sqrt{13}t} \begin{bmatrix} 1 \\ \frac{2}{9}(2 - \sqrt{13}) \end{bmatrix} + c_2 e^{-\sqrt{13}t} \begin{bmatrix} 1 \\ \frac{2}{9}(2 + \sqrt{13}) \end{bmatrix}, \tag{6.86}$$

where c_1 and c_2 are arbitrary constants. Using the conditions $x(1) = 2$ and $\lambda(2) = 0$, we get

$$\begin{bmatrix} 2 \\ 0 \end{bmatrix} = \begin{bmatrix} e^{\sqrt{13}} & e^{-\sqrt{13}} \\ e^{2\sqrt{1}}(2 - \sqrt{13}) & e^{-2\sqrt{1}}(2 + \sqrt{13}) \end{bmatrix} \begin{bmatrix} c_1 \\ c_2 \end{bmatrix}.$$

On solving the above system, the constants c_1 and c_2 can be obtained, and then $x^*(t)$ and $\lambda^*(t)$ can be obtained from (6.86). Finally, the optimal control is obtained from the relation $u^*(t) = -\frac{3}{2}\lambda^*(t)$.

***Example* 6.8.3** Find the optimal control u^* and the corresponding optimal state x^* for the control system $\dot{x} = -2x + u$, which minimizes the performance functional

$$\mathbb{J} = \frac{1}{2}\int_0^\tau (5x^2 + u^2)dt$$

under the following boundary conditions

(*i*) $x(0) = 1, \tau = 1$, and $x(1) = 0$.

(*ii*) $x(0) = 1, \tau = 1$, and $x(1)$ is unspecified.

(*iii*) $x(0) = 1, x(\tau) = 0$, and τ is unspecified.

(*iv*) $x(0) = 1$, but both τ and $x(\tau)$ are unspecified.

Solution The given problem can be converted into the problem of minimizing the functional $\mathbb{J}_2$ as given in equation (6.70), where the Lagrangian is

$$L(t, x, \dot{x}, u, \lambda) = f(t, x, u) + \lambda'(g(t, x, u) - \dot{x}).$$

In this example, $f(t, x, u) = \frac{1}{2}(5x^2 + u^2)$ and $g(t, x, u) = -2x + u$. Hence, the Lagrangian is

$$L = \frac{1}{2}(5x^2 + u^2) + \lambda(-2x + u - \dot{x}).$$

From (6.71), we obtain the following system of Euler–Lagrange equations to find the optimal pair

$$L_x - \frac{d}{dt}L_{\dot{x}} = 0 \implies 5x - 2\lambda + \dot{\lambda} = 0, \tag{6.87}$$

$$L_u = 0 \implies u + \lambda = 0, \tag{6.88}$$

$$L_\lambda = 0 \implies -2x + u - \dot{x} = 0. \tag{6.89}$$

Equation (6.88) provides the control u in terms of the co-state λ as $u = -\lambda$. Hence, equations (6.87) and (6.89) yield the system

$$\begin{cases} \dot{x} = -2x - \lambda, \\ \dot{\lambda} = -5x + 2\lambda. \end{cases} \tag{6.90}$$

Solving (6.90) under suitable boundary conditions will provide the optimal pair. In matrix form, the canonical system (6.90) is written as

$$\begin{bmatrix} \dot{x} \\ \dot{\lambda} \end{bmatrix} = \begin{bmatrix} -2 & -1 \\ -5 & 2 \end{bmatrix} \begin{bmatrix} x \\ \lambda \end{bmatrix}. \tag{6.91}$$

The eigenvalues of the matrix $\begin{bmatrix} -2 & -1 \\ -5 & 2 \end{bmatrix}$ are 3 and -3, with corresponding eigenvectors $\begin{bmatrix} 1 \\ -5 \end{bmatrix}$ and $\begin{bmatrix} 1 \\ 1 \end{bmatrix}$. Thus, the solution of the system (6.91) can be written as

$$\begin{bmatrix} x(t) \\ \lambda(t) \end{bmatrix} = c_1 e^{3t} \begin{bmatrix} 1 \\ -5 \end{bmatrix} + c_2 e^{-3t} \begin{bmatrix} 1 \\ 1 \end{bmatrix},$$

where c_1 and c_2 are arbitrary constants. Therefore,

$$\begin{cases} x(t) = c_1 e^{3t} + c_2 e^{-3t}, \\ \lambda(t) = -5c_1 e^{3t} + c_2 e^{-3t}. \end{cases} \tag{6.92}$$

The constants c_1 and c_2 are determined using the boundary conditions in each case.
(i) $x(0) = 1, \tau = 1$, and $x(1) = 0$.

From (6.92), $c_1 + c_2 = 1$ and $c_1 e^3 + c_2 e^{-3} = 0$. Hence, $c_2 = 1 - c_1$ and $c_1(e^3 - e^{-3}) + e^{-3} = 0$. Thus, $c_1 = \frac{e^{-3}}{e^{-3} - e^3}$ and $c_2 = \frac{e^3}{e^3 - e^{-3}}$. Substituting c_1 and c_2 into (6.92), we get the optimal state $x^*(t)$ and co-state $\lambda^*(t)$. The optimal control $u^*(t)$ is obtained from (6.88) as $u^*(t) = -\lambda^*(t)$. Finally the optimal value of the performance functional $\mathbb{J}$ is obtained by evaluating

$$\mathbb{J}^* = \frac{1}{2} \int_0^1 (5(x^*)^2 + (u^*)^2)\, dt.$$

(ii) $x(0) = 1, \tau = 1$, and $x(1)$ is unspecified.

The condition at $t = 1$ is obtained from (6.72), which states $(L_{\dot{x}})_{t=1} = 0$. Hence, $\lambda(1) = 0$ is the required condition. Now, $x(0) = 1 \implies c_2 = 1 - c_1$ and $\lambda(1) = 0 \implies -5c_1 e^3 + (1 - c_1)e^{-3} = 0$. Hence, $c_1 = \frac{e^{-3}}{5e^{-3} + e^3}$ and $c_2 = \frac{5e^3}{5e^3 + e^{-3}}$.
(iii) $x(0) = 1, x(\tau) = 0$, and τ is unspecified.

As τ is unspecified, the condition to obtain the value of τ is (6.73), which states $(L - \dot{x}L_{\dot{x}})_\tau = 0$. Hence,

$$\left(\frac{1}{2}(5x^2 + u^2) + \lambda(-2x + u)\right)_\tau = 0. \tag{6.93}$$

As $u = -\lambda$, the above equation simplifies to

$$\left(\frac{1}{2}(5x^2 + \lambda^2) - 2x\lambda - \lambda^2\right)_\tau = 0$$

$$\implies \frac{5}{2}x^2 - 2x\lambda - \frac{1}{2}\lambda^2 = 0 \text{ at } t = \tau. \tag{6.94}$$

Using the given condition $x(\tau) = 0$ in the above equation, we obtain $\lambda(\tau) = 0$. Now from (6.92), we get

$$\lambda(\tau) = -5c_1e^{3\tau} + (1-c_1)e^{-3\tau} = 0 \implies c_1 = \frac{e^{-3\tau}}{5e^{3\tau} + e^{-3\tau}} \tag{6.95}$$

and

$$x(\tau) = 0 \implies c_1 = \frac{e^{-3\tau}}{e^{-3\tau} - e^{3\tau}}. \tag{6.96}$$

From (6.95) and (6.96), $6e^{3\tau} = 0 \implies \tau = -\infty$. Thus, the extremization problem with $x(\tau) = 0$, when τ is free, has no finite solution for τ. However, if $x(\tau) = x_1 \neq 0$, then

$$x(\tau) = x_1 = c_1e^{3\tau} + (1-c_1)e^{-3\tau} \implies c_1 = \frac{x_1 - e^{-3\tau}}{e^{3\tau} - e^{-3\tau}}. \tag{6.97}$$

From (6.94), we have $5x_1^2 - 4x_1\lambda(\tau) - \lambda^2(\tau) = 0$, which implies $\lambda(\tau) = x_1$ or $-5x_1$. If

$$\lambda(\tau) = x_1 = -5c_1e^{3\tau} + (1-c_1)e^{-3\tau} \implies c_1 = \frac{x_1 - e^{-3\tau}}{-5e^{3\tau} - e^{-3\tau}}. \tag{6.98}$$

From (6.97) and (6.98), we get $\tau = \infty$. If $\lambda(\tau) = -5x_1$, then

$$c_1 = \frac{-5x_1 - e^{-3\tau}}{-5e^{3\tau} - e^{-3\tau}}. \tag{6.99}$$

From (6.97) and (6.98), we get $e^{3\tau} = x_1 \implies \tau = \frac{1}{3}\log x_1$. Then, from (6.99), we get $c_1 = 1$. Hence, $x^*(t) = e^{3t}$, which satisfies $x(0) = 1$ and $x(\tau) = x_1$ when $\tau = \frac{1}{3}\log x_1$. The optimal control is $u^*(t) = -\lambda^*(t) = 5e^{3t}$.

(iv) $x(0) = 1$, but both τ and $x(\tau)$ are free.

In this case, the conditions at τ are $(L_{\dot{x}})_\tau = 0$ and $(L - \dot{x}L_{\dot{x}})_\tau = 0$. Now

$$L_{\dot{x}} = 0 \implies \lambda(\tau) = 0$$

and

$$L - \dot{x}L_{\dot{x}} = 0 \implies \frac{1}{2}(5x^2 + u^2) + \lambda(-2x + u) = 0 \text{ at } t = \tau. \tag{6.100}$$

Since $u(t) = -\lambda(t)$ and $\lambda(\tau) = 0$, from (6.100), we get $x(\tau) = 0$. In *(iii)*, we have seen that the conditions $x(\tau) = 0$ and $\lambda(\tau) = 0$ lead to $\tau = \infty$.

***Example* 6.8.4** Find the optimal control $u(t)$ for the system $\dot{x} = Ax + Bu$, $0 \le t \le 1$, where $A = \begin{bmatrix} 1 & -1 \\ 0 & 2 \end{bmatrix}$ and $B = \begin{bmatrix} 0 \\ 1 \end{bmatrix}$, with the initial condition $x(0) = \begin{bmatrix} -1 \\ 2 \end{bmatrix}$, while $x(1)$ is unspecified such that

$$\mathbb{J} = \int_0^1 (x_2^2 + u^2)dt$$

is minimized.

Solution Note that $x_2^2 = x'Px$, where $P = \begin{bmatrix} 0 & 0 \\ 0 & 1 \end{bmatrix}$, and $u^2 = u'Qu$, where $Q = 1$. Hence, the canonical system is

$$\begin{bmatrix} \dot{x}_1 \\ \dot{x}_2 \\ \dot{\lambda}_1 \\ \dot{\lambda}_2 \end{bmatrix} = \left[\begin{array}{cc|cc} 1 & -1 & 0 & 0 \\ 0 & 2 & 0 & \frac{1}{2} \\ \hline 0 & 0 & -1 & 0 \\ 0 & -2 & 1 & -2 \end{array}\right] \begin{bmatrix} x_1 \\ x_2 \\ \lambda_1 \\ \lambda_2 \end{bmatrix}.$$

The boundary conditions are $x_1(0) = -1$, $x_2(0) = 2$, and $\left(\frac{\partial L}{\partial \dot{x}}\right)_{t=1} = \lambda(1) = 0$, i.e., $\lambda_1(1) = 0$ and $\lambda_2(1) = 0$. The functions $x(t)$ and $\lambda(t)$ can be obtained in terms of the eigen pairs of the system matrix as in (4.56). Finally, the optimal control is given by

$$u(t) = -\frac{1}{2}Q^{-1}B'\lambda = \begin{bmatrix} 1 & \frac{1}{2} \end{bmatrix} \begin{bmatrix} \lambda_1(t) \\ \lambda_2(t) \end{bmatrix}.$$

Exercise 6.8.1 Show that the Hamiltonian at the optimal pair is constant on the time interval $[0, 2]$ for the following optimal control problem:

$$\text{Minimize } \mathbb{J} = \int_0^2 u^2(t)dt$$

for the control system $\dot{x} = -2x + u$ with the boundary conditions $x(0) = 2$ and $x(2) = 1$.

6.9 Cost Functional Dependent on Final Condition

In this section we consider the cost functional that depends on the final time τ and/or final condition $x(\tau)$, state x and control u. Let $C^1([t_0,\tau],\mathbb{R}^n)$ denote the set of all continuously differentiable vector valued functions defined on $[t_0,\tau]$. Let $h:[t_0,\tau]\times C^1([t_0,\tau],\mathbb{R}^n)\to C^1([t_0,\tau])$ be a real valued function. For every $x\in C^1([t_0,\tau],\mathbb{R}^n)$, assume that $h(t,x(t))$ is differentiable with respect to $t, t_0<t<\tau$. Then,

$$\frac{dh}{dt}=\left(\frac{\partial h}{\partial x}\right)'\frac{dx}{dt}+\frac{\partial h}{\partial t}=h_x'\dot{x}+h_t \tag{6.101}$$

and

$$\int_{t_0}^{\tau}\frac{dh}{dt}dt=h(\tau,x(\tau))-h(t_0,x(t_0)). \tag{6.102}$$

Using (6.101) and (6.102), we can write

$$h(\tau,x(\tau))=h(t_0,x(t_0))+\int_{t_0}^{\tau}\frac{dh}{dt}dt=h(t_0,x(t_0))+\int_{t_0}^{\tau}\left(h_x'\dot{x}+h_t\right)dt. \tag{6.103}$$

Problem V. Consider the problem of finding the optimal control u^* and corresponding state x^* for the system

$$\dot{x}=g(t,x(t),u(t)),\quad t_0\le t\le\tau, \tag{6.104}$$

with the initial condition $x(t_0)=x_0\in\mathbb{R}^n$ such that the optimal pair x^* and u^* minimizes the cost functional

$$\mathbb{J}(\tau,x(\tau),x,u)=h(\tau,x(\tau))+\int_{t_0}^{\tau}f(t,x(t),u(t))dt. \tag{6.105}$$

Remark 6.9.1 If both τ and $x(\tau)$ are specified, then $h(\tau,x(\tau))$ is a constant. Hence (6.105) is only meaningful when either $\tau, x(\tau)$, or both are unspecified.

Using equation (6.103), the cost functional (6.105) can be written as

$$\mathbb{J}_0=h(t_0,x(t_0))+\int_{t_0}^{\tau}\left(f(t,x(t),u(t))+\left(\frac{\partial h}{\partial x}\right)'\dot{x}+\frac{\partial h}{\partial t}\right)dt. \tag{6.106}$$

Since $h(t_0,x(t_0))$ is a constant, it is sufficient to minimize only the integral term in (6.106) to minimize $\mathbb{J}_0$. Using the Lagrange multiplier method described in Section 6.7, the cost functional (6.106) and the constraint (6.104) are combined into a single performance functional $\mathbb{J}_1$ for minimization:

$$\mathbb{J}_1(x,u,\lambda)=\int_{t_0}^{\tau}\left(f+h_x'\dot{x}+h_t+\lambda'(g-\dot{x})\right)dt, \tag{6.107}$$

where $\lambda(t) = \begin{bmatrix} \lambda_1(t) & \lambda_2(t) & \cdots & \lambda_n(t) \end{bmatrix}'$ is the Lagrange multiplier vector function.

Defining the Lagrangian $L(x, \dot{x}, u, \lambda)$ as:

$$L = f + h_x'\dot{x} + h_t + \lambda'(g - \dot{x}) \tag{6.108}$$

and the Hamiltonian $H(x, u, \lambda)$ as:

$$H = f + \lambda' g, \tag{6.109}$$

we get

$$L = H + h_x'\dot{x} + h_t - \lambda'\dot{x}. \tag{6.110}$$

To minimize $\mathbb{J}_1$, the necessary condition for obtaining the optimal functions x^*, u^*, and λ^* are given (see Section 6.7) by the Euler–Lagrange system:

$$\begin{cases} L_x - \dfrac{d}{dt} L_{\dot{x}} = 0, \\ L_u = 0, \\ L_\lambda = 0 \end{cases} \tag{6.111}$$

along with the suitable boundary conditions at t_0 and τ. Using (6.110) in the Euler–Lagrange system, we express it in terms of the Hamiltonian H as:

$$\begin{cases} H_x + \dot{\lambda} = 0, \\ H_u = 0, \\ H_\lambda - \dot{x} = 0. \end{cases} \tag{6.112}$$

The first equation in (6.112) is obtained as follows. From (6.110),

$$\begin{cases} L_x = H_x + h_{xx}\dot{x} + h_{tx}, \\ L_{\dot{x}} = h_x - \lambda, \\ \dfrac{d}{dt} L_{\dot{x}} = h_{xx}\dot{x} + h_{xt} - \dot{\lambda}. \end{cases}$$

Hence

$$L_x - \frac{d}{dt} L_{\dot{x}} = H_x + \dot{\lambda} = 0.$$

The optimal solution is obtained by solving (6.111) or (6.112) along with the boundary conditions at t_0 and τ. As $x(t_0) = x_0$ is assumed to be known, the end condition is obtained as follows (if either $\tau, x(\tau)$, or both are unspecified).

Case I. τ **and** $x(\tau) = x_1$ **are specified.**

Case II. If τ is specified but $x(\tau)$ is unspecified, then

$$(L_{\dot{x}})_\tau = 0. \tag{6.113}$$

In terms of H, (6.113) is written as

$$(h_x - \lambda)_\tau = 0. \tag{6.114}$$

Case III. If $x(\tau)$ is given but τ is free, then

$$\left(L - \dot{x}' L_{\dot{x}}\right)_\tau = 0. \tag{6.115}$$

In terms of H, (6.115) is written as

$$(H + h_t)_\tau = 0. \tag{6.116}$$

Case IV. If both τ and $x(\tau)$ are free, then both (6.113) and (6.115) must be satisfied.

Time-invariant Optimal Control Problem

If, in Problem V, the control system (6.104) is time-invariant, i.e., $\dot{x} = g(x,u)$ for $t_0 \leq t \leq \tau$, and the performance index is of the form

$$\mathbb{J}(x,u) = h(x(\tau)) + \int_{t_0}^{\tau} f(x(t),u(t))dt,$$

then it is called a time-invariant optimal control problem. In such problems, the Hamiltonian has an important property, as shown in the following theorem:

Theorem 6.9.2 If Problem V is time-invariant, then the Hamiltonian at the optimal functions x^*, u^*, and λ^* is constant throughout the interval $[t_0, \tau]$.

Proof. As the problem is time-invariant, the Hamiltonian $H(t,x,u,\lambda)$ simplifies to $H(x,u,\lambda) = f(x,u) + \lambda' g(x,u), t_0 \leq t \leq \tau$. Hence,

$$\frac{dH}{dt} = \left(\frac{\partial H}{\partial x}\right)' \dot{x} + \left(\frac{\partial H}{\partial u}\right)' \dot{u} + \left(\frac{\partial H}{\partial \lambda}\right)' \dot{\lambda}.$$

As the optimal functions x^*, u^*, and λ^* satisfy the equations

$$\frac{\partial H}{\partial x} = -\dot{\lambda}, \frac{\partial H}{\partial \lambda} = \dot{x}, \text{ and } \frac{\partial H}{\partial u} = 0 \quad \text{(form (6.112))},$$

it follows that $\frac{dH}{dt} = 0$ for $t_0 \leq t \leq \tau$. Therefore, the Hamiltonian is constant throughout the interval. □

Corollary 6.9.3 For Cases III and IV of the time-invariant Problem III and IV, the condition (6.116) at τ is

$$(H(x^*,u^*,\lambda^*))_\tau = 0,$$

since $h_t = 0$, where x^*, u^*, and λ^* are the optimal functions. Since $H(x^*,u^*,\lambda^*)$ is constant for all $t_0 \le t \le \tau$ (Theorem 6.9.2), we have

$$H(x^*,u^*,\lambda^*) = 0, \quad t_0 \le t \le \tau.$$

Linear Quadratic Optimal Control Problem

In Section 6.8, we have considered the LQ problem of finding the optimal pair x^* and u^* for a linear control system that extremizes a quadratic performance index. Now we consider cost functionals that depend on the final time and/or final condition.

LQ optimal control problem. Find the optimal control u^* and the corresponding optimal state x^* for the linear time-varying system

$$\begin{cases} \dot{x}(t) = A(t)x(t) + B(t)u(t), & t_0 \le t \le \tau, \\ x(t_0) = x_0, \end{cases} \tag{6.117}$$

which minimizes the performance index

$$\mathbb{J}(\tau, x(\tau), x, u) = x'(\tau)M(\tau)x(\tau) + \int_{t_0}^{\tau} (x'(t)P(t)x(t) + u'(t)Q(t)u(t))\, dt. \tag{6.118}$$

Here, $A(t)$ and $B(t)$ are $n \times n$ and $n \times m$ matrices, respectively. $P(t)$ and $M(t)$ are $n \times n$ RSPSD matrices, and $Q(t)$ is an $m \times m$ RSPD matrix. By substituting $x'(\tau)M(\tau)x(\tau) = h(\tau, x(\tau))$, $x'(t)P(t)x(t) + u'(t)Q(t)u(t) = f(t, x(t), u(t))$, and $A(t)x(t) + B(t)u(t) = g(t, x(t), u(t))$ into (6.117) and (6.118), we observe that the LQ optimal control problem is a particular case of the general problem discussed in Section 6.9 (Problem V). Hence, using (6.108), the Lagrangian L for the LQ problem is given by

$$\begin{aligned} L(t,x,\dot{x},u,\lambda) &= f(t,x,u) + \frac{dh}{dt} + \lambda'(g(t,x,u) - \dot{x}) \\ &= f(t,x,u) + \left(\frac{\partial h}{\partial x}\right)' \dot{x} + \frac{\partial h}{\partial t} + \lambda'(g(t,x,u) - \dot{x}) \\ &= x'Px + u'Qu + 2(Mx)'\dot{x} + x'\dot{M}x + \lambda'(Ax + Bu - \dot{x}). \end{aligned} \tag{6.119}$$

In the above equation, we have used the expression

$$\frac{dh}{dt} = \frac{d}{dt}(x'Mx) = \dot{x}'Mx + x'\dot{M}x + x'M\dot{x} = 2(Mx)'\dot{x} + x'\dot{M}x \quad (\text{as } M = M').$$

The Hamiltonian is

$$H(x,u,\lambda) = f + \lambda' g = x'Px + u'Qu + \lambda'(Ax + Bu). \tag{6.120}$$

The Euler–Lagrange system for obtaining u^* and x^* in terms of H is (see equation (6.112))

$$\begin{cases} \dot{x} = H_\lambda, \\ \dot{\lambda} = -H_x, \\ H_u = 0 \end{cases} \tag{6.121}$$

for $t_0 \le t \le \tau$. From (6.120) and (6.121), we obtain

$$\dot{x} = Ax + Bu, \tag{6.122}$$

$$\dot{\lambda} = -2Px - A'\lambda, \tag{6.123}$$

and

$$2Qu + B'\lambda = 0 \quad (\text{since } \frac{\partial}{\partial u}(\lambda' Bu) = B'\lambda). \tag{6.124}$$

From (6.124), the optimal control u in terms of λ is

$$u = -\frac{1}{2}Q^{-1}B'\lambda. \tag{6.125}$$

Using (6.125) in (6.122), the equations (6.122) and (6.123) can be written as

$$\begin{bmatrix} \dot{x} \\ \dot{\lambda} \end{bmatrix} = \begin{bmatrix} A & -\frac{1}{2}BQ^{-1}B' \\ -2P & -A' \end{bmatrix} \begin{bmatrix} x \\ \lambda \end{bmatrix}. \tag{6.126}$$

Solving this canonical system along with the boundary conditions as in Case I–Case IV, the pair x^* and λ^* can be obtained. Finally, the optimal control u^* is obtained from (6.125).

***Example* 6.9.4** Find the optimal pair u^* and x^* that satisfies $\dot{x} = -2x + u$, $0 \le t \le \tau$, with initial condition $x(0) = 1$, and

$$\text{minimizes } \mathbb{J} = 4x^2(\tau) + \int_0^\tau u^2(t)dt,$$

when

(*i*) $\tau = 1$ and $x(1) = 0$.

(*ii*) $\tau = 1$, but $x(1)$ is unspecified.

Solution Here $A = -2$, $B = 1$, $M = 4$, $P = 0$, and $Q = 1$. From (6.122), $\dot{x} = -2x + u$. From (6.125), $u = -\frac{1}{2}\lambda$. Hence, $\dot{x} = -2x - \frac{1}{2}\lambda$. From (6.123), $\dot{\lambda} = 2\lambda$. Therefore,

$$\begin{bmatrix} \dot{x} \\ \dot{\lambda} \end{bmatrix} = \begin{bmatrix} -2 & -\frac{1}{2} \\ 0 & 2 \end{bmatrix} \begin{bmatrix} x \\ \lambda \end{bmatrix}, \quad x(0) = 1.$$

The eigenvalues are -2 with eigenvector $\begin{bmatrix} 1 \\ 0 \end{bmatrix}$ and 2 with eigenvector $\begin{bmatrix} 1 \\ -8 \end{bmatrix}$. Hence,

$$\begin{bmatrix} x(t) \\ \lambda(t) \end{bmatrix} = c_1 e^{-2t} \begin{bmatrix} 1 \\ 0 \end{bmatrix} + c_2 e^{2t} \begin{bmatrix} 1 \\ -8 \end{bmatrix} \implies x(t) = c_1 e^{-2t} + c_2 e^{2t} \text{ and } \lambda(t) = -8c_2 e^{2t}.$$

Since $x(0) = 1$, we have $c_1 + c_2 = 1$.
(*i*) If $\tau = 1$ and $x(1) = 0$, then $x(1) = c_1 e^{-2} + c_2 e^2 = 0 \implies (1 - c_2)e^{-2} + c_2 e^2 = 0$. Therefore,

$$x^*(t) = \frac{e^{2(1-t)} - e^{-2(1-t)}}{e^2 - e^{-2}}, \lambda^*(t) = \frac{8e^{-2(1-t)}}{e^2 - e^{-2}}, \text{ and } u^*(t) = -\frac{1}{2}\lambda^*(t).$$

(*ii*) If $\tau = 1$ and $x(1)$ is unspecified, then from (6.114), $\lambda(\tau) = 2M(\tau)x(\tau) = 8x(\tau)$. This implies $8c_2 e^2 = 8\left[(1 - c_2)e^{-2} + c_2 e^2\right]$. Therefore,

$$x^*(t) = \frac{2e^{2(1-t)} - e^{-2(1-t)}}{2e^2 - e^{-2}}, \lambda^*(t) = \frac{8e^{-2(1-t)}}{2e^2 - e^{-2}}, \text{ and } u^*(t) = -\frac{4e^{-2(1-t)}}{2e^2 - e^{-2}}. \tag{6.127}$$

Example **6.9.5** In the above example, the control system and the performance index are time-invariant. Hence, by Theorem 6.9.2, the Hamiltonian is constant on $[0, \tau]$, as shown below. The Hamiltonian is given by

$$H(x, u, \lambda) = u^2 - 2\lambda x + \lambda u$$

for $0 \le t \le 1$. For the boundary conditions $x(0) = 1$ and $x(1)$ unspecified, the optimal function x^*, λ^*, and u^* are given by (6.127). By substituting these functions into $H(x, u, \lambda)$, we obtain

$$H(x^*, u^*, \lambda^*) = (u^*)^2 - 2\lambda^* x^* + \lambda^* u^* = -16.$$

Linear Quadratic Regulator (LQR) Problem

Consider the control system

$$\dot{z} = Az + Bu, \quad t_0 \le t \le \tau, \tag{6.128}$$

with initial data $z(t_0) = z_0$. If it is required to find an optimal control $u(t)$ such that the state $z(t)$ is kept near a preassigned point z_τ, we can define a quadratic performance index for minimization as

$$\mathbb{J} = (z(\tau) - z_\tau)' M (z(\tau) - z_\tau) + \int_{t_0}^{\tau} \left((z(t) - z_\tau)' P (z(t) - z_\tau) + u'(t) Q u(t) \right) dt. \tag{6.129}$$

Now, by defining the state variable $x(t)$ as $x(t) = z(t) - z_\tau$, the equations (6.128) and (6.129) become

$$\begin{cases} \dot{x} = Ax + Az_\tau + Bu, \\ x(t_0) = x_0 = z_0 - z_\tau \end{cases} \tag{6.130}$$

and

$$\mathbb{J} = x'(\tau) M x(\tau) + \int_{t_0}^{\tau} \left(x'(t) P x(t) + u'(t) Q u(t) \right) dt. \tag{6.131}$$

Following the procedure described in Section 6.9, we define the Hamiltonian as $H(x, u, \lambda) = x'Px + u'Qu + \lambda'(Ax + Az_\tau + Bu)$ and obtain the Euler–Lagrange equation for optimal state x^* and co-state λ^*, which is the same as (6.121), and u^* is the same as in (6.125). By substituting $u^*(t)$ into (6.130), the optimal state x^* is obtained.

***Example* 6.9.6** Consider the following equation of motion of an object P of mass m moving under the influence of a force $f(t)$:

$$\begin{cases} m\ddot{x} = f, \\ x(0) = x_0, \quad \dot{x}(0) = v_0, \end{cases}$$

where $x(t)$ and $\dot{x}(t)$ are the displacement and velocity of P at time t. Find an optimal force function $f(t)$ (control) such that at $t = \tau$, $x(\tau)$ and $\dot{x}(\tau)$ are close to the values x_τ and v_τ, respectively.

Solution To convert the above problem into an LQR optimal control problem, let $x_1 = x$, $x_2 = \dot{x}$, and $u = \frac{f}{m}$. Then the equation of motion becomes $\dot{X} = AX + Bu$, where

$$X = \begin{bmatrix} x_1 \\ x_2 \end{bmatrix}, A = \begin{bmatrix} 0 & 1 \\ 0 & 0 \end{bmatrix}, B = \begin{bmatrix} 0 \\ 1 \end{bmatrix}, \text{ and } X(0) = \begin{bmatrix} x_0 \\ v_0 \end{bmatrix} = X_0.$$

To find an optimal control to regulate the solution $X(t)$ to be in the neighborhood of $X_\tau = \begin{bmatrix} x_\tau \\ v_\tau \end{bmatrix}$, let the performance index be

$$\mathbb{J} = (X(\tau) - X_\tau)' M (X(\tau) - X_\tau) + \int_0^\tau \left[(X(t) - X_\tau)' P (X(\tau) - X_\tau) + Qu^2(t) \right] dt, \tag{6.132}$$

where $Q > 0$, and M and P are suitable positive semi-definite matrices. For example, if x_τ is specified and v_τ is unspecified, then the performance index can be taken as

$$\mathbb{J} = \alpha (x_1(\tau) - x_\tau)^2 + \int_{t_0}^\tau \left[\beta (x_1(t) - x_\tau)^2 + Qu^2(t) \right] dt, \tag{6.133}$$

where $\alpha > 0, \beta > 0$, and $Q > 0$. To write (6.133) in the form of (6.132), the matrices M and P are taken as $M = \begin{bmatrix} \alpha & 0 \\ 0 & 0 \end{bmatrix}$, $P = \begin{bmatrix} \beta & 0 \\ 0 & 0 \end{bmatrix}$, and $X_\tau = \begin{bmatrix} x_\tau \\ k \end{bmatrix}$, where k is an arbitrary real number.

Exercise 6.9.1 Find $x^*(t)$ that minimizes $\mathbb{J} = \tau^2 + \int_0^\tau \dot{x}^2(t)dt$ subject to the conditions $x(0) = 1, x(\tau) = 2$, with τ unspecified.

Exercise 6.9.2 Find the optimal pair $x^*(t)$ and $u^*(t)$ for the following optimal control problem:

$$\text{Minimize } \mathbb{J}(x,u) = x^2(\tau) + \int_0^\tau (3x^2(t) + u^2(t))dt$$

subject to the constraint $\dot{x} = -x + u$ with the following boundary conditions:

(*i*) $x(0) = 1, x(\tau) = 2$, and $\tau = 3$.

(*ii*) $x(0) = 1, x(\tau) = 2$, and τ is unspecified.

(*iii*) $x(0) = 1, \tau = 3$, but $x(3)$ is unspecified.

(*iv*) $x(0) = 1$, τ and $x(\tau)$ are unspecified.

6.10 Summary

In this chapter, the following problems have been discussed. (i) Finding the extremum of a functional J under four types of boundary conditions. (ii) Finding the extremum of J under some constraints. (iii) Optimal control problems. (iv) Linear Quadratic Optimal Control. (v) Optimal control involving a cost functional J that depends on end conditions.

Problem I: Extremizing a functional .

$$\text{Extremize } \mathbb{J}(x) = \int_{t_0}^{\tau} f(t, x(t), \dot{x}(t))dt,$$

where x is a differentiable function on $[t_0, \tau]$ satisfying one of the following boundary conditions: (I) $x(t_0) = x_0$ and $x(\tau) = x_1$; (II) $x(t_0) = x_0$ but $x(\tau)$ is unspecified; (III) $x(t_0) = x_0$, $x(\tau) = x_1$, but τ is unspecified; and (IV) $x(t_0) = x_0$, while τ as well as $x(\tau)$ are unspecified.

Solution To find the extremal $x^*(t)$, solve the Euler equation $f_x - \frac{d}{dt} f_{\dot{x}} = 0$, along with the boundary conditions as given below:

- **Case I.** $x(t_0) = x_0$ and $x(\tau) = x_1$.
- **Case II.** $x(t_0) = x_0$ and $f_{\dot{x}}|_{\tau} = 0$.
- **Case III.** $x(t_0) = x_0$, $(f - \dot{x} f_{\dot{x}})|_{\tau} = 0$, and $x(\tau) = x_1$.
- **Case IV.** $x(t_0) = x_0$, $f_{\dot{x}}|_{\tau} = 0$, and $f|_{\tau} = 0$.

If t_0 and/or $x(t_0)$ are unspecified, similar conditions as in Cases II–IV apply by replacing τ with t_0.

In the above problem, if $x(t) \in \mathbb{R}^n$ for $t \in [t_0, \tau]$, then same conditions as above hold, with the understanding that x, $\dot{x}$, f_x, and $f_{\dot{x}}$ are vector functions, and $\dot{x} f_{\dot{x}}$ in Case III is replaced by $\dot{x}' f_{\dot{x}}$.

Problem II: Extremizing $\mathbb{J}$ under constraints.

$$\text{Extremize } \mathbb{J}(x) = \int_{t_0}^{\tau} f(t, x(t), \dot{x}(t))dt,$$

under the constraints $g(t, x) = 0$ (note that x and g are vectors) along with the four cases of boundary conditions as given in Problem I.

Solution let $\lambda(t) = \begin{bmatrix} \lambda_1(t) & \lambda_2(t) & \cdots & \lambda_m(t) \end{bmatrix}'$ be the Lagrange multiplier, and let $L(x, \dot{x}, \lambda) = f(t, x, \dot{x}) + \lambda' g(t, x)$ be the Lagrangian. The Euler–Lagrange system of equations to obtain the extremum is $L_x - \frac{d}{dt} L_{\dot{x}} = 0$ and $g(t, x) = 0$ along with the same boundary conditions as in Problem I by replacing f with L.

Problem III: Optimal Control Problem.

$$\text{Extremize } \mathbb{J}(x) = \int_{t_0}^{\tau} f(t, x(t), u(t))dt,$$

where x and u are state and control of a control system $\dot{x} = g(t, x, u)$ with x satisfying any one of the four cases of boundary conditions as in Problem I (Section 6.7).

Solution Let $L(t, x, \dot{x}, \lambda, u) = f(t, x, u) + \lambda'(g(t, x, u) - \dot{x})$ be the Lagrangian, and $H(t, x, \lambda, u) = f(t, x, u) + \lambda' g(t, x, u)$ be the Hamiltonian. To find the optimal control u^* and optimal state x^*, solve the Euler–Lagrange system

$$L_x - \frac{d}{dt}L_{\dot{x}} = 0, L_u = 0, \text{ and } \dot{x} = g(t, x, u) \tag{6.134}$$

along with the boundary conditions as in Problem I with f replaced by L. In terms of the Hamiltonian H the Euler–Lagrange equations are $\dot{x} = H_\lambda$ $\dot{\lambda} = -H_x$ and $H_u = 0$

Problem IV. Linear Quadratic Optimal Control

$$\text{Extremize } \mathbb{J}(x, u) = \int_{t_0}^{\tau} (x'Px + u'Qu)dt,$$

where x and u are state and control vectors of the linear control system $\dot{x} = Ax + Bu$ with one of the four cases of boundary conditions as in Problem I.

Solution Since this is a particular case of Problem III, following the steps of the solutions to Problem III, we get $L = x'Px + u'Qu + \lambda'(Ax + Bu - \dot{x})$ and $H = x'Px + u'Qu + \lambda'(Ax + Bu)$. The Euler–Lagrange equation provides the state and co-state system

$$\dot{x} = Ax - \frac{1}{2}BQ^{-1}B'\lambda,$$
$$\dot{\lambda} = -2Px - A'\lambda,$$

and

$$u = -\frac{1}{2}Q^{-1}B'\lambda,$$

which can be solved under suitable boundary conditions as stated in Problem III (Section 6.8).

Problem V. Cost function $\mathbb{J}$ depending on τ and/or $x(\tau)$

$$\text{Extremize } \mathbb{J}(x, u, \tau, x(\tau)) = h(\tau, x(\tau)) + \int_{t_0}^{\tau} f(t, x, u)dt, \tag{6.135}$$

where x and u are the state and control variables of the control system $\dot{x} = g(t,x,u)$ along with the four cases of boundary conditions as in Problem I.

Solution Let

$$L(t,x,\dot{x},u,\lambda) = f + h_x'\dot{x} + h_t + \lambda'(g - \dot{x}) \tag{6.136}$$

be the Lagrangian, and

$$H(t,x,u,\lambda) = f + \lambda' g \tag{6.137}$$

be the Hamiltonian. Hence, $L = H + h_x'\dot{x} + h_t - \lambda'\dot{x}$. Using L as given in (6.136), the Euler–Lagrange equation and boundary conditions are the same as in (6.134). In terms of the Hamiltonian H, the Euler–Lagrange equations become

$$\dot{x} = H_\lambda, \dot{\lambda} = -H_x, \text{ and } H_u = 0 \tag{6.138}$$

and the boundary conditions are

- **Case I.** $x(t_0) = x_0$ and $x(\tau) = x_1$.
- **Case II.** $x(t_0) = x_0$ and $(h_x - \lambda)|_\tau = 0$.
- **Case III.** $x(t_0) = x_0, x(\tau) = x_1$, and $(H - h_x)|_\tau = 0$.
- **Case IV.** $x(t_0) = x_0$, $(h_x - \lambda)|_\tau = 0$, and $(H + h_t)|_\tau = 0$.

6.11 Exercise

1. If $f(t) = t^2 + 4t$, find the increment Δf and the differential δf when the increment in t is $\Delta t = 0.1$.

2. Let $x = (x_1,x_2)' \in \mathbb{R}^2$ and $f(x_1,x_2) = x_1^2 + 2x_2^2$. If the increment in x is $\Delta x = (0.1,-0.2)'$, find the increment Δf and the differential δf.

3. Let $x(t) = (x_1(t),x_2(t))' \in \mathbb{R}^2$ for each $t \in [0,1]$, and let $\mathbb{J}(x)$ be the functional given by

$$\mathbb{J}(x) = \int_0^1 (x_1^2(t) + 3x_2(t)x_1(t))dt.$$

   Find the increment $\Delta\mathbb{J}$ and the variation $\delta\mathbb{J}$ if the variation in x is $\delta x = (t(1-t), \sin \pi t)'$.

4. If $f(x,y) = xy - x^2 - y^2 - 2x - 2y + 4$, find the extremum points and extremum values of f using differential calculus.

5. Find the greatest value of $f(x,y) = xy$ on the ellipse $\frac{x^2}{8} + \frac{y^2}{2} = 1$.

6. Maximize $x^2 + y^2 + z^2$ subject to the constraints $2y + 4z = 5$ and $4x^2 + 4y^2 - z^2 = 0$.

7. Find $x^*(t)$, $t \in [0,2]$, which minimizes $\mathbb{J} = \int_0^2 x\sqrt{1+\dot{x}^2}dt$ given that $x^*(0) = x^*(2) = \cosh(1)$.

8. Find $x^*(t)$, which extremizes

$$\mathbb{J} = \int_0^{\frac{\pi}{4}} (x^2 - \dot{x}^2)dt$$

under the following conditions:

(*i*) $x(0) = 1$ and $x(\frac{\pi}{4}) = 2$.

(*ii*) $x(0) = 1$ and $x(\frac{\pi}{4})$ is unspecified.

9. Find $x^*(t)$ and τ^*, which extremize

$$\mathbb{J} = \int_0^{\tau} (x^2 + \dot{x}^2)dt$$

with the boundary conditions $x(0) = 1$ and $x(\tau^*) = 2$.

10. Find the extremum (if it exists) of the functional $\mathbb{J}$, where

$$\mathbb{J} = \int_0^{\tau} (x^2 - 3x\dot{x} + \dot{x}^2)dt$$

under the following boundary conditions:

(*i*) $x(0) = 1$, $\tau = 1$, and $x(1) = 2$.

(*ii*) $x(0) = 1$, $\tau = 1$, and $x(1)$ is unspecified.

(*iii*) $x(0) = 1$, $x(\tau) = 2$, and τ is unspecified.

11. Find $x_1^*(t)$ and $x_2^*(t)$ that minimize the functional

$$\mathbb{J} = \int_0^{\tau} (x_1x_2 + \frac{1}{2}\dot{x}_1^2 + \frac{1}{2}\dot{x}_2^2)dt$$

with the boundary conditions:

(*i*) $x_1(0) = 1, x_1(1) = 0, x_2(0) = 0$, and $x_2(1) = 1$.

(*ii*) $x_1(0) = 1, x_1(\tau) = 0$, and $x_2(0), x_2(\tau)$, and τ are unspecified.

12. Minimize $\mathbb{J} = \int_0^{\tau}(x_1^2 + x_1x_2 + \dot{x}_1^2 + 4\dot{x}_2^2)dt$
under the constraint $x_1 - x_2 = 1$, with the boundary conditions:

    (i) $x_1(0) = 1, x_1(\tau) = 2$, and $\tau = 1$.

    (ii) $x_1(0) = 1, x_1(\tau) = 0$, and τ is unspecified.

13. Write the Euler–Lagrange equations along with the boundary conditions to find the optimal solution of the following:

$$\text{Minimize } \mathbb{J} = \int_{t_0}^{\tau}(x_1^2 + x_2^2 + x_2x_3 - \dot{x}_1^2 - \dot{x}_2^2 + 2\dot{x}_3^2)dt$$

subject to the constraints $x_1^2 - 4x_2^2 = -3, x_1^2 + x_2^2 - x_3^2 = 1$, and $x_1(a) = 1$. Consider the following cases:

    (i) $a = 0, b = 2$, and $x_1(b), x_2(a)$, and $x_2(b)$ are unspecified,

    (ii) $a = 0$ and $x_1(b), x_2(a), x_2(b)$, and b are unspecified,

    (iii) $a = 0, b = 2, x_1(b) = 3$, and $x_2(a)$ and $x_2(b)$ are unspecified.

14. Find the optimal control $u^*(t)$ for the system $\dot{x}_1 = x_2, \dot{x}_2 = -2x_1 - 3x_2 + 5u$, which minimizes

$$\mathbb{J} = \int_0^1 (x_1^2 + u^2)dt,$$

given that $x_1(0) = 3$ and $x_2(0) = 2$.

15. Show that the Euler equation for extremizing the functional

$$\mathbb{J}(x) = \int_{t_0}^{\tau} f(t, x(t), \dot{x}(t), \ddot{x}(t))dt$$

over all the functions $x : [t_0, \tau] \to \mathbb{R}$ with fixed boundary conditions for x and $\dot{x}$, i.e., $x(t_0) = \alpha_0, x(\tau) = \alpha_1, \dot{x}(t_0) = \beta_0$, and $\dot{x}(\tau) = \beta_1$, is

$$f_x - \frac{d}{dt}f_{\dot{x}} + \frac{d^2}{dt^2}f_{\ddot{x}} = 0.$$

16. Write the Euler–Lagrange equations and boundary conditions for finding the optimal solution of the following problem:

$$\text{Minimize } \mathbb{J} = x_2^2(\tau) + \int_0^{\tau}(x_1^2 - x_1x_2 + u^2)dt$$

subject to the constraints

$$\begin{cases} \dot{x}_1 = -2x_1 + x_2^2, \\ \dot{x}_2 = u \end{cases}$$

for $0 \le t \le \tau$, where τ and $x_2(\tau)$ are unspecified, $x_1(0) = 1, x_2(0) = 2$, and $x_1(\tau) = 0$.

17. Write the state and co-state canonical matrix for obtaining the optimal pair $x^*(t)$ and $u^*(t)$, which minimizes the functional

$$\mathbb{J} = x'Mx + \int_0^{\tau} (x'Px + u'Qu)dt,$$

subject to the constraint $\dot{x} = Ax + Bu$, where $A = \begin{bmatrix} 1 & 1 \\ 2 & 1 \end{bmatrix}, B = \begin{bmatrix} 0 \\ 1 \end{bmatrix}, M = \begin{bmatrix} 2 & 0 \\ 0 & 1 \end{bmatrix}$, $P = \begin{bmatrix} 2 & 1 \\ 1 & 2 \end{bmatrix}$, and $Q = 4$. Write the appropriate boundary conditions for the following cases:

(*i*) $x(0) = \begin{bmatrix} 1 \\ 0 \end{bmatrix}$, $\tau = 2$, and $x(\tau)$ is unspecified.

(*ii*) $x(0) = \begin{bmatrix} 1 \\ 0 \end{bmatrix}$, $\tau = 2, x_1(\tau) = 0$, and $x_2(\tau)$ is unspecified.

(*iii*) $x(0) = \begin{bmatrix} 1 \\ 0 \end{bmatrix}$, $x_2(\tau) = 1$, and τ and $x_1(\tau)$ are unspecified.

18. Prove that the Hamiltonian $H(x^*, u^*, \lambda^*)$ is constant for $0 \le t \le \tau$ and find $H(x^*, u^*, \lambda^*)$ for the optimal functions of the system $\dot{x} = 4x + 3u$ such that x^* and u^* minimize $\mathbb{J}$, where

$$\mathbb{J} = \int_0^1 (x^2 + u^2)dt$$

with $x(0) = 1$ and $x(1)$ unspecified.

7

Feedback Control and Constrained Control

7.1 Introduction

In Chapter 6, the state and co-state canonical matrix was used to obtain the optimal functions x^* and λ^*. The optimal control u^* was then derived in terms of λ^*. In this chapter, for the linear quadratic (LQ) optimal control problem, the co-state function λ is considered to be a linear feedback function of x, leading to the matrix Riccati equation and, subsequently, to the optimal pair x^* and u^*. In Section 7.3, the Pontryagin principle for finding the optimal control for systems with constraints on the control function is discussed. Applications are provided in Section 7.4.

7.2 Feedback Control

Consider the following LQ problem: Find the optimal control u^* and the corresponding optimal state x^* for the linear time-varying system

$$\begin{cases} \dot{x}(t) = A(t)x(t) + B(t)u(t), & t_0 \leq t \leq \tau, \\ x(t_0) = x_0, \end{cases} \tag{7.1}$$

which minimizes the performance index

$$\mathbb{J} = x'(\tau)M(\tau)x(\tau) + \int_{t_0}^{\tau} \left(x'(t)P(t)x(t) + u'(t)Q(t)u(t)\right) dt, \tag{7.2}$$

where the matrices $A(t), B(t), M(t), P(t)$, and $Q(t)$ are as discussed in Section 6.9 (the LQ problem). Recall equation (6.125), which provides $u(t)$ in terms of $\lambda(t)$ as

$$u(t) = -\frac{1}{2}Q^{-1}B'\lambda(t). \tag{7.3}$$

Let us assume that the co-state $\lambda(t)$ can be expressed in terms of the state $x(t)$ in a linear relation, so that

$$\lambda(t) = R(t)x(t), \tag{7.4}$$

where $R(t)$ is an $n \times n$ matrix (to be obtained later). If such an $R(t)$ exists, then using (7.4) in (7.3), we obtain the control u in terms of x as

$$u = -\frac{1}{2}Q^{-1}B'Rx. \tag{7.5}$$

Equation (7.5) provides an expression for a **linear feedback control**, assuming $R(t)$ exists.

In what follows, it is shown that the matrix $R(t)$ satisfies a first order nonlinear matrix differential equation called the **matrix Riccati equation**.

Matrix Riccati Equation

Since $\lambda = Rx$, we have

$$\dot{\lambda} = \dot{R}x + R\dot{x}. \tag{7.6}$$

In (7.6) replacing $\dot{x}$ and u using (7.1) and (7.3), we get

$$\dot{\lambda} = \dot{R}x + R\left(Ax - \frac{1}{2}BQ^{-1}B'Rx\right) = \dot{R}x + RAx - \frac{1}{2}RBQ^{-1}B'Rx. \tag{7.7}$$

Substituting for $\dot{\lambda}$ from (6.123) on the LHS of (7.7), we get

$$-2Px - A'Rx = \left(\dot{R} + RA - \frac{1}{2}RBQ^{-1}B'R\right)x, \quad t_0 \le t \le \tau, \tag{7.8}$$

implying

$$\left(\dot{R} + RA + A'R + 2P - \frac{1}{2}RBQ^{-1}B'R\right)x = 0.$$

Since the above equation is valid for all solutions x on $[t_0, \tau]$ with arbitrary initial conditions, and the matrix in the bracket is independent of x, it (the matrix) must be identically equal to zero for all $t \in [t_0, \tau]$. Thus, we obtain the matrix Riccati equation:

$$\dot{R} + RA + A'R + 2P - \frac{1}{2}RBQ^{-1}B'R = 0, \quad t_0 \le t \le \tau. \tag{7.9}$$

Final Condition $R(\tau)$

Now consider the case when τ is specified and $x(\tau)$ is unspecified in the LQ problem (7.1)–(7.2). For this case, the condition at τ is

$$\left(\frac{\partial h}{\partial x}-\lambda\right)_{\tau}=0 \quad \text{(from (6.114))},$$

where $h(t)=x^{'}(t)M(t)x(t)$.Hence,

$$(2Mx-\lambda)_{\tau}=0 \implies \lambda(\tau)=2M(\tau)x(\tau). \tag{7.10}$$

Since, from (7.4), $\lambda(\tau)=R(\tau)x(\tau)$, using (7.10), we get $(R(\tau)-2M(\tau))x(\tau)=0$, which holds for any vector $x(\tau)\in\mathbb{R}^n$, leading to the final condition for $R(t)$:

$$R(\tau)=2M(\tau). \tag{7.11}$$

System of ODEs to Obtain x^*

We can obtain the Riccati matrix $R(t)$ by solving (7.9) under the condition (7.11). If the matrix $R(t)$ with real entries is found, then the feedback control u is obtained from (7.5). Substituting the feedback control u into equation (7.1), we get the following system of ODEs in x:

$$\begin{cases} \dot{x}=Ax-\dfrac{1}{2}BQ^{-1}B'Rx, \\ x(t_0)=x_0. \end{cases} \tag{7.12}$$

Solving (7.12) we obtain the optimal state $x^*(t)$, which in turn gives the optimal control $u^*(t)$ as in (7.5).

***Example* 7.2.1** Find the optimal pair x^* and u^* for the following control system

$$\begin{cases} \dot{x}=-x+u, \\ x(0)=1 \end{cases}$$

with the performance index

$$\mathbb{J}=x^2(5)+\int_0^5 u^2(t)dt,$$

where $x(5)$ is unspecified.

Solution **Method I.** (using the Euler–Lagrange equation) From (6.108), the Lagrangian is given by

$$L(x,\dot{x},u,\lambda)=u^2+\lambda(-x+u-\dot{x})+2x\dot{x},$$

and from (6.109), the Hamiltonian is given by

$$H(x,u,\lambda) = u^2 + \lambda(-x+u).$$

From (6.112), the Euler–Lagrange equations are

$$\begin{cases} \dot{x} = H_\lambda \implies \dot{x} = -x+u, \\ \dot{\lambda} = -H_x \implies \dot{\lambda} = \lambda \implies \lambda = e^t\lambda(0), \\ H_u = 0 \implies 2u+\lambda = 0 \implies u = -\dfrac{1}{2}\lambda = -\dfrac{1}{2}e^t\lambda(0). \end{cases} \tag{7.13}$$

Now, replacing $u(t)$ with $-\frac{1}{2}e^t\lambda(0)$ in the system equations, we get $\dot{x} = -x - \frac{1}{2}e^t\lambda(0)$ with $x(0) = 1$. Solving this gives

$$x(t) = e^{-t} - \frac{1}{4}(e^t - e^{-t})\lambda(0). \tag{7.14}$$

Since $x(5)$ is unspecified, from (6.113), the condition at $t = 5$ is

$$[L_{\dot{x}}]_{t=5} = 0 \implies -\lambda(5) + 2x(5) = 0. \tag{7.15}$$

Using (7.13) and (7.14) in (7.15), we get

$$\lambda(0) = \frac{2e^{-5}}{e^5 + \frac{1}{2}(e^5 - e^{-5})}. \tag{7.16}$$

From (7.14), we obtain

$$x^*(t) = \frac{3e^{-(t-5)} - e^{t-5}}{3e^5 - e^{-5}}.$$

Finally, the optimal control $u^*(t)$ is obtained from (7.13) using $\lambda(0)$.

Method II. (using the Riccati equation) In this example, $A = -1, B = 1, P = 0, Q = 1$, and $M(t) = 1$. The Riccati equation becomes $\dot{R} - 2R - \frac{1}{2}R^2 = 0$ with the end condition $R(5) = 2M(5) = 2$. Solving this differential equation we get

$$\frac{2dR(t)}{R(t)(R(t)+4)} = dt \implies \frac{1}{2}\log\left(\frac{R(t)}{R(t)+4}\right) = t+c.$$

Hence,

$$\frac{R(t)}{R(t)+4} = e^{2(t+c)} \implies R(t) = \frac{4e^{2(t+c)}}{1-e^{2(t+c)}}.$$

Now, $R(5) = 2 \implies e^{2(5+c)} = \frac{1}{3} \implies e^{2c} = \frac{1}{3}e^{-10}$. Hence, $R(t) = \frac{4e^{t-5}}{3e^{-(t-5)} - e^{t-5}}$ is the solution of the Riccati equation. Since $u(t) = -\frac{\lambda(t)}{2} = -\frac{R(t)x(t)}{2}$, the given control system becomes

$$\begin{cases} \dot{x} = -x + u = -\dfrac{3e^{-(t-5)} + e^{t-5}}{3e^{-(t-5)} - e^{t-5}} x, \\ x(0) = 1. \end{cases}$$

Now $-\dfrac{3e^{-(t-5)} + e^{t-5}}{3e^{-(t-5)} - e^{t-5}} x = \dfrac{d}{dt}[\log(3e^{-(t-5)} - e^{t-5})]x$.

Then $\dfrac{dx}{x} = d[\log(3e^{-(t-5)} - e^{t-5})]$. Integrating both sides from 0 to t, the optimal state is obtained as

$$x^*(t) = \frac{3e^{-(t-5)} - e^{t-5}}{3e^5 - e^{-5}}.$$

Finally, the optimal control u^* is obtained from the relation $u^*(t) = -\frac{R(t)x^*(t)}{2}$.

Optimal Value of Performance Functional in Terms of $R(t)$

In this section, we shall prove a simple relation between the performance index $\mathbb{J}$ and the Riccati matrix R. Let u^* and x^* be the optimal pair obtained from (7.5), (7.9), and (7.12). Let the performance index be

$$\mathbb{J}_{t_0} = (x^*(\tau))' M(\tau) x^*(\tau) + \int_{t_0}^{\tau} \left((x^*(t))' P(t) x^*(t) + u^{*'}(t) Q(t) u^*(t) \right) dt, \tag{7.17}$$

where $x^*(t)$ and $u^*(t)$ satisfy the control system dynamics

$$\begin{cases} \dot{x}^* = A(t)x^*(t) + B(t)u^*(t), \quad t_0 \le t \le \tau, \\ x^*(t_0) = x_0 \end{cases} \tag{7.18}$$

and extremize $\mathbb{J}_{t_0}$. Note that the notation $\mathbb{J}_{t_0}$ is used here to indicate that the lower limit of the integral in (7.17) is t_0. Let $R(t)$ be the solution of the Riccati equation (7.9) under the condition (7.11), i.e., $R(\tau) = 2M(\tau)$. Replacing $M(\tau)$ by $\frac{R(\tau)}{2}$ in (7.17) yields

$$\begin{aligned} \mathbb{J}_{t_0} &= \frac{1}{2} x^{*'}(\tau) R(\tau) x^*(\tau) + \int_{t_0}^{\tau} \left(x^{*'} P x^* + u^{*'} Q u^* \right) dt \\ &= \frac{1}{2} x^{*'}(t_0) R(t_0) x^*(t_0) + \int_{t_0}^{\tau} \left[\left(x^{*'} P x^* + u^{*'} Q u^* \right) + \frac{d}{dt} \left(\frac{1}{2} x^{*'} R x^* \right) \right] dt \end{aligned}$$

(using (6.101) and (6.104))

$$= \frac{1}{2}x^{*'}(t_0)R(t_0)x^*(t_0) + \int_{t_0}^{\tau}\Big[\Big(x^{*'}Px^* + u^{*'}Qu^*\Big)$$
$$+ \frac{1}{2}\Big(\dot{x}^{*'}Rx^* + x^{*'}\dot{R}x^* + x^{*'}R\dot{x}^{*'}\Big)\Big]\,dt. \tag{7.19}$$

On substituting $\dot{R}$ from the Riccati equation, the integrand in (7.19) becomes

$$x^{*'}Px^* + u^{*'}Qu^* + \frac{1}{2}\dot{x}^{*'}Rx^* + \frac{1}{2}x^{*'}R\dot{x}^*$$
$$+ \frac{1}{2}x^{*'}\left(-RA - A'R - 2P + \frac{1}{2}RBQ^{-1}B'R\right)x^*. \tag{7.20}$$

Now, substituting $\dot{x}^* = Ax^* + Bu^*$ into this expression, it can be seen that integrand in (7.19) is identically zero for all $t \in [t_0, \tau]$. Hence, the optimal value $\mathbb{J}^*_{t_0}$ of $\mathbb{J}_{t_0}$ is given by

$$\mathbb{J}^*_{t_0} = \frac{1}{2}x^{*'}(t_0)R(t_0)x^*(t_0). \tag{7.21}$$

Note that the optimal value $\mathbb{J}^*_{t_0}$ can be obtained from (7.21) using only the initial condition on x and the Riccati matrix $R(t)$. It does not require the explicit optimal pair x^* and u^*.

However, we need to find $x^*(t)$ to compute $\mathbb{J}^*_s$ as defined below. Let

$$\mathbb{J}_s = x'(\tau)M(\tau)x(\tau) + \int_s^{\tau}(x'Px + u'Qu)\,dt, \tag{7.22}$$

where $s \in [t_0, \tau]$. If we consider the minimization of $\mathbb{J}_s$ under the constraint $\dot{x} = Ax + Bu$ with initial condition $x(t_0) = x_0$, then the optimal solution x^* and u^* obtained for $\mathbb{J}_{t_0}$ must minimize $\mathbb{J}_s$ as well. Hence, replacing $x^*(t_0)$ by $x^*(s)$ and t_0 by s in equations (7.17)–(7.21), the minimum value $\mathbb{J}^*_s$ of $\mathbb{J}_s$ is given by

$$\mathbb{J}^*_s = \frac{1}{2}x^{*'}(s)R(s)x^*(s), \quad s \in [t_0, \tau]. \tag{7.23}$$

***Example* 7.2.2** Find the optimal pair x^* and u^*, which minimizes the cost functional

$$\mathbb{J}_s = 4x^2(1) + \int_s^1\left(3x^2(t) + \frac{1}{4}u^2(t)\right)dt, \quad 0 \le s \le 1,$$

under the constraint $\dot{x} = 2x + u$ for $0 \le t \le 1$, with $x(0) = 1$ and $x(1)$ is unspecified. Also, find $\mathbb{J}_0$ and $\mathbb{J}_s$ for $0 \le s \le 1$.

Solution **Method I.** (Using the Euler–Lagrange equation) Here, $A=2, B=1, P=3$, $Q=\frac{1}{4}$, and $M=4$. Substituting these into equation (6.126), we have

$$\begin{bmatrix} \dot{x} \\ \dot{\lambda} \end{bmatrix} = \begin{bmatrix} 2 & -2 \\ -6 & -2 \end{bmatrix} \begin{bmatrix} x \\ \lambda \end{bmatrix} \tag{7.24}$$

with boundary conditions

$$x(0)=1 \text{ and } \lambda(1)=2M(1)x(1), \text{ i.e., } \lambda(1)=8x(1). \tag{7.25}$$

The solution of (7.24) in terms of the eigenpairs is

$$\begin{bmatrix} x(t) \\ \lambda(t) \end{bmatrix} = c_1 e^{4t} \begin{bmatrix} 1 \\ 1 \end{bmatrix} + c_2 e^{-4t} \begin{bmatrix} 1 \\ 3 \end{bmatrix},$$

where $c_i : i=1,2$ are constants, which can be obtained using the boundary conditions (7.25) as $c_1 = \frac{5e^{-4}}{5e^{-4}-9e^4}$ and $c_2 = -\frac{9e^4}{5e^{-4}-9e^4}$. Now, the optimal control is $u(t) = \frac{1}{2}Q^{-1}B'\lambda(t) = 2\lambda(t)$.

Method II. (Using the Riccati equation) Comparing with equations (6.117) and (6.118), we observe that $A=2$, $B=1$, $M=4$, $P=3$, and $Q=\frac{1}{4}$. Hence, the Riccati equation (7.9) becomes $\dot{R}+4R+6-2R^2=0$ for $0 \le t \le 1$, which implies

$$\frac{dR}{(R-3)(R+1)} = 2dt.$$

Integrating both sides we get

$$\log\left(\frac{R(t)-3}{R(t)+1}\right) = 8(t+c), \quad \text{where } c \text{ is a constant.} \tag{7.26}$$

From (7.11), we obtain the condition at $\tau=1$ as

$$R(1)=2M(1)=8. \tag{7.27}$$

Using (7.27) in (7.26), we get $\frac{R(t)-3}{R(t)+1} = \frac{5}{9}e^{8(t-1)}$, which implies $R(t) = \frac{5e^{8(t-1)}+27}{9-5e^{8(t-1)}}$. Thus, from (7.5), the optimal feedback control is given by

$$u^* = -\frac{1}{2}Q^{-1}B'Rx^* = -2R(t)x^* = -2\left(\frac{5e^{8(t-1)}+27}{9-5e^{8(t-1)}}\right)x^*. \tag{7.28}$$

Now, to find $x^*(t)$, substitute u^* from (7.28) into the system equation $\dot{x} = 2x + u$ to get

$$\begin{cases} \dot{x} = 2x - 2\left(\dfrac{5e^{8(t-1)} + 27}{9 - 5e^{8(t-1)}}\right)x = -2\dfrac{10e^{8(t-1)} + 18}{9 - 5e^{8(t-1)}}x, \\ x(0) = 1. \end{cases} \tag{7.29}$$

By solving the initial value problem (IVP) (7.29), we obtain the optimal state

$$x^*(t) = \frac{9e^{-4(t-1)} - 5e^{4(t-1)}}{9e^4 - 5e^{-4}} \quad \text{for } 0 \le t \le 1.$$

From (7.21), the optimal value of $\mathbb{J}$ is given by

$$\mathbb{J}_0^* = x^{*'}(0)R(0)x^*(0) = R(0) = \frac{5e^{-8} + 27}{9 - 5e^{-8}}$$

and

$$\mathbb{J}_s^* = (x^*)'(s)R(s)x^*(s) = \frac{5e^{8(s-1)} + 27}{9 - 5e^{8(s-1)}}\left[\frac{9e^{-4(s-1)} - 5e^{4(s-1)}}{9e^4 - 5e^{-4}}\right]^2 \quad \text{for } 0 \le s \le 1.$$

Proposition 7.2.3 The solution $R(t)$ of the matrix Riccati equation is real symmetric positive definite (RSPD) for all $t \in [t_0, \tau)$ under the assumptions that (i) $M(t)$ and $P(t)$ are real symmetric positive semi definite (RSPSD) matrices. (ii) $Q(t)$ is an RSPD matrix.

Proof. The Riccati equation is

$$\dot{R} + RA + A'R + 2P - \frac{1}{2}RBQ^{-1}B'R = 0, \quad t_0 \le t \le \tau, \tag{7.30}$$

with the end condition

$$R(\tau) = 2M(\tau). \tag{7.31}$$

We know that $P(t)$, $Q(t)$, and $M(t)$ are symmetric matrices. Taking the transpose on both sides of (7.30) and (7.31), we obtain

$$\dot{R}' + A'R' + R'A + 2P - \frac{1}{2}R'BQ^{-1}B'R' = 0 \tag{7.32}$$

with

$$R'(\tau) = 2M'(\tau) = 2M(\tau) \quad (\text{since } M(t) \text{ is an RSPSD matrix}). \tag{7.33}$$

Note that (7.32) and (7.33) are the same as (7.30) and (7.31), respectively, except that $R(t)$ is replaced by $R'(t)$ in (7.32) and (7.33). Since the end conditions are the same in both cases, their solutions are identical (by Theorem 4.4.1 on the uniqueness of solutions). Hence, $R(t) = R'(t)$. □

Proposition 7.2.4 The matrix $R(t)$ is positive definite for all $t \in [t_0, \tau)$.

Proof. It is known that $R(\tau) = 2M(\tau)$. Since $M(\tau)$ is a positive semi-definite matrix, $R(\tau)$ is also a positive semi-definite matrix. From (7.23), we know that

$$\mathbb{J}^*(s) = \frac{1}{2}x^{*'}R(s)x^*(s).$$

Moreover, since $Q(t)$ is a positive definite matrix, $u'(t)Q(t)u(t) > 0$ for $t_0 \leq t \leq \tau$. Hence, $\mathbb{J}(s) > 0$ and $\mathbb{J}(\tau) = x'(\tau)M(\tau)x(\tau) \geq 0$, implying that $\mathbb{J}^*(s) > 0$ for all $s \in [t_0, \tau)$, i.e., $x^{*'}(s)R(s)x^*(s) > 0$ for all $s \in [t_0, \tau)$. Since this result holds for all $x^*(t)$ with arbitrary initial states, $R(t)$ is positive definite for all $t \in [t_0, \tau]$. □

Exercise 7.2.1 Find the optimal state and control for the system $\dot{x} = -x + u$ with $x(0) = 1$ that minimizes $\mathbb{J}(x) = x^2(2) + \int_0^2 u^2(t)dt$ using the Riccati equation. Also, find the optimal value of $\mathbb{J}_1(x) = x^2(2) + \int_1^2 u^2(t)dt$

7.3 Pontryagin Principle (Constrained Control)

In this section, we consider the optimal control problem IV discussed in Section 6.9, with an additional restriction on the control function $u(t)$. The problem is stated as follows:

Find the optimal control $u^*(t)$ and the corresponding optimal state $x^*(t)$, which minimize the functional

$$\mathbb{J} = h(\tau, x(\tau)) + \int_{t_0}^{\tau} f(t, x(t), u(t))dt \tag{7.34}$$

for the system

$$\dot{x}(t) = A(t)x(t) + B(t)u(t) = g(t, x, u) \tag{7.35}$$

with the initial condition $x(t_0) = x_0$. Additionally, the control function u must satisfy $\|u(t)\| \leq M$ for all $t \in [t_0, \tau]$. The Pontryagin principle provides a condition for obtaining the optimal control $u^*(t)$ under this boundedness condition on the control function.

Using the Lagrange multiplier $\lambda(t)$ as discussed in Section 6.9, the problem can be converted into the minimization of $\mathbb{J}_1$, where

$$\mathbb{J}_1 = h(t_0, x(t_0)) + \int_{t_0}^{\tau} \left(f + \frac{dh}{dt} + \lambda'(g - \dot{x}) \right) dt. \tag{7.36}$$

Let $S = \{u \in C^1([t_0, \tau], \mathbb{R}^m) : \|u(t)\| \leq M\}$ be the set of all admissible control functions. If x^*, u^*, and λ^* are the optimal solutions with $u^* \in S$, then

$$\underset{u \in S}{\text{minimum}}\ \mathbb{J}_1(x,u,\lambda) = \mathbb{J}_1(x^*,u^*,\lambda^*), \tag{7.37}$$

Let $L(x,\dot{x},u,\lambda)$ and $H(x,u,\lambda)$ be the Lagrangian and Hamiltonian, respectively. Then $L = f + \frac{dh}{dt} + \lambda'(g - \dot{x}) = H + \frac{dh}{dt} - \lambda'\dot{x}$, where $H(t,x,u,\lambda) = f(t,x,u) + \lambda' g(t,x,u)$. Let $u^* \in S$ be the optimal control and x^* and λ^* the optimal state and co-state functions.

We know from Sections 6.5–6.8 that the variation in $\mathbb{J}_1$ at x^*, u^*, and λ^* is

$$\begin{aligned}\delta \mathbb{J}_1 = &\int_{t_0}^{\tau} \left[\left(\frac{\partial L}{\partial x} - \frac{d}{dt}\left(\frac{\partial L}{\partial \dot{x}}\right)\right) \delta x + \left(\frac{\partial L}{\partial u}\right) \delta u + \left(\frac{\partial L}{\partial \lambda}\right) \delta\lambda(t) \right] dt \\ &+ \left[\frac{\partial L}{\partial \dot{x}}\right]_{\tau} \delta x_{\tau} + \left[L - \left(\frac{\partial L}{\partial \dot{x}}\right)' \dot{x} \right] \Delta\tau. \end{aligned} \tag{7.38}$$

If δu is a variation in u^* and $u = u^* + \delta u \in S$, then $\|u^*(t)\| \leq M$ and $\|u(t)\| \leq M$ for $t_0 \leq t \leq \tau$. Hence, the variation $\delta u(t)$ is not an arbitrary function as in Chapter 6, but is instead restricted by the boundedness condition. From (7.37), the increment $\Delta\mathbb{J}_1$ over $u \in S$ is

$$\Delta\mathbb{J}_1 = \mathbb{J}_1(x,u,\lambda) - \mathbb{J}_1(x^*,u^*,\lambda^*) \geq 0 \implies \delta\mathbb{J}_1 \geq 0 \text{ at } (x^*,u^*,\lambda^*). \tag{7.39}$$

In terms of the Hamiltonian H, equation (7.38) can be written as

$$\begin{aligned}\delta \mathbb{J}_1 = &\int_{t_0}^{\tau} \left[\left(\frac{\partial H}{\partial x} + \dot{\lambda}\right)_* \delta x + \left(\frac{\partial H}{\partial u}\right)_* \delta u(t) + \left(\frac{\partial H}{\partial \lambda} - \dot{x}\right)_* \delta\lambda(t)\right] dt \\ &+ \left[\frac{\partial h}{\partial x} - \lambda(t)\right]_{\tau} \delta x_{\tau} + \left[H + \frac{\partial h}{\partial t}\right]_{\tau} \Delta\tau. \end{aligned} \tag{7.40}$$

Since $\frac{\partial H}{\partial \lambda} = g(x,u)$ and x^* and u^* satisfy (7.35), we obtain $\left(\frac{\partial H}{\partial \lambda} - \dot{x}\right)_* = 0$. Moreover, $\lambda^*(t)$ can be chosen to satisfy the system $\dot{\lambda} = -\frac{\partial H}{\partial x} = -\frac{\partial f}{\partial x} - \lambda' \frac{\partial g}{\partial x}$, along with the end condition $\left(\frac{\partial h}{\partial x} - \lambda(t)\right)_{\tau} = 0$ for the optimal pair u^* and x^*. If τ is unspecified, then τ, $x^*(\tau)$, $u^*(\tau)$, and $\lambda^*(\tau)$ are assumed to satisfy $\left(H + \frac{\partial h}{\partial t}\right)_{\tau} = 0$. Hence, equation (7.40) reduces to

$$\delta\mathbb{J}_1 = \int_{t_0}^{\tau} \left(\frac{\partial H}{\partial u}\right)_{x^*} \delta u(t) dt. \tag{7.41}$$

From (7.39) and (7.41),

$$\int_{t_0}^{\tau} \left(\frac{\partial H}{\partial u}\right)_{x^*} \delta u(t) dt \geq 0,$$

which can be written as

$$\int_{t_0}^{\tau} (H(x^*, u^* + \delta u, \lambda^*)) - H(x^*, u^*, \lambda^*)dt \geq 0 \tag{7.42}$$

for all variations δu such that $u^* + \delta u \in S$. Inequality (7.42) implies that for all admissible variations $\delta u(t)$,

$$H(x^*, u^* + \delta u, \lambda^*) \geq H(x^*, u^*, \lambda^*), \quad t \in [t_0, \tau]. \tag{7.43}$$

The proof for the above inequality is as follows.

If (7.43) is not true, then assume that $H(x^*(t), (u^* + \delta u)(t), \lambda^*(t)) < H(x^*(t), u^*(t), \lambda^*(t)), t \in (a,b) \subset [t_0, \tau]$, for a variation $\delta u(t) = v(t)$ (say). Since both $u^*(t)$ and $u^*(t) + \delta u(t)$ belong to S, $|u^*(t) + v(t)| \leq M$ for all $t \in [t_0, \tau]$. Now consider another variation

$$(\delta u)(t) = \begin{cases} v(t), & t \in (a,b), \\ 0, & \text{elsewhere.} \end{cases}$$

Then

$$u^*(t) + (\delta u)(t) = \begin{cases} u^*(t) + v(t), & t \in (a,b), \\ u^*(t), & \text{elsewhere.} \end{cases}$$

Hence $|(u^* + \delta u)(t)| \leq M$ and $\int_{t_0}^{\tau} [H(x^*, u^* + \delta u, \lambda^*) - H(x^*, u^*, \lambda^*)]\, dt = \int_a^b [H(x^*, u^* + v, \lambda^*) - H(x^*, u^*, \lambda^*)]\, dt < 0$, which contradicts (7.42). Thus, the inequality (7.43) provides the necessary condition for obtaining the solution to the minimization problem (7.37). Hence the optimal solution to the constrained control problem can be obtained from the following.

$$H(x^*, u^*, \lambda^*) = \min_{\|u(t)\| \leq M} H(x^*, u, \lambda^*). \tag{7.44}$$

Equation (7.44) implies that the optimal control $u^*(t)$ minimizes the Hamiltonian. This is known as the **Pontryagin minimum principle**. The inequality (7.43) provides the condition on the variations δu.

Remark 7.3.1 To maximize the functional $\mathbb{J}$ in (7.34) under the constraint $\|u(t)\| \leq M$ for $t \in [t_0, \tau]$, the Pontryagin maximum principle is derived using arguments similar to the previous discussion, with the appropriate adjustment of replacing $\geq$ with $\leq$. In this case, equation (7.44) becomes

$$H(x^*, u^*, \lambda^*) = \max_{\|u(t)\| \leq M} H(x^*, u, \lambda^*).$$

***Example* 7.3.2** Find the optimal control $u^*(t)$ and the corresponding optimal state $x^*(t)$ for the system $\dot{x} = -2x + u, 0 \le t \le \tau$, such that $|u^*(t)| \le 1$ for $0 \le t \le \tau$, and the optimal pair minimizes the performance functional

$$\mathbb{J} = \frac{1}{2}\int_0^\tau (5x^2 + u^2)\,dt$$

under the following boundary conditions: (i) $x(0) = 1, \tau = 1$, and $x(1) = 0$; (ii) $x(0) = 1, \tau = 1$, but $x(1)$ is free; (iii) $x(0) = 1, x(\tau) = 0$, but τ is unspecified.

Solution As in Example 6.8.3, the Lagrangian is $L(x, \dot{x}, u, \lambda) = \frac{1}{2}(5x^2 + u^2) + \lambda(-2x + u - \dot{x})$, and the Hamiltonian is

$$H(x, u, \lambda) = \frac{1}{2}(5x^2 + u^2) + \lambda(-2x + u). \tag{7.45}$$

When there is **no restriction** on u, one can find the optimal control u from the equation $\frac{\partial H}{\partial u} = 0$, which implies

$$u(t) = -\lambda(t), \tag{7.46}$$

where $\lambda(t)$ is obtained by solving the canonical system under the boundary conditions. However, according to Pontryagin's minimum principle, the optimal control u^* with the constraint $|u(t)| \le 1$ minimizes the Hamiltonian as in (7.44). For finding this, consider only the terms involving u in $H(x, u, \lambda)$ for minimization, i.e., $\frac{1}{2}u^2 + \lambda u$, which takes its minimum value when $u = -\lambda$. Now, since $|u(t)| \le 1$, we obtain u^* as

$$u^*(t) = \begin{cases} -\lambda(t), & \text{if } |\lambda(t)| \le 1, \\ -1, & \text{if } \lambda(t) \ge 1, \\ 1, & \text{if } \lambda(t) < -1. \end{cases} \tag{7.47}$$

The Euler–Lagrange equations to obtain x^* corresponding to u^* are the same as in Example 6.8.3. Hence, the equations are

$$\begin{cases} \dot{x} = -2x - \lambda, \\ \dot{\lambda} = -5x + 2\lambda. \end{cases} \tag{7.48}$$

(i) The given boundary conditions $x(0) = 1$ and $x(1) = 0$, are the same as in Example 6.8.3. Therefore, the solution to (7.48) is

$$x(t) = c_1 e^{3t} + c_2 e^{-3t}, \tag{7.49}$$

$$\lambda(t) = -5c_1 e^{3t} + c_2 e^{-3t}, \tag{7.50}$$

where $c_1 = \frac{e^{-3}}{e^{-3}-e^3}$ and $c_2 = \frac{e^3}{e^3-e^{-3}}$. Substituting c_1 and c_2 into (7.50), we get

$$\lambda(t) = \frac{e^{-3}}{e^3 - e^{-3}} \left(4e^{3t} + 2\cosh 3t\right) + e^{-3t}, \tag{7.51}$$

which is positive and increasing for all $t \geq 0$. Let t_1 be such that $\lambda(t_1) = 1$. Then, from (7.47),

$$u^*(t) = \begin{cases} -\lambda(t), & 0 \leq t \leq t_1, \\ -1, & t > t_1. \end{cases} \tag{7.52}$$

(Note that if $t_1 \geq \tau$, then $u(t) = -\lambda(t)$ for all $t \in [0, \tau]$.) Using this optimal control, the optimal state $x^*(t)$ can be obtained by solving the equation

$$\dot{x}^* = -2x^* + u^*. \tag{7.53}$$

The solution $x^*(t)$ obtained from (7.53) may differ from the expression in (7.49), which represents the optimal solution when there is **no restriction** on the control $u(t)$. By contrast, (7.53) will give the optimal solution under the condition $|u(t)| \leq 1$. However, the solution $x^*(t)$ obtained from (7.53) may not satisfy the given end condition $x(1) = 0$ due to the restriction on $u(t)$ along with the time restriction (fixed final time).
(*ii*) Since final time $\tau = 1$ is fixed and $x(1)$ is unspecified, using (6.72), the final condition is $\lambda(1) = 0$. Now, $x(t)$ and $\lambda(t)$ can be obtained from (7.49) and (7.50) using the boundary conditions $x(0) = 1$ and $\lambda(1) = 0$. Finally the the optimal control under the given constraint can be obtained as in case (i).
(*iii*) $x(0) = 1$ and $x(\tau) = 0$, where τ is unspecified. From (7.49), $x(0) = 1$ implies $c_2 = 1 - c_2$. $x(\tau) = 0$ provides c_2 in terms of τ. For the **general problem**, where there are **no restrictions on** $u(t)$, τ can be obtained using the condition $H(x(\tau), u(\tau), \lambda(\tau)) = 0$, where $u(t) = -\lambda(t)$ as in (7.46). Thus, from (7.45), we obtain

$$\begin{aligned} &\frac{1}{2}\left(5x^2(\tau) + \lambda^2(\tau)\right) + \lambda(\tau)\left(-2x(\tau) - \lambda(\tau)\right) = 0 \\ \Longrightarrow\ &\frac{5}{2}x^2(\tau) - 2x(\tau)\lambda(\tau) - \frac{1}{2}\lambda^2(\tau) = 0. \end{aligned} \tag{7.54}$$

Substituting $x(\tau)$ and $\lambda(\tau)$ from (7.49) and (7.50), the final time τ is obtained from (7.54). Note that this τ is valid for the general problem. If τ is less than t_1, then $|\lambda(t)| \leq 1$ for all $t \in [0, \tau]$, implying $|u(t)| \leq 1$. Hence this optimal solution is valid for both general problem and constrained problems. If τ obtained from (7.54) is greater than t_1, then $u^*(t)$ must follow (7.52), and $x^*(t)$ is obtained from (7.53), along with initial condition $x(0) = 1$. The final time τ^* can be obtained from the condition $x^*(\tau) = 0$.

7.4 Applications

***Example* 7.4.1** An object of unit mass moves on a straight line under the influence of a force u, without friction. Let the force $u(t)$ have a constraint $|u(t)| \leq M$, where M is a constant. Initially, the object is at a distance x_0 from a point O and is moving with a velocity v_0. If it is required to reach and stop at O in minimum time, find the optimal time τ, the optimal control $u(t), 0 \leq t \leq \tau$, and the optimal state of the system. Assume that the object can move in both forward and in backward directions.

Solution Let $x_1(t)$ and $x_2(t)$ be the distance from O and the velocity of the object at time t, respectively. Then the equation of motion is represented by

$$\begin{cases} \dot{x}_1 = x_2, \\ \dot{x}_2 = u \end{cases} \tag{7.55}$$

with initial conditions $x_1(0) = x_0$ and $x_2(0) = v_0$. As it is required to minimize the travel time, the objective functional for minimization is $\mathbb{J} = \int_0^\tau dt$. Thus, the problem is to find the optimal control $u^*(t)$ for the system (7.55) such that $|u^*(t)| \leq M$ and $\mathbb{J} = \int_0^\tau dt$ is minimized. Now, let us apply Pontryagin's minimum principle to solve the problem.

Let $x = [x_1 \quad x_2]'$ and $\lambda = [\lambda_1 \quad \lambda_2]'$. For the optimal problem, the Lagrangian is (see (6.69))

$$L = 1 + \lambda'(t)(g(x,u) - \dot{x}) = 1 + \lambda_1(t)(x_2 - \dot{x}_1) + \lambda_2(u - \dot{x}_2), \tag{7.56}$$

and the Hamiltonian is (see (6.76))

$$H(x, \lambda, u) = 1 + \lambda_1 x_2 + \lambda_2 u. \tag{7.57}$$

According to Pontryagin's minimum principle, we need to minimize H with respect to u, assuming x^* and λ^* are fixed, i.e.,

$$H(x^*, \lambda^*, u^*) = \min_{\substack{|u(t)| \leq M \\ 0 \leq t \leq \tau}} H(x^*, \lambda^*, u). \tag{7.58}$$

From (7.57),

$$H(x^*, \lambda^*, u) = 1 + \lambda_1^* x_2^* + \lambda_2^* u. \tag{7.59}$$

By varying $u(t)$ and fixing λ_1^*, λ_2^*, and x_2^*, it can be seen that $H(x^*, \lambda^*, u)$ attains its minimum for each $t \in [0, \tau]$ when $u(t) = u^*(t)$, where

$$u^*(t) = \begin{cases} M, & \text{if } \lambda_2^*(t) < 0, \\ -M, & \text{if } \lambda_2^*(t) > 0, \\ 0, & \text{if } \lambda_2^*(t) = 0. \end{cases}$$

Now let us analyze all possibilities.
Case I. Let $u(t) = M$. Then, from (7.55),

$$x_2(t) = Mt + c_1 \text{ and } x_1(t) = \frac{M}{2}t^2 + c_1 t + c_2,$$

where c_1 and c_2 are constants. Using $x_1(0) = x_0$ and $x_2(0) = v_0$, we get $c_2 = x_0$ and $c_1 = v_0$. Eliminating t between x_1 and x_2, we obtain the family of parabolas $x_1 = \frac{1}{2M}x_2^2 + c_3$, where $c_3 = c_2 - \frac{c_1^2}{M}$ (see Figure 7.1).

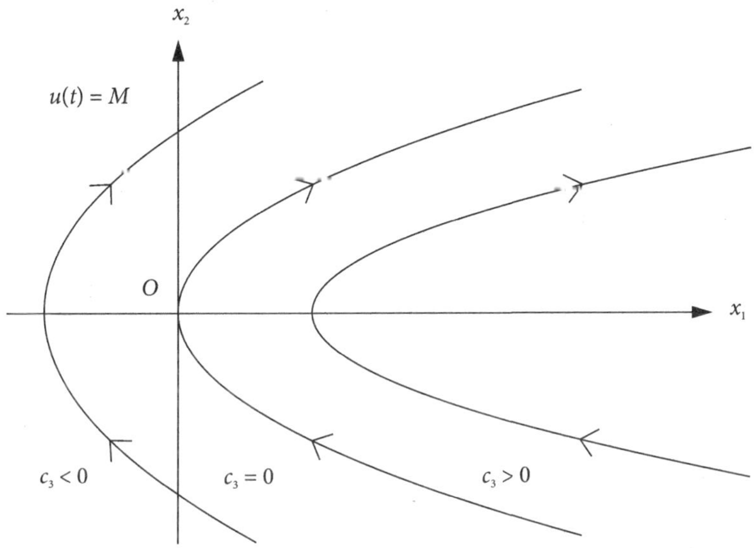

Figure 7.1 Family of parabolas for $u(t) = M$

Case II. Let $u(t) = -M$. Then, from (7.55),

$$x_2(t) = -Mt + c_4 \text{ and } x_1(t) = -\frac{M}{2}t^2 + c_4 t + c_5,$$

where $c_5 = x_0$ and $c_4 = v_0$. Eliminating t from x_1 and x_2, we get

$$x_1 = -\frac{1}{2M}x_2^2 + c_6,$$

where $c_6 = c_5 + \frac{c_4^2}{M}$ is a constant. This gives a family of parabolas in the x_1-x_2 phase plane, as shown in Figure 7.2.

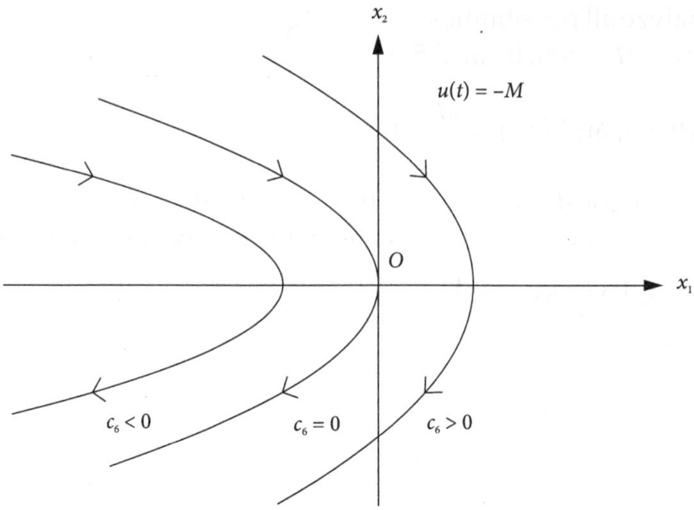

Figure 7.2 Family of parabolas for $u(t) = -M$

If $x_0 > 0$ and $v_0 > 0$, then at $t = 0$, the object is at a distance x_0 and is moving away from O. To bring it towards O, a negative force must be applied. Hence, the choice of $u(t)$ is $u(t) = -M$. Now, from the family of curves corresponding to $u = -M$, choose the parabola I on which (x_0, v_0) lies in the first quadrant (see Figure 7.3).

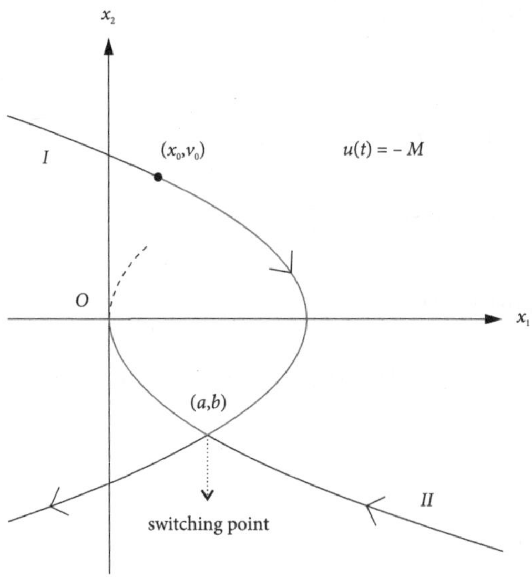

Figure 7.3 Trajectory with the initial state (x_0, v_0)

As the object needs to reach and stop at O, the force $u(t)$ should be applied in such a way that at the final time τ (which is unknown), $x_1(\tau) = 0$ and $x_2(\tau) = 0$.

If the force $u(t) = -M$ is kept unchanged, the required final conditions will never be met, as curve I does not pass through $(0,0)$. Therefore, it is necessary to change the value of $u(t)$ at some time t_1 (say), when curve I intersects curve II (which passes through $(0,0)$). Thus, t_1 is called the switching time, and the force $u(t)$ is switched from $-M$ to $+M$. From time $t = t_1$ to $t = \tau$, $(x_1(t), x_2(t))$ lies on curve II, which leads to the origin at $(x_1(\tau), x_2(\tau))$.

Now, let us illustrate the above discussion with numerical values. Let $x_1(0) = x_0 = 10$, $x_2(0) = v_0 = 2$, and $|u(t)| \le 1$. Then, from Case II, for $u(t) = -1$, we obtain curve I as

$$x_1 = -\frac{1}{2}x_2^2 + 12. \tag{7.60}$$

For $u(t) = 1$, from Case I, the curve II that passes through $(0,0)$ is

$$x_1 = \frac{1}{2}x_2^2 \tag{7.61}$$

(see Figure 7.4).

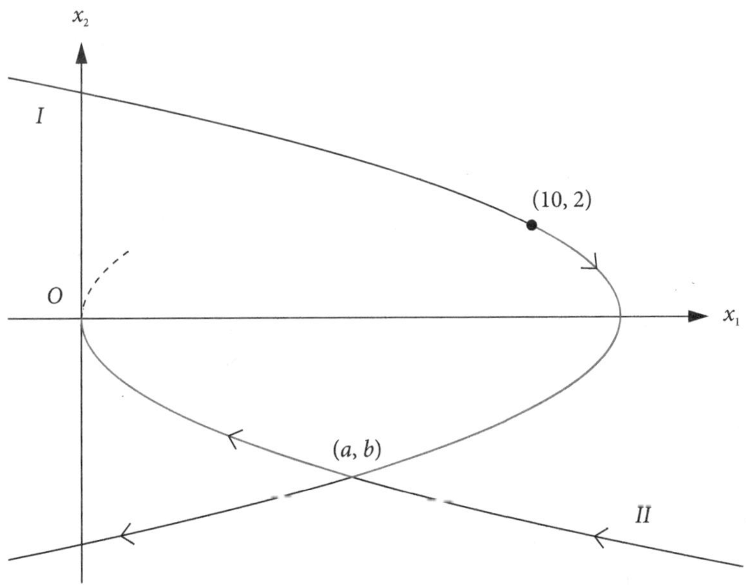

Figure 7.4 Trajectory with the initial state $(10,2)$

Let (a,b) be the point of intersection of curves I and II. From (7.60) and (7.61), we get $(a,b) = (6, -\sqrt{12})$. In Case II, $x_2(t) = -t + c_4$. Since $x_2(0) = 2$ and $x_2(t_1) = -\sqrt{12}$,

$$c_4 = 2 \text{ and } t_1 = 2 + \sqrt{12}. \tag{7.62}$$

As the switching occurs from curve I to curve II at t_1, the point $(x_1(t), x_2(t))$ moves on curve II from $(6, -\sqrt{12})$ to $(0,0)$ in the time interval (t_1, τ).

For curve II, $x_2(t) = t + c_1$. Since $x_2(2+\sqrt{12}) = -\sqrt{12}$, we obtain $c_1 = -2(\sqrt{12}+1)$. Hence, $x_2(t) = t - 2(\sqrt{12}+1)$. Then $x_2(\tau) = 0 \implies \tau - 2(\sqrt{12}+1) = 0 \implies \tau = 2(\sqrt{12}+1)$ is the optimal final time.

***Example* 7.4.2** A rocket starts from the ground from rest and moves vertically upwards under the influence of gravity and a thrust force that is proportional to the fuel mass released by the rocket propellant. Let m_0 be the initial mass of the rocket, and it can lose its mass for propulsion until it reaches a mass of m_1 units. If the vertical thrust force cannot exceed M units, then find the optimal control (thrust force) such that the maximum height is achieved.

Solution The equation of motion of the rocket is

$$\begin{cases} m\ddot{x} = -gm + cu, \\ \dot{m} = -u, \end{cases}$$

where $m(t)$ is the mass at time t, and $x(t), \dot{x}(t)$, and $\ddot{x}(t)$ are the displacement, velocity, and acceleration of the rocket at time t. The term $u(t)$ is the vertical force acting on the rocket, and g is the acceleration due to gravity acting in the downward vertical direction. The constant c is a proportionality constant. The upward vertical direction is assumed to be positive.

It is required that the optimal force $u(t)$ is applied on the rocket so that it attains the maximum height x at an unspecified time τ, when the rocket attains zero velocity and the mass $m(\tau) = m_1$, a fixed value. It is assumed that the force $u(t)$ is restricted by $|u(t)| \leq M$ for all $t \in [0, \tau]$.

To find the optimal control function $u(t)$, let us formulate the problem as follows. Let $x_1 = x$ and $x_2 = \dot{x}$. Then we obtain the control system

$$\begin{cases} \dot{x}_1 = x_2, \\ \dot{x}_2 = -g + \dfrac{cu}{m}, \\ \dot{m} = -u, \end{cases} \tag{7.63}$$

where $u(t)$ must satisfy

$$|u(t)| \leq M. \tag{7.64}$$

The given boundary conditions are $x_1(0) = 0$, $x_2(0) = 0$, $m(0) = m_0$, and $x_2(\tau) = 0$, where τ is unspecified. Since the requirement is to maximize the height using the optimal control $u(t)$, let the performance index be

$$\mathbb{J} = x_1(\tau) + \int_0^\tau u(t)dt,$$

which can be written as

$$\mathbb{J} = \int_0^\tau (u(t) + \dot{x}_1(t))\,dt. \tag{7.65}$$

The Lagrangian is given by (see (6.108))

$$\begin{aligned} L(x_1, x_2, \dot{x}_1, \dot{x}_2, m, \dot{m}, \lambda_1, \lambda_2, \lambda_3) &= u + \dot{x}_1 + \lambda_1(x_2 - \dot{x}_1) \\ &\quad + \lambda_2(-g + \frac{cu}{m} - \dot{x}_2) + \lambda_3(-u - \dot{m}), \end{aligned} \tag{7.66}$$

and the Hamiltonian is given by

$$H(x_1, x_2, m, \lambda_1, \lambda_2, \lambda_3) = u + \lambda_1 x_2 - \lambda_2 g + \lambda_2 \frac{cu}{m} - \lambda_3 u. \tag{7.67}$$

Since $u(t)$ is restricted by $|u(t)| \leq M$, we can apply Pontryagin's maximum principle. The Euler–Lagrange equations are

$$\frac{\partial L}{\partial x_1} - \frac{d}{dt}\left(\frac{\partial L}{\partial \dot{x}_1}\right) = 0 \Longrightarrow \dot{\lambda}_1 = 0 \Longrightarrow \lambda_1 = c_1 \text{ (a constant)}, \tag{7.68}$$

$$\frac{\partial L}{\partial x_2} - \frac{d}{dt}\left(\frac{\partial L}{\partial \dot{x}_2}\right) = 0 \Longrightarrow \lambda_1 + \dot{\lambda}_2 = 0 \Longrightarrow \dot{\lambda}_2 = -\lambda_1 \Longrightarrow \lambda_2 = -c_1 t + c_2, \tag{7.69}$$

$$\frac{\partial L}{\partial m} - \frac{d}{dt}\left(\frac{\partial L}{\partial \dot{m}}\right) = 0 \Longrightarrow -\frac{1}{m^2}\lambda_2 cu + \dot{\lambda}_3 = 0 \Longrightarrow \dot{\lambda}_3 = \frac{1}{m^2}cu(-c_1 t + c_2) \tag{7.70}$$

and

$$\dot{x}_1 = x_2, \tag{7.71}$$

$$\dot{x}_2 = -g + \frac{cu}{m}, \tag{7.72}$$

$$\dot{m} = -u. \tag{7.73}$$

By observing the terms involving $u(t)$ in (7.67), we can say that H attains its maximum value when $u(t) = M$, provided $\lambda_3 < 0$ and $\lambda_2 > 0$. Assuming that $\lambda_3 < 0$ and $\lambda_2 > 0$, let $u(t) = M$. Then, from (7.73), we obtain

$$m(t) = -Mt + c_3 = -Mt + m_0. \tag{7.74}$$

It follows from (7.72) that

$$\dot{x}_2 = -g + \frac{cu}{m} = -g + \frac{cM}{-Mt + m_0} \implies x_2 = -gt - c\log(-Mt + m_0) + c_4.$$

Now, $x_2(0) = 0 \implies c_4 = c\log m_0$. Hence,

$$x_2 = -gt + c\log\frac{m_0}{m_0 - Mt}. \tag{7.75}$$

From (7.71), we obtain

$$\dot{x}_1 = x_2 = -gt + c\log\frac{m_0}{m_0 - Mt}$$
$$\implies x_1 = -\frac{g}{2}t^2 + ct\log m_0 - c\int_0^t \log(m_0 - Mt)dt + c_5.$$

Now, $x_1(0) = 0 \implies c_5 = 0$. Hence,

$$\begin{aligned} x_1 &= -\frac{g}{2}t^2 + ct\log m_0 - c\left[t\log(m_0 - Mt) - t - \frac{m_0}{M}\log(m_0 - Mt)\right] \\ &= -\frac{g}{2}t^2 + ct\log\left(\frac{m_0}{m_0 - Mt}\right) + c\left(t + \frac{m_0}{M}\log(m_0 - Mt)\right). \end{aligned} \tag{7.76}$$

Note that

$$u(t) = \begin{cases} M, & 0 \le t \le t_1, \\ 0, & t_1 \le t \le \tau, \end{cases} \tag{7.77}$$

where t_1 is the switching time when the mass $m(t)$ becomes m_1, i.e., when $m(t_1) = m_1$ which implies that

$$t_1 = (m_0 - m_1)/M \quad \text{(from (7.74))}. \tag{7.78}$$

Hence

$$x_1(t) = -\frac{g}{2}t^2 + ct\log\left(\frac{m_0}{m_0 - Mt}\right) + c\left(t + \frac{m_0}{M}\log(m_0 - Mt)\right) \quad \text{for } 0 \le t \le t_1 \tag{7.79}$$

and

$$x_2(t) = -gt + c\log\frac{m_0}{m_0 - Mt_1} \quad \text{for } 0 \le t \le \tau_1. \tag{7.80}$$

From (7.71), (7.72) and (7.77), we infer that

$$\begin{cases} \dot{x}_1 = x_2, \\ \dot{x}_2 = -g \end{cases} \tag{7.81}$$

for $t_1 \le t \le \tau$ and the solution for (7.80) is obtained using the boundary values $x_1(t_1)$ and $x_2(t_1)$. Finally, the final time τ is determined from the condition $x_2(\tau) = 0$ (i.e., the velocity of the rocket becomes zero in the end).

From (7.80), we have $x_2 = -gt + c_4$ for $t_1 \le t \le \tau$, and from (7.79),

$$x_2(t_1) = -gt_1 + c\log\frac{m_0}{m_0 - Mt_1} = -gt_1 + c\log(m_0/m_1).$$

Hence,

$$c_4 = c\log\frac{m_0}{m_1}$$

and

$$x_2(t) = -gt + c\log\frac{m_0}{m_1} \quad \text{for } t_1 \le t \le \tau. \tag{7.82}$$

From (7.80) and (7.82), we obtain

$$x_1(t) = -\frac{g}{2}t^2 + ct\log\frac{m_0}{m_1} + c_5 \quad \text{for } t_1 \le t \le \tau. \tag{7.83}$$

Equating the value $x_1(t_1)$ from (7.76) and (7.83), we get

$$c_5 = ct_1 + \frac{cm_0}{M}\log(m_1).$$

Thus,

$$x_1(t) = -\frac{g}{2}t^2 + ct\log\frac{m_0}{m_1} + ct_1 + \frac{cm_0}{M}\log(m_1) \quad \text{for } t_1 \le t \le \tau. \tag{7.84}$$

Since $x_2(\tau) = 0$, from (7.82), we obtain the optimal final time as

$$\tau = \frac{c}{g}\log\frac{m_0}{m_1}.$$

The maximum height achieved by the rocket is given by $x_1(\tau)$.

7.5 Summary

In this chapter the following problems were discussed.

- **Linear quadratic optimal control problem:** Find the optimal state x^* and the optimal control u^* that extremize

$$\mathbb{J} = x'(\tau)M(\tau)x(\tau) + \int_{t_0}^{\tau} \left(x'(t)P(t)x(t) + u'(t)Q(t)u(t)\right) dt \tag{7.85}$$

for the control system

$$\begin{cases} \dot{x}(t) = A(t)x(t) + B(t)u(t), & t_0 \le t \le \tau, \\ x(t_0) = x_0. \end{cases} \tag{7.86}$$

To find the optimal pair, it is assumed that the Lagrange multiplier $\lambda(t)$ is a linear feedback function of $x(t)$, i.e.,

$$\lambda(t) = R(t)x(t), \tag{7.87}$$

where $R(t)$ is an $n \times n$ matrix, called the Riccati matrix. Using (7.87) in the Euler–Lagrange system 6.111, we obtain the matrix Riccati equation

$$\dot{R} + RA + A'R + 2P - \frac{1}{2}RBQ^{-1}B'R = 0, \quad t_0 \le t \le \tau, \tag{7.88}$$

with the end condition $R(\tau) = 2M(\tau)$. Then the optimal control u is given in terms of $R(t)$ and x by $u = -\frac{1}{2}Q^{-1}B'Rx$ and the optimal state is obtained by solving $\dot{x} = Ax + Bu = (A - \frac{1}{2}BQ^{-1}B'R)x$ with the given initial condition.

- **Constrained control:** If the control function $u(t)$ is constrained by the boundedness condition $|u(t)| \le M$, $t_0 \le t \le \tau$, for some constant M, then the optimal control is obtained from the Pontryagin minimum (maximum) principle (see 7.44)

$$H(x^*, u^*, \lambda^*) = \min_{\|u(t)\| \le M} H(x^*, u, \lambda^*),$$

where $H(x, u, \lambda)$ is the Hamiltonian of the corresponding optimal control problem.

7.6 Exercise

1. If $x(t)$ and $v(t)$ represent the position and velocity of a particle moving in a straight line during the time interval $0 \le t \le 10$, and if $x(0) = 0$ and $x(10) = 20$, then find the optimal velocity function that

$$\text{minimizes } \mathbb{J} = \int_0^{10} v^2(t)dt.$$

2. The equation of motion of an object of mass m under the influence of a force $u(t)$ is $m\frac{d^2x}{dt^2} = u(t)$, where $x(t)$ is the displacement at time t. Let $x(0) = 0, \dot{x}(0) = 0$, and $m = 1$. Find the optimal control u^*, which minimizes

$$\mathbb{J} = \frac{1}{2}\int_0^\tau (x^2 + u^2)dt$$

under the constraint $|u(t)| \leq 2$ such that $x(\tau) = 100$, $\dot{x}(\tau) = 0$, and τ is unspecified.

3. At time $t = 0$, a spacecraft is at a height $h = 100$ units, moving with a velocity $v(0) = 2$ units. The vertically upward direction is considered positive. The spacecraft moves under the influence of a control force $u(t)$, which is restricted by $|u(t)| \leq 2$. It moves either upward or downward by instantly switching the control. Find the optimal control $u^*(t)$ such that the spacecraft lands smoothly on the ground in the shortest time τ.

4. For the control system

$$\begin{cases} \dot{x}_1 = x_2, \\ \dot{x}_2 = u \end{cases}$$

for $0 \leq t \leq \tau$ (τ is unspecified), find the optimal switching control, which minimizes

$$\mathbb{J} = \int_0^\tau \left(u^2 + 1\right) dt$$

such that $|u(t)| \leq 1, x_1(0) = 100, x_2(0) = 2, x_1(\tau) = 0$, and $x_2(\tau) = 0$.

5. Let $A, C, E \in \mathbb{R}^{n\times n}$, where C and E are symmetric. Let $\Delta = \begin{bmatrix} A & E \\ C & -A' \end{bmatrix}$. Show that if α is an eigenvalue of Δ, then $-\alpha$ is also its eigenvalue.

6. For the functional

$$\mathbb{J} = \frac{1}{2}x_1^2(5) + \int_0^5 u^2(t)dt$$

and the system

$$\begin{cases} \dot{x}_1 = x_2, \\ \dot{x}_2 = u \end{cases}$$

with $x_1(0) = 0$ and $x_2(0) = 1$, find the state and co-state canonical matrix Δ and its eigenpairs. Using the eigenpairs, find the Riccati matrix $R(t)$.

7. (*i*)

$$\text{Minimize } \mathbb{J} = \frac{1}{2}\int_0^5 (5x^2 + u^2)dt$$

for the system $\dot{x} = -2x + u$ with $x(0) = 1$ and $x(5)$ unspecified, using the Riccati matrix solution $R(t)$.

(*ii*) Consider the following problem

$$\text{minimize } \mathbb{J} = \frac{1}{2}\int_0^\tau (5x^2 + u^2)dt$$

under the constraint $\dot{x} = -2x + u$ with $x(0) = 1$ and $x(\tau)$ unspecified. Then:

(*a*) For a fixed τ, find the solution R_τ of the Riccati equation in terms of τ.

(*b*) Show that $\lim_{\tau\to\infty} R_\tau = 1$.

8. If

$$\mathbb{J}_t = \frac{1}{2}x^2(5) + \frac{1}{2}\int_t^5 u^2(t)dt,$$

where $x(t)$ and $u(t)$ satisfy $\dot{x} = -2x + u$ and $x(0) = 1$, find $\text{Min}(\mathbb{J}_0)$ and $\text{Min}(\mathbb{J}_2)$.

8

Stability

8.1 Introduction

In many practical situations, it is desirable for the system under consideration to remain near its equilibrium state. In such cases, we say that the system is stable at the equilibrium position. If a system tends to move away from equilibrium, but it is possible to apply a control to keep it near the equilibrium, this process is called stabilization of the system. In this chapter, we present the stability properties of linear and nonlinear dynamical systems at equilibrium points, as well as the stabilizability of control systems.

In Section 8.2, definitions of equilibrium point, stability, asymptotic stability, and instability of equilibrium points are given. In Section 8.3, stability of the trivial solutions of a linear time-invariant system is studied using the eigenvalues of the coefficient matrix. Some stability results for nonautonomous linear systems are also presented. For some nonlinear systems, stability is analyzed through linearization in Section 8.4. In Section 8.5, Lyapunov stability theorems for time-invariant systems are presented with examples. Stability analysis using matrix equations is also given in this section. Lyapunov theorems for time-varying systems are presented in Section 8.6. The gradient method and the Krasovskii formula for constructing Lyapunov functions are also presented.

8.2 Equilibrium Point and Stability

Consider the dynamical system

$$\dot{x}_i = f_i(t, x_1, x_2, \ldots, x_n) \quad \text{for } t \geq t_0,\ i = 1, 2, \ldots, n,$$

where $f_i : \mathbb{R}^{n+1} \to \mathbb{R}, i = 1, 2, \ldots, n$, are nonlinear functions of t and the state variables $x_i : i = 1, 2, \ldots, n$. The system, in vector form, can be represented as

$$\dot{x} = f(t, x), \tag{8.1}$$

where $x : [t_0, \infty) \to \mathbb{R}^n$ and $x(t) = \begin{bmatrix} x_1(t) & x_2(t) & \cdots & x_n(t) \end{bmatrix}' \in \mathbb{R}^n$ for each t. Let θ denote the zero vector in R^n

Definition 8.2.1 Consider the dynamical system (8.1). Let $f(t, x_e) = \theta \in \mathbb{R}^n$ for all $t \geq t_0$, where x_e is a constant vector in $\mathbb{R}^n$, and $f(t, x) \neq \theta$ for all x in a neighborhood of x_e. Then x_e is said to be an **isolated equilibrium** or a **critical point** of the system (8.1).

If $x(t_0) = x_e$ is the initial condition, then from (8.1), it follows that $x(t) = x_e$ for all $t \geq t_0$. Therefore, any solution starting at the equilibrium point x_e will remain there.

Example **8.2.2** Consider the nonlinear system

$$\begin{cases} \dot{x}_1 = x_1 - 2x_1 x_2, \\ \dot{x}_2 = -2x_2 + x_1 x_2. \end{cases}$$

Equating the RHS of the equations to zero and solving, we find $(0, 0)$ and $(2, \frac{1}{2})$ as equilibrium points.

Definition 8.2.3 An equilibrium state x_e is said to be **stable** if, for any $\varepsilon > 0$, there exists a $\delta > 0$ such that

$$\|x(t_0) - x_e\| < \delta \implies \|x(t) - x_e\| < \varepsilon, \quad t \geq t_0.$$

Definition 8.2.4 An equilibrium state x_e is said to be **asymptotically stable** if x_e is stable and, in addition, $x(t) \to x_e$ as $t \to \infty$.

Definition 8.2.5 An equilibrium state x_e is said to be **unstable** if x_e is not stable. In other words, we can find a fixed $\varepsilon > 0$ satisfying the following: For every $\delta > 0$, there exists an $x(t_0)$ such that $\|x(t_0) - x_e\| < \delta$ and a $t_1 > t_0$ such that $\|x(t) - x_e\| \geq \varepsilon$ for all $t > t_1$. If this holds for every initial condition $x(t_0)$, then the equilibrium is called completely unstable.

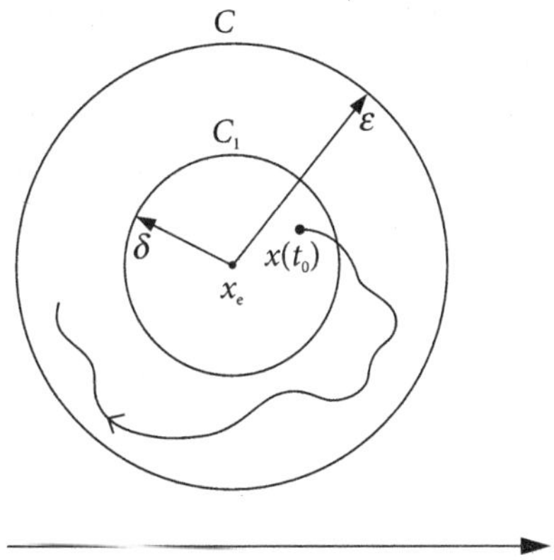

Figure 8.1 Stable equilibrium

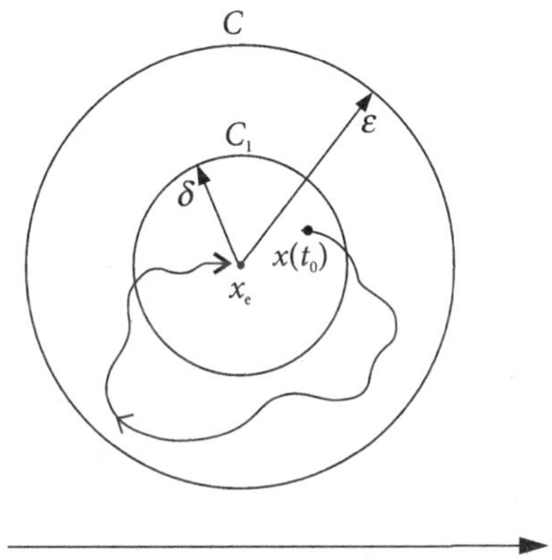

Figure 8.2 Asymptotically stable

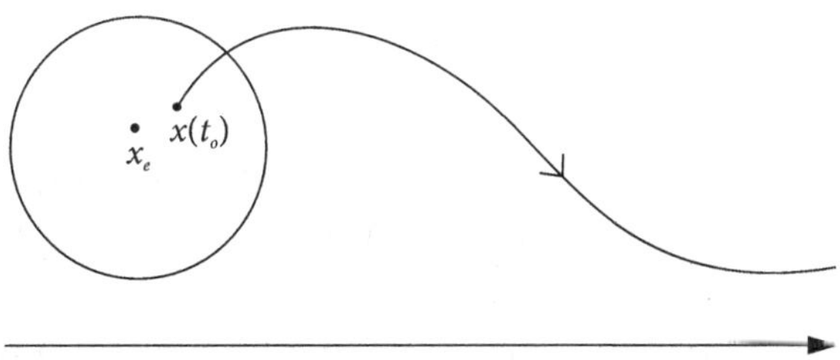

Figure 8.3 Unstable equilibrium

The above definitions are called stability definitions in the sense of Lyapunov.

In two dimensions, a simple geometric interpretation of stability is given below. x_e is stable, if and only if, for a given disc C with center at x_e and radius ε, there exists an inner disc C_1 with center at x_e and radius δ such that trajectories $x(t)$ starting within C_1 never leave C. x_e is asymptotically stable, if and only if, it is stable and trajectories starting inside C_1 tend to x_e as $t \to \infty$.

Clearly, by introducing the new state variable $z = x - x_e$, we can transform the equation (8.1) to $\dot{z} = f(t, z + x_e) = g(t, z)$ (say) and the equilibrium point x_e to the origin θ.Thus a system with a non zero equilibrium point can be transformed to a system with zero equilibrium point. Hence, without loss of generality, we can assume that in (8.1), $f(t, 0) = 0$ for all $t \geq t_0$.

8.3 Linear Time-invariant Systems

Consider the linear time-invariant system

$$\dot{x} = Ax, \quad t \geq 0, \tag{8.2}$$

where A is an $n \times n$ constant **nonsingular** matrix. Then $x = 0$ is the only equilibrium point.

Note that if A is **singular**, then $x = \theta$ is an equilibrium point, but it is not an isolated one since all elements in the null space of A are equilibrium points. .

Theorem 8.3.1 (i) The system (8.2) is asymptotically stable at $x = \theta$ if and only if **all** the eigenvalues $\lambda_i : i = 1, 2, \ldots, n$ of A have negative real parts. (ii) The system (8.2) is unstable at $x = \theta$ if and only if $\text{Re}(\lambda_i) > 0$ for **some** i and is completely unstable if and only if $\text{Re}(\lambda_i) > 0$ for all i.

Proof. (i) Let A be diagonalizable. If $\lambda_i : i = 1, 2, \ldots, n$ are the eigenvalues and X_i are the corresponding eigenvectors of A, then using (4.56), the solution of the system can be written as

$$x(t) = c_1 e^{\lambda_1 t} X_1 + c_2 e^{\lambda_2 t} X_2 + \cdots + c_n e^{\lambda_n t} X_n, \tag{8.3}$$

where $c_i : i = 1, 2, \ldots, n$ are constants (not necessarily real). It can be easily seen that if $\text{Re}(\lambda_i) < 0$ for all $i = 1, 2, \ldots, n$, then $e^{\lambda_i t} \to 0$ as $t \to \infty$. Hence, for any c_i, the vector $x(t) \to \theta$ as $t \to \infty$.

(ii) If λ_j for some j has a positive real part, then $e^{\lambda_j t} \to \infty$ as $t \to \infty$. Now, if $x(0) = X_j$, then

$$c_i = \begin{cases} 0 & \text{for } i \neq j, \\ 1 & \text{for } i = j. \end{cases}$$

Hence, from (8.3), $x(t) = e^{\lambda_j t}X_j$, and $\|x(t)\| = |e^{\lambda_j t}|\|X_j\| \to \infty$ as $t \to \infty$. Therefore, the solution of (8.2) with the initial condition $x(0) = X_j$ is unbounded, implying the instability at the origin. If A is not diagonalizable, the solution $x(t)$ of (8.2) can be written using (4.59) and (4.63). Then the rest of the proof follows from Remark 4.7.1. □

***Example* 8.3.2** Consider the time-invariant linear system $\dot{x} = Ax$, where A is a 2×2 nonsingular matrix. For the following matrices: (i) $A = \begin{bmatrix} -1 & 0 \\ 0 & -2 \end{bmatrix}$, (ii) $A = \begin{bmatrix} 1 & 1 \\ 0 & 2 \end{bmatrix}$, (iii) $A = \begin{bmatrix} 0 & -1 \\ 4 & 0 \end{bmatrix}$, (iv) $A = \begin{bmatrix} 0 & 1 \\ 3 & 2 \end{bmatrix}$, and (v) $A = \begin{bmatrix} -1 & 1 \\ -4 & -1 \end{bmatrix}$, analyze the stability properties of the isolated equilibrium point $x = \theta$ and sketch the solution $x(t)$ in the state space $\mathbb{R}^2$.

Solution (i) $A = \begin{bmatrix} -1 & 0 \\ 0 & -2 \end{bmatrix}$: The eigenvalues are -1 and -2 with corresponding eigenvectors $\begin{bmatrix} 1 \\ 0 \end{bmatrix}$ and $\begin{bmatrix} 0 \\ 1 \end{bmatrix}$, respectively. From Theorem 8.3.1, the equilibrium $x = \theta$ is asymptotically stable. From (8.3), the general solution of the system is

$$\begin{bmatrix} x_1(t) \\ x_2(t) \end{bmatrix} = c_1 e^{-t} \begin{bmatrix} 1 \\ 0 \end{bmatrix} + c_2 e^{-2t} \begin{bmatrix} 0 \\ 1 \end{bmatrix}.$$

Hence, $x_1(t) = c_1 e^{-t}$ and $x_2(t) = c_2 e^{-2t}$. Thus, $x_1^2 = kx_2$, where $k = \frac{c_1^2}{c_2}$, provides a family of parabolas representing the solutions in the x_1x_2-plane (known as the **phase space**). The constant k can be determined using an initial condition $x(0)$. Since $(x_1(t), x_2(t)) \to (0,0)$ as $t \to \infty$, from any initial point P, the solution curve tends towards the origin, as shown in Figure 8.4. The equilibrium point $(0,0)$ is known as a **stable node** or sink.

(ii) $A = \begin{bmatrix} 1 & 1 \\ 0 & 2 \end{bmatrix}$; The eigenvalues are 1 and 2, and the corresponding eigenvectors are $\begin{bmatrix} 1 \\ 0 \end{bmatrix}$ and $\begin{bmatrix} 1 \\ 1 \end{bmatrix}$. As both eigenvalues are positive, from Theorem 8.3.1, the origin is unstable. The general solution of the system is

$$\begin{bmatrix} x_1(t) \\ x_2(t) \end{bmatrix} = c_1 e^t \begin{bmatrix} 1 \\ 0 \end{bmatrix} + c_2 e^{2t} \begin{bmatrix} 1 \\ 1 \end{bmatrix}.$$

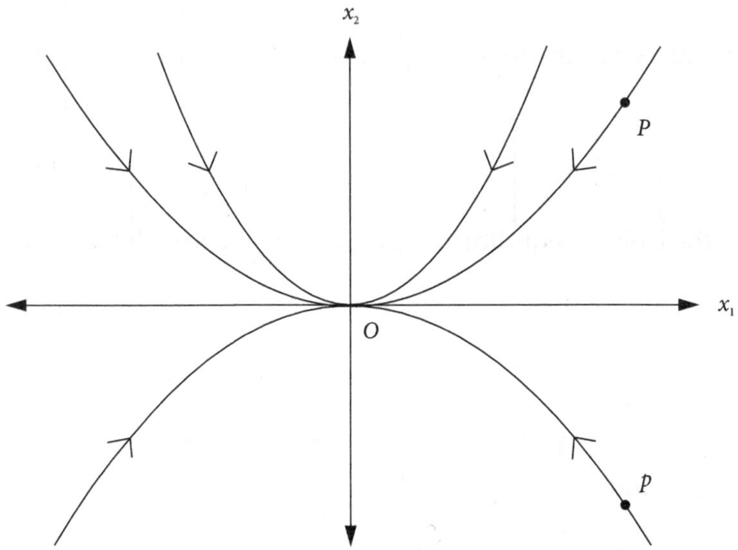

Figure 8.4 Stable node

Hence, $x_1(t) = c_1 e^t + c_2 e^{2t}$ and $x_2(t) = c_2 e^{2t}$, which gives the relation $(x_1 - x_2)^2 = kx_2$, where $k = \frac{c_1^2}{c_2}$. Now, $\frac{dx_2}{dx_1} = \frac{2c_2 e^{2t}}{c_1 e^t + 2c_2 e^{2t}}$, which tends to 1 as $t \to \infty$ and tends to 0 as $t \to -\infty$, implying that the slope of the x_1x_2-curve (phase space diagram) approaches 1 as $t \to \infty$.

Moreover, $(x_1(t), x_2(t)) \to (\infty, \infty)$ as $t \to \infty$. Hence, in the sketch of the general solution $(x_1(t), x_2(t))$ (Figure 8.5), the curves tend towards infinity with the slope tending to 1 (the slope the eigenvector [1 1]) as $t \to \infty$, while the slope at the origin is 0 (the slope of the eigenvector [1 0]) for all curves. Here, the equilibrium point $(0,0)$ is called an **unstable node** or source.

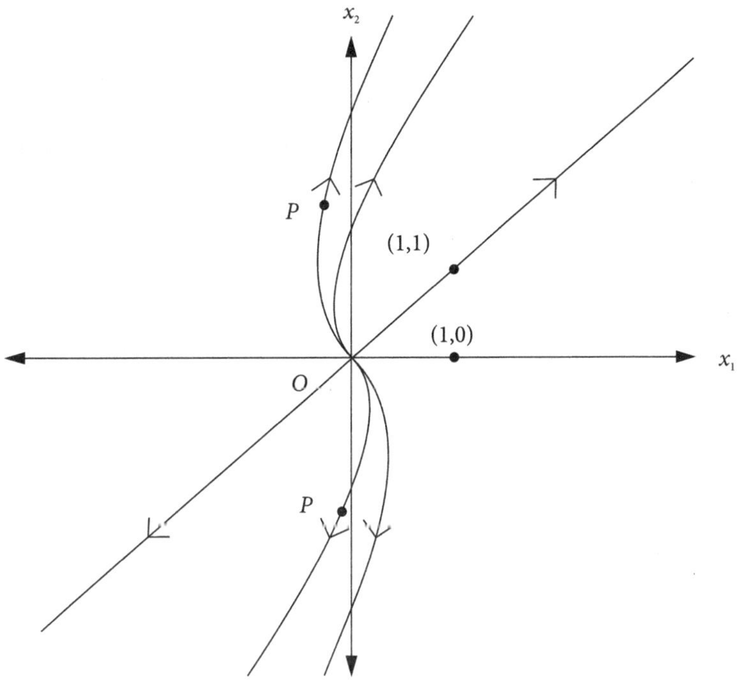

Figure 8.5 Unstable node

$(iii)\ A = \begin{bmatrix} 0 & -1 \\ 4 & 0 \end{bmatrix}$: Then eigenpairs are $\left(2i, \begin{bmatrix} 1 \\ -2i \end{bmatrix}\right)$ and $\left(-2i, \begin{bmatrix} 1 \\ 2i \end{bmatrix}\right)$. The general solution is

$$\begin{bmatrix} x_1(t) \\ x_2(t) \end{bmatrix} = c_1 e^{2it} \begin{bmatrix} 1 \\ -2i \end{bmatrix} + c_2 e^{-2it} \begin{bmatrix} 1 \\ 2i \end{bmatrix}.$$

From Theorem 8.3.3, the system is stable at $(0,0)$. Since we are interested only in real valued solutions, we can set $c_1 = \bar{c}_2$. Hence,

$$x_1(t) = 2(a\cos 2t - b\sin 2t) \text{ and } x_2(t) = 4(a\sin 2t + b\cos 2t),$$

where $c_1 = a + ib$ and $c_2 = a - ib$. Thus, $x_1^2 + \frac{x_2^2}{4} = k$, where $k = 4(a^2 + b^2)$, which represents a family of ellipses with the center at $(0,0)$. The slopes of these curves are given by

$$\frac{dx_2}{dx_1} = \frac{2(a\cos 2t - b\sin 2t)}{(-a\sin 2t - b\cos 2t)} = -4\frac{x_1}{x_2}.$$

Using the slope and the increasing or decreasing nature of the functions $x_1(t)$ and $x_2(t)$ at every point (x_1, x_2), the direction of the trajectory can be decided. For example, at the point $(2,8)$, the slope of the curve is -1. Moreover, $\frac{dx_1}{dt} = -x_2 = -8 < 0$ and $\frac{dx2}{dt} = 4x_1 > 0$, which implies that x_1 is decreasing and x_2 is increasing. The phase space diagram of the solution is shown in Figure 8.6. In this case, the equilibrium $(0,0)$ is referred to as a **center**.

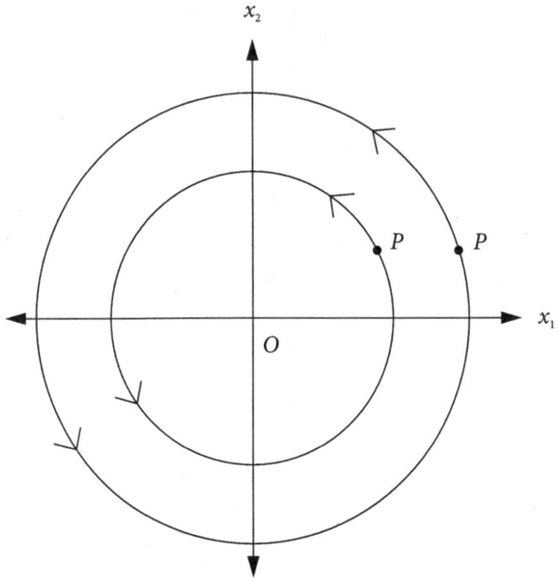

Figure 8.6 Center

(*iv*) $A = \begin{bmatrix} 0 & 1 \\ 3 & 2 \end{bmatrix}$: Then eigenpairs are $\left(3, \begin{bmatrix} 1 \\ 3 \end{bmatrix}\right)$ and $\left(-1, \begin{bmatrix} 1 \\ -1 \end{bmatrix}\right)$. The general solution is

$$x = c_1 e^{3t} \begin{bmatrix} 1 \\ 3 \end{bmatrix} + c_2 e^{-t} \begin{bmatrix} 1 \\ -1 \end{bmatrix} \implies x_1 = c_1 e^{3t} + c_2 e^{-t} \text{ and } x_2 = 3c_1 e^{3t} - c_2 e^{-t}.$$

From Theorem 8.3.1, the equilibrium $(0,0)$ is unstable. Note that $3x_1 - x_2 = 4c_2 e^{-t}$ and $x_1 + x_2 = 4c_1 e^{3t}$. Thus, $(x_1 + x_2)(3x_1 - x_2)^3 = 4c_1 e^{3t}(4c_2 e^{-t})^3 = k$, a constant. Also, $\frac{dx_2}{dx_1} = \frac{9c_1 e^{3t} + c_2 e^{-t}}{3c_1 e^{3t} - c_2 e^{-t}}$ tends to 3 (the slope of the first eigenvector) as $t \to \infty$ and tends to -1 (the slope of the second eigenvector) as $t \to -\infty$. Thus, the slope of every solution curve tends to 3 as $t \to \infty$, as shown in Figure 8.7. In this case, the equilibrium point $(0,0)$ is known as a **saddle point**.

(v) $A = \begin{bmatrix} -1 & 1 \\ -4 & -1 \end{bmatrix}$: Then eigenpairs are $\left(-1+2i, \begin{bmatrix} 1 \\ 2i \end{bmatrix}\right)$ and $\left(-1-2i, \begin{bmatrix} 1 \\ -2i \end{bmatrix}\right)$.

The general solution is

$$\begin{bmatrix} x_1(t) \\ x_2(t) \end{bmatrix} = c_1 e^{-t} e^{2it} \begin{bmatrix} 1 \\ 2i \end{bmatrix} + c_2 e^{-t} e^{-2it} \begin{bmatrix} 1 \\ -2i \end{bmatrix}$$

$$\implies x_1 = c_1 e^{3t} + c_2 e^{-t} \text{ and } x_2 = 3c_1 e^{3t} - c_2 e^{-t}.$$

The origin $(0,0)$ is asymptotically stable (from Theorem 8.3.1). To obtain real

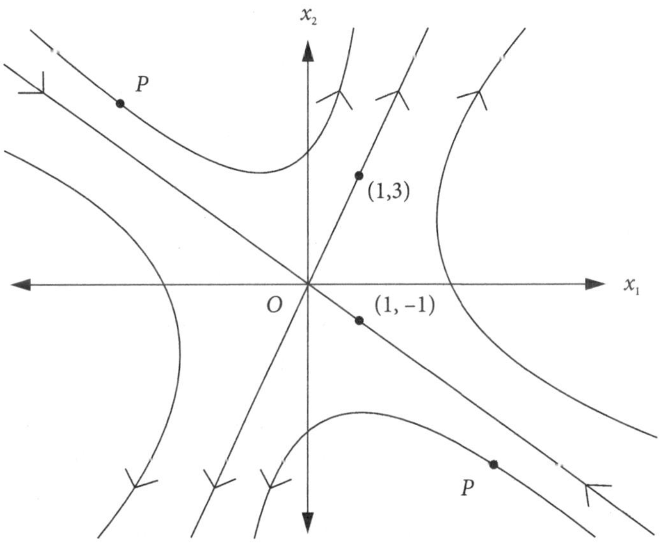

Figure 8.7 Saddle point

solutions, we set $c_1 = \bar{c}_2 = a + ib$. Hence,

$$x_1(t) = e^{-t}(a\cos 2t - b\sin 2t) \text{ and } x_2(t) = -2e^{-t}(a\sin 2t + b\cos 2t).$$

Let $a = r\cos\theta$ and $b = r\sin\theta$. Then $x_1(t) = re^{-t}\cos(2t+\theta)$ and $x_2(t) = -2re^{-t}\sin(2t+\theta)$. These expressions lead to the relation $x_1^2 + \frac{x_2^2}{4} = re^{-2t}$, which represents a spiral around $(0,0)$ (Figure 8.8). Hence, the equilibrium point $(0,0)$ is known as a **spiral**.

The following theorem provides conditions under which the system $\dot{x} = Ax$ is stable but not asymptotically stable.

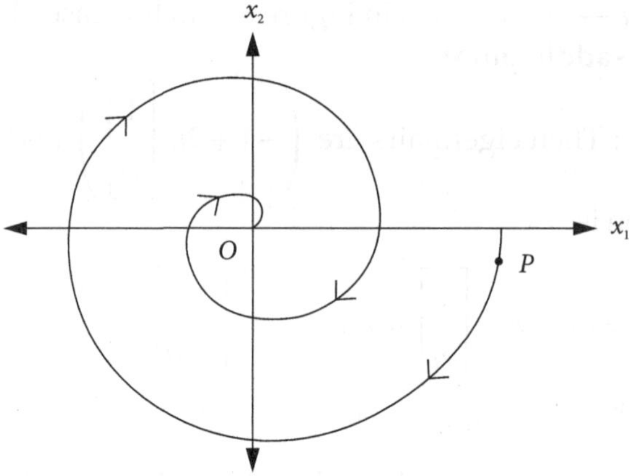

Figure 8.8 Spiral

Theorem 8.3.3 Let A be an $n \times n$ matrix with $l(l \leq n)$ distinct eigenvalues $\lambda_i : i = 1,2,\ldots,l$ satisfying $\text{Re}(\lambda_i) \leq 0$ for $i = 1,2,\ldots,l$. Let $\lambda_i : i = 1,2,\ldots,k$ be the eigenvalues with $\text{Re}(\lambda_i) = 0$ (i.e., each λ_i is purely imaginary or zero), and $\text{Re}(\lambda_i) < 0$ for $i = k+1,\ldots,l$.

(*i*) If the algebraic multiplicity and geometric multiplicity of $\lambda_i : i = 1,2,\ldots,k$ are equal, then the system $\dot{x} = Ax$ is stable but not asymptotically stable at $x = \theta$.

(*ii*) If the algebraic multiplicity $>$ geometric multiplicity for **any** $\lambda_i : i = 1,2,\ldots,k$, then $x = \theta$ is an unstable equilibrium point.

Proof. (*i*) For $\lambda_i : i = 1,2,\ldots,k$, let the algebraic multiplicity = geometric multiplicity $= m_i$. Then, corresponding to λ_i, there are m_i linearly independent eigenvectors, say, $X_1^i, X_2^i, \ldots, X_{m_i}^i$. The remaining eigenvalues for $j = k+1,...,l$ have eigenvectors and generalized eigenvectors depending on their algebraic and geometric multiplicities. The solution of the system can be written as a linear combination of eigenvectors and generalized eigenvectors, as in (4.59), which has terms of the form $t^r e^{\lambda_i t} Y_{ir}$ for $i = k+1,\ldots,l, r \geq 0$, where Y_{ir} is a generalized eigenvector, and terms of the form $e^{\lambda_i t} X_r^i$: $r = 1,2,\ldots,m_i$. Since $\text{Re}(\lambda_i) < 0$ for $i = k+1,\ldots,l$, all the terms involving $t^r e^{\lambda_i t} \to 0$ as $t \to \infty$. Since $\lambda_i = 0$ or is imaginary for $i = 1,2,\ldots,k$, $|e^{\lambda_i t}| = 1$. Hence, $\|e^{\lambda_i t} X_r^i\| = \|X_r^i\|$ is a nonzero finite value. Therefore, $\|x(t)\|$ remains nonzero and bounded. Thus, the solution of $\dot{x} = Ax$ is stable but not asymptotically stable at $x = \theta$.
(*ii*) For some i let $\text{Re}(\lambda_i) = 0$ and algebraic multiplicity $>$ geometric multiplicity then, the solution of the system $\dot{x} = Ax$ can be written as in (4.59),that includes terms of the form $t^r e^{\lambda_i t} Y$, where $r \geq 1$ and Y is a generalized eigenvector corresponding to λ_i.

Since $\lambda_i = 0$ or imaginary $\lim_{t\to\infty} \|t^m e^{\lambda_i t} Y\| = \lim_{t\to\infty} t^m \|Y\| = \infty$. Hence the solution $x(t)$ is unbounded, implying the instability of $x = \theta$. □

***Example* 8.3.4** If $A = \begin{bmatrix} 0 & 1 & 0 \\ 0 & 0 & 1 \\ 0 & 0 & 0 \end{bmatrix}$, then, from Theorem 8.3.3, $\dot{x} = Ax$ is unstable at $x = 0$ since the eigenvalue 0 has algebraic multiplicity 3 and geometric multiplicity 1.

***Example* 8.3.5** If $A = \begin{bmatrix} 0 & 1 & 0 & 0 \\ -1 & 0 & 0 & 0 \\ 0 & 0 & 0 & -1 \\ 0 & 0 & 1 & 0 \end{bmatrix}$, then, from Theorem 8.3.1(i), we can conclude that the system $\dot{x} = Ax$ is stable at $x = 0$ since the eigenvalues i and $-i$ of A have algebraic multiplicity = geometric multiplicity = 2.

***Example* 8.3.6** Consider the system $\dot{x} = Ax$, where $A = \begin{bmatrix} 0 & 0 & 1 & 0 \\ 0 & 0 & 0 & -1 \\ -1 & 0 & 0 & 0 \\ 0 & 1 & 0 & 0 \end{bmatrix}$, then, from Theorem 8.3.3(ii), we conclude that the system is unstable at $x = 0$ since the eigenvalues i and $-i$ of A have algebraic multiplicity 2 and geometric multiplicity 1.

Time-varying Linear Systems

In this section, we shall show that Theorems 8.3.1 and 8.3.3 do not hold for time-varying systems. Consider the nonautonomous linear system

$$\dot{x} = A(t)x,$$

where $A(t)$ is an $n \times n$ matrix with elements which are continuous functions of t. It can be shown that even if all eigenvalues of $A(t)$ have negative real parts for all t, the system may still be unstable at the origin.

***Example* 8.3.7** (i) Consider the system

$$\dot{x} = e^{-t}x, \quad t \geq 0.$$

Here, $x = 0$ is the equilibrium point. Though $e^{-t} \geq 0$ for all $t \geq 0$, the equilibrium $x = 0$ is **stable** (Definition 8.2.3), since the solution is $x(t) = \exp(-e^{-t})ex(0)$ and $|x(t)| \leq ex(0) < \varepsilon$ whenever $|x(0)| < \frac{\varepsilon}{e}$.

(ii) The system

$$\dot{x} = \frac{x}{(t+1)^2}, \quad t \geq 0,$$

is **stable** at $x = 0$, since the solution $e^{-\frac{1}{t+1}}ex(0)$ is bounded by $ex(0)$. Note that $\frac{1}{(t+1)^2} > 0$ for all $t \geq 0$.

(iii) The system

$$\dot{x} = -\frac{1}{(1+t)^{\frac{3}{2}}}x$$

has a bounded solution $x(t) = e^{\frac{2}{\sqrt{1+t}}}e^{-2}x(0)$. Hence, the system is **stable** but not asymptotically stable at the origin, though $-\frac{1}{(1+t)^{\frac{3}{2}}}$ is negative for all $t \geq 0$.

(iv) Consider the time-varying system

$$\begin{cases} \dot{x}_1 = -x_1 + e^{3t}x_2, \\ \dot{x}_2 = -2x_2. \end{cases}$$

The coefficient matrix is $A(t) = \begin{bmatrix} -1 & e^{3t} \\ 0 & -2 \end{bmatrix}$, whose eigenvalues are -1 and -2 for all t. Though the eigenvalues are negative for all t, it can be easily seen that the equilibrium $(0,0)$ is **unstable**, since the solution to the system is $(x_1(t), x_2(t))$, where $x_2(t) = e^{-2t}x_2(0)$ and $x_1(t) = e^{-t}x_1(0) + \frac{e^t}{2}x_2(0) - \frac{e^{-t}}{2}x_2(0)$. Note that $x_1(t) \to \infty$ as $t \to \infty$ whenever $x_2(0) \neq 0$.

The above examples show that the stability properties of a time-varying system at the origin is not directly related to the real parts of the eigenvalues of $A(t)$. The following theorems provide some understanding on the stability of time-varying systems at the origin.

Theorem 8.3.8 Let $A(t)$ be an $n \times n$ continuous matrix on $[t_0, \infty)$ and $M(t)$ be the largest eigenvalue of $A(t) + A'(t)$.

(i) If

$$\int^{\infty} M(t)dt = -\infty, \tag{8.4}$$

then the system $\dot{x} = A(t)x$ is asymptotically stable at the origin.

(*ii*) If

$$\int^{\infty} M(s)ds < \infty,$$

then $\|x(t)\| < \infty$ for all t, i.e., $x(t)$ is bounded for all t.

Proof. (*i*) Let $x(t)$ be a solution of the system $\dot{x} = A(t)x$. Then,

$$\begin{aligned}\frac{d}{dt}\|x(t)\|^2 = \frac{d}{dt}(x'(t)x(t)) &= x'(t)\dot{x}(t) + \dot{x}'(t)x(t) \\ &= x'(t)A(t)x(t) + x'(t)A'(t)x(t) \\ &= x'(t)(A(t) + A'(t))x(t). \end{aligned} \tag{8.5}$$

Noting that $A(t) + A'(t)$ is symmetric and $M(t)$ is its largest eigenvalue, we have (see Exercise 2.9.3)

$$x'(t)(A(t) + A'(t))x(t) \leq M(t)\|x(t)\|^2. \tag{8.6}$$

From (8.5) and (8.6), we obtain $\frac{d}{dt}\|x(t)\|^2 \leq M(t)\|x(t)\|^2$. Integrating both sides from t_0 to t, we get

$$\|x(t)\|^2 - \|x(t_0)\|^2 \leq \int_{t_0}^{t} M(s)\|x(s)\|^2 ds \quad \text{for all } x(t) \in \mathbb{R}^n. \tag{8.7}$$

Now, using Gronwall's inequality (Theorem 4.10.1), we get

$$\|x(t)\|^2 \leq \|x(t_0)\|^2 \exp\left(\int_{t_0}^{t} M(s)ds\right). \tag{8.8}$$

Since $\int_{t_0}^{t} M(s)ds \to -\infty$ as $t \to \infty$, we get $\|x(t)\| \to 0$ as $t \to \infty$, implying the asymptotic stability of the origin.
(*ii*) The boundedness of $x(t)$ follows immediately from (8.8). □

Theorem 8.3.9 Let $A(t)$ be an $n \times n$ continuous matrix for $t \in [t_0, \infty)$, and let $m(t)$ be the smallest eigenvalue of $A(t) + A'(t)$. Then the system $\dot{x}(t) = A(t)x(t)$ is unstable at the origin if

$$\lim_{t \to \infty} \int^{t} m(s)ds = \infty. \tag{8.9}$$

Proof. The proof is similar to that of the previous theorem. Since $m(t)$ is the smallest eigenvalue of $A(t) + A'(t)$, we have

$$x'(t)(A(t) + A'(t))x(t) \geq m(t)\|x(t)\|^2 \quad \text{for all } x(t) \in \mathbb{R}^n. \tag{8.10}$$

Moreover, from (8.5),

$$\frac{d}{dt}\|x(t)\|^2 = x'(t)(A(t)+A'(t))x(t). \tag{8.11}$$

Hence, $\frac{d}{dt}\|x(t)\|^2 \geq m(t)\|x(t)\|^2$. Integrating both sides from t_0 to t, we get

$$\|x(t)\|^2 \geq \|x(t_0)\|^2 + \int_{t_0}^{t} m(s)\|x(s)\|^2 ds.$$

Using Gronwall's inequality, we obtain

$$\|x(t)\|^2 \geq \|x(t_0)\|^2 \exp\left(\int_{t_0}^{t} m(s)ds\right). \tag{8.12}$$

Now, from (8.8) and (8.12), it is clear that $\lim_{t\to\infty}\|x(t)\| = \infty$, implying instability at the origin. □

***Example* 8.3.10** Let $A(t) = \begin{bmatrix} -\frac{1}{t} & t^2 \\ -t^2 & -2 \end{bmatrix}$ for $t \geq 1$. Then $A(t)+A'(t) = \begin{bmatrix} -\frac{2}{t} & 0 \\ 0 & -4 \end{bmatrix}$, and $M(t) = -\frac{2}{t}$ for $t \geq \frac{1}{2}$. Since $-\int_1^\infty \frac{2}{t}dt = -\infty$, from Theorem 8.3.8, the solutions of the system $\dot{x} = A(t)x$ satisfy $\lim_{t\to\infty} x(t) = \theta$. Hence, $x = \theta$ is an asymptotically stable equilibrium point.

***Example* 8.3.11** Let $A(t) = \begin{bmatrix} -1 & e^{3t} \\ 0 & -2 \end{bmatrix}$. Then $A(t)+A'(t) = \begin{bmatrix} -2 & e^{3t} \\ e^{3t} & -4 \end{bmatrix}$, and the largest eigenvalue $M(t)$ of $A(t)+A'(t)$ is $-3+\sqrt{1+e^{6t}}$. Since $M(t) > 0$ for all $t > \log\sqrt{2}$,

$$\lim_{t\to\infty}\int^{t} M(s)ds = \infty. \tag{8.13}$$

Thus, $M(t)$ does not satisfy condition (8.4). Hence, Theorem 8.3.8 is not applicable. From this, we cannot conclude anything about the stability of the trivial solution of the system $\dot{x} = A(t)x$. However, by solving the system directly, we have shown in Example 8.3.7(iv) that the system is unstable, though all the eigenvalues of $A(t)$ are negative for all t.

***Example* 8.3.12** $A(t) = \begin{bmatrix} \frac{1}{(1+t)^2} & t^4 \\ -t^4 & -2 \end{bmatrix}$. Then $A(t)+A'(t) = \begin{bmatrix} \frac{2}{(1+t)^2} & 0 \\ 0 & -4 \end{bmatrix}$. Hence, $M(t) = \frac{2}{(1+t)^2}$ and $\int_0^\infty M(t)dt = 2 < \infty$. Hence, from Theorem 8.3.8(ii), the solution of the system $\dot{x}(t) = A(t)x(t)$ is bounded for all t.

Example **8.3.13** Let $A(t) = \begin{bmatrix} \frac{4t}{(1+t)^2} & t^3 \\ -t^3 & \frac{1}{(1+t)} \end{bmatrix}$ for $t \geq 1$. Then $A(t) + A'(t) = \begin{bmatrix} \frac{8t}{(1+t)^2} & 0 \\ 0 & \frac{2}{1+t} \end{bmatrix}$.
Its minimum eigenvalue is $m(t) = \frac{2}{1+t}$, and $\lim_{t\to\infty} \int^t m(s)ds = \lim_{t\to\infty} 2\log(1+t) = \infty$. Hence, from Theorem 8.3.9, the solution of the system $\dot{x} = A(t)x$ is unbounded, and thus the system is unstable at the origin.

Exercise 8.3.1 Determine the stability properties of the trivial solution of the system $\dot{x} = Ax$ and sketch the phase plane curves representing the solutions of the system, where the system matrix A is given by: *(i)* $\begin{bmatrix} 2 & 3 \\ 0 & 1 \end{bmatrix}$; *(ii)* $\begin{bmatrix} 2 & 3 \\ 0 & -1 \end{bmatrix}$; *(iii)* $\begin{bmatrix} -2 & 3 \\ 0 & -1 \end{bmatrix}$; *(iv)* $\begin{bmatrix} 0 & 2 \\ -2 & 0 \end{bmatrix}$; *(v)* $\begin{bmatrix} 0 & 2 \\ 0 & 0 \end{bmatrix}$; *(vi)* $\begin{bmatrix} 1 & 2 \\ -4 & -3 \end{bmatrix}$.

Exercise 8.3.2 If $A(t) = \begin{bmatrix} -t & e^t \\ -e^t & -t^2 \end{bmatrix}$, $t \geq 1$, discuss the stability of the trivial solution of the system $\dot{x} = A(t)x$.

8.4 Linearization

Consider the nonlinear time-invariant dynamical system of the form

$$\dot{x} = g(x), \quad g(\theta) = \theta, \tag{8.14}$$

where $x(t) = \begin{bmatrix} x_1(t) & x_2(t) & \cdots & x_n(t) \end{bmatrix}'$ and $g(x) = \begin{bmatrix} g_1(x) & g_2(x) & \cdots & g_n(x) \end{bmatrix}'$, with $g_i(x) : i = 1, 2, \ldots, n$ being real valued functions. Assuming that $g(x)$ has higher order derivatives, by Taylor's theorem, $g(x)$ can be expressed as

$$g(x) = g(\theta) + \left(\frac{dg}{dx}\right)_{x=\theta} x + f(x), \tag{8.15}$$

where

$$\left(\frac{dg}{dx}\right)_{x=\theta} = \begin{bmatrix} \frac{\partial g_1}{\partial x_1} & \frac{\partial g_1}{\partial x_2} & \cdots & \frac{\partial g_1}{\partial x_n} \\ \frac{\partial g_2}{\partial x_1} & \frac{\partial g_2}{\partial x_2} & \cdots & \frac{\partial g_2}{\partial x_n} \\ \vdots & & & \\ \frac{\partial g_n}{\partial x_1} & \frac{\partial g_n}{\partial x_2} & \cdots & \frac{\partial g_n}{\partial x_n} \end{bmatrix}_{x=\theta}$$

is the Jacobian of g with respect to x at $x = \theta$, and $f(x)$ is the remainder term of the Taylor expansion, containing higher order terms. Denoting the Jacobian matrix $\left(\frac{dg}{dx}\right)_{x=\theta}$ as A, and substituting (8.15) into (8.14), we get

$$\dot{x} = Ax + f(x). \tag{8.16}$$

Systems represented in the form of equation (8.16) are called **semilinear systems**, as they contain both a linear term and a nonlinear term. By considering only the linear part and neglecting the nonlinear term on the RHS of (8.16), we obtain the linearized form of the system (8.14). Thus, the linearized system is given by

$$\dot{x} = Ax,$$

which is also referred to as the linear system corresponding to the nonlinear system (8.14) or (8.16).

Linearization of a nonlinear system about an equilibrium point is a very useful tool for testing whether the nonlinear system is asymptotically stable or unstable at the equilibrium point. This is due to the fact that the asymptotic stability or instability of the linear system at $x = \theta$ implies the **same** for the nonlinear system under certain assumptions on the nonlinear function. These important results can be found in Theorems 8.4.2, 8.4.10, and 8.5.27. However, **if the linear system is only stable (but not asymptotically stable) at $x = \theta$, the nonlinear system may or may not be stable at** $x = 0$. Example 8.4.14 illustrates this fact.

***Example* 8.4.1** Linearize the following nonlinear system at the equilibrium points

$$\begin{cases} \dot{x}_1 = x_1 - x_1 x_2, \\ \dot{x}_2 = -2x_2 + x_1 x_2. \end{cases}$$

Solution Equating the RHS of the system to zero, two equilibrium points are obtained, namely $x_{e_1} = (0,0)$ and $x_{e_2} = (2,1)$.

I. Linearization at $(0,0)$. Let $g_1(x_1,x_2) = x_1 - x_1x_2$ and $g_2(x_1,x_2) = -2x_2 + x_1x_2$. Then the linearized system at $(0,0)$ is

$$\begin{bmatrix} \dot{x}_1 \\ \dot{x}_2 \end{bmatrix} = A \begin{bmatrix} x_1 \\ x_2 \end{bmatrix}, \quad \text{where } A = \begin{bmatrix} \frac{\partial g_1}{\partial x_1} & \frac{\partial g_1}{\partial x_2} \\ \frac{\partial g_2}{\partial x_1} & \frac{\partial g_2}{\partial x_2} \end{bmatrix}_{(0,0)} = \begin{bmatrix} 1 & 0 \\ 0 & -2 \end{bmatrix},$$

which is unstable at $(0,0)$.

II. Linearization at $(2,1)$. Let $z_1 = x_1 - 2$ and $z_2 = x_2 - 1$. In terms of z_1 and z_2, the given system can be written as

$$\dot{z}_1 = (z_1+2) - (z_1+2)(z_2+1) = -2z_2 - z_1z_2 = h_1(z_1,z_2),$$
$$\dot{z}_2 = -2(z_2+1) + (z_1+2)(z_2+1) = z_1 + z_1z_2 = h_2(z_1,z_2).$$

Now, linearizing the above system at $(0,0)$, we get

$$\begin{bmatrix} \dot{z}_1 \\ \dot{z}_2 \end{bmatrix} = \begin{bmatrix} \frac{\partial h_1}{\partial z_1} & \frac{\partial h_1}{\partial z_2} \\ \frac{\partial h_2}{\partial z_1} & \frac{\partial h_2}{\partial z_2} \end{bmatrix}_{(0,0)} \begin{bmatrix} z_1 \\ z_2 \end{bmatrix} \Longrightarrow \begin{bmatrix} \dot{z}_1 \\ \dot{z}_2 \end{bmatrix} = \begin{bmatrix} 0 & -2 \\ 1 & 0 \end{bmatrix} \begin{bmatrix} z_1 \\ z_2 \end{bmatrix},$$

which is stable at $(0,0)$ according to Theorem 8.3.3. It can be shown that the nonlinear system is also stable at (0,0) (see Exercise 8.4.2).

Let A be an $n \times n$ constant matrix and $f : \mathbb{R}^n \to \mathbb{R}^n$ be a nonlinear function satisfying $f(\theta) = \theta$. Consider the semilinear system

$$\dot{x} = Ax + f(x). \tag{8.17}$$

It is clear that the trivial solution $x \equiv \theta$ is an equilibrium point of (8.17). Consider the linearized system

$$\dot{x} = Ax, \tag{8.18}$$

that corresponds to (8.17). The following theorem provides **a sufficient condition** on f for analyzing the stability of (8.17) at $x = \theta$ using the linear system (8.18). Let the function f satisfy the condition

$$\lim_{x \to \theta} \frac{f(x)}{\|x\|} = \theta. \tag{8.19}$$

Then the following theorem holds.

Theorem 8.4.2 At the equilibrium point $x = \theta$, if the linear system (8.18) is asymptotically stable, then the nonlinear system (8.17) is also asymptotically stable provided that the function f satisfies (8.19).

Proof. Given that the linear system (8.18) is asymptotically stable at $x = 0$, it follows from Theorem 8.3.1 that all the eigenvalues of A have negative real parts. Then, by Lemma 3.2.16, we have

$$\|e^{At}\| \leq Me^{-\mu t}, \quad t > 0 \tag{8.20}$$

for some positive constants M and μ. From condition (8.19), for any $\varepsilon > 0$ there exists a $\delta > 0$ such that

$$\frac{\|f(x)\|}{\|x\|} < \varepsilon \quad \text{whenever } \|x\| < \delta.$$

In other words,

$$\|f(x)\| \leq \varepsilon\|x\| \quad \text{whenever } \|x\| < \delta. \tag{8.21}$$

The solution $x(t)$ of the equation (8.17) is given by

$$x(t) = e^{A(t-t_0)}x(t_0) + \int_{t_0}^{t} e^{A(t-s)}f(x(s))ds, \tag{8.22}$$

where $x(t_0)$ is the initial condition. Now, using (8.20) and (8.21) in (8.22), we get

$$\begin{aligned} \|x(t)\| &\leq \|e^{A(t-t_0)}\|\|x(t_0)\| + \int_{t_0}^{t} \|e^{A(t-s)}\|\|f(x(s))\|ds \\ &\leq Me^{-\mu(t-t_0)}\|x(t_0)\| + M\int_{t_0}^{t} e^{-\mu(t-s)}\|f(x(s))\|ds. \end{aligned} \tag{8.23}$$

Assuming that $\|x(s)\| < \delta$ for all $s > t_0$ (in the end of the proof, we will show that this assumption is valid for a suitable choice of initial condition), from (8.21) and (8.23), we get

$$\|x(t)\| \leq Me^{-\mu(t-t_0)}\|x(t_0)\| + M\varepsilon\int_{t_0}^{t} e^{-\mu(t-s)}\|x(s)\|ds.$$

Hence,

$$e^{\mu(t-t_0)}\|x(t)\| \leq M\|x(t_0)\| + M\varepsilon\int_{t_0}^{t} e^{\mu(s-t_0)}\|x(s)\|ds.$$

Using Gronwall's inequality (Corollary 4.10.3), we obtain

$$e^{\mu(t-t_0)}\|x(t)\| \leq M\|x(t_0)\|e^{M\varepsilon(t-t_0)}, \quad t > t_0.$$

Thus,

$$\|x(t)\| \leq M\|x(t_0)\|e^{(M\varepsilon-\mu)(t-t_0)}, \quad t > t_0. \tag{8.24}$$

By choosing $\varepsilon < \frac{\mu}{M}$ in (8.24), we have $M\varepsilon - \mu < 0$. Hence, $e^{(M\varepsilon-\mu)(t-t_0)} \to 0$ as $t \to \infty$, and $\|x(t)\| \to 0$ as $t \to \infty$. If the initial condition is chosen such that $\|x(t_0)\| < \frac{\delta}{M}$ (where δ corresponds to ε owing to (8.21)), then from (8.24), we get $\|x(t)\| < \delta$ for all $t > t_0$. Thus, we conclude the following:

For a given ε with $0 < \varepsilon < \frac{\mu}{M}$, we can find $\delta > 0$ such that

$$\|f(x)\| < \varepsilon\|x\| \quad \text{whenever } \|x\| < \delta.$$

If the initial condition is chosen such that $\|x(t_0)\| < \frac{\delta}{M}$, then from (8.24), we have $\|x(t)\| \to 0$ as $t \to \infty$. Hence, (8.17) is asymptotically stable at $x = \theta$. □

Consider the system

$$\dot{x} = Ax + b(t), \tag{8.25}$$

where $x(t), b(t) \in \mathbb{R}^n$ for each $t > 0$. For an initial condition $x(t_0) = x_0$, the solution of (8.25) is given by

$$x(t) = e^{A(t-t_0)}x_0 + \int_{t_0}^{t} e^{A(t-s)}b(s)ds. \tag{8.26}$$

Note that the origin $x = \theta$ is not an equilibrium point of the system (8.25). However, the following theorem provides **a sufficient condition** on $b(t)$ for obtaining the solution $x(t)$ such that

$$\lim_{t\to\infty} x(t) = \theta. \tag{8.27}$$

Theorem 8.4.3 The solution (8.26) of the system (8.25) satisfies condition (8.27) if

(*i*) the eigenvalues of A have negative real parts,

(*ii*) $\lim_{t\to\infty} \|b(t)\| = 0$.

Proof. From (8.26), we have

$$\|x(t)\| \leq \|e^{A(t-t_0)}\|\|x_0\| + \int_{t_0}^{t} \|e^{A(t-s)}\|\|b(s)\|ds. \tag{8.28}$$

From condition (*i*) and Lemma 3.2.16, there exist constants $M > 0$ and $\mu > 0$ such that

$$\|e^{A(t-t_0)}\| \leq Me^{-\mu(t-t_0)}.$$

Hence,

$$\|x(t)\| \leq Me^{-\mu(t-t_0)}\|\|x_0\| + M\int_{t_0}^{t} e^{-\mu(t-s)}\|b(s)\|ds. \tag{8.29}$$

Since $\lim_{t\to\infty}\|b(t)\| = 0$, for any $\varepsilon_1 > 0$, there exists $t_1 > t_0$ such that $\|b(t)\| < \varepsilon_1$ for $t > t_1$. Now, for $t > t_1$, (8.29) becomes

$$\begin{aligned}\|x(t)\| &\leq Me^{-\mu(t-t_0)}\|x_0\| + M\int_{t_0}^{t_1} e^{-\mu(t-s)}\|b(s)\|ds + M\int_{t_1}^{t} e^{-\mu(t-s)}\|b(s)\|ds \\ &= Me^{-\mu(t-t_1)}\left[e^{-\mu(t_1-t_0)}\|x_0\| + \int_{t_0}^{t_1} e^{-\mu(t_1-s)}\|b(s)\|ds\right] \\ &\quad + M\int_{t_1}^{t} e^{-\mu(t-s)}\|b(s)\|ds.\end{aligned}$$

Let $e^{-\mu(t_1-t_0)} < \varepsilon_2$ and $\|b(t)\| \leq c$ for $t_0 \leq t \leq t_1$. The above inequality becomes

$$\|x(t)\| \leq Me^{-\mu(t-t_1)}\left[\varepsilon_2\|x_0\| + c\frac{1}{\mu}\right] + M\frac{\varepsilon_1}{\mu}. \tag{8.30}$$

For a given $\varepsilon > 0$, choose $\varepsilon_1 = \frac{\mu}{2M}\varepsilon$ (for sufficiently large t_1) and

$$e^{-\mu(t-t_1)}\left[\varepsilon_2\|x_0\| + \frac{c}{\mu}\right] < \frac{\varepsilon}{2} \quad \text{for sufficiently large } t > t_1.$$

Then, from (8.30), $\|x(t)\| < \varepsilon$ for sufficiently large t, which implies that $\lim_{t\to\infty}\|x(t)\| = 0$. □

***Example* 8.4.4** (*i*) Consider $\dot{x} = -2t - \frac{1}{t}$ for $t > 1$ with the initial condition $x(1) = 0$. From Theorem 8.4.3, we can conclude that $\lim_{t\to\infty} x(t) = \theta$.

(*ii*) Consider the IVP, $\dot{x} = Ax + b(t), x(0) = x_0$, where $A = \begin{bmatrix} -1 & 1 \\ 0 & -2 \end{bmatrix}$, $b(t) = \begin{bmatrix} \frac{1}{t+1} \\ \frac{\sin t}{t} \end{bmatrix}$, and $x_0 = \begin{bmatrix} 1 \\ -1 \end{bmatrix}$. Without solving the system for exact solution, we infer from Theorem 8.4.3 that $\lim_{t\to\infty} x(t) = \theta$.

The following lemmas will be useful in various theorems on the stability of linear systems.

Lemma 8.4.5 (*i*) If A is an $n \times n$ diagonalizable matrix, then there exists a nonsingular matrix P such that

$$P^{-1}AP = \left[\begin{array}{c|c} D_1 & O \\ \hline O & D_2 \end{array}\right], \tag{8.31}$$

where D_1 and D_2 are diagonal matrices such that diagonal entries of D_1 are the eigenvalues of A with positive real parts, and the diagonal entries of D_2 are eigenvalues with non-positive real parts.

(*ii*) If A is not diagonalizable, then for any arbitrary real number $\lambda > 0$, there exists a nonsingular matrix P such that the Jordan form of A is given by

$$J(\lambda) = P^{-1}AP = \left[\begin{array}{c|c} J_1(\lambda) & O \\ \hline O & J_2(\lambda) \end{array}\right], \tag{8.32}$$

where $J_1(\lambda)$ has eigenvalues with positive real parts, and $J_2(\lambda)$ has eigenvalues with non-positive real parts on its diagonal. Moreover, $J_1(\lambda)$ and $J_2(\lambda)$ have λ on the super diagonal (upper diagonal) of every Jordan sub-block of order greater than or equal to 2.

Proof. (*i*) By arranging the eigenvalues as $\lambda_i : i = 1, 2, \ldots, n$ such that the real parts are in decreasing order, and taking the corresponding linearly independent eigenvectors $x_i : i = 1, 2, \ldots, n$ as the ith column of P, it can be shown that

$$P^{-1}AP = \left[\begin{array}{c|c} D_1 & O \\ \hline O & D_2 \end{array}\right].$$

(*ii*) Arrange the eigenvalues in decreasing order with respect to their real parts. If A is not diagonalizable, then for some eigenvalues, the algebraic multiplicity is greater than the geometric multiplicity. In such cases, generalized eigenvectors can be obtained. Let λ_j be one such eigenvalue and x_j be an eigenvector corresponding to λ_j. If x_j has r_j generalized eigenvectors, then the generalized eigenvectors $Y_1, Y_2, \ldots, Y_{r_j}$ can be obtained from the following equations (see equation (2.27))

$$\begin{aligned}
(A - \lambda_j I)Y_1 &= x_j \\
(A - \lambda_j I)Y_2 &= Y_1 \\
&\vdots
\end{aligned}$$

$$(A-\lambda_j I)Y_{r_j} = Y_{r_j-1}.$$

Let $\lambda > 0$ be an arbitrary number. Then $Z_i = \lambda^i Y_i : i = 1,2,\ldots,r_j$, are called the λ-generalized eigenvectors corresponding to x_j. Similarly, λ-generalized eigenvectors corresponding to all eigenvalues can be obtained (if applicable). Then the matrix P is formed by writing the eigenvectors and λ-generalized eigenvectors as columns (see Section 2.7), and it can be shown that

$$P^{-1}AP = \left[\begin{array}{c|c} J_1(\lambda) & O \\ \hline O & J_2(\lambda) \end{array}\right].$$

□

Remark 8.4.6 In the Jordan-type canonical form, note that λ can be chosen as any nonzero real number.

Lemma 8.4.7 Let $J(\lambda) = \begin{bmatrix} \alpha & \lambda & 0 & \cdots & 0 \\ 0 & \alpha & \lambda & \cdots & 0 \\ \vdots & & & & \lambda \\ 0 & 0 & 0 & \cdots & \alpha \end{bmatrix}_{r\times r}$ and $x = \begin{bmatrix} x_1 \\ x_2 \\ \vdots \\ x_r \end{bmatrix}$, where $r \geq 2$. If $\mathrm{Re}(\alpha) > \sigma > 0$ and $0 < \lambda < \sigma$, then $\mathrm{Re}\langle J(\lambda)x, x\rangle > (\sigma-\lambda)\|x\|^2$ for all $x \in \mathbb{C}^n$.

Proof. Since $J(\lambda)x = \sum_{i=1}^{r-1}(\alpha x_i + \lambda x_{i+1}) + \alpha x_r$, we have $\langle J(\lambda)x, x\rangle = \bar{x}' J(\lambda)x = \alpha(\bar{x}_1x_1 + \bar{x}_2x_2 + \cdots + \bar{x}_rx_r) + \lambda(\bar{x}_1x_2 + \bar{x}_2x_3 + \cdots + \bar{x}_{r-1}x_r)$. Now, $2\mathrm{Re}\langle J(\lambda)x,x\rangle = \langle J(\lambda)x,x\rangle + \overline{\langle J(\lambda)x,x\rangle} = 2\mathrm{Re}(\alpha)\|x\|^2 + 2\lambda\mathrm{Re}(\bar{x}_1x_2 + \bar{x}_2x_3 + \cdots + \bar{x}_{r-1}x_r)$. Since $-\|x\|^2 \leq \mathrm{Re}(\bar{x}_1x_2 + \bar{x}_2x_3 + \cdots + \bar{x}_{r-1}x_r) \leq \|x\|^2$ and $\mathrm{Re}(\alpha) > \sigma$, we obtain $\mathrm{Re}\langle J(\lambda)x,x\rangle > (\sigma-\lambda)\|x\|^2$. □

Lemma 8.4.8 Let $J(\lambda) = \begin{bmatrix} J_1(\lambda) & & & 0 \\ & J_2(\lambda) & & \\ & & \ddots & \\ 0 & & & J_k(\lambda) \end{bmatrix}$ be a Jordan form, where $J_i(\lambda) = \begin{bmatrix} \lambda_i & \lambda & 0 & \cdots & 0 \\ 0 & \lambda_i & \lambda & \cdots & 0 \\ \vdots & & & & \lambda \\ 0 & 0 & 0 & \cdots & \lambda_i \end{bmatrix}_{r_i\times r_i}$ for $i = 1,2,\ldots,k$, and $\lambda > 0$. Let $x = \left[\begin{array}{c} X_1 \\ \hline X_2 \\ \hline \vdots \\ \hline X_k \end{array}\right] \in \mathbb{R}^n$, with blocks

$X_i \in \mathbb{R}^{r_i} : i = 1,2,\ldots,k$. If $\text{Re}(\lambda_i) \geq \sigma > 0$ for $i = 1,2,\ldots,n$ and $0 < \lambda < \sigma$, then

$$\text{Re}\langle J(\lambda)x,x\rangle > (\sigma - \lambda)\|x\|^2 \quad \text{for all } x \in \mathbb{C}^n.$$

Proof. Note that $\langle J(\lambda)x,x\rangle = \sum_{i=1}^{k}\langle J_i(\lambda)X_i,X_i\rangle$. Since $\text{Re}\langle J_i(\lambda)X_i,X_i\rangle > (\sigma-\lambda)\|X_i\|^2$, by Lemma 8.4.7, we obtain $\text{Re}\langle J(\lambda)x,x\rangle > (\sigma-\lambda)\sum_{i=1}^{k}\|X_i\|^2 = (\sigma-\lambda)\|x\|^2$. □

Lemma 8.4.9 Let $J(\lambda) = \begin{bmatrix} J_1(\lambda) & & & 0 \\ & J_2(\lambda) & & \\ & & \ddots & \\ 0 & & & J_k(\lambda) \end{bmatrix}$ be a Jordan form, where $J_i(\lambda) = \begin{bmatrix} \lambda_i & \lambda & 0 & \cdots & 0 \\ 0 & \lambda_i & \lambda & \cdots & 0 \\ \vdots & & & & \lambda \\ 0 & 0 & 0 & \cdots & \lambda_i \end{bmatrix}_{r_i \times r_i}$ for $i = 1,2,\ldots,k$, where $\lambda > 0$. Let $x = \begin{bmatrix} X_1 \\ \hline X_2 \\ \hline \vdots \\ \hline X_k \end{bmatrix} \in \mathbb{R}^n$, with blocks $X_i \in \mathbb{R}^{r_i} : i = 1,2,\ldots,k$. If $\text{Re}(\lambda_i) \leq 0$ for $i = 1,2,\ldots,n$, then

$$\text{Re}\langle J(\lambda)x,x\rangle \leq \lambda\|x\|^2 \quad \text{for all } x \in \mathbb{C}^n.$$

Proof. We know that $2\text{Re}\langle J_i(\lambda)X_i,X_i\rangle = \langle J_i(\lambda)X_i,X_i\rangle + \overline{\langle J_i(\lambda)X_i,X_i\rangle} = 2\text{Re}(\lambda_i)\|X_i\|^2 + 2\lambda\text{Re}(\bar{x}_1x_2 + \bar{x}_2x_3 + \cdots + \bar{x}_{r_i-1}x_{r_i}) \leq 2\lambda\text{Re}(\bar{x}_1x_2 + \bar{x}_2x_3 + \cdots + \bar{x}_{r_i-1}x_{r_i})$. Since $\text{Re}(\bar{x}_1x_2 + \bar{x}_2x_3 + \cdots + \bar{x}_{r_i-1}x_{r_i}) \leq \|X_i\|^2$, we obtain $\text{Re}\langle J_i(\lambda)X_i,X_i\rangle \leq \lambda\|X_i\|^2$. Hence, $\text{Re}\langle J(\lambda)x,x\rangle = \sum_{i=1}^{k}\langle J_i(\lambda)X_i,X_i\rangle \leq \lambda\sum_{i=1}^{k}\|X_i\|^2 = \lambda\|x\|^2$. □

Theorem for Instability

Let

$$J(\lambda) = \left[\begin{array}{c|c} J_1(\lambda) & O \\ \hline O & J_2(\lambda) \end{array}\right],$$

where $J_1(\lambda)$ and $J_2(\lambda)$ are as in Lemma 8.4.5. Let the order of the matrices $J_1(\lambda)$ and $J_2(\lambda)$ be p and q, respectively. Let $f : \mathbb{R}^n \to \mathbb{R}^n$ be a continuous function such that $\lim_{x\to\theta}\frac{f(x)}{\|x\|} = \theta$. In other words, for any given $\varepsilon > 0$, there exists a $\delta > 0$ such that whenever $\|x\| < \delta$,

$$\|f(x)\| < \varepsilon\|x\|. \tag{8.33}$$

Now, consider the system

$$\dot{x} = J(\lambda)x + f(x). \tag{8.34}$$

Let $x = \begin{bmatrix} X_1 \\ \hline X_2 \end{bmatrix}$, where $X_1 \in \mathbb{R}^p$ and $X_2 \in \mathbb{R}^q$, and let $f(x) = \begin{bmatrix} f_1(x) \\ \hline f_2(x) \end{bmatrix}$ be suitable blocks, where $f_1 \in \mathbb{R}^p$ and $f_2 \in \mathbb{R}^q$. Then from (8.34), we have

$$\dot{X}_1 = J_1(\lambda)X_1 + f_1(x), \tag{8.35}$$
$$\dot{X}_2 = J_2(\lambda)X_2 + f_2(x). \tag{8.36}$$

Note that $\|x\|^2 = \|X_1\|^2 + \|X_2\|^2$, where $\|X_1\|^2 = \langle X_1, X_1\rangle_{\mathbb{R}^p}$ and $\|X_2\|^2 = \langle X_2, X_2\rangle_{\mathbb{R}^q}$. Hence,

$$\frac{d}{dt}\|X_1\|^2 = \left\langle \frac{dX_1}{dt}, X_1 \right\rangle + \left\langle X_1, \frac{dX_1}{dt} \right\rangle = 2\text{Re}\left\langle \frac{dX_1}{dt}, X_1 \right\rangle. \tag{8.37}$$

So from (8.35)–(8.37), we obtain

$$\frac{d}{dt}\|X_1\|^2 = 2\|X_1\|\frac{d}{dt}\|X_1\| = 2\big[\text{Re}\,\langle J_1(\lambda)X_1, X_1\rangle + \text{Re}\,\langle f_1(x), X_1\rangle\big]. \tag{8.38}$$

Let $\|x\| < \delta$. Then $\|X_1\| < \delta$ and $\|X_2\| < \delta$. From (8.33) and the Cauchy–Schwarz inequality for inner products,

$$\begin{aligned} |\text{Re}\,\langle f_1(x), X_1\rangle| \le |\langle f_1(x), X_1\rangle| &\le \|f_1(x)\|\|X_1\| \\ &\le \|f(x)\|\|X_1\| \\ &\le \varepsilon\sqrt{\|X_1\|^2 + \|X_2\|^2}\,\|X_1\| \quad \text{(using (8.33))} \\ &\le \varepsilon(\|X_1\| + \|X_2\|)\|X_1\|. \end{aligned}$$

Hence,

$$-\varepsilon(\|X_1\| + \|X_2\|)\|X_1\| \le \text{Re}\langle f_1(x), X_1\rangle \le \varepsilon(\|X_1\| + \|X_2\|)\|X_1\|. \tag{8.39}$$

Using Lemma 8.4.8 and the left side inequality of (8.39) in (8.38), we get

$$\|X_1\|\frac{d}{dt}\|X_1\| > (\sigma - \lambda)\|X_1\|^2 - \varepsilon(\|X_1\| + \|X_2\|)\|X_1\|.$$

Thus,

$$\frac{d}{dt}\|X_1\| > (\sigma - \lambda)\|X_1\| - \varepsilon(\|X_1\| + \|X_2\|). \tag{8.40}$$

Similarly, for X_2, we deduce

$$2\|X_2\|\frac{d}{dt}\|X_2\| = \frac{d}{dt}\|X_2\|^2 = \frac{d}{dt}\langle X_2, X_2\rangle = 2\mathrm{Re}\left\langle \frac{dX_2}{dt}, X_2 \right\rangle.$$

By following the steps of (8.37)–(8.39), we get

$$\mathrm{Re}\langle f_2(x), x\rangle \leq \varepsilon(\|X_1\| + \|X_2\|)\|X_2\|. \tag{8.41}$$

From Lemma 8.4.9 and inequality (8.41),

$$\begin{aligned}2\|X_2\|\frac{d}{dt}\|X_2\| &= 2\big[\mathrm{Re}\,\langle J_2(\lambda)X_2, X_2\rangle + \mathrm{Re}\,\langle f_2(x), X_2\rangle\big]\\ &\leq 2\big(\lambda\|X_2\|^2 + \varepsilon(\|X_1\| + \|X_2\|)\|X_2\|\big).\end{aligned}$$

Hence,

$$\frac{d}{dt}\|X_2\| \leq \lambda\|X_2\| + \varepsilon(\|X_1\| + \|X_2\|). \tag{8.42}$$

From (8.40) and (8.42),

$$\begin{aligned}\frac{d}{dt}(\|X_1\| - \|X_2\|) &\geq (\sigma - \lambda)\|X_1\| - \lambda\|X_2\| - 2\varepsilon(\|X_1\| + \|X_2\|)\\ &= \|X_1\|(\sigma - \lambda - 2\varepsilon) - \|X_2\|(\lambda + 2\varepsilon).\end{aligned}$$

Since λ and ε can be chosen arbitrarily, let $\lambda = \frac{\sigma}{4}$ and $\varepsilon = \frac{\sigma}{8}$. Then,

$$\begin{aligned}&\frac{d}{dt}(\|X_1\| - \|X_2\|) \geq \frac{\sigma}{2}(\|X_1\| - \|X_2\|)\\ \implies\ &\|X_1(t)\| - \|X_2(t)\| \geq e^{\frac{\sigma}{2}t}(\|X_1(0)\| - \|X_2(0)\|).\end{aligned} \tag{8.43}$$

As $\sigma > 0$, it is clear from (8.43) that $\|X_1(t)\| - \|X_2(t)\| \to \infty$ as $t \to \infty$ when $\|X_1(0)\| > \|X_2(0)\|$. Hence $\|x\| = \sqrt{\|x_1\|^2 + \|X_2\|^2}$ is unbounded, implying that the equilibrium point $x = \theta$ is unstable. Thus, we obtain the following result.

Theorem 8.4.10 The nonlinear system $\dot{x} = J(\lambda)x + f(x), f(\theta) = \theta$, is unstable at the equilibrium point $x = \theta$ under the following sufficient conditions:

(*i*) f satisfies (8.19), i.e., $\lim_{x\to\theta} \frac{f(x)}{\|x\|} = \theta$.

(*ii*) The matrix $J(\lambda)$ has at least one eigenvalue with positive real part.

Remark 8.4.11 Let A be any $n \times n$ matrix with Jordan-type canonical form $J(\lambda), \lambda > 0$ such that $P^{-1}AP = J(\lambda)$ for a nonsingular matrix P. Consider the nonlinear system

$\dot{z} = Az + g(z)$, where $g(\theta) = \theta$. Then it is clear that $z = \theta$ is an equilibrium point. To analyze the stability at $z = \theta$, we transform the system to $\dot{x} = J(\lambda)x + f(x)$, where $z = Px$ and $P^{-1}g(z) = P^{-1}g(Px) = f(x)$, and then examine the stability at $x = \theta$ for this transformed system.

Remark 8.4.12 Note that the above theorem does not cover the case when all eigenvalues of $J(\lambda)$ have non-positive real parts and some eigenvalues have zero real parts.

***Example* 8.4.13** Consider the prey–predator population model

$$\begin{cases} \dot{x}_1 = ax_1 - bx_1x_2 = x_1(a - bx_2), \\ \dot{x}_2 = -cx_2 + dx_1x_2 = x_2(-c + dx_1), \end{cases}$$

where a, b, c, and d are positive constants. Equating the RHSs to zero, we obtain the equilibrium points $(0,0)$ and $(\frac{c}{d}, \frac{a}{b})$. First, let us analyze the stability property at $(0,0)$. The above system can be written as

$$\dot{x} = Ax + f(x), \tag{8.44}$$

where

$$A = \begin{bmatrix} a & 0 \\ 0 & -c \end{bmatrix} \text{ and } f(x) = \begin{bmatrix} -bx_1x_2 \\ dx_1x_2 \end{bmatrix}.$$

It can be seen that

$$\frac{f(x)}{\|x\|} = \begin{bmatrix} -b\frac{x_1x_2}{\sqrt{x_1^2+x_2^2}} \\ d\frac{x_1x_2}{\sqrt{x_1^2+x_2^2}} \end{bmatrix} \to \begin{bmatrix} 0 \\ 0 \end{bmatrix} \quad \text{as } \|x\| \to 0, \quad \text{where } \|x\| = \sqrt{x_1^2 + x_2^2}.$$

Since a and c are positive, the linear system $\dot{x} = Ax$ is unstable at $x = \theta$ (Theorem 8.3.1). Hence, from Theorem 8.4.10, the nonlinear system (8.44) is also unstable at $(0,0)$.

At a later stage, we shall show that the other equilibrium point $(\frac{c}{d}, \frac{a}{b})$ is stable. This cannot be demonstrated using linearization and the above theorems.

***Example* 8.4.14** Consider the following systems

(i)

$$\begin{cases} \dot{x}_1 = -x_2 - x_1^3, \\ \dot{x}_2 = x_1 + x_2^3. \end{cases}$$

(ii)

$$\begin{cases} \dot{x}_1 = -x_2 - x_1^3, \\ \dot{x}_2 = x_1 - x_2^3. \end{cases}$$

In the following section, using the Lyapunov theory, it is shown that the system (i) is unstable at (0,0) and the system (ii) is asymptotically stable at (0,0). However, for both systems, the corresponding linear system is stable but not asymptotically stable at (0,0).

Example 8.4.14 illustrates that we cannot infer the stability property of an equilibrium point of a nonlinear system solely from the corresponding linear system if the linear system is stable (but not asymptotically stable) at that equilibrium point.

Exercise 8.4.1 Find all equilibrium points of the system

$$\begin{cases} \dot{x}_1 = 3x_1 + x_2 - x_1 \cos x_2, \\ \dot{x}_2 = 2x_2 \sin x_1 + x_2(1 + \cos x_1). \end{cases}$$

Using linearization, identify which equilibrium points are: (i) unstable, or (ii) asymptotically stable.

Exercise 8.4.2 Show that the equation of the phase-space curve of the system

$$\dot{x}_1 = -2\,x_2 - x_1\,x_2$$

$$\dot{x}_2 = x_1 + x_1\,x_2$$

with initial condition $(x_1(0), x_2(0)) = (a, b)$ is $(x_2 - b) - \log(1 + x_2) = \log(2 + x_1) - (x_1 - a) - \log(2 + a)^2 - \log(1 + b)$.

8.5 Lyapunov Theory for Time-invariant Systems

In previous sections, we discussed the stability theory for autonomous linear systems of the form $\dot{x} = Ax$ and semi-linear systems of the form $\dot{x} = Ax + f(x)$. For autonomous linear system, the stability of the equilibrium point $x = \theta$ is analyzed with the help of the eigenvalues of the matrix A, as given in Theorem 8.3.1. For semilinear systems with $f(\theta) = \theta$, stability at $x = 0$ can be analyzed using Theorems 8.4.2 and 8.4.10. For nonlinear systems of the form $\dot{x} = f(x)$ with $f(0) = 0$, stability at $x = \theta$ can be determined by linearizing the system. However, as illustrated in Example 8.4.14, the aforementioned methods do not cover all the cases.

In this section, we introduce Lyapunov's approach to stability analysis, which addresses most situations involving the stability properties of an equilibrium point in nonlinear systems.

Before introducing Lyapunov theory, we define certain terms. Let S be an open set of $\mathbb{R}^n$ containing the origin.

Definition 8.5.1 A function $V : S \to \mathbb{R}$ is said to be **positive (negative) definite** on S if $V(x) > 0$ $(V(x) < 0)$ for all nonzero $x \in S$ and $V(\theta) = 0$.

Definition 8.5.2 A function $V : S \to \mathbb{R}$ is said to be **positive (negative) semi-definite** if $V(x) \geq 0$ $(V(x) \leq 0)$ for all nonzero $x \in S$ and $V(\theta) = 0$.

Example **8.5.3** (a) The function $V : \mathbb{R}^2 \to \mathbb{R}$ defined by

(i) $V(x_1, x_2) = 2x_1^2 + 3x_2^2$ is positive definite on $\mathbb{R}^2$.

(ii) $V(x_1, x_2) = (x_1 - x_2)^2$ is positive semi-definite on $\mathbb{R}^2$.

(iii) $V(x_1, x_2) = (x_1^2 + 2x_2^2)(1 - x_1^2 - x_2^2)$ is positive definite in the disc $x_1^2 + x_2^2 < 1$.

(b) The function $V : \mathbb{R}^3 \to \mathbb{R}$ defined by $V(x_1, x_2, x_3) = -2x_1^2 - 3x_2^2$ is negative semi-definite on $\mathbb{R}^3$.

Definition 8.5.4 Let $\mathbb{R}_0^+ = \{x \in \mathbb{R} : x \geq 0\}$. A function $\phi : \mathbb{R}_0^+ \to \mathbb{R}_0^+$ is said to be of **class-*K*** if $\phi(0) = 0$, $\phi(x) > 0$ for $x \neq 0$, and ϕ is continuous and strictly increasing.

Example **8.5.5** (i) $\phi(|x|) = |x|^2$, where $x \in \mathbb{R}$ is a function of class-K.

(ii) For $x = (x_1, x_2, \ldots, x_n) \in \mathbb{R}^n$, let $\|x\| = \sqrt{x_1^2 + x_2^2 + \ldots + x_n^2}$. Then $\phi(\|x\|) = \|x\|^m$, m is a positive integer, is of class-K.

Let $B_r(\theta) = \{x \in \mathbb{R}^n : \|x\| \leq r, r > 0\}$ be the ball of radius r with center at the origin.

Lemma 8.5.6 (i) If $V : B_r(\theta) \to [0, \infty)$ is continuous and positive definite, then there exist functions ϕ and ψ of class-K such that

$$\phi(\|y\|) \leq V(y) \leq \psi(\|y\|) \quad \text{for all } y \in B_r(\theta).$$

(ii) If $V : B_r(\theta) \to (-\infty, 0]$ is continuous and negative definite, then there exist functions ϕ and ψ of class-K such that

$$-\phi(\|y\|) \geq V(y) \geq -\psi(\|y\|) \quad \text{for all } y \in B_r(0).$$

Proof. (i) Let $r > 0$ be a constant and $V : B_r(\theta) \to [0, \infty)$ be a continuous and positive definite function. Since V is continuous on the closed ball $B_r(\theta)$, it is bounded. Let $0 < s \leq a \leq r$. Then, for all y on the hypersphere $\|y\| = a$,

$$V(y) \leq \max_{\|x\| \leq a} V(x) = \psi(a) \text{ (say)}.$$

Note that $\psi(0) = 0$ and $\psi(a)$ is monotonically increasing with respect to a. Hence ψ is a function of class-K. Similarly,

$$V(y) \geq \min_{a \leq \|x\| \leq r} V(x) = \phi(a) \text{ (say)}$$

on the hypersphere $\|y\| = a$. Note that ϕ is also of class-K. Thus, if V is continuous and positive definite on $B_r(\theta)$, there exist functions ϕ and ψ of class-K such that

$$\phi(\|y\|) \leq V(y) \leq \psi(\|y\|) \quad \text{for all } y \in B_r(\theta). \tag{8.45}$$

(*ii*) If V is negative definite, then $-V$ is positive definite. By applying (8.45) to $-V$, we obtain the desired result. □

Example **8.5.7** Let $V(x_1, x_2) = x_1^4 + x_2^2$. Find a function ϕ of class-K such that $V(x) \geq \phi(\|x\|)$.

Solution For $\|x\| < 1$, $V(x) = x_1^4 + x_2^2 \geq x_1^4 + x_2^4$. Let $x_1 = r\cos\theta$ and $x_2 = r\sin\theta$. Then $x_1^4 + x_2^4 = r^4(\cos^4\theta + \sin^4\theta) = r^4(1 - 2\cos^2\theta\sin^2\theta) = r^4(1 - \frac{\sin^2 2\theta}{2}) \geq \frac{r^4}{2} = \frac{(x_1^2+x_2^2)^2}{2}$. Thus, $V(x) = x_1^4 + x_2^2 \geq \frac{\|x\|^4}{2}$, a function of class-$K$. Here, $\phi(\|x\|) = \frac{\|x\|^4}{2}$.

Definition 8.5.8 (Lyapunov function) Let $V : B_r(\theta) \to \mathbb{R}$ be such that

(*i*) $V(\theta) = 0$,

(*ii*) V is continuous on $B_r(\theta)$,

(*iii*) $\frac{\partial V}{\partial x_i} : i = 1, 2, \ldots, n$ are continuous on $B_r(\theta)$.

Then V is called a **Lyapunov function**.

The following three theorems provide sufficient conditions on Lyapunov functions for establishing (*i*) stability (*ii*) asymptotic stability, or (*iii*) instability of the equilibrium $x = \theta$ for autonomous systems of the form

$$\dot{x} = f(x), \quad f(\theta) = \theta. \tag{8.46}$$

Lyapunov Theorems on Stability

Theorem 8.5.9 For the system $\dot{x} = f(x)$, $f(\theta) = \theta$, the equilibrium $x = \theta$ is stable if there exists a Lyapunov function $V(x)$ satisfying

(*i*) V is positive definite on $B_r(\theta)$ for some $r > 0$,

(*ii*) $\frac{dV}{dt}$ is negative semi-definite on $B_r(\theta)$, where $\frac{dV}{dt} = \frac{\partial V}{\partial x_1}\dot{x}_1 + \cdots + \frac{\partial V}{\partial x_n}\dot{x}_n$.

Example **8.5.10** Consider the linear system $\dot{x} = Ax$, where $x(t) = \begin{bmatrix} x_1(t) \\ x_2(t) \end{bmatrix}$ and $A = \begin{bmatrix} 0 & -1 \\ 1 & 0 \end{bmatrix}$. As the eigenvalues of A are distinct with zero real parts, from Theorem 8.3.3, the equilibrium $x = \theta$ is stable.

Now we shall establish the stability of $x = \theta$ using a Lyapunov function. Consider the Lyapunov function

$$V(x_1, x_2) = x_1^2 + x_2^2,$$

which satisfies condition (i) of Theorem 8.5.9 on $\mathbb{R}^2$. V fulfills condition (ii) also since

$$\frac{dV}{dt} = \frac{\partial V}{\partial x_1}\frac{dx_1}{dt} + \frac{\partial V}{\partial x_2}\frac{dx_2}{dt} = 2x_1(-x_2) + 2x_2(x_1) = 0.$$

Hence, by Theorem 8.5.9, the trivial solution $x = \theta$ is stable.

Proof (of Theorem 8.5.9). Since $V(x)$ is positive definite on $B_r(\theta)$, from Lemma 8.5.6, there exists a function ϕ of class-K such that

$$V(x) \geq \phi(\|x\|) \text{ for all } x \in B_r(\theta). \tag{8.47}$$

To prove stability at $x = \theta$, we show that for any $\varepsilon, 0 < \varepsilon < r$, there exists a $\delta > 0$ such that

$$\|x(t_0)\| < \delta \text{ implies } \|x(t)\| < \varepsilon \text{ for all } t > t_0.$$

Let $\varepsilon < r$ be given. If $x = \theta$ is not stable, then for any initial state $x(t_0)$, there exists a time instant $t_1 > t_0$ such that

$$\|x(t_1)\| = \varepsilon. \tag{8.48}$$

Since $V(\theta) = 0$, V is continuous and $V(x) > 0$ on $B_r(\theta)$, there exists a $\delta > 0$ such that $x(t_0) \in B_\delta(\theta)$ implies

$$V(x(t_0)) < \phi(\varepsilon), \tag{8.49}$$

as shown in Figure 8.9. Since $\frac{dV}{dt} \leq 0$ on B_r, $V(x(t))$ is monotonically non-increasing as a function of t. Hence,

$$t_1 > t_0 \text{ implies } V(x(t_1)) \leq V(x(t_0)). \tag{8.50}$$

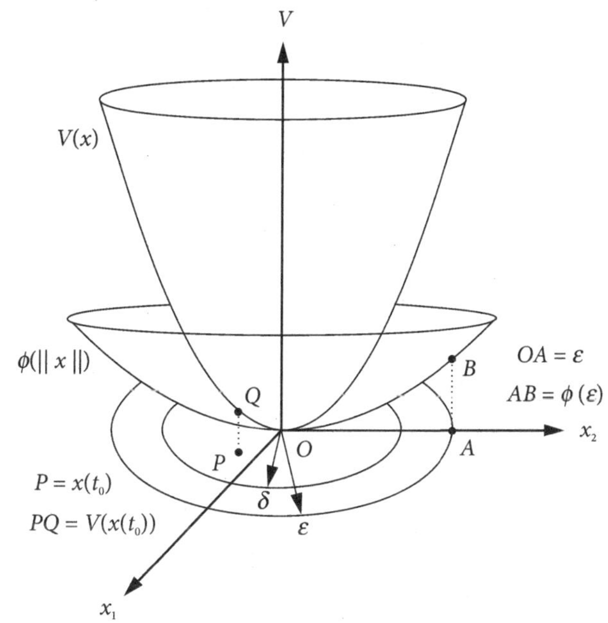

Figure 8.9 Lyapunov function

From (8.47)–(8.50), we have

$$\phi(\varepsilon) = \phi(\|x(t_1)\|) \leq V(x(t_1)) \leq V(x(t_0)) < \phi(\varepsilon),$$

which is a contradiction. Hence, if $\|x(t_0)\| < \delta$, then $\|x(t)\| < \varepsilon$ for all $t > t_0$. This completes the proof. □

Remark 8.5.11 In conditions (i) and (ii) of Theorem 8.5.9, the terms "positive definite" and "negative semi-definite" may be replaced by "negative definite" and "positive semi-definite", respectively. The proof will be similar with suitable changes in the inequality sign.

Theorem 8.5.12 The trivial solution $x = \theta$ of the system $\dot{x} = f(x)$, $f(\theta) = \theta$, is asymptotically stable if there exists a Lyapunov function $V : B_r(\theta) \to \mathbb{R}$ for some $r > 0$ that satisfies the following:

(i) V is positive definite on $B_r(\theta)$,

(ii) $\frac{dV}{dt}$ is negative definite on $B_r(\theta)$.

Proof. It is clear that $x = \theta$ is stable since V satisfies the conditions of Theorem 8.5.9. Therefore, for any $\varepsilon > 0$, there exists $\delta > 0$ such that $\|x(t)\| < \varepsilon$ whenever

$\|x(t_0)\| < \delta$. We now prove asymptotic stability by showing that $\lim_{t\to\infty}\|x(t)\| = 0$. If $\lim_{t\to\infty}\|x(t)\| \neq 0$, then there exists $\omega > 0$ such that $\omega < \|x(t)\| < \varepsilon$ for all $t \geq t_0$. Since $\frac{dV}{dt} < 0$, according to Lemma 8.5.6, there exists a function β of class-K such that

$$\frac{d}{dt}V(x(t)) \leq -\beta(\|x(t)\|) \quad \text{for all } x \in B_r(0).$$

Integrating both sides from t_0 to t, we obtain

$$V(x(t)) - V(x(t_0)) \leq -\int_{t_0}^{t} \beta(\|x(s)\|)ds$$
$$\implies V(x(t)) \leq V(x(t_0)) - \beta(\omega)(t-t_0) \quad (\text{since } \beta(\|x(t)\|) > \beta(\omega) > 0).$$

For large t, the RHS of the inequality becomes negative, which contradicts the fact that $V(x(t))$ is positive. Hence, $\omega = 0$. Moreover, since $\frac{dV(x)}{dt} < 0$ for $x \neq 0$ and $\frac{dV}{dt} = \theta$ at $x = \theta$, $V(x(t))$ is decreasing, and $\lim_{t\to\infty} V(x(t)) = 0$, which implies $\lim_{t\to\infty} x(t) = \theta$. □

Remark 8.5.13 In conditions (i) and (ii), "positive definite" and "negative definite" may be replaced with "negative definite" and "positive definite," respectively.

***Example* 8.5.14** Analyze the stability of the equilibrium $(0,0)$ of the system

$$\begin{cases} \dot{x}_1 = -x_2 - x_1^3, \\ \dot{x}_2 = x_1 - x_2^3. \end{cases}$$

Solution Let $V(x_1,x_2) = x_1^2 + x_2^2$ be a Lyapunov function. Then $\frac{dV}{dt} = -2(x_1^4 + x_2^4)$. From Theorem 8.5.12, the system is asymptotically stable at $(0,0)$. **Note that the corresponding linear system is stable but not asymptotically stable at (0,0).**

***Example* 8.5.15** Consider the system

$$\begin{cases} \dot{x}_1 = -kx_1 - 3x_2, \\ \dot{x}_2 = kx_1 - 2x_2. \end{cases}$$

Using $V(x_1,x_2) = -(x_1^2 + x_2^2)$ as a Lyapunov function, find all possible values of k for which the system is asymptotically stable at $(0,0)$.

Solution The given function $V(x_1,x_2)$ is negative definite. Hence, if $\frac{dV}{dt}$ is positive definite in a neighborhood of $(0,0)$, then the system is asymptotically stable at $(0,0)$. Now

$$\frac{dV}{dt} = \frac{\partial V}{\partial x_1}\dot{x}_1 + \frac{\partial V}{\partial x_2}\dot{x}_2 = -2x_1(-kx_1 - 3x_2) - 2x_2(kx_1 - 2x_2)$$

$$= 2kx_1^2 + 4x_2^2 + (6-2k)x_1x_2$$
$$= 3x_2^2 + ((3-k)x_1 + x_2)^2 - (9 + k^2 - 8k)x_1^2.$$

Hence, $\frac{dV}{dt} > 0$ if $k^2 - 8k + 9 \leq 0$, i.e., $\frac{8-\sqrt{28}}{2} < k < \frac{8+\sqrt{28}}{2}$.

Note that the given system is linear and it is asymptotically stable for all $k > 0$. This can be shown by finding the eigenvalues of the coefficient matrix. However the Lyapunov function $V(x_1, x_2) = -(x_1^2 + x_2^2)$ provides only a restricted region for the values of k. Instead, if we consider the Lyapunov function $V(x_1, x_2) = -(kx_1^2 + 3x_2^2)$, then $V < 0$ and $\frac{dV}{dt} > 0$ for all $k > 0$. This example illustrates that different Lyapunov functions may provide different conditions on the parameters involved in the system equations.

Theorem 8.5.16 The system $\dot{x} = f(x)$, $f(\theta) = \theta$, is unstable at $x = \theta$ if there exists a Lyapunov function $V(x)$ satisfying the following:

(*i*) $\frac{dV}{dt}$ is positive definite in a neighborhood $B_r(\theta)$,

(*ii*) In every neighborhood of the origin contained in $B_r(\theta)$, there exists at least one x such that $V(x)$ takes a positive value.

Proof. Let $x(t)$ be a solution of the given system with initial condition $x(t_0)$ such that $V(x(t_0)) > 0$. Such a choice is possible due to condition (*ii*). If we assume that the trivial solution $x = \theta$ is stable, then for some $\varepsilon > 0$,

$$\|x(t)\| \leq \varepsilon \quad \text{for all } t \geq t_0. \tag{8.51}$$

Since V is a continuous function, $V(x(t))$ is bounded. Since $\frac{dV}{dt}$ is positive definite in $B_r(\theta)$, there exists a function ϕ of class-K such that

$$\frac{d}{dt}V(x(t)) \geq \phi(\|x(t)\|). \tag{8.52}$$

Hence by integration ,

$$V(x(t)) \geq V(x(t_0)) + \int_{t_0}^{t} \phi(\|x(s)\|)ds.$$

Since $V(x(t_0)) > 0$ and $\phi(\cdot) \geq 0$, it follows from (8.52) that $V(x(t)) \geq V(x(t_0)) > 0$ for all t. Hence, there exists an $m > 0$ such that

$$\|x(t)\| \geq m \quad \text{for all } t \geq t_0. \tag{8.53}$$

From (8.51) and (8.53), we infer that for all t,

$$m \leq \|x(t)\| \leq \varepsilon. \tag{8.54}$$

From (8.52) and (8.53), we get $\frac{dV(x(t))}{dt} \geq \phi(\|x(t)\|) \geq \phi(m)$. Hence, $V(x(t)) > V(x(t_0)) + \phi(m)(t - t_0) \to \infty$ as $t \to \infty$, which contradicts the fact that $V(x(t))$ is bounded on $\|x\| \leq \varepsilon$. Hence, the trivial solution is unstable. □

Remark 8.5.17 In Theorem 8.5.16, the terms "positive definite" and "positive value" in (i) and (ii) may be replaced by "negative definite" and "negative value," respectively. The proof will be similar, with the inequalities suitably reversed.

***Example* 8.5.18** Using the Lyapunov function $V(x_1, x_2) = x_1^2 - 3x_2^2$, show that the system

$$\begin{cases} \dot{x}_1 = 3x_1^2 x_2 + x_1 x_2^2 - x_1^3, \\ \dot{x}_2 = x_1^3 + x_2 \end{cases}$$

is unstable at $(0,0)$.

Solution Note that $V(x_1, x_2) = x_1^2 - 3x_2^2$ takes both positive and negative values in every neighborhood of $(0,0)$. It can be seen that $\frac{dV}{dt} = 2x_1\dot{x}_1 - 6x_2\dot{x}_2 = x_2^2(2x_1^2 - 3) - 2x_1^4$ is negative definite in the disc $x_1^2 + x_2^2 < \frac{3}{2}$. Hence, from Theorem 8.5.16 and Remark 8.5.17, the system is unstable at $(0,0)$.

***Example* 8.5.19** Show that the following system

$$\begin{cases} \dot{x}_1 = -x_2 + x_1^3, \\ \dot{x}_2 = x_1 + x_2^3 \end{cases}$$

is unstable at $(0,0)$.

Solution Let $V = x_1^2 + x_2^2$. Then $\frac{dV}{dt} = -2x_1x_2 + 2x_2x_1 + 2x_1(x_1^3) + 2x_2(x_2^3) = 2(x_1^4 + x_2^4)$. Therefore, both V and $\frac{dV}{dt}$ are positive definite. Hence, from Theorem 8.5.16, the system is unstable at $(0,0)$. Note that the linearized system is stable but not asymptotically stable at $(0,0)$.

***Example* 8.5.20** Consider the following system, which represents the equation of motion of a simple pendulum (see Section 1.3, equation (1.14)):

$$\begin{cases} \dot{x}_1 = x_2, \\ \dot{x}_2 = -k \sin x_1. \end{cases}$$

For this system, $(0,0)$ is an equilibrium point. Consider the Lyapunov function

$$V(x_1, x_2) = \frac{x_2^2}{2} + k(1 - \cos x_1).$$

It can be easily seen that V is positive definite when $\sqrt{x_1^2+x_2^2} < 2\pi$. Now,

$$\frac{dV}{dt} = x_2(-k\sin x_1) + (k\sin x_1)x_2 = 0.$$

Hence, by Theorem 8.5.9, the system is stable at $(0,0)$. Since $\frac{dV}{dt} = 0$ along any solution $(x_1(t), x_2(t))$, $V(x_1(t), x_2(t))$ is constant along any solution. If $x_1(0) = \frac{\pi}{3}$ and $x_2(0) = 0$, then $V(x_1,x_2) = \frac{K}{2}$, which implies $x_2^2 = K(2\cos x_1 - 1)$. Note that this represents a closed curve on the x_1-x_2 phase plane around the equilibrium $(0,0)$. Moreover, $(0,0)$ does not lie on this curve. Hence, the equilibrium point $(0,0)$ is stable but not asymptotically stable. Similarly for any initial point in the neighborhood of (0, 0), the solution can be shown to be a closed curve containing the origin. This proves the stability of the equilibrium point (0, 0).

It can be seen that the linearized system

$$\begin{cases} \dot{x}_1 = x_2, \\ \dot{x}_2 = -kx_1, \end{cases}$$

is also stable at $(0,0)$.

***Example* 8.5.21** Analyze the stability of the system

$$\begin{cases} \dot{x}_1 = x_2, \\ \dot{x}_2 = -k\sin x_1 - k_1x_2 \end{cases}$$

at the equilibrium point $(0,0)$.

Solution Consider the Lyapunov function

$$V(x_1,x_2) = \frac{x_2^2}{2} + k(1-\cos x_1).$$

Since $V(x_1,x_2)$ is positive definite and $\frac{dV}{dt} = -k_1x_2^2$ is negative semi-definite in the region $\sqrt{x_1^2+x_2^2} < 2\pi$, from Theorem 8.5.9, the system is stable at $(0,0)$. However, we can establish asymptotic stability by linearizing the system.

Using Taylor series expansion, the system can be written as

$$\begin{cases} \dot{x}_1 = x_2, \\ \dot{x}_2 = -kx_1 - k_1x_2 + g(x_1), \end{cases}$$

where $g(x_1) = k(\frac{x_1^3}{3!} - \frac{x_1^5}{5!} + \cdots)$. Note that the linearized system is asymptotically stable at $(0,0)$. Hence, by Theorem 8.4.2, the given system is also asymptotically stable at $(0,0)$.

In the above example, even though $\frac{dV}{dt} = -kx_2^2$ is negative semi-definite, we can conclude that (0,0) is asymptotically stable using the following argument. If $x_2(t) \equiv 0$, then $\dot{x}_2(t) \equiv 0$. Hence, from the second equation of the system, we get $\sin x_1(t) = 0$, implying $x_1(t) \equiv 0$ for all t. Thus, $\frac{dV}{dt} = 0$ only when $(x_1, x_2) = (0,0)$ in the region $\sqrt{x_1^2 + x_2^2} < 2\pi$. If $(x_1(t), x_2(t))$ is a nontrivial solution, then $x_2(t)$ cannot be identically zero unless $x_1(t) \equiv 0$ for all t. Hence, $\frac{dV(t)}{dt} = -k_1 x_2^2(t)$ implies that $V(x_1(t), x_2(t))$ is decreasing until it reduces to 0. Hence, from Theorem 8.5.12, the system is asymptotically stable at $(0,0)$.

The above result follows directly from LaSalle's theorem, which is stated below without proof.

Theorem 8.5.22 (LaSalle's theorem) The equilibrium point $x = \theta$ of the system $\dot{x} = f(x)$ is asymptotically stable if $V(x) > 0$ for all $x \in B_r(0)$, $\frac{dV(x)}{dt} \leq 0$ in a bounded region containing θ, and $\frac{dV(x)}{dt}$ does not vanish along any trajectory other than $x = \theta$.

Stability Analysis Using Matrix Equation

Consider the linear autonomous system

$$\dot{x} = Ax, \tag{8.55}$$

where A is an $n \times n$ nonsingular matrix. Then $x = 0$ is an isolated equilibrium point. Let P be a real symmetric positive definite (RSPD) $n \times n$ constant matrix, and let

$$V(x) = x'Px. \tag{8.56}$$

If x is a solution of the system (8.55), then

$$\frac{dV(x)}{dt} = \dot{x}'Px + x'P\dot{x} = (x'A')Px + x'P(Ax) = x'(A'P + PA)x = -x'Qx, \tag{8.57}$$

where

$$-Q = A'P + PA. \tag{8.58}$$

If Q is an RSPD matrix, then (8.56) and (8.57) imply that $V(x)$ is positive definite and $\frac{dV}{dt}$ is negative definite. Hence from Theorem 8.5.12, the asymptotic stability of the system (8.55) at $x = \theta$ is guaranteed. If Q is real symmetric positive semi definite (RSPSD) and P is RSPD, then from Theorem 8.5.9, it follows that $x = \theta$ is stable.

If P is a real symmetric matrix with at least one negative eigenvalue and Q is an RSPD matrix, then (8.55) is unstable at $x = \theta$. This follows from Theorem 8.5.16.

To find a Lyapunov function for a linear system of the form (8.55) to determine the stability property of the equilibrium $x = \theta$, one can follow the procedure stated in the theorem below.

Theorem 8.5.23 Let Q be a known matrix and $P = \{P_{ij}\}_{i,j=1,2,\ldots,n}$ be a real matrix that satisfies the matrix equation (8.58), i.e., $A'P + PA = -Q$. Then the following statements hold.

(*i*) If Q is RSPD and the solution P of the equation (8.58) turns out to be RSPD, then the system (8.55) is asymptotically stable at $x = \theta$.

(*ii*) If Q is RSPD and the solution P of the equation (8.58) has at least one negative eigenvalue, then $x = \theta$ is unstable.

(*iii*) If Q is RSPSD and the solution P of the equation (8.58) is RSPD, then by Theorem 8.5.9, the system (8.55) is stable at $x = \theta$.

Proof. The proof follows directly from equations (8.55)–(8.58) and Theorems 8.5.9, 8.5.12, and 8.5.16. □

Remark 8.5.24 (*i*) $V(x) = x'Px$ can be considered as a Lyapunov function for the above cases.

(*ii*) In Theorem 8.5.23(*i*), "RSPD matrices" can be replaced by "real symmetric negative definite (RSND) matrices". In Theorem 8.5.23(*ii*), "Q is RSPD" can be replaced by "Q is RSND". In that case, "at least one negative eigenvalue" must be replaced by "at least one positive eigenvalue". In Theorem 8.5.23(*iii*), "RSPSD" can be replaced by "real symmetric negative semi definite (RSNSD)" and "RSPD" can be replaced by "RSND".

***Example* 8.5.25** Show that the system given below is asymptotically stable at $(0, 0)$ using a suitable Lyapunov function:

$$\begin{cases} \dot{x}_1 = -3x_1 + x_2, \\ \dot{x}_2 = -2x_1. \end{cases}$$

Solution Here, the system matrix is $A = \begin{bmatrix} -3 & 1 \\ -2 & 0 \end{bmatrix}$. Let us choose an RSPD matrix Q as $Q = \begin{bmatrix} 1 & 0 \\ 0 & 2 \end{bmatrix}$, and an unknown real symmetric matrix P as $P = \begin{bmatrix} a & b \\ b & c \end{bmatrix}$. The matrix equation $A'P+PA=-Q$ gives $\begin{bmatrix} -6a-4b & a-3b-2c \\ a-3b-2c & 2b \end{bmatrix} = \begin{bmatrix} -1 & 0 \\ 0 & -2 \end{bmatrix}$. Solving this,

we get $P = \begin{bmatrix} \frac{5}{6} & -1 \\ -1 & \frac{23}{12} \end{bmatrix}$. It can be verified that P is RSPD. Hence, by Theorems 8.5.12 and 8.5.23, the system is asymptotically stable at $(0,0)$, and the Lyapunov function is

$$V(x_1,x_2) = \begin{bmatrix} x_1 & x_2 \end{bmatrix} P \begin{bmatrix} x_1 \\ x_2 \end{bmatrix} = \frac{5}{6}x_1^2 - 2x_1x_2 + \frac{23}{12}x_2^2.$$

Note. By choosing a different RSPD matrix Q, another Lyapunov function can be found that guarantees the asymptotic stability of the system at $(0,0)$.

***Example* 8.5.26** Find a Lyapunov function for the system

$$\begin{cases} \dot{x}_1 = 2x_1 + 4x_2, \\ \dot{x}_2 = x_1 - x_2, \end{cases}$$

and show that the equilibrium $(0,0)$ is unstable.

Solution Let $Q = \begin{bmatrix} 2 & 0 \\ 0 & 4 \end{bmatrix}$ and $P = \begin{bmatrix} a & b \\ b & c \end{bmatrix}$. Solving the matrix equation $A'P + PA = -Q$, we get $P = \begin{bmatrix} -\frac{1}{2} & 0 \\ 0 & 2 \end{bmatrix}$, which has a negative eigenvalue. Hence, by Theorem 8.5.16, the equilibrium $(0,0)$ is unstable.The required Lyapunov function is $V(x_1,x_2) = -\frac{1}{2}x_1^2 + 2x_2^2$ with $\frac{dV}{dt} = 2x_1^2 + 4x_2^2$.

Theorem 8.5.27 Consider the nonlinear system

$$\begin{cases} \dot{x} = Ax + f(x), \\ f(\theta) = \theta \end{cases} \tag{8.59}$$

and the corresponding linear system

$$\dot{x} = Ax. \tag{8.60}$$

Let $f : \mathbb{R}^n \to \mathbb{R}^n$ satisfy the condition

$$\lim_{x \to \theta} \frac{f(x)}{||x||} = \theta \in \mathbb{R}^n. \tag{8.61}$$

Then

(*i*) If $x = \theta$ is asymptotically stable for the linear system (8.60), then (8.59) is also asymptotically stable at $x = \theta$.

(*ii*) If $x = \theta$ is unstable for (8.60), then (8.59) is also unstable at $x = \theta$.

Remark 8.5.28 Note that statement (*i*) is the same as that of Theorem 8.4.2. Here, an alternative proof is provided. Statement (*ii*) cannot be proved using the steps of Theorem 8.4.2. Also see Theorem 8.4.10 and Remark 8.4.12 for more details.

Proof (of Theorem 8.5.27). (*i*) If the system (8.60) is asymptotically stable at $x = \theta$, then from Theorem 8.5.23, for any RSPD matrix Q, there exists an RSPD matrix P such that $A'P + PA = -Q$. A Lyapunov function for (8.60) is

$$V(x) = x'Px \text{ with } \frac{dV}{dt} = -x'Qx. \tag{8.62}$$

Considering $V(x)$ as the Lyapunov function for the nonlinear system (8.59) as well, we have

$$\begin{aligned}
\frac{dV}{dt} &= \dot{x}'Px + x'P\dot{x} \\
&= (Ax + f(x))'Px + x'P(Ax + f(x)) \\
&= x'A'Px + [f(x)]'Px + x'PAx + x'Pf(x) \\
&= x'(A'P + PA)x + 2x'Pf(x) \quad (\text{since } (f(x))'Px = x'Pf(x)) \\
&= -x'Qx + 2x'Pf(x).
\end{aligned} \tag{8.63}$$

Since Q is positive definite, there exists an $\alpha > 0$ such that

$$x'Qx > \alpha\|x\|^2. \tag{8.64}$$

Condition (8.61) implies that for any $\varepsilon > 0$, there exists a $\delta > 0$ such that

$$\frac{\|f(x)\|}{\|x\|} < \varepsilon \quad \text{whenever } \|x\| < \delta.$$

Thus,

$$|2x'Pf(x)| \leq 2\|P\|\|x\|\|f(x)\| \leq 2\|P\|\varepsilon\|x\|^2 \quad \text{when } \|x\| < \delta. \tag{8.65}$$

From (8.63), (8.64), and (8.65), we get

$$\frac{dV}{dt} < -\alpha\|x\|^2 + 2\|P\|\varepsilon\|x\|^2 < -(\alpha - 2\|P\|\varepsilon)\|x\|^2. \tag{8.66}$$

The RHS is negative when ε is chosen such that $\varepsilon < \frac{\alpha}{2\|P\|}$. Since $V > 0$ and $\frac{dV}{dt} < 0$ for $\|x\| < \delta$, from Theorem 8.5.12, the system (8.59) is asymptotically stable at $x = \theta$.
(ii) If (8.60) is unstable at $x = \theta$, then from Theorem 8.5.23, it follows that for a given positive definite matrix Q, the solution P to the equation (8.58) is either negative definite or indefinite (having both positive and negative eigenvalues). By taking $V(x)$ given in (8.62) as the Lyapunov function for (8.59) and following the same steps as in (8.63)–(8.66), we obtain that $\frac{dV}{dt} < 0$ while either $V < 0$ or V is indefinite. Thus, the instability of $x = \theta$ is established (using Theorem 8.5.16). □

***Example* 8.5.29** Consider the nonlinear system

$$\begin{cases} \dot{x}_1 = x_2, \\ \dot{x}_2 = -x_1 - x_2 + (x_1 + 2x_2)(x_2^2 - 1). \end{cases}$$

Show that the system is asymptotically stable at $(0,0)$ using the Lyapunov function $V(x_1,x_2) = 5x_1^2 + 2x_1x_2 + 2x_2^2$.

Solution The given Lyapunov function $V = 4x_1^2 + x_2^2 + (x_1 + x_2)^2$ is positive definite. If we show that $\frac{dV}{dt}$ is negative definite, then the system is asymptotically stable at $(0,0)$. Now $\frac{dV}{dt} = 10x_1\dot{x}_1 + 2x_1\dot{x}_2 + 2x_2\dot{x}_1 + 4x_2\dot{x}_2 = -12x_2^2 - 4x_1^2 + 8x_2^4 + 8x_1x_2^3 + 2x_1^2x_2^2$. If $\frac{dV}{dt} < 0$, then $8x_2^4 + 8x_1x_2^3 + 2x_1^2x_2^2 < 12x_2^2 + 4x_1^2$. Set $x_1 = r\cos\phi$ and $x_2 = r\sin\phi$. Then the inequality becomes $r^4(8\sin^4\phi + 8\cos\theta\sin^3\phi + 2\cos^2\phi\sin^2\theta) < r^2(4 + 8\sin^2\phi)$. If $r < \frac{1}{4}$, then $\frac{1}{2}\sin^4\phi + \frac{1}{2}\cos\phi\sin^3\phi + \frac{1}{8}\cos^2\phi\sin^2\phi < 4 + 8\sin^2\phi$, which is true for all ϕ. Hence, in the neighborhood $x_1^2 + x_2^2 < \frac{1}{16}$, $V(x_1,x_2) > 0$ and $\frac{d}{dt}V(x_1,x_2) < 0$, which imply the asymptotic stability at $(0,0)$.

Stabilizability

If a dynamical system is unstable at an equilibrium point, and if it is possible to introduce a control on the system such that the resulting solution becomes stable at that equilibrium, then the system is called **stabilizable**. For example, consider a linear system $\dot{x} = Ax$, where A is an $n \times n$ matrix. If any of the eigenvalues of A has a positive real part, then the equilibrium $x = \theta$ is unstable. To stabilize the system at $x = \theta$, let us introduce a control $u(t) \in \mathbb{R}^m$ for each t and a suitable control matrix B such that the resulting control system is $\dot{x} = Ax + Bu$. If this system is controllable, then it is stabilizable, as we can find a feedback control $u = Kx$ such that $A + BK$ has negative eigenvalues (see Theorem 5.5.1), which guarantees that $x(t) \to \theta$ as $t \to \infty$.

***Example* 8.5.30** Consider the equation of motion of a harmonic oscillator (see Section 1.3)

$$\frac{d^2x}{dt^2} = -kx, \tag{8.67}$$

where $k > 0$ is a constant, with the initial condition $x(0) = L$ and $\frac{dx}{dt} = 0$ at $t = 0$. Show that the system is stable but not asymptotically stable at $x = \dot{x} = 0$. Find a suitable control $u(t)$ to make this system asymptotically stable at $x = \dot{x} = 0$.

Solution Let $x = x_1$ and $\dot{x} = x_2$. Then, from (8.67) we obtain the system

$$\begin{bmatrix} \dot{x}_1 \\ \dot{x}_2 \end{bmatrix} = \begin{bmatrix} 0 & 1 \\ -k & 0 \end{bmatrix} \begin{bmatrix} x_1 \\ x_2 \end{bmatrix}, \tag{8.68}$$

whose solution is

$$\begin{bmatrix} x_1(t) \\ x_2(t) \end{bmatrix} = \begin{bmatrix} a\cos(\sqrt{k}t) - b\sin(\sqrt{k}t) \\ -a\sqrt{k}\sin(\sqrt{k}t) - b\sqrt{k}\cos(\sqrt{k}t) \end{bmatrix},$$

where a and b are real constants such that $a = x_1(0) = L$ and $b = -\frac{x_2(0)}{\sqrt{k}} = 0$. From this, we obtain $x_1^2(t) + \frac{x_2^2(t)}{k} = L^2$, which implies that any trajectory (solution) of (8.68) forms an ellipse with the center $(0,0)$. Hence, the system (8.68) is stable but not asymptotically stable.

Now, let $u(t)$ be a control (force) applied on the system so that the RHS of the equation (8.67) becomes $-kx + u$. Then, the resulting control system is

$$\begin{bmatrix} \dot{x}_1 \\ \dot{x}_2 \end{bmatrix} = \begin{bmatrix} 0 & 1 \\ -k & 0 \end{bmatrix} \begin{bmatrix} x_1 \\ x_2 \end{bmatrix} + \begin{bmatrix} 0 \\ 1 \end{bmatrix} u(t). \tag{8.69}$$

Since the above system $\dot{X} = AX + Bu$ is controllable (from the Kalman rank condition), it is possible to find a linear feedback control $u(t) = KX = \alpha x_1(t) + \beta x_2(t)$, where $K = \begin{bmatrix} \alpha & \beta \end{bmatrix}$ for some suitable constants α and β such that $A + BK$ has negative eigenvalues (see Theorem 5.5.1). For example, by choosing $\alpha = k - 5$ and $\beta = -6$, the eigenvalues of $A + BK$ are -1 and -5, and hence the solution of (8.69) tends to the zero vector as $t \to \infty$. Thus, the control function $u(t) = (k-5)x_1 - 6x_2$ makes the system (8.69) asymptotically stable at the origin. Physically, this means that the control force $u(t)$ causes the oscillator to stop at $x = 0$ as $t \to \infty$.

***Example* 8.5.31** Consider the control system

$$\begin{cases} \dot{x}_1 = x_2, \\ \dot{x}_2 = -k\sin x_1 - k_1 x_2 + u, \end{cases}$$

representing the equation of motion of a simple pendulum under the torque u. If, at time $t = 0$, the pendulum is at rest with $x_1(0) = 0$ and $x_2(0) = 0$, and if the desired motion is $x_{1d}(t) = \frac{\pi}{4}t$ for $0 \le t \le 4$, find a suitable control $u(t)$ such that the pendulum tracks the desired trajectory.

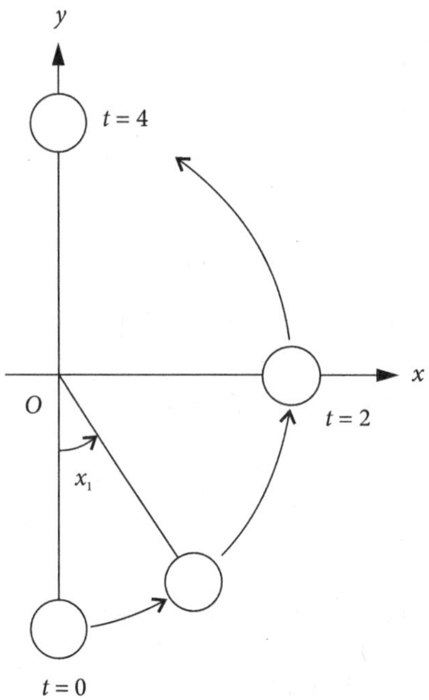

Figure 8.10 The desired trajectory of the pendulum

Solution Let $e_1 = x_1 - x_{1d}$ and $e_2 = x_2 - x_{2d}$ be the errors representing the difference between the actual state and the desired state at time t. To track the desired states x_{1d} and x_{2d}, we need to find a control $u(t)$ such that both $e_1(t)$ and $e_2(t)$ tend to zero. In terms of e_1 and e_2, the system can be expressed as

$$\begin{cases} \dot{e}_1 = e_2, \\ \dot{e}_2 = -k\sin(e_1 + x_{1d}) - k_1(e_2 + x_{2d}) + u - \dot{x}_{2d}. \end{cases}$$

Choosing the control u as $u = \dot{x}_{2d} + k_1 x_{2d} + k\sin(e_1 + x_{2d}) - ae_1$ for some constant $a > 0$, we obtain the linear system

$$\begin{cases} \dot{e}_1 = e_2, \\ \dot{e}_2 = -ae_1 - k_1 e_2. \end{cases}$$

It can be shown that the above system is asymptotically stable.

The following result demonstrates an application of Theorem 8.5.23 for finding a feedback control $u(t)$ to stabilize a linear control system. Assume that the following linear time-invariant system

$$\dot{x} = Ax + Bu \tag{8.70}$$

is controllable. Then, the controllability Gramian matrix

$$W_T = \int_0^T e^{-At} BB^* e^{-A^*t} dt \tag{8.71}$$

is positive definite for any $T > 0$. Define the linear feedback control law

$$u = -B^* W_T^{-1} x = K_T x. \tag{8.72}$$

It can be shown that this feedback control law stabilizes the system (8.70). From (8.71), we have

$$\frac{dW_T}{dT} = e^{-AT} BB^* e^{-A^*T}$$

and

$$-AW_T - W_T A^* = \int_0^T \frac{d}{dt} \left(e^{-At} BB^* e^{-A^*t} \right) dt = e^{-AT} BB^* e^{-A^*T} - BB^*.$$

Thus,

$$\begin{aligned} AW_T + W_T A^* - 2BB^* &= \left(A - BB^* W_T^{-1}\right) W_T + W_T \left(A - BB^* W_T^{-1}\right)^* \\ &= -\left(e^{-AT} BB^* e^{-A^*T}\right) - BB^* < 0. \end{aligned}$$

Let us denote W_T by P and the RHS by $-Q$. Then, from Theorem 8.5.23(i), we infer that the matrix $\left(A - BB^* W_T^{-1}\right)^*$ has eigenvalues with negative real parts, which implies the same for the matrix $(A - BB^* W_T^{-1})$. Hence, the feedback control $u(t) = -B^* W_T^{-1} x(t)$ for some fixed $T > 0$ renders the system (8.70) asymptotically stable at $x = \theta$.

Exercise 8.5.1 If $A = \begin{bmatrix} 0 & 0 \\ 1 & 0 \end{bmatrix}$ and $B = \begin{bmatrix} 1 \\ 0 \end{bmatrix}$, find a feedback control $u(t) = Kx(t)$ using (8.72) with $T = 1$ such that the system $\dot{x} = Ax + Bu$ is asymptotically stable at $x = 0$.

Exercise 8.5.2 Using a Lyapunov function, show that the system

$$\begin{cases} \dot{x}_1 = x_1 + x_1 x_2, \\ \dot{x}_2 = x_2 - x_1^2 \end{cases}$$

is unstable at $(0,0)$.

Exercise 8.5.3 Show that the system

$$\begin{cases} \dot{x}_1 = x_2, \\ \dot{x}_2 = -\sin x_1 + \alpha x_2 \end{cases}$$

is asymptotically stable at $(0,0)$ if $\alpha < 0$ and unstable at $(0,0)$ if $\alpha > 0$, using: (i) linearization, (ii) the Lyapunov function $V(x_1,x_2) = (1 - \cos x_1) + \frac{x_2^2}{2}$.

Exercise 8.5.4 For the dynamical system

$$\begin{cases} \dot{x}_1 = 4x_2, \\ \dot{x}_2 = -x_1 - 2x_2^3, \end{cases}$$

which of the following Lyapunov functions provides results on stability at $(0,0)$: (i) $x_1^2 + x_2^2$; (ii) $x_1^2 + 4x_2^2$; (iii) $4x_1^2 + x_2^2$.

8.6 Lyapunov Theory for Time-varying Systems

Let S be an open set in $\mathbb{R}^n$ containing the origin. Let $\mathbb{R}_0^+ = \{t \in \mathbb{R} : t \geq 0\}$.

Definition 8.6.1 A function $V : \mathbb{R}_0^+ \times S \to \mathbb{R}$ is **positive definite (negative definite)** if $V(t,\theta) = 0$ for all $t \geq 0$, and there exists a function ϕ of class-K such that

$$V(t,x) \geq \phi(\|x\|)\ (V(t,x) \leq -\phi(\|x\|)) \quad \text{for all } (t,x) \in \mathbb{R}_0^+ \times S. \tag{8.73}$$

Definition 8.6.2 A function $V : \mathbb{R}_0^+ \times S \to \mathbb{R}$ is called **positive semi-definite (negative semi-definite)** if $V(t,\theta) = 0$ for all $t \geq 0$, and $V(t,x) \geq 0$ (≤ 0) for all $(t,x) \in \mathbb{R}_0^+ \times S$.

Definition 8.6.3 A function $V : \mathbb{R}_0^+ \times S \to \mathbb{R}$ is said to be **decrescent** if there exists a function ψ of class-K such that

$$V(t,x) \leq \psi(\|x\|) \quad \text{for all } (t,x) \in \mathbb{R}_0^+ \times S. \tag{8.74}$$

***Example* 8.6.4** For $x = (x_1, \dot{x}_2) \in \mathbb{R}^2$ and $t \geq 0$, the function

(*i*) $V(t,x) = (t+2)x_1^2 + 3x_2^2$ is positive definite since $V(t,x) \geq 2(x_1^2 + x_2^2) = \phi(r)$, where $r = \sqrt{x_1^2 + x_2^2}$ and $\phi(r) = 2r^2$ is of class-K.

(*ii*) $V(t,x) = t^2x_1^2 + x_2^2$ is positive but **not** positive definite, since there is no function of class-K that satisfies inequality (8.73). Additionally, $V(t,x)$ is not decrescent, as there does not exist any function ψ of class-K satisfying (8.74).

(*iii*) $V(t,x) = (1+\cos t)x_1^2 + x_2^2$ is positive definite in $[0, \pi/2) \times \mathbb{R}^2$, since $(1+\cos t)x_1^2 + x_2^2 \geq x_1^2 + x_2^2$ on $[0, \pi) \times \mathbb{R}^2$. However, for $t \in \mathbb{R}_0^+$, $V(t,x)$ is positive semi-definite, as $V(t,x)$ can take zero values at points other than the origin. For example, $V(\pi, (x_1, 0)) = 0$ for any x_1.

(*iv*) $V(t,x) = 4x_1^2 - x_1^4 + (1+\sin^2 t)x_2^2$ is positive definite and decrescent in $\mathbb{R}_0^+ \times S$, where $S = \{x \in \mathbb{R}^2 : \|x\| < 1\}$. Here, $x_1^2 + x_2^2 \leq V(t,x) \leq 4(x_1^2 + x_2^2)$.

(*v*) $V(t,x) = \frac{1}{(t+1)^2}x_1^2 + 2x_2^2$ is decrescent since $V(t,x) \leq 2(x_1^2 + x_2^2) = \psi(r)$, where $\psi(r) = 2r^2$ and $r = \sqrt{x_1^2 + x_2^2}$. However, $V(t,x)$ is not positive definite because $1/(t+1)$ tends to zero as t tends to infinity.

(*vi*) $V(t,x) = (1+\sin^2 t)x_1^2 + (1+\cos^2 t)x_2^2$ is both positive definite and decrescent since $x_1^2 + x_2^2 \leq V(t,x) \leq 2(x_1^2 + x_2^2)$.

Lyapunov Theorems for Stability

Consider the time-varying system

$$\dot{x} = f(t,x) \quad \text{on } \mathbb{R}_0^+ \times \mathbb{R}^n. \tag{8.75}$$

Let $x = \theta$ be an isolated equilibrium point, i.e., $f(t,\theta) = \theta$ for all $t \in \mathbb{R}_0^+$.

Theorem 8.6.5 For the system (8.75), the trivial solution $x = \theta$ is stable if there exists a Lyapunov function $V(t,x)$ on $\mathbb{R}_0^+ \times S$ satisfying:

(*i*) $V(t,x)$ is positive definite,

(*ii*) $\frac{dV}{dt}$ is negative semi-definite.

Proof. The proof follows the same reasoning as that of Theorem 8.5.9. □

Note. If $x = (x_1, x_2, \ldots, x_n)' \in \mathbb{R}^n$, then

$$\frac{dV}{dt} = \frac{\partial V}{\partial t} + \frac{\partial V}{\partial x_1}\dot{x}_1 + \frac{\partial V}{\partial x_2}\dot{x}_2 + \cdots + \frac{\partial V}{\partial x_n}\dot{x}_n. \tag{8.76}$$

Theorem 8.6.6 The time-varying system (8.75) is asymptotically stable at $x = \theta$ if there exists a Lyapunov function $V(t,x)$ on $\mathbb{R}_0^+ \times S$ satisfying:

(*i*) $V(t,x)$ is **positive definite and decrescent,**

(*ii*) $\frac{dV}{dt}$ is negative definite.

Proof. From Theorem 8.6.5, the stability at $x = \theta$ follows. Hence, for any $\varepsilon > 0$, there exists a $\delta > 0$ such that

$$\|x(t)\| < \varepsilon \quad \text{for } t \geq t_0 \text{ whenever } \|x(t_0)\| < \delta.$$

If $x = \theta$ is not asymptotically stable, then there exists a constant $\omega > 0$ such that

$$\omega < \|x(t)\| < \varepsilon \quad \text{for all } t \geq t_0.$$

Now, following the steps in Theorem 8.5.12, we obtain $\omega = 0$, and hence,

$$0 < \|x(t)\| < \varepsilon \quad \text{for all } t \geq t_0. \tag{8.77}$$

It remains to show that $\lim_{t\to\infty} \|x(t)\| = 0$. Since $V(t,x)$ is decrescent, there exists a function ψ of class-K such that

$$V(t,x(t)) \leq \psi(\|x(t)\|) \text{ and } V(t,0) = \psi(0) = 0 \quad \text{for all } t \geq t_0. \tag{8.78}$$

Since $\frac{dV}{dt} < 0$, $V(t,x(t))$ decreases with respect to t. Hence $\lim_{t\to\infty} V(t,x(t)) = 0$, which implies that $\lim_{t\to\infty} x(t) = \theta$. Thus, the asymptotic stability at $x = \theta$ is established. □

Remark 8.6.7 The decrescent property (equation (8.78)) of $V(t,x)$ in time-varying system is useful for establishing the asymptotic stability. However, it is not a necessary condition (see Lemma 8.6.10 and Example 8.6.11). In the time-invariant case, if $V(x)$ is positive definite, then it is also decrescent.

***Example* 8.6.8** Consider the system

$$\begin{cases} \dot{x}_1 = x_2, \\ \dot{x}_2 = -a(t)x_1, \end{cases}$$

where $a(t) \geq c > 0$ and $\dot{a}(t) \leq 0$ for $t \in [0,\infty)$. Show that the equilibrium $(0,0)$ is stable.

Solution Let $V(t,x_1,x_2) = a(t)x_1^2 + x_2^2$. Then V is positive-definite and

$$\frac{dV}{dt} = \dot{a}(t)x_1^2 + 2a(t)x_1x_2 + 2x_2(-a(t)x_1) = \dot{a}(t)x_1^2,$$

which is negative semi-definite. Hence, by Theorem 8.6.5, the equilibrium $(0,0)$ is stable.

Theorem for Instability

Theorem 8.6.9 The system $\dot{x} = f(t,x)$, $f(t,0) = 0$ for all $t \geq 0$, is unstable at $x = 0$ if there exists a Lyapunov function $V(t,x)$ satisfying the following:

(*i*) $\frac{dV}{dt}$ is positive definite in S, a neighborhood of the origin,

(*ii*) In every neighborhood of the origin contained in S, there exists at least one x such that $V(t,x) > 0$ for all $t \geq 0$.

(*iii*) $|v(t,x)|$ is decrescent.

Proof. The proof is similar to that of Theorem 8.5.16. Condition (*iii*) is required to establish the boundedness of $V(t,x(t))$ using (8.51). □

The following Lemma provides sufficient conditions for non-decrescent Lyapunov function $V(t,x)$ that guarantees asymptotic stability.

Lemma 8.6.10 Let $V(t,x)$ be a Lyapunov function for the system $\dot{x} = f(t,x)$, $f(t,\theta) = \theta$, for all t. If $\frac{d}{dt}V(t,x) \leq -cV(t,x)$ for $t \geq c$ for some $c > 0$, then the equilibrium $x = \theta$ is asymptotically stable.

Proof. From the given differential inequality, we have $V(t,x) \leq e^{-c(t-t_0)}V(t_0,x)$. Hence, $\lim_{t\to\infty} V(t,x) = 0$ implies $\lim_{t\to\infty} x(t) = \theta$, which shows that $x = \theta$ is asymptotically stable. □

Example **8.6.11** Show that $\dot{x} = (t\sin t - 2t)x$ is asymptotically stable at $x = 0$, using the Lyapunov function

$$V(t,x) = x^2 \exp\left(\int_0^t (2s - s\sin s)ds\right).$$

Solution The function $V(t,x)$ is positive definite since $V(t,x) > x^2$ for all $t \geq 0$ and $x \in \mathbb{R}^+$. Moreover, $V(t,0) = 0$ for all $t \geq 0$. Now

$$\begin{aligned} \frac{dV}{dt} &= -V(t,x)(2t - t\sin t) \\ &\leq -tV(t,x) \quad (\text{as } t(2-\sin t) \geq t) \\ &\leq -cV(t,x) \quad \text{for a positive constant } c \text{ such that } t > c > 0. \end{aligned}$$

Hence, from Lemma 8.6.10, the asymptotic stability of the trivial solution $x = 0$ follows.

Note that $V(t,x)$ is not decrescent.

Gradient Method to Find a Lyapunov Function

Consider the time-invariant system

$$\dot{x} = f(x), \quad f(\theta) = \theta. \tag{8.79}$$

If $x = \theta$ is asymptotically stable, then one can find a Lyapunov function $V(x)$ such that (i) $V(x)$ is positive definite and (ii) $\frac{dV(x)}{dt}$ is negative definite. To construct such a Lyapunov function (if it exists), the gradient method outlined below is useful in some cases. Now,

$$\frac{dV}{dt} = \left(\frac{dV}{dx}\right)' \frac{dx}{dt} = G(x)'\dot{x}, \tag{8.80}$$

where $\frac{dV}{dx} = G(x) = \begin{bmatrix} \frac{\partial V}{\partial x_1} \\ \frac{\partial V}{\partial x_2} \\ \vdots \\ \frac{\partial V}{\partial x_n} \end{bmatrix}$ is the gradient of $V(x)$. Integrating (8.80) from 0 to x, we get

$$V(x) = \int_0^x G(x)'dx = \int_0^x dV.$$

The above integral is a line integral in $\mathbb{R}^n$ from the origin $(0,0,\ldots,0)$ to any $x = (x_1, x_2, \ldots, x_n)$. Since $G(x)$ is the gradient of a scalar function $V(x)$, the integration is path-independent. Hence, it can be evaluated as

$$V(x) = \int_{(0,0,\ldots,0)}^{(x_1,0,\ldots,0)} (G(z_1,0,\ldots,0))'e_1 dz_1 + \int_{(x_1,0,\ldots,0)}^{(x_1,x_2,0,\ldots,0)} (G(x_1,z_2,0,\ldots,0))'e_2 dz_2$$
$$+ \cdots + \int_{(x_1,x_2,\ldots,x_{n-1},0)}^{(x_1,x_2,\ldots,x_n)} (G(x_1,x_2,\ldots,x_{n-1},z_n))'e_n dz_n, \tag{8.81}$$

where $e_i : i = 1,2,\ldots,n$ are the standard basis vectors in R^n. In order to find $V(x)$ using the above step, we need to determine $G(x)$, whose components are $\frac{\partial V}{\partial x_i} : i = 1,2,\ldots,n$. Assume that $\frac{\partial V}{\partial x_i}$ are of the form

$$\frac{\partial V}{\partial x_i} = a_{i1}x_1 + a_{i2}x_2 + \cdots + a_{in}x_n, \quad i = 1,2,\ldots,n, \tag{8.82}$$

where a_{ij} can be suitably chosen to be either constants or functions of $x_i : i = 1,2,\ldots,n$. Since $\frac{\partial}{\partial x_j}\left(\frac{\partial V}{\partial x_i}\right) = \frac{\partial}{\partial x_i}\left(\frac{\partial V}{\partial x_j}\right)$, the coefficients a_{ij} must satisfy

$$\frac{\partial}{\partial x_j}(a_{i1}x_1 + a_{i2}x_2 + \cdots + a_{in}x_n) = \frac{\partial}{\partial x_i}(a_{j1}x_1 + a_{j2}x_2 + \cdots + a_{jn}x_n). \tag{8.83}$$

Now choose a_{ij} such that equation (8.83) is satisfied and $\frac{dV}{dt} = G(x)'\dot{x}$ is negative (or positive or semi-definite). If such a_{ij} are found, then $G(x)$ can be obtained using (8.82). Consequently, $V(x)$ can be obtained using (8.81). It remains to check whether $V(x)$ is positive definite (or negative definite or indefinite). If $V(x)$ turns out to be positive definite (or negative definite or indefinite), then it is one of the solutions to the problem. Then, Theorems 8.5.9, 8.5.12, and 8.5.16 can be used to determine the stability of the equilibrium $x = \theta$. If not, other choices of a_{ij} can be explored to find a suitable Lyapunov function $V(x)$.

Hence, the gradient method is a **trial-and-error method** for obtaining a suitable Lyapunov function.

***Example* 8.6.12** Show that the system

$$\begin{cases} \dot{x}_1 = -x_2 + x_1^3, \\ \dot{x}_2 = x_1 + x_2^3, \end{cases}$$

is unstable at $(0,0)$ by finding a Lyapunov function using the gradient method.

Solution Let $V(x_1, x_2)$ denote a Lyapunov function for the given system. Let

$$\frac{dV}{dx} = \begin{bmatrix} \frac{\partial V}{\partial x_1} \\ \frac{\partial V}{\partial x_2} \end{bmatrix} = \begin{bmatrix} a_{11}x_1 + a_{12}x_2 \\ a_{21}x_1 + a_{22}x_2 \end{bmatrix}, \tag{8.84}$$

where a_{ij} are suitable functions to be determined. Now,

$$\begin{aligned} \frac{dV}{dt} = \left[\frac{dV}{dx}\right]' \dot{x} &= (a_{11}x_1 + a_{12}x_2)(-x_2 + x_1^3) + (a_{21}x_1 + a_{22}x_2)(x_1 + x_2^3) \\ &= a_{11}x_1^4 + a_{21}x_1^2 + a_{22}x_2^4 - a_{12}x_2^2 + x_1x_2(-a_{11} + a_{12}x_1^2 + a_{22} + a_{21}x_2^2). \end{aligned}$$

By choosing

$$a_{12} = 0 = a_{21} \text{ and } a_{11} = a_{22} = 1, \tag{8.85}$$

we get

$$\frac{dV}{dt} = x_1^4 + x_2^4. \tag{8.86}$$

Moreover, from (8.84) and (8.85), we obtain

$$\frac{\partial V}{\partial x_1} = x_1 \text{ and } \frac{\partial V}{\partial x_2} = x_2, \tag{8.87}$$

which satisfy the conservative condition

$$\frac{\partial}{\partial x_2}\left(\frac{\partial V}{\partial x_1}\right)=\frac{\partial}{\partial x_1}\left(\frac{\partial V}{\partial x_2}\right).$$

Hence, from (8.81) and (8.87), we get

$$V(x_1,x_2)=\int x_1 dx_1+\int x_2 dx_2=\frac{1}{2}(x_1^2+x_2^2). \tag{8.88}$$

From (8.87) and (8.88), we observe that both V and $\frac{dV}{dt}$ are positive definite, which imply that the equilibrium $(0,0)$ is unstable (from Theorem 8.5.16).

***Example* 8.6.13** For the system

$$\begin{cases}\dot{x}_1=-4x_1+x_2^3,\\ \dot{x}_2=-x_1-x_2^3,\end{cases}$$

find a Lyapunov function $V(x_1,x_2)$ using the gradient method and analyze the stability at $(0,0)$.

Solution Let

$$\frac{dV}{dx}=\begin{bmatrix}\frac{\partial V}{\partial x_1}\\ \frac{\partial V}{\partial x_2}\end{bmatrix}=\begin{bmatrix}a_{11}x_1+a_{12}x_2\\ a_{21}x_1+a_{22}x_2\end{bmatrix}.$$

Then

$$\begin{aligned}\frac{dV}{dt}&=(a_{11}x_1+a_{12}x_2)(-4x_1+x_2^3)+(a_{21}x_1+a_{22}x_2)(-x_1-x_2^3)\\&=-4a_{11}x_1^2-4a_{12}x_1x_2+a_{11}x_1x_2^3+a_{12}x_2^4-a_{21}x_1^2-a_{21}x_1x_2^3\\&\quad-a_{22}x_1x_2-x_1^2(4a_{11}+a_{21})+x_2^4(a_{12}-a_{22}).\end{aligned}$$

Let $a_{12}=a_{21}=0$, $a_{11}>0$, and $a_{11}x_2^2=a_{22}$. Then $\frac{dV}{dt}=-4a_{11}x_1^2-a_{22}x_2^4$, $V_{x_1}=a_{11}x_1$, and $V_{x_2}=a_{22}x_2=a_{11}x_2^3$. Therefore, from (8.81), $V=\frac{a_{11}}{2}x_1^2+\frac{a_{11}}{4}x_2^4>0$. Since $V>0$ and $\frac{dV}{dt}<0$, the origin is asymptotically stable.

Other than the methods discussed above, stability analysis for a system can also be performed by directly finding the solution of the system, as illustrated in the following example.

***Example* 8.6.14** Show that the following system and the corresponding linear system are stable but not asymptotically stable at $(0,0)$

$$\begin{cases}\dot{x}_1=-x_2-x_1^3,\\ \dot{x}_2=x_1+x_2^3.\end{cases} \tag{8.89}$$

Solution For the system (8.89), the linearized system is

$$\begin{cases} \dot{x}_1 = -x_2, \\ \dot{x}_2 = x_1. \end{cases} \tag{8.90}$$

Clearly, for the linear system (8.90), the origin is stable but not asymptotically stable. Now, let us test the stability of the nonlinear system (8.89) at $(0,0)$ by finding its solution. From (8.89), we get

$$\begin{aligned}
&\frac{dx_1}{-x_2 - x_1^3} = \frac{dx_2}{x_1 + x_2^3} \\
\Longrightarrow\ &\frac{x_1 dx_1 + x_2 dx_2}{x_2^4 - x_1^4} = \frac{x_2 dx_1 + x_1 dx_2}{(x_2^2 - x_1^2)(x_1 x_2 - 1)} \\
\Longrightarrow\ &\frac{1}{2}\frac{d(x_1^2 + x_2^2)}{x_2^2 + x_1^2} = \frac{d(x_1 x_2 - 1)}{(x_1 x_2 - 1)} \\
\Longrightarrow\ &\log\sqrt{x_1^2 + x_2^2} = \log c(x_1 x_2 - 1), \quad \text{where } c \text{ is a constant} \\
\Longrightarrow\ &x_1^2 + x_2^2 = c^2(x_1 x_2 - 1)^2.
\end{aligned} \tag{8.91}$$

If (x_{01}, x_{02}) is a suitable initial point such that $x_{01}x_{02} \neq 1$, then $c^2 = \frac{x_{01}^2 + x_{02}^2}{(x_{01}x_{02} - 1)^2}$ is finite.

Set $x_1 = r\cos\phi$ and $x_2 = r\sin\phi$. Let the initial condition (x_{01}, x_{02}) be chosen such that $\phi_0 = 0$ and $x_{01}^2 + x_{02}^2 = r_0^2 < 1$. Then from (8.91), we obtain

$$c^2 = \frac{r_0^2}{(\frac{1}{2}r_0^2 \sin 2\phi_0 - 1)^2} = r_0^2. \tag{8.92}$$

Now

$$\frac{x_1^2(t) + x_2^2(t)}{(x_1(t)x_2(t) - 1)^2} = \frac{r^2}{(\frac{1}{2}r^2 \sin 2\phi - 1)^2} = c^2 \tag{8.93}$$

$$\Longrightarrow\ \frac{1}{4}r^4 c^2 \sin^2 2\phi - r^2(1 + c^2 \sin 2\phi) + c^2 = 0$$

$$\Longrightarrow\ r^2 = \frac{1 + c^2 \sin 2\phi \pm \sqrt{1 + 2c^2 \sin 2\phi}}{\frac{1}{2}c^2 \sin^2 2\phi}. \tag{8.94}$$

From (8.93), it can be seen that

$$\lim_{\phi \to 0} r^2(\phi) = c^2. \tag{8.95}$$

Hence, in (8.94), we discard the $+$ sign and take

$$r^2(\phi) = \frac{1 + c^2 \sin 2\phi - \sqrt{1 + 2c^2 \sin 2\phi}}{\frac{1}{2}c^2 \sin^2 2\phi}, \tag{8.96}$$

which satisfies (8.95). Now using Binomial expansion for $(1+2c^2\sin 2\phi)^{\frac{1}{2}}$ in (8.96), we obtain

$$\begin{aligned}
r^2 &= \frac{(1+c^2\sin 2\phi)-(1+c^2\sin 2\phi-\frac{1}{8}(2c^2\sin 2\phi)^2+\frac{1}{2}c^6\sin^3 2\phi+\cdots)}{\frac{1}{2}c^2\sin^2 2\phi} \\
&= \frac{\frac{1}{2}c^4\sin^2 2\phi-\frac{1}{2}c^6\sin^3 2\phi+\cdots}{\frac{1}{2}c^2\sin^2 2\phi} \\
&= c^2-c^4\sin 2\phi+\cdots \\
&\leq c^2(1+c^2+c^4+\cdots) \\
&= \frac{c^2}{1-c^2} \quad \text{for all } \phi
\end{aligned} \tag{8.97}$$

From (8.97) we infer that if $c^2<\frac{1}{2}$ then $r<1$. This implies that the solution $(x_1(t),x_2(t))$ lies within the unit disk for all t, if the initial condition lies within a disk of radius $<1/2$. Hence, the origin is stable. Moreover, from (8.96), it is clear that the value of r at $\phi=0$ and $\phi=2\pi$ are equal, which implies that the solution curve is a closed curve. Therefore, $r(t)$ can be zero only when $r_0=0$, i.e., the initial condition is zero. Hence, the origin is stable but NOT asymptotically stable.

Krasovskii's Method

Consider the system

$$\dot{x}=f(x), \quad f(\theta)=\theta. \tag{8.98}$$

Let $\mathbf{J}(x)$ be the Jacobian of f with respect to x given by

$$\mathbf{J}(x)=\frac{df}{dx}=\left[\frac{\partial f_i}{\partial x_j}\right]_{i,j=1,2,\ldots,n}.$$

Let $H(x)$ be the $n\times n$ matrix defined as

$$H(x)=\mathbf{J}'(x)P+P\mathbf{J}(x), \tag{8.99}$$

where P is an $n\times n$ matrix.

Theorem 8.6.15 If there exists an RSPD matrix P such that $H(x)$ defined in (8.99) is negative definite for all x, then the origin is asymptotically stable for the system (8.98). Moreover, no other vector is an equilibrium point of the system other than the origin.

Proof. Let $V(x) = f'(x)Pf(x)$. Since P is positive definite, $V(x) > 0$ for $x \neq 0$ and $V(\theta) = 0$. Moreover, we have $\frac{dV}{dt} = \left[\frac{dV}{dx}\right]' \dot{x} = \left[\frac{dV}{dx}\right]' f(x)$ and $\frac{dV}{dx} = \left[\frac{df}{dx}\right]' \frac{dV}{df} = J'(2Pf)$.

$$\text{Hence,} \frac{dV}{dt} = (2\,J'P\,f)'\,f = f'P\,J\,f + (J\,f)'\,P\,f = f'P\,J\,f + f'J'P\,f$$

$$= f'(x)(P\mathbf{J}(x) + \mathbf{J}'(x)P)f(x) = f'(x)H(x)f(x) < 0,$$

since $H(x)$ is negative definite. Thus, $x = \theta$ is asymptotically stable.

It remains to show that the origin is the only equilibrium point of the system under the assumption $H(x) < 0$. If $z \neq \theta$ is an equilibrium point, then $f(z) = \theta$. For $0 \leq a \leq 1$, consider the function $f(ax)$. Then $\frac{df(ax)}{da} = \frac{df(ax)}{d(ax)}x$ and $\int_0^1 \frac{df(ax)}{da}da = f(x) - f(0) = f(x)$. Hence,

$$x'Pf(x) = \int_0^1 x'P\frac{df(ax)}{da}da = \int_0^1 x'P\frac{df(ax)}{d(ax)}xda$$
$$= \frac{1}{2}\int_0^1 x'\left[P\frac{df(ax)}{d(ax)} + \left[\frac{df(ax)}{d(ax)}\right]'P\right]xda.$$

Thus,

$$x'Pf(x) = \frac{1}{2}\int_0^1 (x'H(ax)x)\,da. \tag{8.100}$$

Evaluating the above expression at $x = z$, we observe that the LHS is 0 as $f(z) = 0$, while the RHS is negative since $(ax)'\left[H(ax)\right](ax) < 0$ for any x. Hence, (8.100) is valid only when $x = 0$. □

***Example* 8.6.16** Consider the system

$$\begin{cases} \dot{x}_1 = x_2, \\ \dot{x}_2 = -x_2 - kg(x_1), \end{cases}$$

where $k > 0$ and $g(0) = 0$. Then $(0,0)$ is an equilibrium point. Let $x = \begin{bmatrix} x_1 \\ x_2 \end{bmatrix}$ and $f(x) = \begin{bmatrix} x_2 \\ -x_2 - kg(x_1) \end{bmatrix}$. The system can be represented as

$$\dot{x} = f(x), \quad f(\theta) = \theta. \tag{8.101}$$

Let

$$J(x) = \frac{df}{dx} = \begin{bmatrix} \frac{\partial f_1}{\partial x_1} & \frac{\partial f_1}{\partial x_2} \\ \frac{\partial f_2}{\partial x_1} & \frac{\partial f_2}{\partial x_2} \end{bmatrix} = \begin{bmatrix} 0 & 1 \\ -k\frac{dg}{dx_1} & -1 \end{bmatrix}.$$

Let $P = \begin{bmatrix} \alpha & \beta \\ \beta & \gamma \end{bmatrix}$ be positive and $H(x) = J'(x)P + PJ(x)$. Then

$$H(x) = \begin{bmatrix} -2\beta k\frac{dg}{dx_1} & \alpha - \beta - \gamma k\frac{dg}{dx_1} \\ \alpha - \beta - \gamma k\frac{dg}{dx_1} & 2(\beta - \gamma) \end{bmatrix}.$$

From Theorem 8.6.15, if $H(x)$ is negative definite, then the system is asymptotically stable at $(0,0)$. For $H(x)$ to be negative definite, the following conditions must be satisfied:

$$2\beta k\frac{dg}{dx_1} > 0 \tag{8.102}$$

and

$$det(H(x)) = -4\beta k\frac{dg}{dx_1}(\beta - \gamma) - \left(\alpha - \beta - \gamma k\frac{dg}{dx_1}\right)^2 > 0. \tag{8.103}$$

Let us choose $\alpha = \beta > 0$ and $\gamma = a\alpha$ for some $a > 1$. Then, we obtain

$$4(a-1) > ka^2\frac{dg}{dx_1}. \tag{8.104}$$

Thus, $\frac{dg}{dx_1} > 0$ along with (8.104) are sufficient conditions for the asymptotic stability at $(0,0)$. As an example, let $g(x_1) = x_1^3$, then $\frac{dg}{dx_1} = 3x_1^2 \geq 0$. Let $\alpha = \beta = 1, \gamma = 2$, and $a = 2$. Then $4(a-1) > ka^2\frac{dg}{dx_1} \Longrightarrow 4 > 4k\frac{dg}{dx_1} \Longrightarrow k\frac{dg}{dx_1} < 1 \Longrightarrow 3kx_1^2 < 1 \Longrightarrow x_1^2 < \frac{1}{3k}$. Thus, inside the circle

$$x_1^2 + x_2^2 < \frac{1}{3k}, \tag{8.105}$$

if we consider the Lyapunov function

$$V(x) = f'(x)Pf(x), \tag{8.106}$$

then by Theorem 8.6.15, $V(x)$ is positive definite, and $\frac{dV}{dt} = f'(x)H(x)f(x)$ is negative definite, which imply the asymptotic stability at $(0,0)$. From (8.106), it can be seen that

$$V(x) = -kx_1^3x_2 + (-x_2 - 2kx_1^3)(-x_2 - kx_1^3) = k^2x_1^6 + (x_2 + kx_1^3)^2,$$

is a Lyapunov function. It is positive definite, and it can be verified that $\frac{dV}{dt}$ is negative definite in the region $x_1^2 + x_2^2 < \frac{1}{3k}$.

8.7 Summary

- For the time-invariant system $\dot{x} = Ax$, the equilibrium point $x = \theta$ is
  - (i) asymptotically stable if and only if all the eigenvalues of A have negative real parts (Theorem 8.3.1).
  - (ii) unstable if and only if at least one of the eigenvalues has positive real part (Theorem 8.3.1).
- If all the eigenvalues of A have non-positive real parts and the eigenvalues with zero real parts have equal algebraic and geometric multiplicities, then $x = \theta$ is stable (Theorem 8.3.3).
- Consider the nonlinear system

$$\dot{x} = Ax + f(x), \quad f(\theta) = \theta, \tag{8.107}$$

  and the corresponding linear system

$$\dot{x} = Ax. \tag{8.108}$$

  Let f satisfy the condition (8.19). Then, if (8.108) is asymptotically stable at $x = \theta$, so is (8.107) at $x = \theta$ (Theorem 8.4.2).
- If (8.108) is unstable at $x = \theta$, then so is (8.107) at $x = \theta$ (Theorem 8.4.10).
- Consider the nonlinear autonomous system $\dot{x} = f(x)$, $f(\theta) = \theta$. Let $V : B_r(\theta) \to \mathbb{R}$ be a Lyapunov function for the system. Then the system is stable at $x = \theta$ if (i) V is positive definite and (ii) $\dot{V}$ is negative semi-definite (Theorem 8.5.9).
- The system is asymptotically stable at $x = \theta$ if (i) V is positive definite and (ii) $\dot{V}$ is negative definite (Theorem 8.5.12).
- The system is unstable at $x = \theta$ if (i) $\dot{V}$ is positive definite and (ii) V is positive definite or indefinite (Theorem 8.5.16).
- The time-varying system $\dot{x} = f(t,x)$, $f(t,\theta) = \theta$ for all $t \geq 0$, is asymptotically stable at $x = \theta$ if there exists a Lyapunov function $V(t,x)$ such that (i) V is positive definite and decrescent and (ii) $\dot{V}$ is negative definite (Theorem 8.6.6).

- The conditions on V for the stability of time-varying systems at $x = \theta$ are similar to those for time-invariant systems.

8.8 Exercise

1. If all solutions of the system $\dot{x} = Ax$ are bounded, show that all solutions of $\dot{x} = (A + B(t))x$ are also bounded, provided $\int^{\infty} \|B(t)\| dt < \infty$.

2. Let $A = \begin{bmatrix} 1 & 3 \\ 3 & 1 \end{bmatrix}$ and $B = \begin{bmatrix} 1 \\ 0 \end{bmatrix}$. Find a feedback control $u(t) = Kx(t)$ such that $\dot{x} = Ax + Bu$ is asymptotically stable at $x = 0$.

3. Analyze the stability at the equilibrium points of the system

$$\begin{cases} \dot{x} = x - xy, \\ \dot{y} = -x + xy, \end{cases}$$

using the linearization method. If the system is unstable at the critical point $(0,0)$, apply a linear feedback control $u(t)$ to the first equation to stabilize the system.

4. (*i*) Analyze the stability at $(0,0)$ of the system

$$\begin{cases} \dot{x} = y - x^3, \\ \dot{y} = -x - y^3, \end{cases}$$

using the Lyapunov function $V(x,y) = x^2 + y^2$.

(*ii*) Analyze the stability at the equilibrium point $(0,0)$ of the system

$$\begin{cases} \dot{x} = 2y^3 + x, \\ \dot{y} = -xy^2 + y, \end{cases}$$

using the Lyapunov function $V(x,y) = \frac{x^2}{2} + y^2$.

(*iii*) Analyze the stability at $(0,0)$ of the system

$$\begin{cases} \dot{x} = -3x + y - x^3, \\ \dot{y} = -5x - y - y^3, \end{cases}$$

using a Lyapunov function $V(x,y) = ax^2 + by^2$ for some suitable constants a and b.

5. Analyze the stability of the system

$$\begin{cases} \dot{x} = y, \\ \dot{y} = -2x - y \end{cases}$$

at $(0,0)$ using the Lyapunov functions:

(*i*) $V(x,y) = 2x^2 + y^2$,
(*ii*) $V(x,y) = 7x^2 + 2xy + 3y^2$.

6. Using a Lyapunov function of the form $V(x_1,x_2) = ax_1^2 + bx_1x_2 + cx_2^2$, with suitable values of a, b and c, show that the system

$$\begin{cases} \dot{x}_1 = 2x_1 - 4x_2, \\ \dot{x}_2 = 2x_1 - 2x_2 \end{cases}$$

is stable at $(0,0)$.

7. Using the Lyapunov function $V(x,y) = y^2 - x^2$, investigate the stability at $(0,0)$ of the system

$$\begin{cases} \dot{x} = x - 2xy^2, \\ \dot{y} = -2y + x^2y. \end{cases}$$

8. Using the Lyapunov function $V(x,y) = 3xy^2 - x^3$, investigate the stability of the equilibrium point $(0,0)$ of the system

$$\begin{cases} \dot{x} = x^2 - y^2, \\ \dot{y} = -2xy. \end{cases}$$

9. Using the Lyapunov function $V(x,y) = 5x^2 + 2xy + 2y^2$, show that the origin $(0,0)$ is asymptotically stable for the system

$$\begin{cases} \dot{x} = y, \\ \dot{y} = -x - y + (x+2y)(y^2 - 1). \end{cases}$$

10. Using the Lyapunov function $V(x_1,x_2) = x_2^2 + (1 - \cos x_1)$ and LaSalle's theorem, show that the system

$$\begin{cases} \dot{x}_1 = x_2, \\ \dot{x}_2 = -\dfrac{1}{2}\sin x_1 - \dfrac{1}{4}x_2^3 \end{cases}$$

is asymptotically stable at $(0,0)$.

11. The equation of motion of a controlled pendulum is $\frac{d^2\theta}{dt^2} + a\frac{d\theta}{dt} + k\sin\theta = \tau$ where $\tau(t)$ is the torque (control) applied on the pendulum. If the desired trajectory for $\theta(t)$ is $\theta_d(t) = \frac{\pi}{6} + \frac{\pi}{36}t^3$ for $t \in [0,3]$, and if the pendulum is initially at rest with $\theta(0) = 0$, find a suitable control function $\tau(t)$.

12. The equation of motion of a damped harmonic oscillator is $\ddot{x} + a\dot{x} + kx = f$, where a and k are positive constants. If $x(0) = 5$ and $\dot{x}(0) = 0$, and it is required to stop the oscillator at $x = -3$, then find a suitable control force $f(t)$.

13. Consider the following control system representing the motion of a controlled pendulum:

$$\begin{cases} \dot{x}_1 = x_2, \\ \dot{x}_2 = -K\sin x_1 - K_1 x_2 + \tau, \end{cases}$$

where K and K_1 are positive constants, $x_1(t)$ is the angle between the pendulum and the negative y-axis, and $x_2(t)$ is the angular velocity at time t. $\tau(t)$ is the torque applied at the pendulum joint. If the pendulum is held motionless at $x_1 = \frac{\pi}{6}$, find the constant torque to be applied (in terms of K and K_1).

14. Show that

    (*i*) $(x_1^2 + 2x_2^2 + 3x_3^2)\sin^2 t$ is not positive definite but is decrescent on $\mathbb{R}_0^+ \times \mathbb{R}^3$.

    (*ii*) $tx_1^2 + e^t x_2^2 + x_3^2$ is neither positive definite nor decrescent on $\mathbb{R}_0^+ \times \mathbb{R}^3$.

    (*iii*) $x_1^2 + e^{-t}x_2^2$ is decrescent on $\mathbb{R}_0^+ \times \mathbb{R}^2$.

    (*iv*) $x_1^2 + \frac{1}{1+t}x_2^2$ is decrescent on $\mathbb{R}_0^+ \times \mathbb{R}^2$ and positive on $\mathbb{R}_0^+ \times B_1(0)$.

15. Show that $V(x_1, x_2) = -2x_1^4 + x_2^2(2x_1^2 - 3)$ is negative definite in $\|x\| \leq 1$. Find a function ψ of class-K such that $V(x) \leq -\psi(\|x\|)$.

9

Discrete-time Systems

9.1 Introduction

Various results pertaining to discrete-time systems are similar to those of continuous-time systems. In this chapter, results on controllability, observability, stability, stabilizability, and optimal control for discrete-time systems are discussed. These results are analogous to those of continuous-time systems presented in Chapters 5–8, with minor changes. For the computational study of a continuous-time system, it is often necessary to discretize the system equations and analyze the equivalent discrete-time system. Thus, the study of discrete-time systems becomes important in applications.

In Section 9.2, the state transition matrix and solutions to discrete-time linear autonomous and nonautonomous systems are defined. Sections 9.3 and 9.4 provide results on controllability and observability for discrete-time linear systems. A method for obtaining a discrete-time system from a continuous-time system while preserving the controllability property is presented in Section 9.5. In Section 9.6, the dual system and the duality theorem are briefly discussed. In Section 9.7, stability and stabilizability of discrete-time systems using eigenvalues are analyzed, and in Section 9.8, Lyapunov theory is explained in detail. Optimal control problems and the matrix Riccati equation analogous to those for continuous-time systems are presented in Sections 9.9–9.11.

9.2 Discrete-time Linear System

In this section, we shall define the state transition matrix and the solution of a discrete-time linear system in terms of the state transition matrix. Consider the discrete-time linear system

$$x(k+1) = A(k)x(k) \quad \text{for } k = k_0, k_0+1, \ldots, k_0+N-1 \tag{9.1}$$

with the initial condition $x(k_0) = x_0$. Here, k_0 is a positive integer, $x(k) \in \mathbb{R}^n$ for each k, and $A(k)$ is an $n \times n$ **nonsingular** matrix whose elements are functions of k. The sequence $x(k).k = k_0, k_0+1, \ldots, k_0+N$, is called the solution of (9.1). From (9.1), it can be seen that

$$\begin{aligned} x(k_0+1) &= A(k_0)x_0, \\ x(k_0+2) &= A(k_0+1)x(k_0+1) = A(k_0+1)A(k_0)x_0. \end{aligned}$$

Proceeding as above, we have

$$x(k_0+r) = A(k_0+r-1)A(k_0+r-2)\cdots A(k_0)x_0, \quad r = 1,2,\ldots,N. \tag{9.2}$$

Let $m = k_0 + r$. Then,

$$x(m) = A(m-1)A(m-2)\cdots A(k_0)x_0, \quad m = k_0+1, k_0+2, \ldots, k_0+N. \tag{9.3}$$

Define the **state transition matrix** $\phi(m,l)$ of $A(m)$ as

$$\phi(m,l) = A(m-1)A(m-2)\cdots A(l), \quad l < m. \tag{9.4}$$

For $l < m$, define $\phi(l,m,)$ as $(\phi(m,l))^{-1}$. Then,

$$\phi(l,m) = (A(l))^{-1}(A(l+1))^{-1}\cdots(A(m-1))^{-1}, \quad m > l. \tag{9.5}$$

From (9.4) and (9.5), we have

$$\begin{aligned} \phi(m,l)\phi(l,m) &= A(m-1)A(m-2)\cdots A(l)(A(l))^{-1}(A(l+1))^{-1}\cdots(A(m-1))^{-1} \\ &= I = \phi(l,m)\phi(m,l). \end{aligned}$$

For $m > l > r$, from (9.4), we have

$$\phi(m,l)\phi(l,r) = \phi(m,r).$$

Moreover, for any three integers m, l, and r, using (9.4) and (9.5), we have

$$\phi(m,l)\phi(l,r) = \phi(m,r) \text{ and } \phi(m+1,l) = A(m)A(m-1)\ldots A(l) = A(m)\phi(m,l) \text{ for } m \geq l.$$

Thus, the state transition matrix ϕ defined in (9.4) and (9.5) satisfies the following properties.

For any positive integers m, l, and r, the following hold:

(1) $\phi(m,m) = I$,

(2) $\phi(m,l)\phi(l,r) = \phi(m,r)$,

(3) $\phi(m,l) = (\phi(l,m))^{-1}$,

(4) $\phi(m+1,l) = A(m)\phi(m,l)$.

From (9.3), the solution of (9.1) is given by

$$x(m) = \phi(m,k_0)x(k_0) \quad \text{for } m = k_0+1, k_0+2, \ldots. \tag{9.6}$$

Autonomous System

If $A(k) = A$, a constant nonsingular matrix for all k, then from (9.4) and (9.5),

$$\phi(m,l) = A^{m-l} \quad \text{for any integers } m \text{ and } l.$$

***Example* 9.2.1** Consider the discrete-time system $x(k+1) = A(k)x(k)$ for $k \geq 3$, where $A(k) = \begin{bmatrix} k & 1 \\ 2 & 1 \end{bmatrix}$. If $x(3) = \begin{bmatrix} 1 \\ 0 \end{bmatrix}$, find $x(5)$.

Solution Note that for $k = 2$, $A(2)$ is singular. Therefore, we must consider $A(k)$ for $k \geq 3$. From (9.6), we have $x(5) = \phi(5,3)x(3)$, where $\phi(5,3) = A(4)A(3)$. Hence, $x(5) = \begin{bmatrix} 4 & 1 \\ 2 & 1 \end{bmatrix}\begin{bmatrix} 3 & 1 \\ 2 & 1 \end{bmatrix}\begin{bmatrix} 1 \\ 0 \end{bmatrix} = \begin{bmatrix} 14 \\ 8 \end{bmatrix}$.

Nonhomogeneous System

Consider the nonhomogeneous time-varying system

$$\begin{cases} x(k+1) = A(k)x(k) + f(k), & k = k_0, k_0+1, \ldots, k_0+N-1, \\ x(k_0) = x_0. \end{cases} \tag{9.7}$$

To find the solution $x(k_0+r)$, $r = 1, 2, \ldots$, to (9.7), we follow the following steps.

$$\begin{aligned} x(k_0+1) &= A(k_0)x(k_0) + f(k_0), \\ x(k_0+2) &= A(k_0+1)x(k_0+1) + f(k_0+1) \\ &= A(k_0+1)A(k_0)x_0 + A(k_0+1)f(k_0) + f(k_0+1). \end{aligned}$$

Proceeding as above, we have

$$\begin{aligned}x(k_0+r) &= A(k_0+r-1)A(k_0+r-2)\cdots A(k_0)x_0 \\ &\quad + A(k_0+r-1)\cdots A(k_0+1)f(k_0)+\cdots \\ &\quad + A(k_0+r-1)f(k_0+r-2)+f(k_0+r-1) \quad \text{for } r=1,2,\ldots.\end{aligned}$$

Let $m = k_0 + r$. Then using (9.4) and (9.5), we obtain

$$\begin{aligned}x(m) &= \phi(m,k_0)x_0+\phi(m,k_0+1)f(k_0)+\cdots+\phi(m,m-1)f(m-2)+f(m-1) \\ &= \phi(m,k_0)\left[x_0+\phi(k_0,k_0+1)f(k_0)+\phi(k_0,k_0+2)f(k_0+1)+\cdots\right. \\ &\quad \left. +\,\phi(k_0,m-1)f(m-2)+\phi(k_0,m)f(m-1)\right] \\ &= \phi(m,k_0)\left[x_0+\sum_{i=k_0}^{m-1}\phi(k_0,i+1)f(i)\right] \quad \text{(using the properties of } \phi). \qquad (9.8)\end{aligned}$$

***Example* 9.2.2** Consider the time-invariant system $x(k+1) = Ax(k) + f(k)$, $k = 3,4,5$, where $A = \begin{bmatrix} 3 & 1 \\ 0 & 3 \end{bmatrix}$, $f(k) = \begin{bmatrix} k^2 \\ 3 \end{bmatrix}$, and $x(3) = \begin{bmatrix} 0 \\ 1 \end{bmatrix}$. Find $x(6)$.

Solution From (9.8), it follows that

$$\begin{aligned}x(6) &= \phi(6,3)\left[x(3)+\phi(3,4)f(3)+\phi(3,5)f(4)+\phi(3,6)f(5)\right] \\ &= A^3\left(\begin{bmatrix}0\\1\end{bmatrix}+A^{-1}\begin{bmatrix}9\\3\end{bmatrix}+A^{-2}\begin{bmatrix}16\\3\end{bmatrix}+A^{-3}\begin{bmatrix}25\\3\end{bmatrix}\right) \\ &= A^3\begin{bmatrix}0\\1\end{bmatrix}+A^2\begin{bmatrix}9\\3\end{bmatrix}+A\begin{bmatrix}16\\3\end{bmatrix}+\begin{bmatrix}25\\3\end{bmatrix}.\end{aligned}$$

Exercise 9.2.1 Find the solution $x(k) : k = 1,2,3,4$ of the system $x(r+1) = A(r)x(r) + f(r)$, $r = 0,1,2,3$, with the initial condition $x(0) = \begin{bmatrix} 1 \\ 2 \end{bmatrix}$, where $A(r) = \begin{bmatrix} 1 & r \\ 0 & 1 \end{bmatrix}$ and $f(r) = \begin{bmatrix} 2 \\ r \end{bmatrix}$.

9.3 Controllability

Consider the linear control system

$$\begin{cases} x(k+1) = A(k)x(k) + B(k)u(k), & k = k_0, k_0+1, \dots, k_0+N-1, \\ x(k_0) = x_0, \end{cases} \tag{9.9}$$

where $A(k)$ and $B(k)$ are $n \times n$ and $n \times m$ matrices, respectively, and the control $u(k) \in \mathbb{R}^m$ for each k. It is assumed that $A(k)$ is nonsingular for each k. If the control sequence $u(i)$ for $i = k_0, k_0+1, \dots, k_0+N-1$ is given, then by replacing $f(k)$ with $B(k)u(k)$ in (9.8), the solution $x(r) : r = k_0+1, k_0+2, \dots, k_0+N$ is written as

$$x(r) = \phi(r, k_0) \left[x_0 + \sum_{i=k_0}^{r-1} \phi(k_0, i+1)B(i)u(i) \right]. \tag{9.10}$$

However, in many control problems, the control sequence may not be given, and one needs to determine a control sequence under certain conditions on the solution $x(k)$, as discussed in the next section on controllability.

Controllability

The system (9.9) is said to be controllable in discrete time between k_0 and k_0+N if, for arbitrary vectors x_0 and x_1, we can find a sequence of control $u(i) : i = k_0, k_0+1, \dots, k_0+N-1$ such that the solution of (9.9) satisfies $x(k_0) = x_0$ and $x(k_0+N) = x_1$.

Condition for Controllability

As in the case of continuous-time system, we can define the controllability Gramian matrix W (analogous to (5.15)) as

$$W = \sum_{i=k_0}^{k_0+N-1} \phi(k_0, i+1)B(i)B'(i)\phi'(k_0, i+1). \tag{9.11}$$

Note that W is constant symmetric matrix of order n. The following result is similar to Theorem 5.3.3 concerning the controllability of continuous-time system.

Theorem 9.3.1 The discrete-time system (9.9) is controllable if and only if W is nonsingular.

Proof. **Sufficiency.** Let W be invertible. Consider the control sequence

$$u(i) = B'(i)\phi'(k_0, i+1)W^{-1}\left(\phi(k_0, k_0+N)x_1 - x_0\right), \quad i = k_0, k_0+1, \dots, k_0+N-1. \tag{9.12}$$

Substituting (9.12) into (9.10), we find that $x(k_0) = x_0$ and

$$x(k_0+N) = \phi(k_0+N,k_0)\left[x_0 + \sum_{i=k_0}^{k_0+N-1} \phi(k_0,i+1)B(i) \right.$$
$$\left. \times \left[B'(i)\phi'(k_0,i+1)W^{-1}\left(\phi(k_0,k_0+N)x_1 - x_0\right)\right]\right]$$
$$= \phi(k_0+N,k_0)\left[x_0 + WW^{-1}\left(\phi(k_0,k_0+N)x_1 - x_0\right)\right] = x_1.$$

Hence, the system is controllable.

Note that the control sequence in (9.12) is one possible control sequence. However, there are infinitely many other control sequences that can steer $x(k)$ from x_0 to x_1 if the system is controllable. If $v(i) \in \text{Null}(\phi(k_0,i+1)B(i))$, then $w(i) = u(i) + v(i)$: $i = k_0, k_0+1, ..., k_0+N-1$ is also a control sequence that steers $x(k)$ from x_0 to x_1.

Necessity. Let the system (9.9) be controllable. We need to prove that W is nonsingular. If not, then there exists a **nonzero vector** $\alpha \in \mathbb{R}^n$ such that

$W\alpha = \theta$ (the zero vector).

Hence,

$$\alpha' W\alpha = 0.$$

From (9.11), it follows that

$$\alpha' \sum_{i=k_0}^{k_0+N-1} \phi(k_0,i+1)B(i)B'(i)\phi'(k_0,i+1)\alpha = 0,$$

which is equivalent to

$$\sum_{i=k_0}^{k_0+N-1} \left(B'(i)\phi'(k_0,i+1)\alpha\right)'\left(B'(i)\phi'(k_0,i+1)\alpha\right) = 0.$$

Hence,

$$\sum_{i=k_0}^{k_0+N-1} \|B'(i)\phi'(k_0,i+1)\alpha\|^2 = 0. \tag{9.13}$$

Note that $B'(i)\phi'(k_0,i+1)\alpha \in \mathbb{R}^m$ for each i, and the norm in the LHS of (9.13) is the Euclidean norm in $\mathbb{R}^m$. From (9.13), it is clear that

$$B'(i)\phi'(k_0,i+1)\alpha = \theta \quad \text{for } i = k_0, k_0+1, \ldots, k_0+N-1. \tag{9.14}$$

Since the system is controllable, we can find a control sequence $u(i) : i = k_0, k_0 + 1, \ldots, k_0 + N - 1$ such that $x(k_0) = \theta$ and $x(k_0+N) = \phi(k_0+N, k_0)\alpha$. Hence, from (9.10),

$$x(k_0+N) = \phi(k_0+N,k_0)\alpha = \phi(k_0+N,k_0)\left[x(k_0) + \sum_{i=k_0}^{k_0+N-1} \phi(k_0,i+1)B(i)u(i)\right].$$

Since $(\phi(k_0+N,k_0))^{-1} = \phi(k_0,k_0+N)$ and $x(k_0) = \theta$, multiplying both sides by $\phi(k_0,k_0+N)$, the above equation becomes

$$\alpha = \sum_{i=k_0}^{k_0+N-1} \phi(k_0,i+1)B(i)u(i).$$

Multiplying both sides by α', we obtain

$$\|\alpha\|^2 = \alpha'\alpha = \sum_{i=k_0}^{k_0+N-1} \alpha'\phi(k_0,i+1)B(i)u(i) = 0 \quad \text{(from(9.14))},$$

implying $\alpha = \theta$. This contradicts the assumption that α is a nonzero vector. Hence, W is nonsingular. □

***Example* 9.3.2** For the discrete-time control system $x(k+1) = A(k)x(k) + B(k)u(k)$, $k = 1,2,3$, where $A(k) = \begin{bmatrix} k^2 & 0 \\ 1 & k \end{bmatrix}$ and $B(k) = \begin{bmatrix} k \\ 1 \end{bmatrix}$, test the system's controllability. If it is controllable, find $u(1)$ and $u(2)$ such that $x_0 = x(1) = \begin{bmatrix} 1 \\ 0 \end{bmatrix}$ and $x_1 = x(3) = \begin{bmatrix} 0 \\ 1 \end{bmatrix}$.

Solution The controllability Gramian matrix of the system is

$$\begin{aligned} W &= \phi(1,2)B(1)B'(1)\phi'(1,2) + \phi(1,3)B(2)B'(2)\phi'(1,3) \\ &= (A(1))^{-1}B(1)B'(1)(A'(1))^{-1} + (A(2)A(1))^{-1}B(2)B'(2)(A'(1)A'(2))^{-1}. \end{aligned}$$

Let us check if W is nonsingular. For $A(k) = \begin{bmatrix} k^2 & 0 \\ 1 & k \end{bmatrix}$, we have $A(1) = \begin{bmatrix} 1 & 0 \\ 1 & 1 \end{bmatrix}$, $A(2) = \begin{bmatrix} 4 & 0 \\ 1 & 2 \end{bmatrix}$, $A^{-1}(1) = \begin{bmatrix} 1 & 0 \\ -1 & 1 \end{bmatrix} = \phi(1,2)$, and $A^{-1}(2) = \frac{1}{8}\begin{bmatrix} 2 & 0 \\ -1 & 4 \end{bmatrix}$. Therefore, $(A(2)A(1))^{-1} = \frac{1}{8}\begin{bmatrix} 2 & 0 \\ -3 & 4 \end{bmatrix} = \phi(1,3)$. Hence, the Gramian matrix is

$$W = \phi(1,2)B(1)B'(1)\phi(1,2) + \phi(1,3)B(2)B'(2)\phi'(1,3),$$

which simplifies to

$$W = \begin{bmatrix} 1 & 0 \\ -1 & 1 \end{bmatrix} \begin{bmatrix} 1 \\ 1 \end{bmatrix} \begin{bmatrix} 1 & 1 \end{bmatrix} \begin{bmatrix} 1 & -1 \\ 0 & 1 \end{bmatrix} + \frac{1}{8} \begin{bmatrix} 2 & 0 \\ -3 & 4 \end{bmatrix} \begin{bmatrix} 2 \\ 1 \end{bmatrix} \begin{bmatrix} 2 & 1 \end{bmatrix} \frac{1}{8} \begin{bmatrix} 2 & -3 \\ 0 & 4 \end{bmatrix}$$

$$= \begin{bmatrix} 1 \\ 0 \end{bmatrix} \begin{bmatrix} 1 & 0 \end{bmatrix} + \frac{1}{64} \begin{bmatrix} 4 \\ -2 \end{bmatrix} \begin{bmatrix} 4 & -2 \end{bmatrix} = \begin{bmatrix} 1 & 0 \\ 0 & 0 \end{bmatrix} + \frac{1}{64} \begin{bmatrix} 16 & -8 \\ -8 & 4 \end{bmatrix} = \begin{bmatrix} \frac{5}{4} & -\frac{1}{8} \\ -\frac{1}{8} & \frac{1}{16} \end{bmatrix}.$$

Hence, W is nonsingular, and from Theorem 9.3.1, the system is controllable. Now,

$$W^{-1} = 16 \begin{bmatrix} \frac{1}{16} & \frac{1}{8} \\ \frac{1}{8} & \frac{5}{4} \end{bmatrix} = \begin{bmatrix} 1 & 2 \\ 2 & 20 \end{bmatrix}.$$

From (9.12), the required control sequence is

$$u(1) = B'(1)\phi'(1,2)W^{-1}(\phi(1,3)x_1 - x_0),$$
$$u(2) = B'(2)\phi'(1,3)W^{-1}(\phi(1,3)x_1 - x_0).$$

Substituting the matrices appropriately, we obtain

$$u(1) = \begin{bmatrix} 1 & 1 \end{bmatrix} \begin{bmatrix} 1 & -1 \\ 0 & 1 \end{bmatrix} 16 \begin{bmatrix} \frac{1}{16} & \frac{1}{8} \\ \frac{1}{8} & \frac{5}{4} \end{bmatrix} \left(\frac{1}{8} \begin{bmatrix} 2 & 0 \\ -3 & 4 \end{bmatrix} \begin{bmatrix} 0 \\ 1 \end{bmatrix} - \begin{bmatrix} 1 \\ 0 \end{bmatrix} \right)$$

$$= \begin{bmatrix} 1 & 0 \end{bmatrix} \begin{bmatrix} 1 & 2 \\ 2 & 20 \end{bmatrix} \left(\frac{1}{8} \begin{bmatrix} 0 \\ 4 \end{bmatrix} - \begin{bmatrix} 1 \\ 0 \end{bmatrix} \right) = \begin{bmatrix} 1 & 2 \end{bmatrix} \left(\begin{bmatrix} 0 \\ \frac{1}{2} \end{bmatrix} - \begin{bmatrix} 1 \\ 0 \end{bmatrix} \right) = 0$$

and

$$u(2) = \begin{bmatrix} 2 & 1 \end{bmatrix} \frac{1}{8} \begin{bmatrix} 2 & -3 \\ 0 & 4 \end{bmatrix} \begin{bmatrix} 1 & 2 \\ 2 & 20 \end{bmatrix} \begin{bmatrix} -1 \\ \frac{1}{2} \end{bmatrix} = \frac{1}{8} \begin{bmatrix} 4 & -2 \end{bmatrix} \begin{bmatrix} 0 \\ 8 \end{bmatrix} = -2.$$

If A and B are constant matrices of order $n \times n$ and $n \times m$, respectively, then the following result on controllability holds.

Theorem 9.3.3 The autonomous system

$$x(k+1) = Ax(k) + Bu(k), \quad k = k_0, k_0+1, \ldots, k_0+N-1 \tag{9.15}$$

is controllable if and only if $\text{rank}(V) = n$, where

$$V = \begin{bmatrix} B & AB & A^2B & \cdots & A^{N-1}B \end{bmatrix}_{n \times Nm}.$$

Proof. **Sufficiency.** Let $\text{rank}(V) = n$, and let x_0 and x_1 be arbitrary vectors in $\mathbb{R}^n$. From (9.10), for a given control sequence, the solution of the system (9.15) is

$$x(r) = A^{r-k_0}\left[x(k_0) + \sum_{i=k_0}^{r-1} A^{k_0-i-1}Bu(i)\right], \quad r = k_0+1,\ldots,k_0+N. \tag{9.16}$$

At the final time $r = k_0 + N$,

$$x(k_0+N) = A^N\left[x(k_0) + \sum_{i=k_0}^{k_0+N-1} A^{k_0-i-1}Bu(i)\right]. \tag{9.17}$$

Let $x(k_0) = x_0$ and $x(k_0+N) = x_1$. The above equation then becomes

$$\begin{aligned} x_1 - A^N x_0 &= \sum_{i=k_0}^{k_0+N-1} A^{N+k_0-i-1}Bu(i) \\ &= A^{N-1}Bu(k_0) + A^{N-2}Bu(k_0+1) + \cdots + Bu(k_0+N-1) \\ &= \begin{bmatrix} B & AB & A^2B & \cdots & A^{N-1}B \end{bmatrix} \begin{bmatrix} u(k_0+N-1) \\ u(k_0+N-2) \\ \vdots \\ u(k_0) \end{bmatrix}. \end{aligned} \tag{9.18}$$

Let

$$p = x_1 - A^N x_0 \text{ and } q = \begin{bmatrix} u(k_0+N-1) \\ u(k_0+N-2) \\ \vdots \\ u(k_0) \end{bmatrix}.$$

Then, (9.18) becomes $p = Vq$, where $p \in \mathbb{R}^n$, $V \in \mathbb{R}^{n\times Nm}$, and $q \in \mathbb{R}^{Nm\times 1}$. If $\text{rank}(V) = n$, then for any arbitrary vector p, there exists a solution q. Thus, for any x_0 and x_1, there exists a control sequence $u(i) : i = k_0, k_0+1, \ldots, k_0+N-1$ satisfying (9.17). Hence, the system is controllable if $\text{rank}(V) = n$. Note that if $Nm < n$, then $\text{rank}(V) < n$, and hence, (9.18) may not have a solution.

Necessity. Let the system be controllable. We shall show that the matrix V has rank n. Suppose, for contradiction, that $\text{rank}(V) < n$. Then, there exists a nonzero vector α in $\mathbb{R}^n$ such that

$$\alpha' V = \theta \in \mathbb{R}^{Nm}. \tag{9.19}$$

As the system is controllable, by choosing $x_0 = \theta$ and $x_1 = \alpha$, we can find a control sequence $u(i) : i = k_0, k_0+1, \ldots, k_0+N-1$ such that the solution x to (9.15) satisfies $x(k_0) = \theta$ and $x(k_0+N) = \alpha$. Substituting in (9.17), we get

$$\alpha = \sum_{i=k_0}^{k_0+N-1} A^{N+k_0-i-1} Bu(i)$$

$$= \begin{bmatrix} B & AB & A^2B & \cdots & A^{N-1}B \end{bmatrix} \begin{bmatrix} u(k_0+N-1) \\ u(k_0+N-2) \\ \vdots \\ u(k_0) \end{bmatrix} = Vq, \tag{9.20}$$

where

$$q = \begin{bmatrix} u(k_0+N-1) \\ u(k_0+N-2) \\ \vdots \\ u(k_0) \end{bmatrix}.$$

Multiplying both sides of (9.20) by α' and using (9.19), we get $\alpha'\alpha = \alpha' Vq = 0$, implying that $\|\alpha\|^2 = 0$, which is a contraction to the assumption that $\alpha \neq \theta$. The proof is complete. □

Remark 9.3.4 If $N \geq n$, then $\text{rank}(U) = \text{rank}(V)$, where $U = \begin{bmatrix} B & AB & A^2B & \cdots & A^{n-1}B \end{bmatrix}$. Therefore, in this case, the rank condition (the Kalman condition) for the controllability of a discrete-time system is the same as that of a continuous-time system (Theorem 5.3.4). If $Nm < n$, then the system is not controllable, since rank $V < n$.

If the system (9.9) is not controllable (i.e., if the controllability Gramian W in (9.11) is not invertible), then a result similar to that of the continuous-time system (solvability of (5.33)) holds.

Let x_0 and x_1 be given. Then, we can find a control sequence $u(i) : i = k_0, k_0 + 1, \ldots, k_0+N-1$ that steers the system from $x(k_0) = x_0$ to $x(k_0+N) = x_1$ provided the following system is solvable for $\gamma \in \mathbb{R}^n$:

$$W\gamma = \phi(k_0, k_0+N)x_1 - x_0. \tag{9.21}$$

In that case, the control sequence is given by

$$u(i) = B'(i)\phi'(k_0, i+1)\gamma, \quad i = k_0, k_0+1, \ldots, k_0+N-1. \tag{9.22}$$

By substituting $u(i)$ into (9.10), it can be shown that the solution

$$x(k) = \phi(k, k_0)\left[x_0 + \sum_{i=k_0}^{k-1} \phi(k_0, i+1)B(i)u(i)\right], \quad k = k_0+1, k_0+2, \ldots, k_0+N-1$$

satisfies the conditions $x(k_0) = x_0$ and $x(k_0+N) = x_1$ since

$$\begin{aligned} x(k_0+N) &= \phi(k_0+N, k_0)\left[x_0 + \sum_{i=k_0}^{k_0+N-1} \phi(k_0, i+1)B(i)B'(i)\phi'(k_0, i+1)\gamma\right] \\ &= \phi(k_0+N, k_0)\left[x_0 + W\gamma\right] \\ &= \phi(k_0+N, k_0)x_0 + \phi(k_0+N, k_0)\left(\phi(k_0, k_0+N)x_1 - x_0\right) = x_1. \end{aligned}$$

***Example* 9.3.5** Consider the system $x(k+1) = A(k)x(k) + B(k)u(k)$, $k = 1, 2$, where $A(k) = \begin{bmatrix} k^2 & 0 \\ 1 & k \end{bmatrix}$ and $B(k) = \begin{bmatrix} k \\ \frac{k+1}{2} \end{bmatrix}$. Show that the system is not controllable. If $x(1) = \begin{bmatrix} 1 \\ 3 \end{bmatrix}$, find the set of all possible final points $x(3) \in \mathbb{R}^2$ that can be reached using suitable control sequences.

Solution We know that $\phi(1,2) = A^{-1}(1)$ and $\phi(1,3) = A^{-1}(1)A^{-1}(2)$. From (9.11), the controllability Gramian matrix is

$$\begin{aligned} W &= \begin{bmatrix} 1 & 0 \\ -1 & 1 \end{bmatrix}\begin{bmatrix} 1 \\ 1 \end{bmatrix}\begin{bmatrix} 1 & 1 \end{bmatrix}\begin{bmatrix} 1 & -1 \\ 0 & 1 \end{bmatrix} + \frac{1}{64}\begin{bmatrix} 2 & 0 \\ -3 & 4 \end{bmatrix}\begin{bmatrix} 2 \\ \frac{3}{2} \end{bmatrix}\begin{bmatrix} 2 & \frac{3}{2} \end{bmatrix}\frac{1}{8}\begin{bmatrix} 2 & -3 \\ 0 & 4 \end{bmatrix} \\ &= \begin{bmatrix} 1 \\ 0 \end{bmatrix}\begin{bmatrix} 1 & 0 \end{bmatrix} + \frac{1}{64}\begin{bmatrix} 4 \\ 0 \end{bmatrix}\begin{bmatrix} 4 & 0 \end{bmatrix} \\ &= \begin{bmatrix} 1 & 0 \\ 0 & 0 \end{bmatrix} + \frac{1}{64}\begin{bmatrix} 16 & 0 \\ 0 & 0 \end{bmatrix} = \begin{bmatrix} \frac{5}{4} & 0 \\ 0 & 0 \end{bmatrix}. \end{aligned}$$

Hence, W is singular implying that the system is not controllable. Note that the range of W is generated by $\begin{bmatrix} 1 \\ 0 \end{bmatrix}$. From (9.21), it follows that

$$\phi(1,3)x_1 - x_0 = W \cdot \gamma = W \begin{bmatrix} \gamma_1 \\ \gamma_2 \end{bmatrix} = \begin{bmatrix} \frac{5}{4}\gamma_1 \\ 0 \end{bmatrix}, \tag{9.23}$$

where

$$\phi(1,3) = (A(2)A(1))^{-1} = \frac{1}{8}\begin{bmatrix} 2 & 0 \\ -3 & 4 \end{bmatrix}.$$

Let $x_0 = x(1) = \begin{bmatrix} \alpha \\ \beta \end{bmatrix}$ and $x_1 = x(3) = \begin{bmatrix} a \\ b \end{bmatrix}$. Then, from (9.23), we obtain

$$\frac{1}{8}\begin{bmatrix} 2 & 0 \\ -3 & 4 \end{bmatrix}\begin{bmatrix} a \\ b \end{bmatrix} - \begin{bmatrix} \alpha \\ \beta \end{bmatrix} = \begin{bmatrix} \frac{1}{4}a - \alpha \\ \frac{4b-3a}{8} - \beta \end{bmatrix} = \begin{bmatrix} \frac{5}{4}\gamma_1 \\ 0 \end{bmatrix}$$
$$\implies 5\gamma_1 = a - 4\alpha \text{ and } 4b - 3a - 8\beta = 0$$
$$\implies \gamma_1 = \frac{a-4\alpha}{5}, \gamma_2 = \text{arbitrary, and } 4b - 3a - 8\beta = 0$$

are the conditions on a, b, α, and β for the initial state $x_0 = \begin{bmatrix} \alpha \\ \beta \end{bmatrix}$ to be steered to the final state $x_1 = \begin{bmatrix} a \\ b \end{bmatrix}$. It is given that $x_0 = x(1) = \begin{bmatrix} 1 \\ 3 \end{bmatrix}$. Therefore, $4b - 3a - 24 = 0$, which implies $b = \frac{24+3a}{4}$. Hence, the set of all possible final points is $x_1 = \begin{bmatrix} a \\ \frac{24+3a}{4} \end{bmatrix}$, where a is arbitrary. From (9.22), the required control sequence is given by $u(i) = B'(i)\phi'(1,i+1)\gamma, i = 1,2$, where $\gamma = \begin{bmatrix} \frac{a-4}{5} \\ \gamma_2 \end{bmatrix}$.

Exercise 9.3.1 Consider the control system $x(k+1) = Ax(k) + Bu(k)$, $k = 0,1,2,\ldots,N-1$, where $A = \begin{bmatrix} 0 & 1 & 0 & 0 \\ 0 & 0 & 1 & 0 \\ 1 & 1 & 0 & 0 \\ 0 & 0 & 0 & 1 \end{bmatrix}$ and $B = \begin{bmatrix} 1 \\ 0 \\ 0 \\ 1 \end{bmatrix}$. Show that the system is not controllable if $N \leq 3$ and controllable if $N \geq 4$.

9.4 Observability

In many real-life problems, it is not possible to measure the state variables directly. However, one can measure or observe certain quantities related to the state variables. These observations or measurements are considered useful only if the state of the system can be uniquely determined from the observations. A system with such useful observations is called an observable system. A formal definition of observability for a discrete-time system is given below.

Definition 9.4.1 Consider the discrete-time system

$$x(k+1) = A(k)x(k) + B(k)u(k), \quad k = k_0, k_0+1, \ldots, k_0+N-1 \tag{9.24}$$

with the observation

$$y(k) = C(k)x(k), \quad k = k_0+1, \ldots, k_0+N, \tag{9.25}$$

where $x(k) \in \mathbb{R}^n$, $u(k) \in \mathbb{R}^m$, and $y(k) \in \mathbb{R}^p$ for each k. The system (9.24) along with (9.25) is said to be **observable** if it is possible to **uniquely** determine the **initial state** $x(k_0)$ from the knowledge of $y(k) : k = k_0+1, \ldots, k_0+N$ and $u(k) : k = k_0, k_0+1, \ldots, k_0+N-1$.

Note that if $x(k_0)$ is uniquely determined, then $x(k)$ for $k = k_0+1, \ldots, k_0+N$ can also be uniquely obtained using (9.10). Thus, the observability of the system (9.24) along with (9.25) implies that from the knowledge of $y(k) : k = k_0+1, \ldots, k_0+N$ and $u(k) : k = k_0, k_0+1, \ldots, k_0+N-1$, one can uniquely obtain the state $x(k)$ of the system for all $k = k_0, k_0+1, \ldots, k_0+N$.

The following theorem provides necessary and sufficient conditions for the observability of the system (9.24) with the observation (9.25). Let $\phi(k, k_0)$ be the state transition matrix of $A(k)$, and let $M[k_0, k_0+N]$ be the $n \times n$ matrix defined as

$$M[k_0, k_0+N] = \sum_{r=k_0+1}^{k_0+N} \phi'(r, k_0) C'(r) C(r) \phi(r, k_0). \tag{9.26}$$

The matrix $M[k_0, k_0+N]$ is called the **observability Gramian matrix**. Hereafter, we denote $M[k_0, k_0+N]$ by M for simplicity.

Theorem 9.4.2 The discrete-time system (9.24) along with (9.25) is observable if and only if M is nonsingular.

Proof. **Sufficiency.** Let M be nonsingular. To prove observability, let $y(k) : k = k_0 + 1, \ldots, k_0+N$ and $u(k) : k = k_0, k_0+1, \ldots, k_0+N-1$ be given. We know that

$$x(r) = \phi(r, k_0)x_0 + \sum_{i=k_0}^{r-1} \phi(r, i+1)B(i)u(i), \quad r = k_0+1, \ldots, k_0+N \tag{9.27}$$

and

$$y(r) = C(r)x(r).$$

Thus, from (9.27), it follows that

$$y(r) = C(r)\phi(r, k_0)x_0 + \sum_{i=k_0}^{r-1} C(r)\phi(r, i+1)B(i)u(i),$$

which is equivalent to

$$y(r) - \sum_{i=k_0}^{r-1} C(r)\phi(r, i+1)B(i)u(i) = C(r)\phi(r, k_0)x_0, \quad r = k_0+1, \ldots, k_0+N. \tag{9.28}$$

Denoting the LHS as $z(r)$ and multiplying both sides by $\phi'(r, k_0)C'(r)$, we obtain

$$\phi'(r, k_0)C'(r)z(r) = \phi'(r, k_0)C'(r)C(r)\phi(r, k_0)x_0.$$

Summing over $r = k_0+1, \ldots, k_0+N$, we get

$$\sum_{r=k_0+1}^{k_0+N} \phi'(r, k_0)C'(r)z(r) = \left[\sum_{r=k_0+1}^{k_0+N} \phi'(r, k_0)C'(r)C(r)\phi(r, k_0)\right] x_0.$$

Keeping the LHS as $P \in \mathbb{R}^n$ and the RHS as Mx_0, where M is the observability Gramian matrix, we have

$$P = Mx_0. \tag{9.29}$$

Note that P is known, and M can be obtained from the matrices $A(k)$ and $C(k)$ for $k = k_0, k_0+1, \ldots, k_0+N$. If the matrix M is nonsingular, then from (9.29), the initial condition x_0 can be uniquely determined. Hence, the system is observable.

Necessity. Suppose the system is observable but M is singular. Then, there exists a nonzero vector $\alpha \in \mathbb{R}^n$ such that $\alpha' M \alpha = 0$, i.e.,

$$\alpha' \sum_{i=k_0+1}^{k_0+N} \phi'(i,k_0)C'(i)C(i)\phi(i,k_0)\alpha = 0,$$

which can be written as

$$\sum_{i=k_0+1}^{k_0+N} (C(i)\phi(i,k_0)\alpha)'(C(i)\phi(i,k_0)\alpha) = 0,$$

implying that

$$\sum_{i=k_0+1}^{k_0+N} \|C(i)\phi(i,k_0)\alpha\|^2 = 0.$$

Hence,

$$C(i)\phi(i,k_0)\alpha = \theta \in \mathbb{R}^p \quad \text{for each } i = k_0+1,\ldots,k_0+N. \tag{9.30}$$

Let $u(i) = \theta$ for all i. Then, from (9.24), (9.25), and (9.28), we get

$$y(i) = C(i)\phi(i,k_0)x(k_0) \quad \text{for } i = k_0+1,\ldots,k_0+N. \tag{9.31}$$

Now, consider two different initial states, $x(k_0) = \alpha$ and $x(k_0) = \theta$. If $x(k_0) = \alpha$, then from (9.30) and (9.31),

$$y(i) = \theta \quad \text{for all } i = k_0+1,\ldots,k_0+N, \tag{9.32}$$

and if $x(k_0) = \theta$, then (9.31) again implies that

$$y(i) = \theta \quad \text{for all } i = k_0+1,\ldots,k_0+N. \tag{9.33}$$

If $\alpha \neq \theta$, then (9.32) and (9.33) contradict the observability of the system, as two different initial conditions result in the same observation $y(i)$ for all i. Hence, α must be θ, which implies M is nonsingular. □

***Example* 9.4.3** Consider the system $x(k+1) = A(k)x(k) + B(k)u(k)$ with observation $y(k) = C(k)x(k)$ for $k = 1,2,3$, where $A(k) = \begin{bmatrix} k^2 & 0 \\ 1 & k \end{bmatrix}$, $B(k) = \begin{bmatrix} k & 0 \end{bmatrix}$, and $C(k) = \begin{bmatrix} k & 1 \end{bmatrix}$.

Test the observability of the system.

Solution Here, $k_0 = 1$ and $k_0 + N = 3$ are the initial and final time instants. The observability Gramian matrix is

$$M = \phi'(2,1)C'(2)C(2)\phi(2,1) + \phi'(3,1)C'(3)C(3)\phi(3,1),$$

where

$$\phi(2,1) = A(1) = \begin{bmatrix} 1 & 0 \\ 1 & 1 \end{bmatrix} \text{ and } \phi(3,1) = A(2)A(1) = \begin{bmatrix} 4 & 0 \\ 3 & 2 \end{bmatrix}.$$

Thus, the observability Gramian matrix is $M = \begin{bmatrix} 234 & 33 \\ 33 & 5 \end{bmatrix}$. Since it is nonsingular, the system is observable.

Example **9.4.4** In the above example, if $u(i) = 0$ for $i = 1, 2, y(2) = 1$, and $y(3) = 4$, find the initial state $x(1)$.

Solution From (9.29), the initial state $x_0 = x(1) = M^{-1}P$, where

$$P = \phi'(2,1)C'(2)y(2) + \phi'(3,1)C'(3)y(3) = \begin{bmatrix} 63 \\ 9 \end{bmatrix}.$$

Thus, $x(1) = \begin{bmatrix} \frac{2}{9} \\ \frac{1}{3} \end{bmatrix}$.

Time-invariant System

Let A, B, and C be constant matrices of order $n \times n$, $n \times m$, and $p \times n$, respectively. Consider the system

$$x(k+1) = Ax(k) + Bu(k)$$

with the observation

$$y(k) = Cx(k) \quad \text{for } k = k_0 + 1, \ldots, k_0 + N.$$

From (9.25), the observability Gramian matrix is

$$M = \sum_{i=k_0+1}^{k_0+N} (A')^{i-k_0} C'CA^{i-k_0}. \tag{9.34}$$

According to Theorem 9.4.2, the system is observable if and only if M is nonsingular. If M is singular, then there exists a nonzero $\alpha \in \mathbb{R}^n$ such that

$$\alpha' M \alpha = 0.$$

This implies

$$\sum_{i=k_0+1}^{k_0+N} \alpha' \left(A'\right)^{i-k_0} C' C A^{i-k_0} \alpha = 0.$$

Thus, as in (9.30), we obtain $CA^{i-k_0}\alpha = \theta \in \mathbb{R}^p$ for $i = k_0+1, \ldots, k_0+N$, which can be represented as

$$\begin{bmatrix} C \\ CA \\ CA^2 \\ \vdots \\ CA^{N-1} \end{bmatrix} A\alpha = \begin{bmatrix} 0 \\ 0 \\ 0 \\ \vdots \\ 0 \end{bmatrix}. \tag{9.35}$$

Since A is assumed to be nonsingular and $\alpha \neq \theta, A\alpha \neq \theta$ in $\mathbb{R}^n$. From (9.35), it follows that

$$\operatorname{rank} \begin{bmatrix} C \\ CA \\ CA^2 \\ \vdots \\ CA^{N-1} \end{bmatrix}_{Np\times n} < n. \tag{9.36}$$

This shows that if M is singular, then rank $\begin{bmatrix} C \\ CA \\ \vdots \\ CA^{N-1} \end{bmatrix} < n$. The converse can be proved by reversing the above steps. Hence,

$$M \text{ is nonsingular} \iff \text{rank} \begin{bmatrix} C \\ CA \\ \vdots \\ CA^{N-1} \end{bmatrix} = n, \tag{9.37}$$

which is the necessary and sufficient condition for the observability of time-invariant systems. Note that if $Np < n$, then rank $\begin{bmatrix} C \\ CA \\ \vdots \\ CA^{N-1} \end{bmatrix} < n$. In this case, the system is not observable. If $N \geq n$, then

$$\text{rank} \begin{bmatrix} C \\ CA \\ \vdots \\ CA^{N-1} \end{bmatrix} = \text{rank} \begin{bmatrix} C \\ CA \\ \vdots \\ CA^{n-1} \end{bmatrix}.$$

In this case, the observability rank condition is the same as that of continuous-time systems (Corollary 5.6.7).

***Example* 9.4.5** Show that the system

$$\begin{cases} x_1(k+1) = 2x_1(k) - x_2(k), \\ x_2(k+1) = x_1(k) + x_2(k) \end{cases}$$

for $k = 1, 2$ with the observation $y(k) = x_1(k) - 2x_2(k)$ for $k = 2, 3$ is observable. If $y(2) = 1$ and $y(3) = 4$, find $x(k) : k = 1, 2, 3, 4$.

Solution Here, $A = \begin{bmatrix} 2 & -1 \\ 1 & 1 \end{bmatrix}$ and $C = \begin{bmatrix} 1 & -2 \end{bmatrix}$. It is clear that rank $\begin{bmatrix} C \\ CA \end{bmatrix} =$ rank $\begin{bmatrix} 1 & -2 \\ 0 & -3 \end{bmatrix} = 2$. Hence, the system is observable. From (9.29), we obtain

$$P = Mx_0, \tag{9.38}$$

where $P = \sum_{r=2}^{3} \phi'(r,1)C'(r)y(r)$ and $M = \sum_{r=2}^{3} \phi'(r,1)C'(r)C(r)\phi(r,1)$. First, let us find the initial state x_0. Since $\phi(2,1) = A, \phi(3,1) = A^2, M = A'C'CA + (A^2)'C'CA^2 = \begin{bmatrix} 9 & 9 \\ 9 & 18 \end{bmatrix}$, and $P = A'C'y(2) + (A^2)'C'y(3) = \begin{bmatrix} -12 \\ -15 \end{bmatrix}$, from (9.38), we obtain $x_0 = M^{-1}P = \begin{bmatrix} \frac{2}{9} & -\frac{1}{9} \\ -\frac{1}{9} & \frac{1}{9} \end{bmatrix} \begin{bmatrix} -12 \\ -15 \end{bmatrix}$. Thus, $x_0 = x(1) = \begin{bmatrix} x_1(1) \\ x_2(1) \end{bmatrix} = \begin{bmatrix} -1 \\ -\frac{1}{3} \end{bmatrix}$. Using $x(1), x(k) : k = 2,3,4$ can be obtained with the help of (9.2).

9.5 Discrete-time System Obtained from Continuous-time System

In this section, we shall see how an appropriate discrete-time system can be derived from an autonomous continuous-time system. Let A and B be constant matrices of orders $n \times n$ and $n \times m$, respectively. Consider the continuous-time control system

$$\begin{cases} \dot{x} = Ax + Bu, & t \in [t_0, \tau], \\ x(t_0) = x_0. \end{cases} \tag{9.39}$$

The solution of the above IVP can be written as

$$x(t) = e^{A(t-t_0)}x_0 + \int_{t_0}^{t} e^{A(t-s)}Bu(s)ds. \tag{9.40}$$

Let $h = \frac{\tau - t_0}{N}$, where N is a positive integer. Consider the partition of the interval $[t_0, \tau]$: $t_0, t_0 + h, t_0 + 2h, \ldots, t_0 + Nh$. From (9.40), we obtain

$$x(t_0 + h) = e^{Ah}x(t_0) + \int_{t_0}^{t_0+h} e^{A(t_0+h-s)}Bu(s)ds. \tag{9.41}$$

Similarly,

$$x(t_0+kh) = e^{Ah}x(t_0+(k-1)h) + \int_{t_0+(k-1)h}^{t_0+kh} e^{A(t_0+kh-s)}Bu(s)ds, \quad k=1,2,\ldots,N.$$

Let us approximate the function $u(s)$ in the discrete form as

$$u(s) = u(t_0+(k-1)h), \quad s \in [t_0+(k-1)h, t_0+kh), k=1,2,\ldots,N.$$

Substituting the above values for $u(s)$ in (9.41), we get

$$x(t_0+h) = e^{Ah}x(t_0) + \int_{t_0}^{t_0+h} e^{A(t_0+h-s)}Bu(t_0)ds,$$

$$x(t_0+2h) = e^{Ah}x(t_0+h) + \int_{t_0+h}^{t_0+2h} e^{A(t_0+2h-s)}Bu(t_0+h)ds,$$

and in general,

$$x(t_0+kh) = e^{Ah}x(t_0+(k-1)h) + \int_{t_0+(k-1)h}^{t_0+kh} e^{A(t_0+kh-s)}Bu(t_0+(k-1)h)ds \tag{9.42}$$

for $k=1,2,\ldots,N$. Since the discretized function $u(s)$ is a constant in each subinterval, (9.42) becomes

$$x(t_0+kh) = e^{Ah}x(t_0+(k-1)h) + \left(\int_{t_0+(k-1)h}^{t_0+kh} e^{A(t_0+kh-s)}Bds\right)u(t_0+(k-1)h), \tag{9.43}$$

for $k=1,2,\ldots,N$. Now, let $r = t_0+kh-s$. Then,

$$\int_{t_0+(k-1)h}^{t_0+kh} e^{A(t_0+kh-s)}Bds = \int_0^h e^{Ar}Bdr. \tag{9.44}$$

Using (9.44) in (9.43), we obtain

$$x(t_0+kh) = e^{Ah}x(t_0+(k-1)h) + \left(\int_0^h e^{Ar}Bdr\right)u(t_0+(k-1)h) \tag{9.45}$$

for $k=1,2,\ldots,N$. Now, let us denote $x(t_0+kh)$ by $X(k)$ and $u(t_0+kh)$ by $U(k)$, where $k=1,2,\ldots,N$. Then (9.45) can be written as

$$X(k) = e^{Ah}X(k-1) + \left(\int_0^h e^{Ar}Bdr\right)U(k-1), \quad k=1,2,\ldots,N. \tag{9.46}$$

Thus, (9.46) can be considered as the discrete-time control system corresponding to the continuous-time system (9.39). The system (9.46) can be written as

$$X(k+1) = FX(k) + GU(k), \quad k = 0,1,2,\ldots,N-1, \tag{9.47}$$

where $X(0) = x(t_0), F = e^{Ah}$, and $G = \left(\int_0^h e^{Ar}dr\right)B$.

***Example* 9.5.1** The equation of motion of an object of mass m under the influence of a force $u(t)$ and a frictional force proportional to the velocity is written as $m\ddot{x} = u - \alpha\dot{x}$, where $\alpha > 0$. Considering $u(t)$ as the control function, write the continuous-time and the corresponding discrete-time systems for the time interval $0 \le t \le \tau$.

Solution Let $x_1 = x$ and $x_2 = \dot{x}$ be the state variables representing displacement and velocity. Then, the continuous-time control system is

$$\dot{x} = Ax + Bu, \quad 0 \le t \le \tau, \tag{9.48}$$

where $A = \begin{bmatrix} 0 & 1 \\ 0 & -\alpha \end{bmatrix}$ and $B = \begin{bmatrix} 0 \\ 1 \end{bmatrix}$. Now, by partitioning $[0,\tau]$ into N equal subintervals, we get $h = \frac{\tau}{N}$. Then, following (9.47), the discrete-time system corresponding to (9.48) is

$$\begin{bmatrix} X_1(k+1) \\ X_2(k+1) \end{bmatrix} = F \begin{bmatrix} X_1(k) \\ X_2(k) \end{bmatrix} + GU(k), \quad k = 0,1,2,\ldots,N-1,$$

where

$$F = e^{Ah} = \begin{bmatrix} 1 & 1-e^{-\alpha h} \\ 0 & e^{-\alpha h} \end{bmatrix}, G = \int_0^h e^{Ar}Bdr = \begin{bmatrix} \frac{h\alpha - 1 + e^{-\alpha h}}{\alpha} \\ \frac{1-e^{-\alpha h}}{\alpha} \end{bmatrix}, \text{ and } U(k) = u(kh).$$

In the above discussion, we obtained a discrete-time system (9.47) from the continuous-time system (9.39). For sufficiently small values of $h > 0$, one can obtain approximate solution for (9.39) by solving (9.47). However, for certain values of h, the controllability property of (9.39) may differ from that of (9.47). Since it is desirable to maintain the controllability property of both (9.39) and the corresponding discrete-time system (9.47), the choice of h is very important. The following result provides the procedure for choosing h so that the controllability property is preserved.

Let the Jordan canonical form of the matrix A be (see Section 2.7)

$$J = \begin{bmatrix} J_1 & & & \\ & J_2 & & \\ & & \ddots & \\ & & & J_p \end{bmatrix},$$

with Jordan blocks

$$J_i = \begin{bmatrix} J_{i1} & & & \\ & J_{i2} & & \\ & & \ddots & \\ & & & J_{il_i} \end{bmatrix}_{m_i \times m_i}, \quad i = 1, 2, \ldots, p,$$

where m_i and l_i are the algebraic and geometric multiplicities of λ_i, respectively, and the sub-blocks

$$J_{ij} = \begin{bmatrix} \lambda_i & 1 & & & 0 \\ & \lambda_i & 1 & & \\ & & \ddots & \ddots & \\ & & & \ddots & 1 \\ 0 & & & & \lambda_i \end{bmatrix}, \quad j = 1, 2, \ldots, l_i.$$

The Jordan sub-block J_{ij} is an $r_{ij} \times r_{ij}$ matrix, where $(r_{ij} - 1)$ is the number of generalized eigenvectors corresponding to the eigenpair (λ_i, x_{ij}). Let $J = P^{-1}AP$, where P is a nonsingular matrix as described in Section 2.7 (equation (2.29)). Then, $\dot{x} = Ax + Bu \implies P^{-1}\dot{x} = P^{-1}Ax + P^{-1}Bu$. Let $P^{-1}x = z$. Then,

$$\dot{z} = P^{-1}APz + P^{-1}Bu \implies \dot{z} = Jz + Du, \tag{9.49}$$

where $D = P^{-1}B$. Note that

$$\begin{aligned} \begin{bmatrix} B & AB & A^2B & \cdots & A^{n-1}B \end{bmatrix} &= P \begin{bmatrix} P^{-1}B & JP^{-1}B & \cdots & J^{n-1}P^{-1}B \end{bmatrix} \\ &= P \begin{bmatrix} D & JD & \cdots & J^{n-1}D \end{bmatrix}. \end{aligned} \tag{9.50}$$

From (9.50), it can be concluded that the system (9.39) and the corresponding canonical system (9.49) have the same controllability property. Therefore, we can study the canonical system (9.49) instead of (9.39). Now, let the discrete-time system corresponding to (9.49) be

$$Z(k+1) = e^{Jh}Z(k) + \left(\int_0^h e^{Jr}dr\right)DU(k) = \bar{F}Z(k) + \bar{G}U(k) \tag{9.51}$$

for $k = 0,1,2,\ldots,N-1$, where $Z(k) = z(t_0 + kh)$. Thus, by analyzing the canonical system (9.49) and the corresponding discrete-time system (9.51), we can obtain the controllability result for the system (9.39) and the corresponding discrete-time system (9.47). Now, (9.49) can be written as

$$\dot{z}(t) = \begin{bmatrix} \dot{z}_1 \\ \dot{z}_2 \\ \vdots \\ \dot{z}_p \end{bmatrix} = \begin{bmatrix} J_1 & & & \\ & J_2 & & \\ & & \ddots & \\ & & & J_p \end{bmatrix} \begin{bmatrix} z_1 \\ z_2 \\ \vdots \\ z_p \end{bmatrix} + \begin{bmatrix} D_1 \\ D_2 \\ \vdots \\ D_p \end{bmatrix} u(t), \tag{9.52}$$

and (9.51) can be written as

$$\begin{bmatrix} Z_1(k+1) \\ Z_2(k+1) \\ \vdots \\ Z_p(k+1) \end{bmatrix} = \begin{bmatrix} e^{J_1h} & & & \\ & e^{J_2h} & & \\ & & \ddots & \\ & & & e^{J_ph} \end{bmatrix} \begin{bmatrix} Z_1(k) \\ Z_2(k) \\ \vdots \\ Z_p(k) \end{bmatrix} + \begin{bmatrix} \overline{G}_1 \\ \overline{G}_2 \\ \vdots \\ \overline{G}_p \end{bmatrix} U(k), \tag{9.53}$$

where $\overline{G}_i = \left(\int_0^h e^{J_ir}dr\right)D_i$. Note that z_i and Z_i are in $\mathbb{R}^{m_i}$. From 9.52 and 9.53, we obtain the following relations:

$$z_i = J_iz_i + D_iu \text{ and } Z_i(k+1) = e^{J_ih}Z_i(k) + \bar{G}_iU(k) \quad \text{for } i = 1,2,\ldots,p,$$

respectively. Further, using the sub-blocks of J_i and the corresponding sub-blocks of D_i, we obtain

$$z_{ij} = J_{ij}z_{ij} + D_{ij}u(t) \quad \text{for } j = 1,2,\ldots,l_i.$$

Let us now consider the following theorem and the corollaries.

Theorem 9.5.2 The rank $\begin{bmatrix} B & AB & A^2B & \cdots & A^{n-1}B \end{bmatrix} = n$ if and only if the n rows of $e^{At}B$ are linearly independent as functions of t.

Corollary 9.5.3 The system (9.39) is controllable if and only if the n rows of the matrix $e^{At}B$ are linearly independent as functions of t.

The proofs of both Theorem 9.5.2 and Corollary 9.5.3 follow from Lemma 3.3.1 and Corollary 3.3.2.

Corollary 9.5.4 The system (9.39) is controllable if and only if the rows of $(sI - A)^{-1}B$ are linearly independent as function of s.

Proof. The Laplace transform of $e^{At}B$ is $(sI - A)^{-1}B$ (Lemma 2.8.10). Since the Laplace transform is a one-to-one linear transformation, the linearly independent rows of $e^{At}B$ are mapped into linearly independent rows of $(sI - A)^{-1}B$. Hence, by Corollary 9.5.3, the system (9.39) is controllable if and only if the rows of $(sI - A)^{-1}B$ are linearly independent as functions of s. □

As system (9.39) can be transformed into the canonical form (9.49), by Corollary 9.5.4, it follows that the system (9.39) is controllable if and only if the rows of $(sI - A)^{-1}B$ are linearly independent as functions of s. Equivalently, the system (9.49) is controllable if and only if the rows of $(sI - J)^{-1}D$ are linearly independent as functions of s (since $A = PJP^{-1}$ and $B = PD$).

Let

$$J_{ij} = \begin{bmatrix} \lambda_i & 1 & & & 0 \\ & \lambda_i & 1 & & \\ & & \ddots & \ddots & \\ & & & \ddots & 1 \\ 0 & & & & \lambda_i \end{bmatrix}_{r_{ij} \times r_{ij}}.$$

Then,

$$(sI - J_{ij})^{-1} = \begin{bmatrix} \frac{1}{s-\lambda_i} & \frac{1}{(s-\lambda_i)^2} & & \frac{1}{(s-\lambda_i)^a} \\ & \frac{1}{s-\lambda_i} & & \frac{1}{(s-\lambda_i)^{a-1}} \\ & & \ddots & \\ 0 & & & \frac{1}{s-\lambda_i} \end{bmatrix}. \tag{9.54}$$

Theorem 9.5.5 Consider the sub-blocks D_{ij} of the matrix D_i for $i = 1,2,\ldots,p$ and $j = 1,2,\ldots,l_i$. Let D_{ij} be $r_{ij} \times m$ matrix defined by

$$D_{ij} = \begin{bmatrix} d_{ij1} \\ d_{ij2} \\ \vdots \\ d_{ijr_{ij}} \end{bmatrix}_{r_{ij}\times m}.$$

Let $DL(i)$ be a matrix having l_i rows, where the kth row is the last row of the matrix D_{ik}, i.e.,

$$DL(i) = \begin{bmatrix} D_{i1r_{i1}} \\ D_{i2r_{i2}} \\ \vdots \\ D_{il_ir_{il_i}} \end{bmatrix}_{l_i\times m}. \quad (9.55)$$

Then, the system (9.49) is controllable if and only if $\text{rank}(DL(i)) = l_i$ for each $i = 1,2,\ldots,p$.

Before proving the theorem, let us consider some examples.

***Example* 9.5.6** If in (9.49) and (9.50), the matrices J and D are given by

$$J = \left[\begin{array}{c|cc|c|cc|ccc} 2 & & & & & & & & \\ \hline & 2 & 1 & & & & & & \\ & 0 & 2 & & & & & & \\ \hline & & & 3 & & & & & \\ \hline & & & & 3 & 1 & & & \\ & & & & 0 & 3 & & & \\ \hline & & & & & & 3 & 1 & 0 \\ & & & & & & 0 & 3 & 1 \\ & & & & & & 0 & 0 & 3 \end{array}\right]_{9\times 9} = \left[\begin{array}{c|c} J_1 & O \\ \hline O & J_2 \end{array}\right] \text{ and } D = \left[\begin{array}{cc} 2 & 1 \\ \hline 0 & 1 \\ 3 & 2 \\ \hline 1 & 1 \\ \hline 1 & 0 \\ 2 & 4 \\ \hline 0 & 2 \\ 1 & -1 \\ -2 & 1 \end{array}\right]_{9\times 2},$$

then

$$J_1 = \left[\begin{array}{c|cc} 2 & & \\ \hline & 2 & 1 \\ & 0 & 2 \end{array}\right]_{3\times 3} \text{ and } J_2 = \left[\begin{array}{c|cc|ccc} 3 & & & & & \\ \hline & 3 & 1 & & & \\ & 0 & 3 & & & \\ \hline & & & 3 & 1 & 0 \\ & & & 0 & 3 & 1 \\ & & & 0 & 0 & 3 \end{array}\right]_{6\times 6}.$$

The blocks of J_1 are

$$J_{11} = \begin{bmatrix} 2 \end{bmatrix} \text{ and } J_{12} = \begin{bmatrix} 2 & 1 \\ 0 & 2 \end{bmatrix} \text{ with } l_1 = 2,$$

and the blocks of J_2 are

$$J_{21} = \begin{bmatrix} 3 \end{bmatrix}, J_{22} = \begin{bmatrix} 3 & 1 \\ 0 & 3 \end{bmatrix}, \text{ and } J_{23} = \begin{bmatrix} 3 & 1 & 0 \\ 0 & 3 & 1 \\ 0 & 0 & 3 \end{bmatrix} \text{ with } l_2 = 3.$$

Then, from (9.52), the blocks of D are

$$D_1 = \left[\begin{array}{cc} 2 & 1 \\ \hline 0 & 1 \\ 3 & 2 \end{array}\right] \text{ and } D_2 = \left[\begin{array}{cc} 1 & 1 \\ \hline 1 & 0 \\ 2 & 4 \\ \hline 0 & 2 \\ 1 & -1 \\ -2 & 1 \end{array}\right],$$

where the subblocks are $D_{11} = \begin{bmatrix} 2 & 1 \end{bmatrix}$, $D_{12} = \begin{bmatrix} 0 & 1 \\ 3 & 2 \end{bmatrix}$, $D_{21} = \begin{bmatrix} 1 & 1 \end{bmatrix}$, $D_{22} = \begin{bmatrix} 1 & 0 \\ 2 & 4 \end{bmatrix}$,

and $D_{23} = \begin{bmatrix} 0 & 2 \\ 1 & -1 \\ -2 & 1 \end{bmatrix}$. From (9.55), $DL(1) = \begin{bmatrix} 2 & 1 \\ 3 & 2 \end{bmatrix}$ and $DL(2) = \begin{bmatrix} 1 & 1 \\ 2 & 4 \\ -2 & 1 \end{bmatrix}$. The rows of $DL(1)$ are linearly independent, but the rows of $DL(2)$ are linearly dependent. Hence, by Theorem 9.5.5, the system $\dot{x} = Jx + Du$ is not controllable.

Remark 9.5.7 We know that the $\text{rank}(DL(i)) \leq \min\{l_i, m\}$ for $i = 1, 2, \ldots, p$, where l_i is the geometric multiplicity of the eigenvalue λ_i, and m is the number of columns of the matrix B. From Theorem 9.5.5, we infer the following:

(*i*) If $l_i \leq m$ and $\text{rank}(DL(i)) = l_i$ for each $i = 1, 2, \ldots, m$, then the system is controllable.

(*ii*) If $l_i > m$ for any $i = 1, 2, \ldots, p$, then the system is not controllable.

***Example* 9.5.8** Consider the system $\dot{x} = Jx + Bu$, where

$$J = \begin{bmatrix} 3 & 0 & 0 \\ 0 & 2 & 0 \\ 0 & 0 & 2 \end{bmatrix} \text{ and } B = \begin{bmatrix} 1 \\ 2 \\ 3 \end{bmatrix}.$$

Here, $\lambda_1 = 3, m_1 = l_1 = 1, \lambda_2 = 2, m_2 = l_2 = 2, n = 3$, and $m = 1$. Since $l_2 > m$, by Theorem 9.5.2, the system is not controllable.

Proof (of Theorem 9.5.5). According to Corollary 9.5.4, the system (9.49) is controllable if and only if $\text{rank}((sI - J)^{-1}D) = n$, where

$$(sI - J)^{-1}D = \begin{bmatrix} (sI - J_1)^{-1} & & & 0 \\ & (sI - J_2)^{-1} & & \\ & & \ddots & \\ 0 & & & (sI - J_p)^{-1} \end{bmatrix} \begin{bmatrix} D_1 \\ D_2 \\ \vdots \\ D_p \end{bmatrix} \tag{9.56}$$

with

$$(sI-J_i)^{-1}D_i = \begin{bmatrix} (sI-J_{i1})^{-1} & & & 0 \\ & (sI-J_{i2})^{-1} & & \\ & & \ddots & \\ 0 & & & (sI-J_{il_i})^{-1} \end{bmatrix} \begin{bmatrix} D_{i1} \\ D_{i2} \\ \vdots \\ D_{il_i} \end{bmatrix}. \tag{9.57}$$

Now,

$$(sI-J_{ij})^{-1}D_{ij} = \begin{bmatrix} \frac{1}{s-\lambda_i} & \frac{-1}{(s-\lambda_i)^2} & & \frac{(-1)^{n+1}}{(s-\lambda_i)^{r_{ij}}} \\ & \frac{1}{s-\lambda_i} & & \frac{(-1)^n}{(s-\lambda_i)^{r_{ij}-1}} \\ & & \ddots & \vdots \\ 0 & & & \frac{1}{s-\lambda_i} \end{bmatrix} \begin{bmatrix} d_{ij1} \\ d_{ij2} \\ \vdots \\ d_{ijr_{ij}} \end{bmatrix}_{r_{ij}\times m}. \tag{9.58}$$

The rows of $(sI-J_{ij})^{-1}D_{ij}$ are as follows:

$$\begin{cases} \text{first row} = \frac{1}{s-\lambda_i}d_{ij1} + \frac{1}{(s-\lambda_i)^2}d_{ij2} + \cdots + \frac{1}{(s-\lambda_i)^{r_{ij}}}d_{ijr_{ij}}, \\ \text{second row} = \frac{1}{s-\lambda_i}d_{ij2} + \frac{1}{(s-\lambda_i)^2}d_{ij3} + \cdots + \frac{1}{(s-\lambda_i)^{r_{ij}-1}}d_{ijr_{ij}}, \\ \quad\vdots \\ r_{ij}^{\text{th}}\text{ row} = \frac{1}{s-\lambda_i}d_{ijr_{ij}}. \end{cases} \tag{9.59}$$

It is clear that if the row $d_{ijr_{ij}}$ (the last row of D_{ij}) is nonzero, then the rows mentioned above are linearly independent as functions of s, i.e., the rows of $(sI-J_{ij})^{-1}D_{ij}$ are linearly independent as functions of s. Now, consider $DL(i)$ as in (9.55)

$$DL(i) = \begin{bmatrix} D_{i1r_{i1}} \\ D_{i2r_{i2}} \\ \vdots \\ D_{il_ir_{il_i}} \end{bmatrix}_{l_i\times m}. \tag{9.60}$$

From (9.57)–(9.60), one can conclude that if the rows of $DL(i)$ are linearly independent, then the m_i rows of $(sI-J_i)^{-1}D_i$ are linearly independent as functions of s.

Thus, if the rank of $DL(i) = l_i$ for each $i = 1, 2, \ldots, p$, then from (9.56) and (9.57), we infer that the n rows of $(sI - J)^{-1}D$ are linearly independent as functions of s, which implies the controllability of the system (9.49). □

Now consider the discrete-time system (9.53), which is obtained from the continuous-time system (9.52). We know that the system (9.52) is controllable if and only if $\text{rank}\begin{bmatrix} D & JD & J^2D & \cdots & J^{n-1}D \end{bmatrix} = n$, which is equivalent to saying that the rows of $(sI - J)^{-1}D$ are linearly independent as functions of s. Similarly, the discrete-time system (9.53) is controllable if and only if $\text{rank}\begin{bmatrix} \overline{G} & e^{Jh}\overline{G} & e^{2Jh}\overline{G} & \cdots & e^{(N-1)Jh}\overline{G} \end{bmatrix} = n$, which is also equivalent to saying that the $n \times m$ matrix $(sI - e^{Jh})^{-1}\overline{G}$ has linearly independent rows as functions of s. Recall from (9.51) and (9.52) that

$$\overline{G} = \left(\int_0^h e^{Jr} dr \right) D \text{ and } \overline{G}_i = \left(\int_0^h e^{J_i r} dr \right) D_i \quad \text{for } i = 1, 2, \ldots, p.$$

Now, using the Cayley–Hamilton theorem, it is known that e^{Jh} can be written as

$$e^{Jh} = a_0 I + a_1 J + a_2 J^2 + \cdots + a_{n-1} J^{n-1}$$

for some suitable constants $a_i : i = 0, 1, 2, \ldots, n-1$. Hence,

$$(sI - e^{Jh})^{-1} \left(\int_0^h e^{Jr} dr \right) = \left(\int_0^h e^{Jr} dr \right) (sI - e^{Jh})^{-1},$$

since $(sI - e^{Jh})^{-1}$ commutes with J^{α} for $\alpha = 1, 2, \ldots$. Therefore,

$$(sI - e^{Jh})^{-1}\overline{G} = (sI - e^{Jh})^{-1} \left[\left(\int_0^h e^{Jr} dr \right) D \right] = \left(\int_0^h e^{Jr} dr \right) (sI - e^{Jh})^{-1} D.$$

As $\int_0^h e^{Jr} dr$ is a nonsingular matrix (for $h > 0$), the number of linearly independent rows of $(sI - e^{Jh})^{-1}\overline{G}$ and $(sI - e^{Jh})^{-1}D$ are equal.

Now consider the matrices $P(s) = (sI - J)^{-1}D$ and $Q(s) = (sI - e^{Jh})^{-1}D$. Let the rows of $P(s)$ be linearly independent. Then the controllability of the continuous-time system (9.49) holds.

Let the jth and k^{th} eigenvalues of J be α and $\alpha + \frac{2\pi i}{h}$, respectively. In this case, both $(sI - J_{jl_j})^{-1}D_{jl_j}$ and $(sI - J_{kl_k})^{-1}D_{kl_k}$ have linearly independent rows as functions of s (from (9.59). However, the rows of $(sI - e^{Jh})^{-1}D$ are **linearly dependent** as the jth and k^{th} eigenvalues $e^{\alpha h}$ and $e^{\left(\alpha + \frac{2\pi i}{h}\right)h}$ are the same. Hence, the discrete-time system is not controllable, while the continuous-time system is controllable. Thus, it is important to choose the time increment h such that both λ and $\lambda + \frac{2q\pi i}{h}$ together are not eigenvalues of the matrix A, for any constant λ (real or complex) and any integer q.

Example **9.5.9** Consider the system $\dot{x} = Ax + Bu$, $0 \le t \le 10$. Let $h = 0.1$ be the length of each subinterval. For the matrix $A = \begin{bmatrix} 1 & -10\pi \\ 10\pi & 1 \end{bmatrix}$, the eigenvalues are $\lambda_1 = 1 + 10\pi i = 1 + \frac{\pi i}{h}$ and $\lambda_2 = 1 - \frac{\pi i}{h}$ which implies that $\lambda_1 = \lambda_2 + \frac{2\pi i}{h}$. From the above discussion, it is clear that if the continuous-time system $\dot{x} = Ax + Bu$ is controllable, then the discrete-time system obtained with $h = 0.1$ is not controllable. Hence, in this case, if we choose any $h \neq \frac{q}{10}$ for any integer q, the controllability properties of the continuous-time and corresponding discrete-time systems will be the same.

Exercise 9.5.1 For the continuous-time system $\dot{x} = Ax + Bu$, $0 \le t \le 1$, where $A = \begin{bmatrix} 2 & 1 \\ 0 & 2 \end{bmatrix}$ and $B = \begin{bmatrix} 0 \\ 1 \end{bmatrix}$, write the corresponding discrete-time system $z(k+1) = Fz(k) + Gv(k), k = 0, 1, 2, \ldots, 10$. If $x(0) = \begin{bmatrix} 2 \\ 1 \end{bmatrix}$ and $u(t) = 1, 0 \le t \le 1$, find $x(t)$, $0 \le t \le 1$, $z(k)$, $k = 0, 1, 2, \ldots, 10$, and the modulus of the difference between the continuous and discrete solutions. $|x(\frac{k}{10}) - z(k)|, k = 0, 1, 2, \ldots, 10$. Also show that both the systems are controllable.

9.6 Dual System

Consider the following discrete-time control system

$$x(k+1) = A(k)x(k) + B(k)u(k) \quad \text{for } k = k_0, k_0+1, \ldots, k_0+N-1 \tag{9.61}$$

with the observation

$$y(k) = C(k)x(k) \quad \text{for } k = k_0+1, \ldots, k_0+N \tag{9.62}$$

and the control system

$$z(k+1) = (A')^{-1}(k)z(k) + C'(k+1)v(k) \quad \text{for } k = k_0, k_0+1, \ldots, k_0+N-1 \tag{9.63}$$

with the observation

$$w(k) = B'(k-1)z(k) \quad \text{for } k = k_0+1, \ldots, k_0+N. \tag{9.64}$$

From (9.4), the state transition matrix of system (9.61) is

$$\phi(r,k_0) = A(r-1)A(r-2)\ldots A(k_0), \quad r = k_0+1,\ldots,k_0+N, \tag{9.65}$$

and the state transition matrix of system (9.63) is

$$\psi(r,k_0) = (A')^{-1}(r-1)(A')^{-1}(r-2)\ldots(A')^{-1}(k_0). \tag{9.66}$$

Then, (9.65) and (9.66) imply

$$\psi(r,k_0) = \phi'(k_0,r). \tag{9.67}$$

From (9.11), the controllability Gramian matrix for system (9.61) is

$$W = \sum_{i=k_0}^{k_0+N-1} \phi(k_0,i+1)B(i)B'(i)\phi'(k_0,i+1), \tag{9.68}$$

and from (9.25), the observability Gramian matrix for system (9.63) is

$$\begin{aligned} M &= \sum_{i=k_0+1}^{k_0+N} \psi'(i,k_0)B(i-1)B'(i-1)\psi(i,k_0) \\ &= \sum_{i=k_0}^{k_0+N-1} \psi'(i+1,k_0)B(i)B'(i)\psi(i+1,k_0). \end{aligned} \tag{9.69}$$

Using (9.67) in (9.69), it is clear that $M = W$, which indicates that the observability Gramian matrix of system (9.63) is the same as the controllability Gramian matrix of system (9.61). Similarly, it can be shown that the observability Gramian of system (9.61) is the same as the controllability Gramian matrix of system (9.63). Hence, the following theorem holds.

Theorem 9.6.1 The system (9.61) is controllable (observable) if and only if the system (9.63) is observable (controllable).

Remark 9.6.2 The above theorem provides an alternative proof for the observability rank condition for **time-invariant** systems.

By applying Theorem 9.6.1, the rank condition for the observability of time-invariant systems can be obtained from the controllability condition of the dual system.

Let the time-invariant system be

$$\begin{cases} x(k+1) = Ax(k) + Bu(k), & k = k_0, k_0+1,\ldots,k_0+N-1, \\ y(k) = Cx(k), & k = k_0+1, k_0+2,\ldots,k_0+N. \end{cases} \tag{9.70}$$

Then its dual system is

$$\begin{cases} z(k+1) = (A')^{-1}z(k) + C'v(k), & k = k_0, k_0+1, \ldots, k_0+N-1, \\ w(k) = B'z(k), & k = k_0+1, k_0+2, \ldots, k_0+N. \end{cases} \tag{9.71}$$

From Theorem 9.3.3, the system (9.70) is observable if and only if the system (9.71) is controllable. System (9.71) is controllable if and only if (Theorem 9.33)

$$\operatorname{rank}\begin{bmatrix} C' & (A')^{-1}C' & \cdots & (A')^{1-N}C' \end{bmatrix} = n$$

$$\Longrightarrow \operatorname{rank}\begin{bmatrix} (A')^{N-1}C' & (A')^{N-2}C' & \cdots & C' \end{bmatrix} = n \quad \text{(since } A \text{ is nonsingular)}$$

$$\Longrightarrow \operatorname{rank}\begin{bmatrix} C \\ CA \\ \vdots \\ CA^{N-1} \end{bmatrix} = n,$$

which is the same as the condition given in (9.37).

9.7 Stability and Stabilizability of Discrete-time Systems

Consider the linear discrete-time system

$$x(k+1) = Ax(k), \quad k = 0, 1, 2, \ldots, \tag{9.72}$$

where A is an $n \times n$ constant nonsingular matrix. The trivial solution $x(k) = \theta$ for all k is the **equilibrium** for the system (9.72). Let $x(0) = x_0 \in \mathbb{R}^n$ be the initial condition. Then the solution of (9.72) is

$$x(r) = A^r x_0, \quad r = 1, 2, \ldots. \tag{9.73}$$

Definition 9.7.1 The trivial solution θ of the system (9.72) is called **stable** if, for any given $\varepsilon > 0$, there exists a $\delta > 0$ such that, whenever $\|x_0\| < \delta$,

$$\|x(k)\| < \varepsilon \quad \text{for } k = 1, 2, \ldots.$$

Let λ_j and x_j for $j = 1, 2, \ldots, n$ be the eigenvalues and corresponding eigenvectors of A.

If A is diagonalizable, then $\{x_1, x_2, \ldots, x_n\}$ forms a basis of $\mathbb{R}^n$. The initial condition x_0 can be uniquely expressed as

$$x_0 = \alpha_1 x_1 + \alpha_2 x_2 + \cdots + \alpha_n x_n, \tag{9.74}$$

where $\alpha_{i_j} : j = 1, 2, \ldots, n$ are real constants. From (9.73) and (9.74), we have

$$x(r) = \alpha_1 \lambda_1^r x_1 + \alpha_2 \lambda_2^r x_2 + \cdots + \alpha_n \lambda_n^r x_n. \tag{9.75}$$

If λ_j is complex, then let $\lambda_j = R_j e^{i\phi}$, where $i = \sqrt{-1}$. Then $\lambda_j^r = R_j^r e^{ir\phi}$. If $|\lambda_j| = R_j < 1$, then $\lambda_j^r \to 0$ as $r \to \infty$. If $|\lambda_j| = R_j > 1$, then λ_j^r becomes unbounded as $r \to \infty$. Hence, from (9.75), we can conclude the following:

(*i*) $x(r) \to \theta$ as $r \to \infty$ if all the eigenvalues λ_j lie in the interior of the unit circle. Therefore, if $|\lambda_j| < 1$ for all j, then the trivial solution is asymptotically stable.

(*ii*) If $|\lambda_j| > 1$ for some j, then $R_j^r \to \infty$ as $r \to \infty$. Consequently, $x(r)$ becomes unbounded as $r \to \infty$. Therefore, if $|\lambda_j| > 1$ for some j, then the zero solution is unstable.

(*iii*) If $|\lambda_j| = 1$ for some j and the remaining eigenvalues have moduli < 1, then in (9.75), only the jth term has modulus $|\alpha_j R_j^r e^{ir\phi}| = |\alpha_j|$. Hence, $|x(r)| < |\alpha_j|$, a finite value for all r, implying stability.

Thus, we obtain the following theorem.

Theorem 9.7.1 If A is an $n \times n$ **nonsingular diagonalizable** matrix, then the system $x(k+1) = Ax(k)$ for $k = 0, 1, 2, \ldots$ is

(*i*) asymptotically stable at the origin if and only if all the eigenvalues of A lie in the interior of the unit circle in the complex plane, i.e., $|\lambda_j| < 1$ for all $j = 1, 2, \ldots, n$,

(*ii*) unstable at the origin if and only if $|\lambda_j| > 1$ for any j,

(*iii*) stable but not asymptotically stable at the origin if and only if $|\lambda_j| \leq 1$ for all $j = 1, 2, \ldots, n$ and at least one eigenvalue has modulus 1.

Alternative proof of Theorem 9.7.1. Since A is diagonalizable, we can write $A = PDP^{-1}$, where D is the diagonal matrix with the eigenvalues of A on the diagonal and P is the matrix with the corresponding eigenvectors as columns. Then the solution of (9.72) is

$$x(r) = A^r x_0 = PD^r P^{-1} x_0. \tag{9.76}$$

(*i*) If the eigenvalues $\lambda_j : j = 1, 2, \ldots, n$ of A lie inside the unit circle in the complex plane, i.e., if $|\lambda_j| < 1$, then $\lim_{r\to\infty} \lambda_j^r = 0$. Hence, $\lim_{r\to\infty} D^r = \mathbb{O}$. From (9.76), we infer that $\lim_{r\to\infty} x(r) = \theta \in \mathbb{R}^n$.
(*ii*) If $|\lambda_j| > 1$ for some $j = 1, 2, \ldots, n$, then $\lim_{r\to\infty} \lambda_j^r = \infty$ in the complex plane. Thus, D^r is a diagonal matrix whose j^{th} diagonal element λ_j^r, is unbounded as $r \to \infty$.

Hence, from (9.76), we can conclude that $\lim_{r\to\infty} x(r) = \infty$, implying instability at the origin.

(iii) If A is **not diagonalizable**, then A has a Jordan canonical form J, where $J = P^{-1}AP$. The solution of the IVP

$$\begin{cases} x(k+1) = Ax(k), & k = 0, 1, \ldots, \\ x(k_0) = x_0 \end{cases}$$

is given by

$$x(r) = A^r x_0 = PJ^r P^{-1} x_0. \tag{9.77}$$

Let J_i be the ith Jordan block of the form

$$J = \begin{bmatrix} \lambda & 1 & 0 & \cdots & & 0 \\ 0 & \lambda & 1 & \cdots & & 0 \\ \vdots & & & & & \\ 0 & 0 & 0 & \cdots & \lambda & 1 \\ 0 & 0 & 0 & \cdots & & \lambda \end{bmatrix}_{l_i \times l_i}. \tag{9.78}$$

Then,

$$J_i^r = \begin{bmatrix} \lambda^r & \begin{pmatrix} r \\ 1 \end{pmatrix} \lambda^{r-1} & \begin{pmatrix} r \\ 2 \end{pmatrix} \lambda^{r-2} & \cdots & \\ 0 & \lambda^r & \begin{pmatrix} r \\ 1 \end{pmatrix} \lambda^{r-1} & \cdots & \\ \vdots & & & & \\ 0 & 0 & 0 & & \lambda^r \end{bmatrix}.$$

If $|\lambda| < 1$, then

$$\lim_{r\to\infty} \begin{pmatrix} r \\ k \end{pmatrix} |\lambda|^{r-k} = 0, \quad k = 0, 1, 2, \ldots, n. \tag{9.79}$$

If $|\lambda| > 1$, then

$$\lim_{r\to\infty} \begin{pmatrix} r \\ k \end{pmatrix} |\lambda|^{r-k} \to \infty, \quad k = 0, 1, 2, \ldots, n. \tag{9.80}$$

If $|\lambda| = 1$, then

$$\lim_{r\to\infty} \binom{r}{k} |\lambda|^r = \lim_{r\to\infty} \binom{r}{k} = \infty, \quad k = 1, 2, \ldots, n. \tag{9.81}$$

Thus, (i) If all the eigenvalues have moduli less than 1, then from (9.77) and (9.79), we infer that $\lim_{r\to\infty} x(r) = 0$, implying the trivial solution is asymptotically stable. (ii) If at least one of the eigenvalues has modulus greater than 1, then from (9.77) and (9.80), we can conclude that $\|x(r)\|$ is unbounded as r increases, implying that the origin is unstable. (iii) If any of the eigenvalues has modulus 1 and its algebraic multiplicity is greater than its geometric multiplicity, then it has a Jordan block of the form (9.78) with order ≥ 2. From (9.77) and (9.81), we can show that $\lim_{r\to\infty} \|x(r)\| = \infty$, which implies that the origin is unstable. □

The summary of the above discussion is given in the following theorem.

Theorem 9.7.2 Let A be any $n \times n$ matrix.

(i) If all the eigenvalues of A have modulus less than 1, then the trivial solution of (9.72) is asymptotically stable.

(ii) If any of the eigenvalues of A has modulus greater than 1, then the equilibrium $x = \theta$ is unstable.

(iii) If A is diagonalizable and the eigenvalues have modulus ≤ 1, with at least one eigenvalue having modulus 1, then $x = \theta$ is stable but not asymptotically stable.

(iv) If A is not diagonalizable and has eigenvalues with modulus ≤ 1, then $x = \theta$ is stable but not asymptotically stable provided that every eigenvalue with modulus 1 has algebraic multiplicity equal to its geometric multiplicity.

(v) If any eigenvalue with modulus 1 has algebraic multiplicity greater than its geometric multiplicity, then $x = \theta$ is unstable.

9.8 Lyapunov Theorem for Discrete-time Systems

As in the case of continuous-time systems, the stability properties of discrete-time systems can be obtained using discrete Lyapunov functions. Consider the time-invariant discrete-time system

$$x(k+1) = F(x(k)), \quad k = 0, 1, 2, \ldots, \tag{9.82}$$

where $x(k) \in \mathbb{R}^n$ and $F : \mathbb{R}^n \to \mathbb{R}^n$ is a function satisfying $F(\theta) = \theta$. The following theorem for the system (9.82) is analogous to Theorem 8.5.9.

Theorem 9.8.1 The discrete-time system (9.82) is stable at the origin θ if there exist a constant $r > 0$ and a Lyapunov function $V : B_r(\theta) \subset \mathbb{R}^n \to \mathbb{R}$ satisfying the conditions: (i) $V(\theta) = 0$, (ii) $V(x(k)) > 0$, and (iii) $V(x(k+1)) \leq V(x(k))$, $k = 0, 1, 2, \ldots$, for all solutions $x(k)$ of the system (9.82).

Theorem 9.8.2 The system (9.82) is asymptotically stable at the origin θ if the conditions (i) and (ii) of Theorem 9.8.1, along with the condition (iii) $V(x(k+1)) < V(x(k)), k = 0, 1, 2, \ldots$, hold for all the solutions $x(k)$ of (9.82).

Theorem 9.8.3 The system (9.82) is unstable at the origin if (i) $V(\theta) = 0$, (ii) $V(x(k+1)) > V(x(k))$, $k = 0, 1, 2, \ldots$, for all the solutions of (9.82), and (iii) in every neighborhood of the origin, there exists a $z \in \mathbb{R}^n$ such that $V(z) > 0$.

The proofs of the above theorems are similar to those of Theorems 8.5.9, 8.5.12, and 8.5.16 with suitable modifications to account for the discrete-time system.

The stability results stated in Theorems (9.8.1) - (9.8.3) hold even if we reverse the inequalities.

Analogous to Theorem 8.5.23 on the stability of continuous-time linear autonomous systems at the origin, the following theorem provides a method for identifying the stability properties of discrete-time autonomous linear systems of the form (9.72).

Theorem 9.8.4 Let A be an $n \times n$ matrix and Q be an $n \times n$ real symmetric negative definite matrix. Let P be the matrix satisfying the matrix equation

$$A^* PA - P = Q. \tag{9.83}$$

Then,

(i) If P is real symmetric positive definite matrix, then the origin is asymptotically stable for the system (9.72).

(ii) If P is real symmetric negative definite, then the origin is unstable for the system (9.72).

Proof. (i) Let P be the solution of (9.83). Let λ be an eigenvalue of A with corresponding eigenvector x. Then $Ax = \lambda x$. From (9.83), we obtain $x^* A^* PAx - x^* Px = x^* Qx$. Hence, $(\lambda x)^* P(\lambda x) - x^* Px = x^* Qx$. Thus,

$$(|\lambda|^2 - 1)x^* Px = x^* Qx, \tag{9.84}$$

which implies $|\lambda|^2 - 1 = \frac{x^*Qx}{x^*Px} < 0$ since Q is negative definite and P is positive definite. Hence $|\lambda| < 1$, implying asymptotic stability of the system (9.72) at the origin.
(*ii*) Since P and Q both are negative definite, from (9.84), we get $|\lambda| > 1$, implying instability at the origin θ. □

Remark 9.8.5 If P and Q are defined as in Theorem 9.8.4, then the function

$$V(x) = x'Px \tag{9.85}$$

acts as a Lyapunov function for the linear system $x(k+1) = Ax(k), k = 0, 1, 2, \ldots$.

***Example* 9.8.6** Using a Lyapunov function, discuss the stability at the equilibrium $(0,0)$ of the system $x(k+1) = Ax(k), k = 0, 1, 2, \ldots$, where $A = \begin{bmatrix} 0 & -\frac{1}{2} \\ \frac{1}{3} & 0 \end{bmatrix}$.

Solution Let $P = \begin{bmatrix} a & b \\ b & c \end{bmatrix}$ and $Q = \begin{bmatrix} -1 & 0 \\ 0 & -3 \end{bmatrix}$. Substituting A, P, and Q into (9.83) and solving, we obtain $P = \begin{bmatrix} \frac{1}{5} & 0 \\ 0 & \frac{4}{5} \end{bmatrix}$, which is positive definite. Hence, from (9.85), the Lyapunov function is $V(x_1, x_2) = \frac{1}{5}x_1^2 + \frac{4}{5}x_2^2$. From Theorem 9.8.4($i$), the given system is asymptotically stable. Note that Q can be any negative definite matrix.

The following theorem is analogous to Theorem 8.4.2, which relates the stability properties of a semilinear system to those of the corresponding linear system. Let A be an $n \times n$ matrix and $f : \mathbb{R}^n \to \mathbb{R}^n$ be a function that satisfy the following conditions:

(a) $\|A\| < 1$,

(b) $f(\theta) = \theta$,

(c) for any $\varepsilon > 0$, there exists a $\delta > 0$ such that if $z \in \mathbb{R}^n$ satisfies $\|z\| < \delta$, then $\|f(z)\| \leq \varepsilon \|z\|$.

Theorem 9.8.7 Consider the semilinear discrete-time system

$$x(k+1) = Ax(k) + f(x(k)), \quad k = 0, 1, 2, \ldots, \tag{9.86}$$

where A and f satisfy the conditions (a)–(c). Then, the trivial solution of (9.86) is asymptotically stable.

Proof. Since $\|A\| < 1$, we can choose $\varepsilon > 0$ such that $\|A\| + \varepsilon < 1$. From (9.8), the solution of the system (9.86) is given by

$$x(m) = A^m x(0) + \sum_{k=0}^{m-1} A^{m-k-1} f(x(k)), \quad m = 1, 2, \ldots. \tag{9.87}$$

From condition (c), it follows that if $\|x(0)\| < \delta$, then $\|f(x(0))\| < \varepsilon\|x(0)\|$. Hence,

$$\|x(1)\| \le (\|A\| + \varepsilon)\|x(0)\|. \tag{9.88}$$

Furthermore,

$$\begin{aligned}\|x(2)\| &\le \|A\|^2\|x(0)\| + \|A\|\|f(x(0))\| + \|f(x(1))\| \\ &\le (\|A\|^2 + \varepsilon\|A\|)\|x(0)\| + \varepsilon\|x(1)\| \le (\|A\| + \varepsilon)^2\|x(0)\| \quad \text{(using (9.88))}.\end{aligned}$$

Similarly, we can show that $\|x(m)\| \le (\|A\| + \varepsilon)^m\|x(0)\|$ for $m = 1, 2, \ldots$. Since $\|A\| + \varepsilon < 1$, we have $\lim_{m\to\infty}\|x(m)\| = 0$, which implies the asymptotic stability of the semilinear system (9.86) at the origin. □

Stabilizability

As in the case of continuous-time dynamical systems, if a discrete-time dynamical system is not asymptotically stable, then a suitable control sequence can be found to make the system asymptotically stable at the origin.

Let A be an $n \times n$ matrix with at least one eigenvalue λ such that $|\lambda| \ge 1$. Then the system

$$x(r+1) = Ax(r), \quad r = 0, 1, 2, \ldots \tag{9.89}$$

is not asymptotically stable at the origin (Theorem 9.7.2). In order to obtain asymptotic stability of the system, we can introduce a suitable control matrix B and a control sequence $u(r)$, $r = 0, 1, 2, \ldots$, in (9.89) so that the resulting system

$$x(r+1) = Ax(r) + Bu(r) \tag{9.90}$$

is asymptotically stable at the origin, i.e., $\lim_{r\to\infty} x(r) = \theta \in \mathbb{R}^n$. For this, B must be chosen such that the system (9.90) is controllable, i.e., rank $\begin{bmatrix} B & AB & \cdots & A^{n-1}B \end{bmatrix} = n$. Then from Theorems 5.5.1 and 5.5.3, for any arbitrary set $S = \{\mu_1, \mu_2, \ldots, \mu_n\}$ of scalars, one can find a matrix K such that the matrix $A + BK$ has $\mu_i : i = 1, 2, \ldots, n$ as its eigenvalues. Thus, by considering the feedback control sequence $u(r) = Kx(r)$, $r = 0, 1, 2, \ldots$, in (9.90), we get

$$x(r+1) = (A + BK)x(r) \quad r = 0, 1, 2, \ldots. \tag{9.91}$$

If the set S is chosen such that $|\mu_i| < 1$, $i = 1, 2, \ldots$, then from Theorem 9.7.2, the solution $x(r) \to 0$ in $\mathbb{R}^n$ as $r \to \infty$, which implies asymptotic stability.

***Example* 9.8.8** The discrete-time dynamical system

$$\begin{cases} x_1(r+1) = -2x_1(r) + x_2(r), \\ x_2(r+1) = -\dfrac{1}{2}x_2(r) \end{cases}$$

for $r = 0, 1, 2, \ldots$ is unstable, as one of the eigenvalues of the coefficient matrix $A = \begin{bmatrix} -2 & 1 \\ 0 & -\frac{1}{2} \end{bmatrix}$ has modulus > 1 (by Theorem 9.7.1(ii)). By introducing the control matrix $B = \begin{bmatrix} 0 \\ 1 \end{bmatrix}$ and an (unknown) control sequence $u(r)$, $r = 0, 1, 2, \ldots$, we obtain the control system

$$\begin{bmatrix} x_1(r+1) \\ x_2(r+1) \end{bmatrix} = \begin{bmatrix} -2 & 1 \\ 0 & -\frac{1}{2} \end{bmatrix} \begin{bmatrix} x_1(r) \\ x_2(r) \end{bmatrix} + \begin{bmatrix} 0 \\ 1 \end{bmatrix} u(r).$$

As rank $\begin{bmatrix} B & AB \end{bmatrix} = 2$, the above system is controllable. Now, let $S = \{0, \frac{1}{2}\}$ and $u(r) = Kx(r) = k_2x_1(r) + k_1x_2(r)$, where $K = \begin{bmatrix} k_2 & k_1 \end{bmatrix}$. It can be easily seen that if $K = \begin{bmatrix} -5 & 3 \end{bmatrix}$, then $A + BK$ has eigenvalues 0 and $\frac{1}{2}$. Thus, the feedback control $u(r) = -5x_1(r) + 3x_2(r)$ makes the system asymptotically stable at the origin (by Theorem 9.7.1(i)).

Exercise 9.8.1 Consider the discrete-time control system $x(k+1) = Ax(k) + Bu(k)$, $k = 0, 1, 2, \ldots$, where $A = \begin{bmatrix} 2 & 1 \\ 0 & 2 \end{bmatrix}$ and $B = \begin{bmatrix} 0 \\ 1 \end{bmatrix}$, and let $u(k)$ be a feedback control of the form $u(k) = Lx(k)$, where L is a 1×2 matrix. Suppose $L = \begin{bmatrix} l_1 & l_2 \end{bmatrix}'$. Find all possible values of l_1 and l_2 such that the system is asymptotically stable at the origin.

Exercise 9.8.2 Using Theorem 9.8.7, show that the trivial solution of the discrete system $x(k+1) = \frac{1}{2}x(k) + x^3(k)$, $k = 0, 1, 2 \ldots$, is asymptotically stable. Also, find a Lyapunov function $V(x)$ using equation (9.83) by taking $Q = -1$.

Show that $V(x)$ establishes the asymptotic stability of the nonlinear system at $x = \theta$ in the neighborhood $|x| < \frac{7}{10}$.

9.9 Discrete-time Optimal Control Systems

In Chapter 6, optimal control problems for continuous-time systems were introduced. Necessary conditions have been derived using Lagrangian and Hamiltonian functions. In this section we shall present analogous results for discrete-time systems.

Let $S = \{k_0, k_0+1, \ldots, k_0+N\} \subset \mathbb{R}$ be a discrete set of positive integers, and let $x : S \to \mathbb{R}^n$ be a function. Consider the real valued function $f : \mathbb{R}^n \times \mathbb{R}^n \times S \to \mathbb{R}$, and define the functional $\mathbb{J}(x, k_0, N)$ by

$$\mathbb{J}(x, k_0, N) = \sum_{k=k_0}^{k_0+N-1} f(x(k), x(k+1), k). \tag{9.92}$$

Hereafter, we denote $\mathbb{J}(x, k_0, N)$ by $\mathbb{J}(x)$ or simply by $\mathbb{J}$. First, let us consider the following minimization problems.

Case I. (Fixed boundary)

$$\text{Minimize } \mathbb{J} = \sum_{k=k_0}^{k_0+N-1} f(x(k), x(k+1), k) \tag{9.93}$$

under the boundary conditions

$$x(k_0) = x_0 \text{ and } x(k_0+N) = x_1. \tag{9.94}$$

Case II. (Free boundary)

$$\text{Minimize } \mathbb{J} = \sum_{k=k_0}^{k_0+N-1} f(x(k), x(k+1), k)$$

under the boundary condition $x(k_0) = x_0$, while $x(k_0+N)$ is unspecified.

In the above problems, we need to find the optimal sequence of vectors $x(k)$ for $k = k_0, k_0+1, \ldots, k_0+N$, which minimizes $\mathbb{J}$ while satisfying the given boundary conditions. The solution to the above problems can be obtained from the following theorems.

Theorem 9.9.1 If $x^*(k) : k = k_0, k_0 + 1, \ldots, k_0 + N$ is the required optimal sequence of vectors that minimizes $\mathbb{J}$ given in (9.93), then $x^*(k)$ satisfies the **Euler's equation**

$$\left[\frac{\partial}{\partial x(k)} f(x(k), x(k+1), k) + \frac{\partial}{\partial x(k)} f(x(k-1), x(k), k-1)\right]_* = \theta \tag{9.95}$$

for $k = k_0 + 1, k_0 + 2, \ldots, k_0 + N - 1$, along with the boundary conditions.

Proof. Let $x(k) = x^*(k) + \delta x(k)$ for $k = k_0, k_0 + 1, \ldots, k_0 + N$ be any arbitrary sequence in the neighborhood of $x^*(k)$. Here, $\delta x(k)$ is known as the incremental sequence or variation in $x^*(k)$. Now, let

$$\mathbb{J} = \sum_{k=k_0}^{k_0+N-1} f(x(k), x(k+1), k)$$

and

$$\mathbb{J}^* = \sum_{k=k_0}^{k_0+N-1} f(x^*(k), x^*(k+1), k), \tag{9.96}$$

the required minimum value of $\mathbb{J}$. Let $\Delta\mathbb{J} = \mathbb{J} - \mathbb{J}^*$ be the increment in $\mathbb{J}$, which is given by

$$\Delta\mathbb{J} = \sum_{k=k_0}^{k_0+N-1} \left[f(x(k), x(k+1), k) - f(x^*(k), x^*(k+1), k)\right].$$

Using the Taylor series expansion of the first term about $(x^*(k), x^*(k+1))$ and neglecting second and higher order terms, we obtain the first variation in $\mathbb{J}$ as

$$\begin{aligned}\delta\mathbb{J} = \sum_{k=k_0}^{k_0+N-1} &\left[\left(\frac{\partial}{\partial x(k)} f(x(k), x(k+1), k)\right)_* \delta x(k)\right.\\ &\left. + \left(\frac{\partial}{\partial x(k+1)} f(x(k), x(k+1), k)\right)_* \delta x(k+1)\right].\end{aligned} \tag{9.97}$$

We know that the necessary condition for minimization of $\mathbb{J}$ is $\delta\mathbb{J} = 0$ (Lemma 6.4.1). Now, rewrite the second term in (9.97) as

$$\begin{aligned}&\sum_{k=k_0}^{k_0+N-1} \left(\frac{\partial}{\partial x(k+1)} f(x(k), x(k+1), k)\right)_* \delta x(k+1)\\ &= \sum_{i=k_0+1}^{k_0+N} \left(\frac{\partial}{\partial x(i)} f(x(i-1), x(i), i-1)\right)_* \delta x(i) \quad \text{(by setting } k = i-1\text{)}\end{aligned}$$

$$= \sum_{i=k_0+1}^{k_0+N-1} \left(\frac{\partial}{\partial x(i)} f(x(i-1), x(i), i-1) \right)_* \delta x(i)$$

$$+ \left[\left(\frac{\partial}{\partial x(i)} f(x(i-1), x(i), i-1)\, \delta x(i) \right)_* \right]_{i=k_0+N}. \tag{9.98}$$

Substituting (9.98) into (9.97) (after replacing the dummy variable i with k), we get

$$\delta \mathbb{J} = \sum_{k=k_0+1}^{k_0+N-1} \left(\frac{\partial}{\partial x(k)} f(x(k), x(k+1), k) \right)_* \cdot \delta x(k)$$

$$+ \sum_{k=k_0+1}^{k_0+N-1} \left(\frac{\partial}{\partial x(k)} f(x(k-1), x(k), k-1) \right)_* \delta x(k)$$

$$+ \left[\left(\frac{\partial}{\partial x(k)} f(x(k-1), x(k), k-1) \right)_* \delta x(k) \right]_{k=k_0+N}$$

$$+ \left[\left(\frac{\partial}{\partial x(k)} f(x(k), x(k+1), k) \right)_* \delta x(k) \right]_{k=k_0}. \tag{9.99}$$

Using (9.99), we can obtain the necessary conditions for cases I and II as follows.

Case I. As $x(k_0) = x_0$ and $x(k_0+N) = x_f$ are fixed, the variations at $k = k_0$ and k_0+N are zero, i.e., $\delta x(k_0) = \theta$ and $\delta x(k_0+N) = \theta$. Hence, the last two terms of (9.99) are zeros. Therefore, the necessary condition $\delta \mathbb{J} = 0$ implies that

$$\delta \mathbb{J} = \sum_{k=k_0+1}^{k_0+N-1} \left[\left(\frac{\partial}{\partial x(k)} f(x(k), x(k+1), k) \right)_* + \left(\frac{\partial}{\partial x(k)} f(x(k-1), x(k), k-1) \right)_* \right] \delta x(k) = 0. \tag{9.100}$$

Since $\delta x(k)$ are arbitrary for each k, the required necessary conditions for the optimal solution are given by (9.95) for $k = k_0+1, \ldots, k_0+N-1$, along with the given boundary conditions.

Case II. In this case, $\delta x(k_0) = \theta$, and $\delta x(k_0+N)$ is arbitrary. Moreover, since Case I is a particular case of Case II, $x^*(k)$ must satisfy Euler's equation (9.95) for $k = k_0+1, \ldots, k_0+N-1$. Using (9.95) and $\delta x(k_0) = 0$ in (9.99), we get

$$\delta \mathbb{J} = \left[\left(\frac{\partial}{\partial x(k)} f(x(k-1), x(k), k-1) \right)_* \delta x(k) \right]_{k=k_0+N} = 0.$$

As $\delta x(k_0+N)$ is arbitrary, we obtain the condition

$$\frac{\partial}{\partial x(k)} f(x(k-1), x(k), k-1) = 0 \text{ at } k = k_0+N, \tag{9.101}$$

which is the required boundary condition for the optimal solution $x^*(k)$. □

Remark 9.9.2 In Theorem 9.9.1, "minimizes $\mathbb{J}$" can be replaced by "maximizes $\mathbb{J}$."

***Example* 9.9.3** Let $x(k)$ for $k = 0,1,2$ be real valued functions, and consider the function $f(x(k),x(k+1),k) = x^2(k) - x^2(k+1) + (x(k) - x(k+1))^2$. Define the functional

$$\mathbb{J} = \sum_{k=0}^{1} f(x(k),x(k+1),k).$$

If $x(0) = 1$ and $x(2) = 5$, find $x(1)$ such that $\mathbb{J}$ is minimized.

Solution **Method I.** This problem can be solved using basic calculus for maxima/minima. Here,

$$\mathbb{J} = x^2(0) - x^2(1) + (x(0) - x(1))^2 + x^2(1) - x^2(2) + (x(1) - x(2))^2.$$

Using the given boundary values, we obtain

$$\mathbb{J} = (1 - x(1))^2 + (x(1) - 5)^2 - 25.$$

As $x(1)$ is an unknown real value, let $x(1) = y$. Then

$$\mathbb{J} = (1-y)^2 + (y-5)^2 - 25,$$

and the minimum of $\mathbb{J}$ occurs at $y = 3$.

Method II. We now solve the problem using Euler's equation (9.95) along with the boundary conditions. Here, $k_0 = 0$ and $N = 2$. From (9.95), we get

$$\left(\frac{\partial}{\partial x(k)} f(x(k),x(k+1),k) + \frac{\partial}{\partial x(k)} f(x(k-1),x(k),k-1)\right)_{k=1} = 0,$$

which simplifies to

$$[2x(k) + 2(x(k) - x(k+1))]_{k=1} + [-2x(k) - 2(x(k-1) - x(k))]_{k=1} = 0.$$

Substituting $x(0) = 1$ and $x(2) = 5$, we obtain

$$2(x(1) - 5) + 2(x(1) - 1) = 0,$$

which gives $x(1) = 3$.

The standard procedure for maxima/minima, as shown in **Method I**, can become cumbersome when dealing with vector valued functions $x(k)$ for $k = k_0, k_0 + 1, \ldots$. In such cases, using Euler's equation (9.100) can lead to a simpler solution.

Example **9.9.4** For the functional

$$\mathbb{J} = \sum_{k=0}^{1} f(x(k), x(k+1), k),$$

with

$$f(x(k), x(k+1), k) = \langle x(k), x(k+1) \rangle + \|x(k)\|^2,$$

where $\langle \cdot, \cdot \rangle$ denotes the dot product, find the optimal solution $x^*(k) = \begin{bmatrix} x_1^*(k) \\ x_2^*(k) \end{bmatrix}$ with the boundary conditions $x^*(0) = \begin{bmatrix} 1 \\ 0 \end{bmatrix}$ and $x^*(2) = \begin{bmatrix} 0 \\ 1 \end{bmatrix}$.

Solution Differentiating $f(x(k), x(k+1), k)$ and $f(x(k-1), x(k), k-1)$ with respect to $x(k)$, we get (see Section 6.3 (III))

$$\frac{\partial}{\partial x(k)} f(x(k), x(k+1), k) = x(k+1) + 2x(k)$$

and

$$\frac{\partial}{\partial x(k)} f(x(k-1), x(k), k-1) = x(k-1),$$

respectively. For the minimization of $\mathbb{J}$, from (9.95), the Euler equation is given by

$$\left(\frac{\partial}{\partial x(k)} f(x(k), x(k+1), k) + \frac{\partial}{\partial x(k)} f(x(k-1), x(k), k-1) \right)_{k=1} = 0,$$

which yields

$$x(2) + x(0) + 2x(1) = 0 \implies x(1) = -\frac{x(0) + x(2)}{2} = \begin{bmatrix} -\frac{1}{2} \\ -\frac{1}{2} \end{bmatrix}.$$

Example **9.9.5** Define the functional

$$\mathbb{J} = \sum_{k=0}^{2} f(x(k), x(k+1), k),$$

where

$$f(x(k), x(k+1), k) = x(k)x(k+1) + x^2(k) + x^2(k+1).$$

Find the optimal function $x^*(k)$ with the boundary conditions $x^*(0) = 1$ and $x^*(3) = 5$ that minimizes $\mathbb{J}$.

Solution **Method I.** Let $x(1) = y$ and $x(2) = z$ be real variables. Then,

$$\mathbb{J} = (y + 1 + y^2) + (yz + y^2 + z^2) + (5z + z^2 + 25).$$

As $\mathbb{J}$ is a function of y and z, to find the minimum of $\mathbb{J}$, solve the following equations

$$\begin{cases} \dfrac{\partial \mathbb{J}}{\partial y} = 1 + 4y + z = 0, \\ \dfrac{\partial \mathbb{J}}{\partial z} = y + 4z + 5 = 0, \end{cases} \tag{9.102}$$

which gives $y = \frac{51}{5}$ and $z = -\frac{19}{5}$.

Method II. Using (9.95), we get

$$\left(\frac{\partial}{\partial x(k)} f(x(k), x(k+1), k) + \frac{\partial}{\partial x(k)} f(x(k-1), x(k), k-1) \right) = 0 \quad \text{for } k = 1, 2,$$

which leads to

$$\begin{aligned} &[(x(k+1) + 2x(k)) + (x(k-1) + 2x(k))]_{k=1,2} = 0 \\ \Longrightarrow &\begin{cases} x(2) + 4x(1) + x(0) = 0, \\ x(3) + 4x(2) + x(1) = 0, \end{cases} \end{aligned}$$

and it is same as (9.102).

Optimal Control

In this subsection, necessary conditions, known as Euler–Lagrange equations, for obtaining the optimal control sequence for discrete-time control systems are presented. This section is analogous to Section 6.7 on continuous systems.

Case I. (Fixed boundary conditions). Let $S = \{k_0, k_0 + 1, \ldots, k_0 + N\}$ be a discrete set of integers, and $x : S \to \mathbb{R}^n$ and $u : S \to \mathbb{R}^m$ be the state and control functions, respectively. Let $f : \mathbb{R}^n \times \mathbb{R}^n \times \mathbb{R}^m \times \mathbb{S} \to \mathbb{R}$ be a continuous function with respect to the first three variables and with continuous partial derivatives. Consider the functional

$$\mathbb{J}(x(k), x(k+1), u(k), k) = \sum_{k=k_0}^{k_0+N-1} f(x(k), x(k+1), u(k), k).$$

The problem is to find the optimal control sequence $u^*(k) : k = k_0, k_0 + 1, \ldots, k_0 + N - 1$ and the optimal states $x^*(k) : k = k_0 + 1, \ldots, k_0 + N - 1$ for the control system

$$x(k+1) = g(x(k), u(k), k),$$

with boundary conditions $x(k_0) = x_0$ and $x(k_0+N) = x_1$ such that $\mathbb{J}$ is minimized, i.e.,

$$\mathbb{J}(x^*(k), x^*(k+1), u^*(k), k) = \min_u \mathbb{J}(x(k), x(k+1), u(k), k).$$

Case II: In the above problem $x(k_0) = x_0$ is specified but $x(k_0+N)$ is unspecified.

Solution **Case I:** Following the procedure similar to that for continuous-time systems (see Section 6.7, equation (6.68)), a Lagrange multiplier function $\lambda(k+1) \in R^n$ is introduced to obtain the augmented functional $\mathbb{J}_1(x(k), x(k+1), u(k), \lambda(k+1), k)$ as follows:

$$\mathbb{J}_1 = \sum_{k=k_0}^{k_0+N-1} f(x(k), x(k+1), u(k), k) + \lambda'(k+1)\,[g(x(k), u(k), k) - x(k+1)]\,. \quad (9.103)$$

Thus, the given problem of extremizing $\mathbb{J}$ with constraints is converted into a problem of extremizing $\mathbb{J}_1$ without any constraints. By defining the Lagrangian function L as

$$\begin{aligned} L(x(k), x(k+1), u(k), \lambda(k+1), k) = {} & f(x(k), x(k+1), u(k), k) \\ & + \lambda'(k+1)\,[g(x(k), u(k), k) - x(k+1)]\,, \end{aligned} \quad (9.104)$$

the equation (9.103) can be written as

$$\mathbb{J}_1 = \sum_{k=k_0}^{k_0+N-1} L(x(k), x(k+1), u(k), \lambda(k+1), k). \quad (9.105)$$

Since $\mathbb{J}_1$ and the Lagrangian L depend on the variables $x(k), x(k+1), u(k)$, and $\lambda(k)$ for $k = k_0, k_0+1, \ldots, k_0+N$, the Euler–Lagrange equations are obtained from (9.95) (replacing f with L) with respect to the variables $x(k), u(k)$, and $\lambda(k)$ as given below.

$$\begin{cases} \left(\dfrac{\partial}{\partial x(k)} L(x(k), x(k+1), u(k), \lambda(k+1), k) \right. \\ \left. + \dfrac{\partial}{\partial x(k)} L(x(k-1), x(k), u(k-1), \lambda(k), k-1) \right)_* = 0 \text{ for } k = k_0+1, k_0+2, \ldots, k_0+N-1., \\ \left(\dfrac{\partial}{\partial u(k)} L(x(k), x(k+1), u(k), \lambda(k+1), k) \right. \\ \left. + \dfrac{\partial}{\partial u(k)} L(x(k-1), x(k), u(k-1), \lambda(k), k-1) \right)_* = 0 \text{ for } k = k_0, k_0+1, \ldots, k_0+N-1., \\ \left. \dfrac{\partial}{\partial \lambda(k)} L(x(k-1), x(k), u(k-1), \lambda(k), k-1) \right)_* = 0 \text{ for } k = k_0+1, k_0+2, \ldots, k_0+N-1 \end{cases}$$

(9.106)

under the given boundary conditions.
Case II: In this case the solution may be obtained by solving (9.106) along with the boundary conditions : $x(k_0) = x_0$ and $(\partial/\partial x(k))(L(k-1)) = 0$ at $k = k_0 + N$. The notation $L(k)$ is used here to denote the LHS of (9.104).

Remark 9.9.6 In (9.103), the Lagrangian L is formed using the Lagrange multiplier $\lambda(k+1)$. This choice is made only for convenience in numerical computation. Alternatively, one could use $\lambda(k)$ in place of $\lambda(k+1)$, in which case a one step shift would be needed at a later stage.

***Example* 9.9.7** Find the states $x(1), x(2), x(3), x(4)$ and the control sequence $u(0), u(1), u(2), u(3), u(4)$, which minimize the functional $\mathbb{J}$, where

$$\mathbb{J} = \sum_{k=0}^{4} u^2(k)$$

under the constraint

$$\begin{cases} x(k+1) = x(k) + u(k), & k = 0, 1, 2, 3, 4, \\ x(0) = 1, \quad x(5) = 0. \end{cases}$$

Solution From (9.103), the minimization of $\mathbb{J}$ under the given constraint can be written as

$$\text{minimize } \mathbb{J}_1 = \sum_{k=0}^{4} \left[u^2(k) + \lambda(k+1)\left[x(k) + u(k) - x(k+1)\right]\right],$$

where $\lambda(k+1), k = 0, 1, 2, 3,$ and 4 are Lagrange multipliers.

From (9.106), the Euler–Lagrange equations are

$$\begin{cases} \lambda(k+1) - \lambda(k) = 0, \\ 2u(k) + \lambda(k+1) = 0, \\ x(k+1) = x(k) + u(k) \end{cases}$$

for $k = 1, 2, 3, 4$. The first equation implies that $\lambda(k)$ is constant for all k, denoted by c. The second equation yields $u(k) = -\frac{\lambda(k+1)}{2} = -\frac{c}{2}$ for $k = 0, 1, 2, 3, 4$. The third equation becomes $x(k+1) = x(k) + u(k) - x(k) - \frac{c}{2}$ for $k = 0, 1, 2, 3, 4$. Now, using $x(0) = 1$ and $x(5) = 0$, we obtain

$$x(1) = x(0) - \frac{c}{2} = 1 - \frac{c}{2}.$$

$$x(2) = 1 - \frac{c}{2} - \frac{c}{2} = 1 - c.$$

$$x(3) = 1 - \frac{3}{2}c.$$
$$x(4) = 1 - 2c.$$
$$x(5) = 1 - \frac{5}{2}c = 0 \Longrightarrow c = \frac{2}{5}.$$

Hence, the optimal control is $u(k) = -\frac{c}{2} = -\frac{1}{5}$ for $k = 0,1,2,3,4$, and the optimal value of $\mathbb{J}$ is given by $\mathbb{J} = \sum_{k=0}^{4} u^2(k) = \frac{1}{5}$.

***Example* 9.9.8** Find the optimal control sequence $u(k) : k = 1,2,3$ that extremizes

$$\mathbb{J} = \sum_{k=1}^{3} (x^2(k) + 2x^2(k+1) + u^2(k))$$

under the constraint

$$\begin{cases} x(k+1) = 2x(k) + u(k), & k = 1,2,3, \\ x(1) = 3, \quad x(4) = 1. \end{cases}$$

Solution From (9.104), the Lagrangian is

$$\begin{aligned} L(x(k), x(k+1), u(k), \lambda(k), k) = {} & x^2(k) + 2x^2(k+1) + u^2(k) \\ & + \lambda(k+1)(2x(k) + u(k) - x(k+1)). \end{aligned}$$

From (9.106), the Euler–Lagrange equations are

$$6x(k) + 2\lambda(k+1) - \lambda(k) = 0, \tag{9.107}$$
$$2u(k) + \lambda(k+1) = 0, \tag{9.108}$$
$$x(k+1) - 2x(k) - u(k) = 0 \tag{9.109}$$

for $k = 1,2,3$. From (9.107)–(9.109), we get

$$x(k+1) = 2x(k) - \frac{\lambda(k+1)}{2} = 2x(k) - \frac{1}{4}\lambda(k) + \frac{3}{2}x(k) = \frac{7}{2}x(k) - \frac{1}{4}\lambda(k). \tag{9.110}$$

From (9.110) and (9.107), we obtain

$$\begin{bmatrix} x(k+1) \\ \lambda(k+1) \end{bmatrix} = \begin{bmatrix} \frac{7}{2} & -\frac{1}{4} \\ -3 & \frac{1}{2} \end{bmatrix} \begin{bmatrix} x(k) \\ \lambda(k) \end{bmatrix}$$

for $k = 1,2,3$. The state transition matrix for the above system is $\bar{A}^{k-1} : k = 1,2,3,4$, where $\bar{A} = \begin{bmatrix} \frac{7}{2} & -\frac{1}{4} \\ -3 & \frac{1}{2} \end{bmatrix}$. Hence, the solution of the above system is

$$\begin{bmatrix} x(r) \\ \lambda(r) \end{bmatrix} = \bar{A}^{r-1} \begin{bmatrix} x(1) \\ \lambda(1) \end{bmatrix}, \quad r = 1,2,3,4.$$

In particular, for $r = 4$, we have

$$\begin{bmatrix} x(4) \\ \lambda(4) \end{bmatrix} = \bar{A}^{3} \begin{bmatrix} x(1) \\ \lambda(1) \end{bmatrix}.$$

This yields

$$x(4) = \frac{97}{2}x(1) - 7\lambda(1) \quad \text{and} \quad \lambda(4) = -45x(1) + \frac{7}{2}\lambda(1).$$

Substituting $x(1) = 3$ and $x(4) = 1$, we obtain $\lambda(1) = \frac{289}{14}$ and $\lambda(4) = -\frac{251}{4}$. Furthermore, using $\begin{bmatrix} x(3) \\ \lambda(3) \end{bmatrix} = \bar{A}^2 \begin{bmatrix} x(1) \\ \lambda(1) \end{bmatrix}$ and $\begin{bmatrix} x(2) \\ \lambda(2) \end{bmatrix} = \bar{A} \begin{bmatrix} x(1) \\ \lambda(1) \end{bmatrix}$, the values of $x(2)$, $x(3)$, $\lambda(2)$, and $\lambda(3)$ can be computed explicitly. Finally, from (9.108), the optimal control sequence is given by

$$u(k) = -\frac{\lambda(k+1)}{2}, \quad k = 1,2,3.$$

9.10 Cost Functional Involving End Condition

Consider the problem of finding the optimal pair $u^*(k) : k = k_0, \ldots, k_0 + N - 1$ and $x^*(k) : k = k_0 + 1, \ldots, k_0 + N$ for the discrete-time linear control system

$$\begin{cases} x(k+1) = A(k)x(k) + B(k)u(k), & k = k_0, k_0 + 1, \ldots, k_0 + N - 1, \\ x(k_0) = x_0 \end{cases}$$

that extremizes the quadratic form of cost functional $\mathbb{J}$ involving the final state $x(k_0 + N)$, where

$$\mathbb{J} = x'(k_0 + N)M(k_0 + N)x(k_0 + N) + \sum_{k=k_0}^{k_0+N-1} \left[x'(k)P(k)x(k) + u'(k)Q(k)u(k)\right]. \quad (9.111)$$

Note that $\mathbb{J}$ can be rewritten as

$$\mathbb{J} = x'(k_0)M(k_0)x(k_0) + \sum_{k=k_0}^{k_0+N-1} \big[x'(k)P(k)x(k) + u'(k)Q(k)u(k) + x'(k+1)M(k+1)x(k+1) - x'(k)M(k)x(k)\big]. \tag{9.112}$$

By introducing the Lagrange multiplier vector function $\lambda(k+1)$, as in Section 9.9, the Lagrangian of the extremization problem can be written as

$$\begin{aligned} L(x(k), x(k+1), u(k), \lambda(k+1)) = & x'(k)P(k)x(k) + u'(k)Q(k)u(k) \\ & + x'(k+1)M(k+1)x(k+1) - x'(k)M(k)x(k) \\ & + \lambda'(k+1)\left[A(k)x(k) + B(k)u(k) - x(k+1)\right]. \end{aligned} \tag{9.113}$$

Denote this Lagrangian as $L(k)$, and when k is replaced by $k-1$, denote it by $L(k-1)$. Since the first term in $\mathbb{J}$ is a constant, the extremization of $\mathbb{J}$ is now equivalent to the extremization of the functional

$$\mathbb{J}_1 = \sum_{k=k_0}^{k_0+N-1} L(x(k), x(k+1), u(k), \lambda(k+1)) = \sum_{k=k_0}^{k_0+N-1} L(k). \tag{9.114}$$

This extremization problem is an unconstrained problem, as discussed in Section 9.9. Using Euler's equation (9.106) for (9.114), we obtain the following equations

$$\frac{\partial L(k)}{\partial x(k)} + \frac{\partial L(k-1)}{\partial x(k)} = 0, \quad k = k_0+1, \ldots, k_0+N-1, \tag{9.115}$$

$$\frac{\partial L(k)}{\partial u(k)} = 0, \quad k = k_0, k_0+1, \ldots, k_0+N-1, \tag{9.116}$$

$$\frac{\partial L(k-1)}{\partial \lambda(k)} = 0, \quad k = k_0+1, \ldots, k_0+N. \tag{9.117}$$

From (9.115), we obtain

$$2P(k)x(k) + A'\lambda(k+1) - \lambda(k) = 0. \tag{9.118}$$

From (9.116), we get $2Q(k)u(k) + B'(k)\lambda(k+1) = 0$, which implies

$$u(k) = -\frac{1}{2}Q^{-1}(k)B'(k)\lambda(k+1). \tag{9.119}$$

From (9.117), we obtain

$$x(k+1) = A(k)x(k) + B(k)u(k), \quad k = k_0, k_0+1, \ldots, k_0+N-1. \tag{9.120}$$

Substituting $u(k)$ from (9.119) into (9.120), we obtain

$$x(k+1) = A(k)x(k) - \frac{1}{2}B(k)Q^{-1}(k)B'(k)\lambda(k+1). \tag{9.121}$$

Denoting $\frac{1}{2}B(k)Q^{-1}(k)B'(k)$ by $G(k)$, (9.121) becomes

$$x(k+1) = A(k)x(k) - G(k)\lambda(k+1). \tag{9.122}$$

From (9.118), we obtain

$$\lambda(k+1) = -2(A'(k))^{-1}p(k)x(k) + (A'(k))^{-1}\lambda(k). \tag{9.123}$$

Substituting this expression of $\lambda(k+1)$ into (9.122), we obtain

$$x(k+1) = \left(A(k) + 2G(k)(A'(k))^{-1}P(k)\right)x(k) - G(k)(A'(k))^{-1}\lambda(k). \tag{9.124}$$

Let

$$E(k) = (A'(k))^{-1}p(k) \text{ and } D(k) = A(k) + 2G(k)E(k). \tag{9.125}$$

Then, from (9.122)–(9.125), the system for state $x(k)$ and co-state $\lambda(k)$ can be written as

$$\begin{bmatrix} x(k+1) \\ \lambda(k+1) \end{bmatrix} = \begin{bmatrix} D(k) & -G(k)(A'(k))^{-1} \\ -2E(k) & (A'(k))^{-1} \end{bmatrix} \begin{bmatrix} x(k) \\ \lambda(k) \end{bmatrix}. \tag{9.126}$$

Denoting $\begin{bmatrix} x(k) \\ \lambda(k) \end{bmatrix}$ by $X(k) \in \mathbb{R}^{2n}$ and the $2n \times 2n$ matrix given in (9.126) by $\overline{A}(k)$, this system can be expressed as

$$X(k+1) = \overline{A}(k)X(k), \quad k = k_0, k_0+1, \ldots, k_0+N-1. \tag{9.127}$$

Let $\phi(k, k_0)$ denote the state transition matrix of $\overline{A}(k)$. Then

$$\begin{bmatrix} x(k) \\ \lambda(k) \end{bmatrix} = \phi(k, k_0) \begin{bmatrix} x(k_0) \\ \lambda(k_0) \end{bmatrix}, \tag{9.128}$$

where

$$\phi(k, k_0) = \overline{A}(k-1)\overline{A}(k-2)\cdots\overline{A}(k_0) = \begin{bmatrix} \phi_{11}(k, k_0) & \phi_{12}(k, k_0) \\ \phi_{21}(k, k_0) & \phi_{12}(k, k_0) \end{bmatrix} \text{(say)}.$$

Two cases of boundary conditions are considered.

Case I. (Fixed boundary) $x(k_0) = x_0$ and $x(k_0+N) = x_N$ are specified. In this case, from (9.128), we obtain

$$x_N = x(k_0+N) = \phi_{11}(k_0+N,k_0)x_0 + \phi_{12}(k_0+N,k_0)\lambda_0, \text{ where } \lambda_0 = \lambda(k_0),$$

Hence

$$\lambda_0 = \phi_{12}^{-1}(k_0+N,k_0)\left(x_N - \phi_{11}(k_0+N,k_0)x_0\right). \tag{9.129}$$

Now, using x_0 and λ_0 in (9.128), we can obtain $\begin{bmatrix} x(k) \\ \lambda(k) \end{bmatrix}$ for $k = k_0+1,\ldots,k_0+N$. The optimal control $u(k) : k = k_0, k_0+1,\ldots,k_0+N-1$ is then obtained from (9.119) as

$$u(k) = -\frac{1}{2}Q^{-1}(k)B'(k)\lambda(k+1), \quad k = k_0, k_0+1,\ldots,k_0+N.$$

Case II. (Free boundary) $x(k_0) = x_0$ is specified, but $x(k_0+N)$ is unspecified. Then from (9.101),

$$\frac{\partial L(k-1)}{\partial x(k)} = 0 \text{ at } k = k_0+N,$$

which implies (from (9.113))

$$\lambda(k_0+N) = 2M(k_0+N)x(k_0+N). \tag{9.130}$$

Now, from (9.128) and (9.130),

$$2M(k_0+N)x(k_0+N) = \phi_{21}(k_0+N,k_0)x_0 + \phi_{22}(k_0+N,k_0)\lambda_0,$$

which gives

$$\lambda_0 = \phi_{22}^{-1}(k_0+N,k_0)\left(2M(k_0+N)x(k_0+N) - \phi_{21}(k_0+N,k_0)x_0\right).$$

Using x_0 and λ_0 in (9.128), we can obtain all values of $x(k)$ and $\lambda(k)$ for $k = k_0, k_0+1,\ldots,k_0+N$. The optimal control sequence $u(k) : k = k_0, k_0+1,\ldots,k_0+N-1$ is then obtained from (9.119).

***Example* 9.10.1** Find the optimal control sequence $u(k) : k = 0,1,2$ and the optimal states $x(k) : k = 1,2,3$ for the control system $x(k+1) = 2x(k) + u(k)$ for $k = 0,1,2$ that minimize the functional

$$\mathbb{J} = 4x^2(3) + \frac{1}{2}\sum_{k=0}^{2}(x^2(k) + u^2(k)),$$

given the following boundary conditions: (*i*) $x(0) = 2$ and $x(3) = 4$ and (*ii*) $x(0) = 2$ and $x(3)$ is arbitrary.

Solution In this problem, $A=2, B=1, P=\frac{1}{2}, Q=\frac{1}{2}$, and $M=4$. Then, $G=\frac{1}{2}BQ^{-1}B'=1$. From (9.125), we have $E=(A')^{-1}P=\frac{1}{4}$ and $L=A+2GE=2+2.\frac{1}{4}=\frac{5}{2}$. Substituting these values into (9.126), we get

$$\begin{bmatrix} x(k+1) \\ \lambda(k+1) \end{bmatrix} = \begin{bmatrix} \frac{5}{2} & -\frac{1}{2} \\ -\frac{1}{2} & \frac{1}{2} \end{bmatrix} \begin{bmatrix} x(k) \\ \lambda(k) \end{bmatrix}. \tag{9.131}$$

Case (i). $x(0)=2$ and $x(3)=4$. The solution of (9.131) is

$$\begin{bmatrix} x(k) \\ \lambda(k) \end{bmatrix} = \phi(k,0) \begin{bmatrix} x(0) \\ \lambda(0) \end{bmatrix},$$

where $\phi(k,0)=\overline{A}^k$ for $k=1,2,3$, where

$$\overline{A} = \begin{bmatrix} \frac{5}{2} & -\frac{1}{2} \\ -\frac{1}{2} & \frac{1}{2} \end{bmatrix}.$$

Hence

$$\begin{bmatrix} x(3) \\ \lambda(3) \end{bmatrix} = \overline{A}^3 \begin{bmatrix} x(0) \\ \lambda(0) \end{bmatrix} = \begin{bmatrix} 17 & -4 \\ -4 & 1 \end{bmatrix} \begin{bmatrix} x(0) \\ \lambda(0) \end{bmatrix}, \tag{9.132}$$

which implies $x(3)=17x(0)-4\lambda(0)$, leading to $\lambda(0)=\frac{15}{2}$. Then,

$$\begin{bmatrix} x(2) \\ \lambda(2) \end{bmatrix} = \overline{A}^2 \begin{bmatrix} x(0) \\ \lambda(0) \end{bmatrix} = \begin{bmatrix} \frac{13}{2} & -\frac{3}{2} \\ -\frac{3}{2} & \frac{1}{2} \end{bmatrix} \begin{bmatrix} 2 \\ \frac{15}{2} \end{bmatrix} = \begin{bmatrix} \frac{7}{4} \\ \frac{3}{4} \end{bmatrix}.$$

Similarly,

$$\begin{bmatrix} x(1) \\ \lambda(1) \end{bmatrix} = \begin{bmatrix} \frac{5}{2} & -\frac{1}{2} \\ -\frac{1}{2} & \frac{1}{2} \end{bmatrix} \begin{bmatrix} x(0) \\ \lambda(0) \end{bmatrix} = \begin{bmatrix} \frac{5}{4} \\ \frac{11}{4} \end{bmatrix}.$$

Thus, the optimal state sequence is $x(0)=2$, $x(1)=\frac{5}{4}$, $x(2)=\frac{7}{4}$, and $x(3)=4$, and the optimal co-state sequence is $\lambda(0)=\frac{15}{2}$, $\lambda(1)=\frac{11}{4}$, $\lambda(2)=\frac{3}{4}$, and $\lambda(3)=-\frac{1}{2}$. The optimal control $u(k)$ is obtained from the relation (9.119) as $u(k)=-\frac{1}{2}Q^{-1}B'\lambda(k+1)$. Hence, the optimal control sequence is $u(0)=-\lambda(1)=-\frac{11}{4}$, $u(1)=-\lambda(2)=-\frac{3}{4}$, and $u(2)=-\lambda(3)=\frac{1}{2}$.

Case (ii). Given that $x(0)=2$ and $x(3)$ is unspecified, we have from (9.130) that $\lambda(3)=2M(3)x(3)=8x(3)$. From (9.132),

$$x(3)=17x(0)-4\lambda(0)=34-4\lambda(0)$$

and

$$8x(3)=\lambda(3)=-4x(0)+\lambda(0)=-8+\lambda(0),$$

which imply $\lambda(0)=\frac{280}{33}$ and $x(3)=34-\frac{1120}{33}$. Using $x(0)$ and $\lambda(0)$, other values of $x(k)$, $\lambda(k)$, and $u(k)$ can be obtained as in Case (i).

Exercise 9.10.1 Consider the discrete-time system $x(k+1)=Ax(k)+Bu(k)$, where $A=\begin{bmatrix}0 & 1\\ 1 & -1\end{bmatrix}$ and $B=\begin{bmatrix}0\\ 1\end{bmatrix}$, $k=1,2,3$. Write the state and co-state system with boundary conditions for finding the optimal state $x^*(t)$, $k=2,3,4$, and the optimal control $u^*(t)$, $k=1,2,3$, which minimize the performance index

$$\mathbb{J}=2x_1^2(4)+\sum_{k=0}^{3}\left(x_1^2(k)+x_2^2(k+1)+u^2(k)\right)$$

subject to the boundary conditions $x(1)=\begin{bmatrix}2\\ 1\end{bmatrix}$ and $x(4)$ unspecified.

9.11 Optimal Feedback Control and Riccati Equation

Riccati equation for discrete-time optimal control problems can be obtained in a similar way as discussed in Section 7.2 for continuous-time systems. Following the notations of the previous section, consider the equations (9.122) and (9.118), as given below:

$$x(k+1)=A(k)x(k)-G(k)\lambda(k+1) \tag{9.133}$$

and

$$\lambda(k)=2P(k)x(k)+A'(k)\lambda(k+1), \tag{9.134}$$

along with the initial condition $x(k_0)=x_0$. Let $R(k)$ be a matrix (to be obtained later) such that

$$\lambda(k)=R(k)x(k). \tag{9.135}$$

Using (9.134) in (9.135), we obtain

$$2P(k)x(k) + A'(k)\lambda(k+1) = R(k)x(k). \tag{9.136}$$

From (9.133) and (9.135), we have

$$\begin{aligned}\lambda(k+1) = R(k+1)x(k+1) &= R(k+1)(Ax(k) + Bu(k)) \\ &= R(k+1)(Ax(k) - G(k)\lambda(k+1)).\end{aligned}$$

Hence,

$$(I + R(k+1)G(k))\lambda(k+1) = R(k+1)Ax(k),$$

implying

$$\lambda(k+1) = (I + R(k+1)G(k))^{-1}R(k+1)A(k)x(k). \tag{9.137}$$

Substitute (9.137) into (9.136) for $\lambda(k+1)$ to obtain

$$\left[2P(k) + A'(k)(I + R(k+1)G(k))^{-1}R(k+1)A(k)\right]x(k) = R(k)x(k). \tag{9.138}$$

As the above equation is valid for any arbitrary state $x(k)$, we obtain the following equation, called the **Matrix Riccati Equation** for discrete-time systems:

$$R(k) = 2P(k) + A'(k)(I + R(k+1)G(k))^{-1}R(k+1)A(k), \tag{9.139}$$

which can also be written as

$$R(k) = 2P(k) + A'(k)\left[R^{-1}(k+1) + G(k)\right]^{-1}A(k). \tag{9.140}$$

Here, we assume that $R^{-1}(k)$ exists for all k. Later, we will prove that $R(k)$ exists and is, in fact, positive definite and hence nonsingular. Equation (9.139) or (9.140) can be solved backward to obtain $R(k)$ using the final condition at $k_0 + N$. To obtain the final condition, from (9.130) we know that $\lambda(k_0+N) = 2M(k_0+N)x(k_0+N)$, and from (9.135), $\lambda(k_0+N) = R(k_0+N)x(k_0+N)$. Since $x(k_0+N)$ can be an arbitrary vector in $\mathbb{R}^n$, we obtain

$$R(k_0+N) = 2M(k_0+N). \tag{9.141}$$

Proposition 9.11.1 The Riccati matrix $R(k)$ is symmetric for all $k = k_0, k_0+1, \ldots, k_0+N$.

Proof. The matrices $P(k), Q(k)$, and $M(k)$ are symmetric for all $k = k_0, k_0+1, \ldots, k_0+N$. In (9.140), set $k = k_0+N-1$, and using $R(k_0+N) = 2M(k_0+N)$, observe that the RHS of (9.140) is symmetric. Hence, the LHS of (9.140), i.e., $R(k_0+N-1)$, is symmetric. By iteratively applying the same process for all $k = k_0+N-1, k_0+N-2, \ldots, k_0+1$, it can be concluded that $R(k)$ is symmetric for all $k = k_0, k_0+1, \ldots, k_0+N$. □

Optimal Value of $\mathbb{J}$ in terms of Riccati Matrix

Consider the performance functional $\mathbb{J}$ as given in equation (9.111):

$$\mathbb{J} = x'(k_0+N)M(k_0+N)x(k_0+N) + \sum_{k=k_0}^{k_0+N-1} \left[x'(k)P(k)x(k) + u'(k)Q(k)u(k)\right]. \tag{9.142}$$

Now, replacing $M(k_0+N)$ with $\frac{1}{2}R(k_0+N)$ (using (9.141)), we have

$$\mathbb{J} = \frac{1}{2}x'(k_0+N)R(k_0+N)x(k_0+N) + \sum_{k=k_0}^{k_0+N-1} \left[x'(k)P(k)x(k) + u'(k)Q(k)u(k)\right].$$

Rearranging the terms suitably, the RHS can be rewritten as

$$\begin{aligned}\mathbb{J} = \frac{1}{2}x'(k_0)R(k_0)x(k_0) + \sum_{k=k_0}^{k_0+N-1} &\left[x'(k)P(k)x(k) + u'(k)Q(k)u(k)\right] \\ &+ \frac{1}{2}\left[x'(k+1)R(k+1)x(k+1) - x'(k)R(k)x(k)\right].\end{aligned} \tag{9.143}$$

Using (9.139), note that

$$\begin{aligned}x'(k)P(k)x(k) - \frac{1}{2}x'(k)R(k)x(k) &= x'(k)\left(P(k) - \frac{1}{2}R(k)\right)x(k) \\ &= -\frac{1}{2}x'(k)\left(A'(k)\left(I+R(k+1)G(k)\right)^{-1}R(k+1)A(k)\right)x(k).\end{aligned} \tag{9.144}$$

Using (9.119) and (9.135), we obtain

$$\begin{aligned}u'(k)Q(k)u(k) &= -\frac{1}{2}u'(k)Q(k)\left(Q^{-1}(k)B'(k)\lambda(k+1)\right) \\ &= -\frac{1}{2}u'(k)B'(k)\lambda(k+1)\end{aligned} \tag{9.145}$$

and

$$\begin{aligned}x'(k+1)R(k+1)x(k+1) &= x'(k+1)\lambda(k+1) \\ &= \left(A(k)x(k) + B(k)u(k)\right)'\lambda(k+1),\end{aligned} \tag{9.146}$$

respectively. Substituting (9.144)–(9.146) into (9.143) and then simplifying, the optimal value of $\mathbb{J}$ is obtained as

$$\mathbb{J}^* = \frac{1}{2}x'(k_0)R(k_0)x(k_0) + \sum_{k=k_0}^{k_0+N-1} \left[-\frac{1}{2}x'(k)A'(k)\left(I+R(k+1)G(k)\right)^{-1}\right.$$

$$\times R(k+1)A(k)x(k) + \frac{1}{2}x'(k)A'(k)\lambda(k+1)\Big]$$
$$= \frac{1}{2}x'(k_0)R(k_0)x(k_0) \quad (\text{using (9.137) for } \lambda(k+1)). \tag{9.147}$$

Now, if $\mathbb{J}(i)$ is defined as

$$\mathbb{J}(i) = x'(k_0+N)M(k_0+N)x(k_0+N) + \sum_{k=i}^{k_0+N-1} \left[x'(k)P(k)x(k) + u'(k)Q(k)u(k)\right], \quad i = k_0, k_0+1, \ldots, k_0+N-1 \tag{9.148}$$

with the initial condition $x(k_0) = x_0$, then by following the same steps from (9.142)–(9.147) (replacing k_0 by i), it can be shown that

$$\mathbb{J}^*(i) = x'(i)R(i)x(i), \quad i = k_0, k_0+1, \ldots, k_0+N-1. \tag{9.149}$$

From (9.148) and (9.149), we can conclude that the Riccati matrix $R(k)$ is positive definite, as stated in the following proposition:

Proposition 9.11.2 The Riccati matrix $R(k)$ is positive definite for $k = k_0, k_0 + 1, \ldots, k_0+N-1$.

Proof. Since the matrices $P(k)$ and $M(k)$ are positive semi-definite and the matrix $Q(k)$ is positive definite for all $k = k_0, k_0+1, \ldots, k_0+N$, it is clear that $\mathbb{J}(i)$ is positive for all $i = k_0, k_0+1, \ldots, k_0+N-1$. Hence, from (9.149), we conclude that $R(i)$ is positive definite for $i = k_0, k_0+1, \ldots, k_0+N-1$. At $k = k_0+N$, $R(k_0+N)$ is positive semi-definite since $R(k_0+N) = 2M(k_0+N)$ and $M(k)$ is positive semi-definite. □

***Example* 9.11.3** Consider the Case II of Example 9.10.1. Using the Riccati equation, find the optimal control sequence $u(i) : i = 0, 1, 2$, the optimal state $x(i) : i = 1, 2, 3$, and the optimal value of $\mathbb{J}_0$ and $\mathbb{J}_1$.

Solution Here, $k_0 = 0$, $N = 3$, $A = 2$, $B = 1$, $P = Q = \frac{1}{2}$, and $M = 4$. Now, $G(k) = \frac{1}{2}BQ^{-1}B' = 1$. From (9.140),

$$R(k) = 2P + A'(R^{-1}(k+1) + G)^{-1}A = 1 + 4(R^{-1}(k+1)+1)^{-1}.$$

Since $R(3) = 2M(3) = 8$, we obtain $R(2) = \frac{41}{9}$, $R(1) = \frac{214}{50}$, and $R(0) = \frac{478}{264}$. From (9.133) and (9.135),

$$x(k+1) = Ax(k) - G(k)\lambda(k+1) = Ax(k) - G(k)R(k+1)x(k+1),$$

implying

$$x(k+1) = (I + G(k)R(k+1))^{-1}Ax(k), \quad k = 0,1,2.$$

Now, we get

$$x(k+1) = 2(1 + R(k+1))^{-1}x(k).$$

Hence, we obtain $x(1) = 2(1+\frac{214}{50})^{-1}2 = \frac{25}{33}$, $x(2) = \frac{9}{33}$, and $x(3) = \frac{2}{33}$.

For optimal control: $u(k) = -\frac{1}{2}Q^{-1}(k)B'(k)\lambda(k+1)$ for $k = 0,1,2$. Hence, we obtain $u(k) = -R(k+1)x(k+1)$. Thus, $u(0) = -R(1)x(1)$, $u(1) = -R(2)x(2)$, and $u(2) = -R(3)x(3)$.

For optimal values of $\mathbb{J}_0$ and $\mathbb{J}_1$: From (9.149), $\mathbb{J}_0 = \mathbb{J} = x'(k_0)R(k_0)x(k_0) = 4R(0) = \frac{239}{33}$ and $\mathbb{J}_1 = x'(k_0+1)R(k_0+1)x(k_0+1) = \left(\frac{25}{33}\right)^2 R(1) = \frac{2675}{1089}$.

Exercise 9.11.1 Consider the discrete-time system $x(k+1) = Ax(k) + Bu(k)$, where $A = \begin{bmatrix} 0 & 1 \\ 1 & -1 \end{bmatrix}$ and $B = \begin{bmatrix} 0 \\ 1 \end{bmatrix}$, $k = 1,2,3$. Write the state and co-state system with boundary conditions for finding the optimal state $x^*(t), k = 2,3,4$, and the optimal control $u^*(t)$, $k = 1,2,3$, which minimize the performance index

$$\mathbb{J} = 2x_1^2(4) + \sum_{k=0}^{3}(x_1^2(k) + x_2^2(k) + u^2(k))$$

subject to the boundary conditions $x(1) = \begin{bmatrix} 2 \\ 1 \end{bmatrix}$ and $x(4)$ unspecified. Also, write the matrix Riccati equation and the boundary condition.

9.12 Summary

- For the linear discrete-time system $x(k+1) = A(k)x(k)$ for $k = k_0, k_0+1, \ldots, k_0+N$, the state transition matrix $\phi(m,l)$ for $m,l \in \mathbb{N} \cup \{0\}$ is given by (see (9.4) and (9.5))

$$\phi(m,l) = \begin{cases} A(m-1)A(m-2)\cdots A(l), & \text{for } l < m, \\ (A(m))^{-1}(A(m+1))^{-1}\cdots(A(l-1))^{-1}, & \text{for } l > m, \\ I, & \text{for } m = l. \end{cases}$$

If A is a constant square matrix, then $\phi(m,l) = A^{m-l}$.

- In Section 9.2, properties of the state transition matrix are given.
- For the control system

$$\begin{cases} x(k+1) = A(k)x(k) + B(k)u(k), & k = k_0, k_0+1, \ldots, k_0+N, \\ x(k_0) = x_0, \end{cases}$$

the controllability Gramian matrix is defined by (see (9.11))

$$W = \sum_{i=k_0}^{k_0+N-1} \phi(k_0, i+1)B(i)B'(i)\phi'(k_0, i+1).$$

- The system is controllable if and only if W is nonsingular, and a control sequence that steers the system from the initial state x_0 to the final state x_1 is

$$u(i) = B'(i)\phi'(k_0, i+1)W^{-1}\left(\phi(k_0, k_0+N)x_1 - x_0\right), \quad i = k_0, k_0+1, \ldots, k_0+N-1$$

(see Theorem 9.3.1).

- The system

$$x(k+1) = A(k)x(k) + B(k)u(k), \quad k = k_0, k_0+1, \ldots, k_0+N-1 \tag{9.150}$$

with the observation

$$y(k) = C(k)x(k), \quad k = k_0, k_0+1, \ldots, k_0+N, \tag{9.151}$$

is observable if and only if the observability Gramian

$$M[k_0, k_0+N] = \sum_{r=k_0+1}^{k_0+N} \phi'(r, k_0)C'(r)C(r)\phi(r, k_0)$$

is nonsingular (see Theorem 9.4.2).

- If A, B, and C are constant matrices, then the system (9.150) is controllable if and only if the rank $\begin{bmatrix} B & AB & A^2B & \cdots & A^{N-1}B \end{bmatrix}_{n\times Nm} = n$ (Theorem 9.3.3).

- The system (9.150)–(9.151) is observable if and only if the rank $\begin{bmatrix} C \\ CA \\ CA^2 \\ \vdots \\ CA^{N-1} \end{bmatrix} = n$ (Section 9.4).
- For the autonomous continuous-time linear system, $\dot{x} = Ax + Bu$, $t_0 \leq t \leq \tau$, an equivalent discrete-time system $X(k+1) = FX(k) + GU(k)$, $k = 0, 1, \ldots, N-1$, is defined in Section 9.5, where $F = e^{Ah}$, $G = \left(\int_0^h e^{Ar} dr\right) B$, $h = \frac{\tau - t_0}{N}$, and $X(k) = x(t_0 + kh)$. The partition length h is chosen such that the controllability property of the continuous-time system is preserved in the corresponding discrete-time system (Example 9.5.9).
- The duality theorem analogous to that for continuous-time systems is proved in Section 9.6.
- In Section 9.7, stability and stabilizability of the trivial solution of discrete-time systems are discussed. For the autonomous linear system $x(k+1) = Ax(k)$, Theorem 9.7.1 shows that the origin is (i) asymptotically stable if and only if all the eigenvalues of A have modulus < 1, (ii) unstable if and only if at least one of the eigenvalues of A has modulus > 1, and (iii) stable (but not asymptotically stable) if and only if all the eigenvalues have modulus ≤ 1 and the eigenvalues with modulus 1 have equal algebraic and geometric multiplicities.
- In Section 9.8, Lyapunov theorems for stability and instability of discrete-time system at the origin are proved. These results are similar to those of continuous-time systems.
- Discrete-time optimal control problem is discussed in Section 9.9. The problem is to find the optimal control sequence $u(k)$, $k = k_0, k_0+1, \ldots, k_0 + N-1$, and the optimal state sequence $x(k)$, $k = k_0, k_0+1, \ldots, k_0+N$, of the control system $x(k+1) = g(x(k), u(k)k)$, $k = k_0, k_0+1, \ldots, k_0+N-1$, that minimizes the functional $\mathbb{J} = \sum_{k=k_0}^{k_0+N-1} f(x(k), x(k+1), u(k), k)$. Two cases, namely fixed and free boundary conditions, are considered. To find the optimal solution, the Lagrangian is defined in equation (9.104). The Euler–Lagrange equations are given in (9.106).

- The linear quadratic optimal control problem for discrete-time linear system is discussed in Section 9.10.
- Optimal feedback control using the Riccati matrix is discussed in Section 9.11.

9.13 Exercise

1. Let $\dot{x} = Ax + Bu$ for $2 \leq t \leq 10$, where $A = \begin{bmatrix} 2 & 1 \\ 0 & 2 \end{bmatrix}$ and $B = \begin{bmatrix} 1 \\ 3 \end{bmatrix}$. Find an equivalent discrete-time system $y(k+1) = Ey(k) + Fv(k)$ for $k = 0, 1, \ldots, N$ such that $y(k) = x(2+kh)$ and $v(k) = u(2+kh)$ for $k = 0, 1, \ldots, N$, where $h = \frac{1}{2}$.

2. Consider the system $\dot{x} = Ax + Bu$ for $0 \leq t \leq 2$, where $A = \begin{bmatrix} 1 & 5\pi \\ -5\pi & 1 \end{bmatrix}$ and $B = \begin{bmatrix} 0 \\ 1 \end{bmatrix}$. Let $z((k+1)h) = Fz(kh) + Gu(kh)$ be the corresponding discrete-time system with $h = 0.2$. Show that the continuous-time system is controllable, but the discrete-time system is not controllable.

3. Show that the discrete-time dynamical system $x(k+1) = Ax(k)$: $k = 0, 1, 2 \ldots$, where $A = \begin{bmatrix} 2 & 3 \\ 3 & 2 \end{bmatrix}$, is unstable at $(0,0)$. By introducing the control matrix $B = \begin{bmatrix} 0 \\ 1 \end{bmatrix}$, find a feedback control $u(k)$ for $k = 0, 1, 2 \ldots$ such that the system $x(k+1) = Ax(k) + Bu(k)$: $k = 0, 1, 2 \ldots$ is asymptotically stable at $(0,0)$.

4. Consider the functional

$$\mathbb{J} = \sum_{k=1}^{3} f(x(k), x(k+1), k),$$

where

$$f(x(k), x(k+1), k) = x(k)x(k+1) + x^2(k) + x^2(k+1).$$

Find the optimal function $x^*(k)$ with the boundary conditions $x^*(1) = 2$ and $x^*(4) = 6$ that minimizes $\mathbb{J}$.

5. For the functional

$$\mathbb{J} = \sum_{k=1}^{3} f(x(k), x(k+1), k),$$

with

$$f(x(k), x(k+1), k) = \langle x(k), x(k+1) \rangle + \|x(k)\|^2,$$

where $\langle \cdot, \cdot \rangle$ denotes the dot product in R^2, find the optimal solution $x^*(k) = \begin{bmatrix} x_1^*(k) \\ x_2^*(k) \end{bmatrix}$ with the boundary conditions $x^*(1) = \begin{bmatrix} 1 \\ 0 \end{bmatrix}$ and $x^*(4) = \begin{bmatrix} 4 \\ -3 \end{bmatrix}$.

6. Find the optimal control sequence $u(k)$ for $k = 2, 3, 4$ that optimizes

$$\mathbb{J} = \sum_{k=2}^{4} (x^2(k) + 2x^2(k+1) + u^2(k))$$

under the constraint

$$\begin{cases} x(k+1) = 2x(k) + u(k), & k = 2, 3, 4, \\ x(2) = 1, \quad x(5) = 3. \end{cases}$$

7. Find the optimal control sequence $u(k)$ for $k = 0, 1, 2, 3$ and the optimal states $x(k)$ for $k = 1, 2, 3, 4$ for the control system $x(k+1) = -2x(k) + 3u(k)$ for $k = 0, 1, 2, 3$ that minimize the functional

$$\mathbb{J} = x^2(4) + \sum_{k=0}^{3} (x^2(k) + u^2(k)),$$

given the following cases: (i) $x(0) = 2$ and $x(4) = 4$ and (ii) $x(0) = 2$ and $x(4)$ is arbitrary.

8. Find the optimal control sequence $u(k)$ for $k = 0, 1, 2, 3$ and the optimal states $x(k)$ for $k = 1, 2, 3, 4$ for the control system $x(k+1) = 2x(k) + u(k)$ for $k = 0, 1, 2, 3$ that minimize the functional

$$\mathbb{J} = 4x^2(3) + \frac{1}{2} \sum_{k=0}^{3} (x^2(k) + u^2(k)),$$

given that $x(0) = 2$ and $x(3)$ is arbitrary. Using the Riccati equation, find the optimal control sequence $u(i)$ for $i = 0, 1, 2, 3$, the optimal state $x(i)$ for $i = 1, 2, 3, 4$, and the optimal value of $\mathbb{J}_0$ and $\mathbb{J}_1$.

9. Consider the system $x(k+1) = Ax(k)$ for $k = 0, 1, 2, \ldots$, where

$$(i)\ A = \begin{bmatrix} 0 & -\frac{1}{2} \\ \frac{1}{3} & 0 \end{bmatrix},$$

$$(ii)\ A = \begin{bmatrix} -1 & 1 \\ -\frac{1}{2} & 1 \end{bmatrix}.$$

Determine the stability properties at $(0,0)$ of the systems using suitable Lyapunov functions.

10
Infinite Dimensional Systems

10.1 Introduction

The aim of this chapter is to introduce the concept of controllability for control systems defined in infinite dimensional spaces. In particular, the controlled heat equation is studied in the framework of semigroup theory. Basic concepts from functional analysis are prerequisites for understanding this chapter.

In Section 10.2, example of an infinite dimensional linear control system is given. In Section 10.3, uniformly continuous and strongly continuous semigroups are defined with examples. The definition of infinitesimal generator of a semigroup and related examples are provided in Section 10.4. In Section 10.5, important properties of C_0-semigroups and uniformly continuous semigroups and their generators are discussed. The mild solution of initial value problems (IVPs) in terms of semigroups is also introduced in this section. In Section 10.6, approximate controllability of infinite dimensional systems and ill-posed problems are defined, and examples are provided. In Section 10.7, computation of a control function for an approximately controllable system as a solution to an ill-posed problem is demonstrated using the linear heat equation. Section 10.8 contains a brief discussion on a semilinear control system and its controllability using the corresponding linear system.

10.2 Infinite Dimensional Control System

Let A and B be constant matrices of order $n \times n$ and $n \times m$, respectively. A finite dimensional time-invariant linear control system is of the form

$$\begin{cases} \dot{x}(t) = Ax(t) + Bu(t) & \text{for } t_0 < t \leq \tau, \\ x(t_0) = x_0, \end{cases} \tag{10.1}$$

whose solution is (see equation (5.13))

$$x(t) = \exp((t - t_0)A)x_0 + \int_{t_0}^{t} \exp((t-s)A)Bu(s)ds \quad \text{for } t_0 \leq t \leq \tau. \tag{10.2}$$

The matrix $\exp((t-t_0)A)$ is known as the **state transition matrix.** In (10.1), the system matrix $A : \mathbb{R}^n \to \mathbb{R}^n$ and the control matrix $B : \mathbb{R}^m \to \mathbb{R}^n$ are **bounded linear** operators.

In this chapter, we shall consider linear control systems of the form (10.1), i.e., $\dot{x} = Ax + Bu$, where A and B are linear operators defined on infinite dimensional spaces. Let X and U be real Banach spaces, and $A : D(A) \subset X \to X$ and $B : U \to X$ be linear operators. Here, $D(A)$ denotes the domain of A. The operator B is assumed to be bounded, whereas A may be bounded or unbounded. If A is a bounded linear operator, then $D(A) = X$.

In many applications, control systems on infinite dimensional spaces are obtained from partial differential equations. The following example illustrates a control system of the form (10.1) defined on an infinite dimensional space. Since absolutely continuous functions play an important role in this chapter, let us first define absolute continuity.

Definition 10.2.1 (Absolute continuity) A function $f : [a,b] \to \mathbb{R}$ is said to be **absolutely continuous** on $[a,b]$ if for any given $\varepsilon > 0$, there exists a $\delta > 0$ such that $\sum_{k=1}^{n} |f(x_k) - f(y_k)| < \varepsilon$ whenever $\sum_{k=1}^{n} |x_k - y_k| < \delta$ for any finite set of mutually disjoint intervals $\{(x_k, y_k) : k = 1, 2, \ldots, n\}$ in $[a,b]$.

The following statements hold.

(*i*) If $f : [a,b] \to \mathbb{R}$ is absolutely continuous, then f is uniformly continuous on $[a,b]$.

(*ii*) If $f : [a,b] \to \mathbb{R}$ is Lipschitz continuous, then f is absolutely continuous on $[a,b]$.

(*iii*) If $f : [a,b] \to \mathbb{R}$ is absolutely continuous, then f is differentiable almost everywhere on $[a,b]$ and the derivative $\frac{df}{dt} \in L_1[a,b]$.

(*iv*) $f : [a,b] \to \mathbb{R}$ is absolutely continuous if and only if there exists a function $g \in L_1[a,b]$ such that $f(t) - f(a) = \int_a^t g(s)ds$, where $\frac{df(t)}{dt} = g(t)$ for a.e. $t \in [a,b]$.

Example **10.2.2** Consider the following equation that represents the heat conduction on a thin rod of length 1

$$\begin{cases} \frac{\partial}{\partial t}y(\rho,t) = \frac{\partial^2}{\partial \rho^2}y(\rho,t) + w(\rho,t), \quad 0 \le \rho \le 1, t > 0, \\ y(\rho,0) = y_0(\rho), \\ \frac{\partial}{\partial \rho}y(0,t) = 0 = \frac{\partial}{\partial \rho}y(1,t). \end{cases} \tag{10.3}$$

Here, $y(\rho,t)$ represents the temperature at a point ρ on the rod at time t, and $w(\rho,t)$ represents the external heat source (control) function.

Let $x(t) = y(\cdot,t)$, $u(t) = w(\cdot,t)$, and $\frac{dx(t)}{dt} = \frac{\partial}{\partial t}y(\cdot,t)$. Let $X = L_2[0,1]$, and let $x(t) \in X$ for each $t \ge 0$. Let $A : D(A) \subset X \to X$ be defined as $Az = \frac{d^2z}{d\rho^2}$. Then (10.3) can be formulated as an **abstract** differential equation in X of the form

$$\begin{cases} \dot{x}(t) = Ax(t) + u(t), \quad t > 0, \\ x(0) = x_0, \end{cases} \tag{10.4}$$

where $x_0 \in X$ and $x_0(\rho) = y(\rho,0)$. To satisfy the given boundary conditions, the domain of the operator A is defined as

$$D(A) = \Big\{ z \in L_2[0,1] : z, \frac{dz}{d\rho} \text{ are absolutely continuous functions of } \rho, \\ \frac{d^2z}{d\rho^2} \in L_2[0,1], \text{ and } \frac{dz}{d\rho}(0) = 0 = \frac{dz}{d\rho}(1) \Big\}.$$

In the next section, we introduce the semigroup of operators defined on a Banach space. One can observe that the semigroup of operators has similar properties to those of the state transition matrix (Section 4.7) and that the role of the semigroup is analogous (with suitable modifications) to that of the state transition matrix in proving certain properties such as controllability and observability of abstract control systems in infinite dimensional setting.

Let X be a Banach space and $BL(X)$ be the set of all bounded linear operators defined from X into itself. From Theorem 3.4.18, it is known that $BL(X)$ is a Banach space under the operator norm. In the following sections we shall use the notation $\|\cdot\|$ to denote the norm of an element in a Banach space Z,instead of $\|\cdot\|_Z$. If $T : Z \to Z$ is a bounded linear operator, then $\|T\|$ represents $\|T\|_{BL(Z)}$.

10.3 Semigroup of Bounded Linear Operators

Consider the finite dimensional dynamical system

$$\begin{cases} \dot{x} = Ax, & t > 0, \\ x(0) = x_0, \end{cases} \tag{10.5}$$

where A is an $n \times n$ matrix. It is well known that its solution can be written in terms of the state transition matrix as $x(t) = e^{At}x_0$ (see (4.54)). It can be observed that the state transition matrix $e^{At}, t \geq 0$, forms a semigroup of operators on $\mathbb{R}^n$ by satisfying the conditions stated in Definition 10.3.1.

In this section, we shall study the solutions of infinite dimensional systems of the form (10.5) in terms of the semigroup of operators associated with the operator A.

Definition 10.3.1 Let X be a Banach space. A **semigroup** on X is a family of **bounded linear** operators $T(t) : X \to X$ for $t \geq 0$ satisfying the following properties

(*a*) $T(0) = I$, where I is the identity operator on X;

(*b*) $T(t+s) = T(t)T(s)$ for all $t, s \geq 0$.

Example **10.3.2** Let $X = \mathbb{R}$ and $T(t) : \mathbb{R} \to \mathbb{R}$ be defined as $T(t)x = e^{at}x$ for $t \geq 0$ and $a \in \mathbb{R}$. Then, according to Definition 10.3.1, $T(t) : t \geq 0$ forms a semigroup.

Definition 10.3.3 A **strongly continuous** semigroup, or C_0**-semigroup**, on X is a semigroup that satisfies the following condition

(*c*) $\lim_{t\to 0^+} \|T(t)x - x\| = 0$ for all $x \in X$.

Definition 10.3.4 A semigroup $T(t) : X \to X$ is called a **uniformly continuous** semigroup if it satisfies

(*d*) $\lim_{t\to 0^+} \|T(t) - I\| = 0$, where I is the identity operator on X.

Note that every uniformly continuous semigroup is a C_0-semigroup, but the converse is not necessarily true.

Theorem 10.3.5 Let X be a Banach space, and $A : X \to X$ be a **bounded linear** operator. Then the family of operators $T(t) : t \geq 0$, defined as $T(t) = e^{At}$, forms a uniformly continuous semigroup on X.

Proof. First, let us prove that e^{At} is a bounded linear operator on X. Consider the operator sequence

$$T_n(t) = I + tA + \frac{t^2A^2}{2!} + \cdots + \frac{t^nA^n}{n!}, \quad n = 0, 1, 2 \ldots.$$

Note that $T_n(t)$ is a bounded linear operator for each $n = 0, 1, 2, \ldots$ and for each $t \geq 0$. (Since it is a finite sum of bounded linear operators.) Hence, $T_n(t) \in BL(X)$ for each $n = 1, 2, \ldots$. If the sequence $\{T_n(t)\}$ converges in $BL(X)$, then its limit is denoted by e^{At}. To prove this, we shall show that the sequence $\{T_n(t)\}$ is Cauchy.

We know that $e^r = \sum_{j=0}^{\infty} \frac{r^j}{j!}$ converges for every $r \in \mathbb{R}$. Hence, for a given $\varepsilon > 0$, there exists $N \in \mathbb{N}$ such that $\sum_{j=m+1}^{n} \frac{r^j}{j!} < \varepsilon$ for all $n > m > N$. Now, for each fixed $t > 0$, we have

$$\|T_n(t) - T_m(t)\| = \left\| \sum_{j=m+1}^{n} \frac{t^j A^j}{j!} \right\| \leq \sum_{j=m+1}^{n} \frac{|t|^j \|A\|^j}{j!} < \varepsilon \quad \text{for } n > m > N$$

(since $\|A\|$ is finite and $\sum_{j=0}^{\infty} \frac{t^j \|A\|^j}{j!}$ is convergent). Hence, $\{T_n(t)\}$ is Cauchy in $BL(X)$ and converges to the limit e^{At}.

Now, we verify the conditions (a) and (b) of the semigroup.
(a) It can be easily seen that $T(0) = I$.
(b) For all $t, r \geq 0$, we have

$$\begin{aligned}
T(t)T(r) &= e^{tA}e^{rA} \\
&= \lim_{n\to\infty} \left(I + tA + \frac{t^2}{2!}A^2 + \cdots + \frac{t^n}{n!}A^n\right)\left(I + rA + \frac{r^2}{2!}A^2 + \cdots + \frac{r^n}{n!}A^n\right) \\
&= \lim_{n\to\infty} \left(I + (t+r)A + \frac{(t+r)^2}{2!}A^2 + \cdots + \frac{(t+r)^n}{n!}A^n\right) \\
&\quad + \lim_{n\to\infty} \left[\left(t\frac{r^n}{n!} + \frac{t^2}{2!}\frac{r^{n-1}}{(n-1)!} + \cdots + \frac{t^n}{n!}r\right)A^{n+1}\right. \\
&\quad \left. + \left(\frac{t^2}{2!}\frac{r^n}{n!} + \frac{t^3}{3!}\frac{r^{n-1}}{(n-1)!} + \cdots + \frac{t^n}{n!}\frac{r^2}{2!}\right)A^{n+2} + \cdots + \left(\frac{t^n}{n!}\frac{r^n}{n!}\right)A^{2n}\right] \\
&= \lim_{n\to\infty} g_n(t,r) + \lim_{n\to\infty} \left[\sum_{k=1}^{n} h_{n+k}(t,r)\frac{A^{n+k}}{(n+k)!}\right],
\end{aligned} \tag{10.6}$$

where

$$g_n(t,r) = I + (t+r)A + \frac{(t+r)^2}{2!}A^2 + \cdots + \frac{(t+r)^n}{n!}A^n$$

and

$$h_{n+k}(t,r) = (n+k)!\left(\frac{t^k}{k!}\frac{r^n}{n!} + \frac{t^{k+1}}{(k+1)!}\frac{r^{n-1}}{(n-1)!} + \cdots + \frac{t^n}{n!}\frac{r^k}{k!}\right).$$

Since $h_{n+k}(t,r) \leq (t+r)^{n+k}$ for $t,r \geq 0$, we have

$$\left\| \sum_{k=1}^{n} h_{n+k}(t,r) \frac{A^{n+k}}{(n+k)!} \right\| \leq \sum_{k=1}^{n} \frac{(t+r)^{n+k}}{(n+k)!} \|A\|^{n+k}$$
$$= \sum_{k=n+1}^{2n} \frac{(t+r)^k}{k!} \|A\|^k \to 0 \text{ as } n \to \infty.$$

Hence, from (10.6),

$$T(t)T(r) = \lim_{n\to\infty} g_n(t,s) = e^{(t+r)A} = T(t+r).$$

Therefore, $e^{At}e^{Ar} = e^{A(t+r)}$, which verifies condition (b).

Now, let us show that $T(t)$ forms a uniformly continuous semigroup. For this, the condition (d) needs to be verified. For all $t > 0$,

$$\|T(t) - I\| = \|e^{tA} - I\| = \lim_{n\to\infty} \left\| \sum_{j=0}^{n} \frac{(tA)^j}{j!} - I \right\| = \lim_{n\to\infty} \left\| \sum_{j=1}^{n} \frac{(tA)^j}{j!} \right\|. \tag{10.7}$$

Note that

$$\left\| \sum_{j=1}^{n} \frac{(tA)^j}{j!} \right\| \leq \sum_{j=1}^{n} \frac{|t|^j \|A\|^j}{j!} \leq e^{\|A\|t} - 1. \tag{10.8}$$

Using (10.8) in (10.7) and then taking the limit as $t \to 0^+$, we get

$$\lim_{t\to 0^+} \|T(t) - I\| = 0.$$

Hence, $T(t)$ is a uniformly continuous semigroup. □

***Example* 10.3.6** Let $T(t) : L_1[0,\infty) \to L_1[0,\infty)$ be defined by $T(t)x = x_t$, where $x_t(s) = x(t+s)$ for $t,s \geq 0$. To prove that $T(t) : t \geq 0$ forms a C_0-semigroup, let us verify conditions (a), (b), and (c).
(a) Let $x \in L_1[0,\infty)$. Then $T(0)x = x_0$, where $x_0(s) = x(s)$. Hence, $T(0) = I$.
(b) We have $T(t+s)x = x_{t+s}$ and $(T(t)T(s))x = T(t)x_s = (x_s)_t$. Now, $x_{t+s}(r) = x(t+s+r) = x_s(t+r) = (x_s)_t(r)$ for all $r > 0$. Hence, condition (b) is satisfied.
(c) We need to show that $\lim_{t\to 0^+} \|T(t)x - x\| = 0$ for every $x \in L_1[0,\infty)$. As every $x \in L_1[0,\infty)$ can be approximated by a continuous function with compact support, for a given $\varepsilon > 0$, one can find a continuous function y with compact support in $[0,\infty)$ such that

$$\|x - y\| < \frac{\varepsilon}{3}, \text{ i.e., } \int_0^\infty |x(s) - y(s)| ds < \frac{\varepsilon}{3}.$$

Hence,

$$\|x_t - y_t\| = \int_t^{\infty} |x(s) - y(s)|ds < \frac{\varepsilon}{3}.$$

For sufficiently small $t > 0$, $\|y_t - y\| < \frac{\varepsilon}{3}$. Now,

$$\begin{aligned}\|T(t)x - x\| = \|x_t - x\| &= \|x_t - y_t + y_t - y + y - x\| \\ &\leq \|x_t - y_t\| + \|y_t - y\| + \|y - x\| < \varepsilon,\end{aligned}$$

for sufficiently small $t > 0$. Hence, $T(t) : t \geq 0$ is a C_0-semigroup.

***Example* 10.3.7** Let $\{\phi_j : j = 1, 2, \ldots\}$ be an orthogonal basis in $L_2[0, l]$, and let $\{\lambda_j : j = 1, 2. \ldots\}$ be a set of real numbers with $\sup_j \lambda_j < \infty$. Consider the operator $T(t) : t \geq 0$ on $L_2[0, l]$ defined as

$$T(t)x = \sum_{j=1}^{\infty} e^{\lambda_j t} \langle x, \phi_j \rangle \phi_j, \tag{10.9}$$

where $\langle \cdot, \cdot \rangle$ denotes the inner product in $L_2[0, l]$. Then, (i) $T(t) : t \geq 0$ is a C_0-semigroup on $L_2[0, l]$, (ii) $T(t)$ is self-adjoint for all $t \geq 0$, and (iii) $T(t)$ is compact for all $t > 0$, provided that $\{\lambda_j\}_{j \in \mathbb{N}}$ is decreasing and $\inf_j \lambda_j = -\infty$.

Solution As $L_2[0, l]$ is a Hilbert space and $\{\phi_j : j = 1, 2, \ldots\}$ is an orthogonal basis,

$$x = \sum_{j=1}^{\infty} \langle x, \phi_j \rangle \phi_j, \ \|x\|^2 = \sum_{j=1}^{\infty} |\langle x, \phi_j \rangle|^2,$$

and

$$T(t)\phi_k = e^{\lambda_k t} \phi_k \quad \text{for } k = 1, 2, \ldots \quad \text{(from (10.9)}.$$

Moreover, $\langle T(t)x, \phi_k \rangle = e^{\lambda_k t} \langle x, \phi_k \rangle$. For each $t \geq 0$, $T(t)$ is bounded, as $\sup_j \lambda_j < \infty$ (from (10.9) and Definition 3.4.7).
Proof of (i): It can be observed that (a) $T(0) = I$, and

$$\begin{aligned}(b) \quad T(t)T(r)x = T(t)\,(T(r)x) &= \sum_{j=1}^{\infty} e^{\lambda_j t} e^{\lambda_r t} \langle x, \phi_j \rangle \phi_j \\ &= \sum_{j=1}^{\infty} e^{(t+r)j} \langle x, \phi_j \rangle \phi_j = T(t+r)x.\end{aligned}$$

Hence, $T(t) : t \geq 0$ forms a semigroup.

To prove that $T(t)$ forms a C_0-semigroup, we verify condition (c): $\lim_{t\to 0^+}\|T(t)x-x\|=0$ for all $x\in L_2[0,l]$. Now,

$$\begin{aligned}\|T(t)x-x\|^2 &= \left\|\sum_{j=1}^{\infty}(e^{t\lambda_j}-1)\langle x,\phi_j\rangle\phi_j\right\|^2\\ &= \sum_{j=1}^{\infty}(e^{t\lambda_j}-1)^2|\langle x,\phi_j\rangle|^2\\ &= \sum_{j=1}^{N}(e^{t\lambda_j}-1)^2|\langle x,\phi_j\rangle|^2+\sum_{j=N+1}^{\infty}(e^{t\lambda_j}-1)^2|\langle x,\phi_j\rangle|^2. \end{aligned} \tag{10.10}$$

Let $\alpha=\sup_{t\in[0,1],j\geq 1}(e^{t\lambda_j}-1)^2$. Now, for any given $x\in L_2[0,l]$ and $\varepsilon>0$, there exists an $N\in\mathbb{N}$ and a sufficiently small $t>0$ such that

$$(e^{t\lambda_j}-1)^2<\frac{\varepsilon}{\|x\|^2},\quad j=1,2,\ldots,N \tag{10.11}$$

and

$$\sum_{j=N+1}^{\infty}\langle x,\phi_j\rangle^2<\frac{\varepsilon}{\alpha}. \tag{10.12}$$

Using (10.11) and (10.12) in (10.10), we obtain

$$\|T(t)x-x\|^2<\frac{\varepsilon}{\|x\|^2}\sum_{j=1}^{N}|\langle x,\phi_j\rangle|^2+\alpha\sum_{j=N+1}^{\infty}|\langle x,\phi_j\rangle|^2<2\varepsilon.$$

Hence, $\lim_{t\to 0^+}\|T(t)x-x\|^2=0$ for every $x\in L_2[0,l]$, implying that $T(t):t\geq 0$ forms a C_0-semigroup.

Proof of (ii): To prove that $T(t)$ is self-adjoint, it is sufficient to show that

$$\langle T(t)x,y\rangle=\langle x,T(t)y\rangle \quad \text{for all } x,y\in L_2[0,l],$$

which can be verified using (10.9) and the properties of the inner product.

Proof of (iii): We now prove that $T(t)$ is compact for each $t>0$, provided that $\{\lambda_j\}_{j\in\mathbb{N}}$ is decreasing and $\inf_j\lambda_j=-\infty$. Note that $T(0)=I$ is not compact on $L_2[0,l]$. Let

$$T_n(t)x=\sum_{j=1}^{n}e^{t\lambda_j}\langle x,\phi_j\rangle\phi_j,\quad n=1,2,\ldots.$$

Since the range of $T_n(t)$ is a finite dimensional subspace of $L_2[0,l]$, $T_n(t)$ is compact for each $n\in\mathbb{N}$ (see Section 3.6). Now,

$$\|(T(t)-T_n(t))x\|^2=\left\|\sum_{j=n+1}^{\infty}e^{t\lambda_j}\langle x,\phi_j\rangle\phi_j\right\|^2.$$

As $\{\lambda_j\}_{j\in\mathbb{N}}$ is decreasing with $\inf_j \lambda_j = -\infty$,

$$\|(T(t)-T_n(t))x\|^2 = \left\| \sum_{j=n+1}^{\infty} e^{t\lambda_j}\langle x,\phi_j\rangle\phi_j \right\|^2 = \sum_{j=n+1}^{\infty} e^{2t\lambda_j}|\langle x,\phi_j\rangle|^2. \tag{10.13}$$

For a fixed $t > 0$, given $\varepsilon > 0$, and a sufficiently large n,

$$e^{2t\lambda_n} < \varepsilon. \tag{10.14}$$

Now, if $\|x\| = 1$, then from Definition 3.4.10, (10.13), and (10.14), we have

$$\|T(t)-T_n(t)\| = \sup_{\|x\|=1} \|(T(t)-T_n(t))x\| < \varepsilon, \quad \text{when } n \text{ is large.}$$

Hence, $\{T_n(t)\}$ converges to $T(t)$ uniformly, implying (from Property 3.6.14(vii)) that, $T(t)$ is compact for each $t > 0$.

Theorem 10.3.8 Suppose that $T(t) : t \geq 0$ is a C_0-semigroup. Then there exist constants $C > 1$ and $\alpha > 0$ such that for $t \geq 0$,

$$\|T(t)\| \leq Ce^{\alpha t}.$$

Proof. First, let us show that there exists a $\tau > 0$ such that $\|T(t)\|$ is bounded on the interval $[0,\tau]$. Suppose this is not true. Then there exists a nonnegative sequence $\{t_n\}$ such that $\lim_{n\to\infty} t_n = 0$ and $\|T(t_n)\| \geq n$, $n = 1,2,\ldots$. Now, by the uniform boundedness theorem (Theorem 3.4.20), for some $x \in X$, $\|T(t_n)x\|$ is unbounded. However, by the definition of a C_0-semigroup, $\lim_{t\to 0^+} T(t)x = x$, which implies that the sequence $\{T(t_n)x\}$ is bounded. Therefore, $\|T(t)\| \leq C$ for $0 \leq t \leq \tau$ for some $\tau > 0$ and finite constant $C > 1$ (since $T(0) = I$).

Next, take $t = n\tau + s$ with $0 < s \leq \tau$. By the semigroup property, we have

$$\begin{aligned}\|T(t)\| = \|T(n\tau+s)\| = \|T(\underbrace{\tau+\tau+\cdots+\tau}_{n \text{ times}}+s\| &= \|T(s)(T(\tau))^n\| \\ &\leq C^{n+1} = Ce^{n\log C}.\end{aligned}$$

Define $\alpha = \frac{\log C}{\tau} > 0$, so that $n\log C = n\alpha\tau \leq \alpha t$ since $t = n\tau + s$ with $0 \leq s \leq \tau$. Thus,

$$\|T(t)\| \leq Ce^{\alpha t}.$$

The function $Ce^{\alpha t}$ is called a **growth bound** for $T(t)$. □

If A is a square matrix, a growth bound for $T(t) = e^{At}$ has been shown in Lemma 3.2.16.

Remark 10.3.9 It is possible to obtain a better growth bound than the one given in Theorem 10.3.8. For example, if $A = \begin{bmatrix} -1 & 1 \\ 0 & -1 \end{bmatrix}$, then

$$T(t) = e^{At} = \begin{bmatrix} e^{-t} & te^{-t} \\ 0 & e^{-t} \end{bmatrix} \text{ and } \|T(t)\| \leq \begin{cases} e^{-t} & \text{for } 0 \leq t \leq \tau \leq 1, \\ \tau e^{-t} & \text{for } 0 \leq t \leq \tau \text{ and } \tau > 1. \end{cases}$$

Exercise 10.3.1 Find a growth bound for the operator $T(t)$ defined in Example 10.3.7.

10.4 Infinitesimal Generator of a Semigroup

Let A be an $n \times n$ matrix. For the state transition matrix $e^{At}, t \geq 0$, it is easy to verify that $\frac{d}{dt} e^{At} = A$ at $t = 0^+$. This finite dimensional concept motivates the following definition of the infinitesimal generator of a semigroup in infinite dimensional spaces.

Definition 10.4.1 The **infinitesimal generator** A of a C_0-semigroup $T(t) : t \geq 0$ on a Banach space X is defined by

$$Ax = \lim_{t \to 0^+} \frac{T(t)x - x}{t},$$

if the limit exists. The domain of A, denoted by $D(A)$, is the set of elements $x \in X$ for which the limit exists. The operator A is represented as $A : D(A) \subset X \to X$.

In Theorem 10.5.5, we shall show that two different semigroups of operators cannot have the same infinitesimal generator.

Theorem 10.4.2 If $A : X \to X$ is a bounded linear operator, then it is the infinitesimal generator of the uniformly continuous semigroup e^{At}.

Proof. Let $T(t) = e^{At}$ for $t > 0$. In Theorem 10.3.5, it was shown that $T(t) : t \geq 0$ forms a uniformly continuous semigroup on X. To show that A is the infinitesimal generator of $T(t)$, consider

$$\left\| \frac{T(t)x - x}{t} - Ax \right\| = \left\| \frac{(I + tA + \frac{t^2A^2}{2!} + \cdots - I)x}{t} - Ax \right\|$$

$$= \lim_{n \to \infty} \left\| \frac{1}{t} \sum_{j=2}^{n} \frac{t^j A^j x}{j!} \right\|$$

$$\leq \lim_{n\to\infty} \sum_{j=2}^{n} \frac{t^{j-1}\|A\|^j}{j!}\|x\|,$$

which tends to 0 as $t \to 0^+$ for all $x \in X$. Hence, $\lim_{t\to 0^+} \frac{T(t)x-x}{t} = Ax$ for all $x \in X$. Therefore, by Definition 10.4.1, A is the infinitesimal generator of $T(t)$. Since A is bounded, its domain $D(A) = X$. □

In Theorem 10.5.4, we shall prove the converse to Theorem 10.4.2, i.e, the infinitesimal generator of a uniformly continuous semigroup is bounded.

***Example* 10.4.3** Let $T(t) : L_1[0,\infty) \to L_1[0,\infty)$ be defined as $T(t)x = x_t$, where $x_t(r) = x(t+r)$. We have seen in Example 10.3.6 that $T(t)$ is a C_0-semigroup. Let A be a function defined from $L_1[0,\infty)$ into itself as

$$Ax = \lim_{t\to 0^+} \frac{T(t)x-x}{t},$$

Then,

$$(Ax)(r) = \lim_{t\to 0^+} \left(\frac{x_t(r)-x(r)}{t}\right) = \lim_{t\to 0^+} \frac{x(r+t)-x(r)}{t} = \frac{dx}{dr}.$$

Hence, the differential operator $\frac{d}{dr}$ is the infinitesimal generator of $T(t)$, and the domain $D(A)$ of A is $C^1[0,\infty)$ with the L_1 norm which is dense in $L_1[0,\infty]$.

***Example* 10.4.4** Let $T(t) : L_2[0,l] \to L_2[0,l]$, $t > 0$, be the operator given in Example 10.3.7, i.e.,

$$T(t)x = \sum_{i=1}^{\infty} e^{\lambda_i t}\langle x, \phi_i\rangle \phi_i : \sup_i \lambda_i < \infty.$$

Then, from Definition 10.4.1, it can be verified that the infinitesimal generator of $T(t)$ is defined on $L_2[0,l]$ as

$$Ax = \sum_{i=1}^{\infty} \lambda_i \langle x, \phi_i\rangle \phi_i$$

with domain

$$D(A) = \left\{x \in L_2[0,l] : \sum_{i=1}^{\infty} \lambda_i^2 |\langle x, \phi_i\rangle|^2 < \infty\right\}.$$

As $\{\phi_i\}$ is an orthogonal basis in $L_2[0,l]$, one can observe that (λ_i, ϕ_i) and $(e^{\lambda_i t}, \phi_i)$, $i = 1, 2, \ldots$, are the eigenpairs of A and $T(t)$, respectively. Moreover, A and $T(t)$ are self-adjoint. Note that the operator A is unbounded if $\sup_i |\lambda_i| = \infty$.

Definition 10.4.5 The semigroup $T(t)$ is said be a **contraction** semigroup if $\|T(t)\| < 1$ for all $t \geq 0$.

The semigroup $T(t)$ given in Example 10.4.4 is a contraction semigroup provided $\lambda_i < 0$ for all i, since $\|T(t)x\| < \|x\|$ for all $x \in L_2[0,l]$.

Exercise 10.4.1 If $T(t)$ is a C_0-semigroup on a Hilbert space X, show that $e^{\alpha t}T(t)$, $\alpha \in \mathbb{R}$, is also a C_0-semigroup on X. If A is the infinitesimal generator of $T(t)$, find the infinitesimal generator of $e^{\alpha t}T(t)$.

Exercise 10.4.2 Let $X = L_1[0,b]$ and $T(t) : X \to X, t \geq 0$, be defined as

$$(T(t)x)(s) = \begin{cases} x(t+s), & 0 \leq s \leq b, t \leq b-s, \\ 0, & 0 \leq s \leq b, t > b-s. \end{cases}$$

Show that:

(*i*) $T(t)$ defines a C_0-semigroup on X.

(*ii*) The infinitesimal generator of $T(t)$ is $A = \frac{d}{ds}$ with domain

$$D(A) = \{x \in L_1[0,b] : x \text{ is absolutely continuous on } [0,b] \text{ and } x(b) = 0\}.$$

10.5 Properties of $T(t)$ and A

To analyze abstract differential equations, the following properties of the semigroup $T(t)$ and the infinitesimal generator A will be useful.

Theorem 10.5.1 Let $T(t) : t \geq 0$ be a C_0-semigroup on a Banach space X. Then the following statements hold.

(*i*) For all $x \in X$, the map $t \to T(t)x$ is continuous.

(*ii*) For all $x \in X$, $\lim_{h\to 0^+} \frac{1}{h}\int_t^{t+h} T(s)x ds = T(t)x$.

(*iii*) If $x \in D(A)$, then $T(t)x \in D(A)$ for $t \geq 0$.

(*iv*) For $x \in X$, $\int_0^t T(s)x ds \in D(A)$, and $A\int_0^t T(s)x ds = T(t)x - x$.

(*v*) $D(A)$ is dense in X.

(*vi*) A is closed.

(vii) For $x_0 \in D(A)$, $\frac{d}{dt}(T(t)x_0) = AT(t)x_0 = T(t)Ax_0$ for $t > 0$.

Proof. (i) For $t, h > 0$, by Theorem 10.3.8,

$$\|T(t+h)x - T(t)x\| \leq \|T(t)\|\|T(h)x - x\| \leq Ce^{\alpha t}\|T(h)x - x\| \to 0 \text{ as } h \to 0^+,$$

and for $t - h > 0$,

$$\begin{aligned} \|T(t-h)x - T(t)x\| &= \|T(t-h)x - T(t-h+h)x\| \\ &\leq \|T(t-h)\|\|(I - T(h))x\| \\ &\leq Ce^{\alpha t}\|x - T(h)x\| \to 0 \text{ as } h \to 0^+. \end{aligned}$$

Hence, $t \to T(t)x$ is continuous for all $x \in X$.
(ii) Since $\|T(s)x - x\| \to 0$ as $s \to 0^+$, we have $\|T(s)x - x\| < \varepsilon$ whenever $0 < s < h$ for sufficiently small $h > 0$. Now,

$$\left\| \frac{1}{h}\int_0^h T(s)xds - x \right\| = \left\| \frac{1}{h}\int_0^h (T(s)x - x)ds \right\| \leq \frac{1}{h}\int_0^h \varepsilon ds = \varepsilon.$$

Hence,

$$\lim_{h \to 0^+} \frac{1}{h}\int_0^h T(s)xds = x. \tag{10.15}$$

Now,

$$\begin{aligned} \left\| \frac{1}{h}\int_t^{t+h} T(s)xds - T(t)x \right\| &= \left\| \frac{1}{h}\int_0^h (T(t+r) - T(t))xdr \right\| \\ &\quad \text{(by substituting } s = t + r) \\ &= \left\| T(t)\left[\frac{1}{h}\int_0^h (T(r)x - x)dr\right] \right\|, \end{aligned}$$

which tends to 0 as $h \to 0^+$ from (10.15). Hence, $\lim_{h\to 0^+} \frac{1}{h}\int_t^{t+h} T(s)xds = T(t)x$ for all $x \in X$.
(iii) Let $x \in D(A)$. Then,

$$\begin{aligned} \lim_{h \to 0^+} \left(\frac{T(h) \quad I}{h}\right) T(t)x &= \lim_{h \to 0^+} \left(\frac{T(t+h) - T(t)}{h}\right) x \\ &= T(t) \lim_{h \to h^+} \frac{T(h)x - x}{h} = T(t)Ax. \end{aligned}$$

Hence, $T(t)x \in D(A)$ for $t \geq 0$.

(iv) Let $x \in X$, and consider

$$\begin{aligned}
\lim_{h\to 0}\frac{T(h)-I}{h}\int_0^t T(s)xds &= \lim_{h\to 0}\frac{1}{h}\int_0^t (T(h+s)-T(s))xds \\
&= \lim_{h\to 0}\frac{1}{h}\left[\int_h^{t+h} T(s)xds - \int_0^t T(s)xds\right] \\
&= \lim_{h\to 0}\frac{1}{h}\left[\int_t^{t+h} T(s)xds - \int_0^h T(s)xds\right] \\
&= T(t)x - x \quad (\text{from } (ii)).
\end{aligned}$$

(v) For each $x \in X$ and $h > 0$, $\int_0^h T(s)xds \in D(A)$ (from (iv)). Hence, $\frac{1}{h}\int_0^h T(s)xds \in D(A)$. By (ii), we have $\lim_{h\to 0}\frac{1}{h}\int_0^h T(s)xds = x$ for all $x \in X$, and hence, $D(A)$ is dense in X.
(vi) Let $\{x_n\} \subset D(A)$ be such that $x_n \to x \in X$ and $Ax_n \to y \in X$ as $n \to \infty$. We need to prove that $x \in D(A)$ and $y = Ax$. For $0 \le t \le \tau$, by Theorem 10.3.8, $\|T(t)\| \le Me^{\alpha\tau} =: C_\tau$, and

$$\|T(t)Ax_n - T(t)y\| \le C_\tau \|Ax_n - y\| \to 0.$$

Hence, $T(t)Ax_n \to T(t)y$ as $n \to \infty$ uniformly with repect to t. From (iv),

$$A\int_0^t T(s)x_n ds = \int_0^t T(s)Ax_n ds = T(t)x_n - x_n. \tag{10.16}$$

As $T(s)Ax_n \to T(s)y$ uniformly on $[0,t]$, $0 \le t \le \tau$, the middle term of (10.16) tends to $\int_0^t T(s)yds$, and its RHS tends to $T(t)x - x$. Hence,

$$\lim_{t\to 0^+}\frac{1}{t}\int_0^t T(s)yds = \lim_{t\to 0^+}\frac{T(t)x-x}{t} \implies y = Ax.$$

Hence, A is closed.
(vii) From the definition of the infinitesimal generator, for $x_0 \in D(A)$, we obtain

$$\frac{d}{dt}(T(t)x_0) = \lim_{h\to 0}\frac{T(t+h)-T(t)}{h}x_0 = \lim_{h\to 0}\frac{T(h)-I}{h}T(t)x_0 = AT(t)x_0.$$

On the other hand, we also have

$$\frac{d}{dt}(T(t)x_0) = \lim_{h\to 0}\frac{T(t+h)-T(t)}{h}x_0 = \lim_{h\to 0}T(t)\frac{T(h)-I}{h}x_0 = T(t)Ax_0.$$

Hence, the identity follows. □

Remark 10.5.2 From Theorem 10.5.1(vii), one can conclude the following: For every $x_0 \in D(A)$, the (classical) solution of the abstract Cauchy problem $\frac{dx}{dt} = Ax$, with the initial condition $x(0) = x_0$, is given by $x(t) = T(t)x_0$. If $x_0 \notin D(A)$, then $x(t)$ is referred to as the **mild solution** of the Cauchy problem. Since $D(A)$ is dense in X and $T(t)$ is bounded, every mild solution can be approximated by a classical solution. If $x_0 \notin D(A)$, then we can find a sequence $\{x_n\} \subset D(A)$ such that $x_n \to x_0$ as $n \to \infty$. Consequently, the sequence of classical solutions $\{T(t)x_n\}$ tends to the mild solution $T(t)x_0$.

Lemma 10.5.3 If $T(t) : t \geq 0$ is a uniformly continuous semigroup on X, then $\frac{1}{h}\int_0^h T(s)ds$ is invertible for sufficiently small $h > 0$, and its inverse is bounded.

Proof. Note that

$$\left\|\frac{1}{h}\int_0^h T(s)ds - I\right\| = \left\|\frac{1}{h}\int_0^h (T(s)-I)ds\right\| \leq \frac{1}{h}\int_0^h \|T(s)-I\|ds.$$

Due to the uniform continuity of $T(s)$, $\|T(s) - I\| < \varepsilon$ for $0 \leq s < h$ when h is sufficiently small. Hence, $\|\frac{1}{h}\int_0^h T(s)ds - I\| < \varepsilon < 1$ for small h. This, together with Lemma 3.4.18, implies that $\frac{1}{h}\int_0^h T(s)ds$ is invertible. Since $\|T(s)\| \leq C$ for $0 \leq s \leq h$, we have $\|\frac{1}{h}\int_0^h T(s)ds\| \leq C$. Therefore, by the open mapping theorem (Theorem 3.6.13), $\left(\frac{1}{h}\int_0^h T(s)ds\right)^{-1}$ is bounded. □

Theorem 10.5.4 If $T(t) : t \geq 0$ is a uniformly continuous semigroup, then the infinitesimal generator A of $T(t)$ is bounded.

Proof. Consider the following operator identity

$$\frac{T(h)-I}{h}\int_0^t T(s)ds = \frac{1}{h}\int_0^t (T(s+h)-T(s))ds = \frac{1}{h}\left[\int_h^{t+h} T(s)ds - \int_0^t T(s)ds\right]$$
$$= \frac{1}{h}\left[\int_t^{t+h} T(s)ds - \int_0^h T(s)ds\right].$$

For sufficiently small $h > 0$, by Lemma 10.5.3, we obtain

$$\frac{T(h)-I}{h} = \frac{1}{h}\left[\int_t^{t+h} T(s)ds - \int_0^h T(s)ds\right]\left(\int_0^t T(s)ds\right)^{-1}.$$

As $h \to 0^+$, the above equality reduces to $A = (T(t)-I)\left(\int_0^t T(s)ds\right)^{-1}$, which shows that A is a bounded linear operator. □

Theorem 10.5.5 If two C_0-semigroups $T_1(t) : t \geq 0$ and $T_2(t) : t \geq 0$ on a Banach space X have the same generator $A : D(A) \subset X \to X$, then $T_1(t) = T_2(t)$ for $t \geq 0$.

Proof. Let $x \in D(A)$, and define $S(r) = T_1(t-r)T_2(r)$ for $t \geq r \geq 0$. Then $S(0) = T_1(t)$ and $S(t) = T_2(t)$. Now, using Theorem 10.5.1(*vii*), we obtain $\frac{d}{dr}S(r)x = \frac{d}{dr}T_1(t-r)(T_2(r)x) + T_1(t-r)\frac{d}{dr}T_2(r)x = -AT_1(t-r)T_2(r)x + T_1(t-r)(AT_2(r)x) = 0$ since T_1 and A are commutative on $D(A)$ (Theorem 10.5.1 (vii)). Hence, $S(r)x$ is constant in X for $0 \leq r \leq t$. This implies $S(0)x = S(t)x$, which gives $T_1(t)x = T_2(t)x$ for all $x \in D(A)$. Since $T(t)$ is continuous and $D(A)$ is dense in X, we conclude that $T_1(t)x = T_2(t)x$ for all $x \in X$. □

Theorem 10.5.6 Consider the C_0-semigroup $T(t) : t \geq 0$ on X, with A as its infinitesimal generator. Then, for any $x_0 \in D(A)$,

(*i*) the function

$$x(t) = T(t)x_0 \quad \text{for } t \geq 0 \tag{10.17}$$

defines a unique solution to the abstract homogeneous IVP

$$\begin{cases} \dot{x} = Ax, & t > 0, \\ x(0) = x_0. \end{cases} \tag{10.18}$$

(*ii*) the function

$$x(t) = T(t)x_0 + \int_0^t T(t-s)w(s)ds \quad \text{for } t \geq 0 \tag{10.19}$$

defines a unique solution to the nonhomogeneous IVP

$$\begin{cases} \dot{x} = Ax + w, & t > 0, \\ x(0) = x_0, \end{cases} \tag{10.20}$$

where w is a given function of t, and $w(t) \in X$ for $t \geq 0$.

Proof. (*i*) Let $x(t) = T(t)x_0$, where $x_0 \in D(A)$. From Definition 10.4.1 and 10.3.1(b) we know that $AT(t)x_0 = T(t)Ax_0$, and the map $t \to T(t)x_0$ is continuous and differentiable function from $[0,\infty)$ into $D(A)$. Now, from Theorem 10.5.1(vii) $\frac{dx}{dt} = \frac{d}{dt}T(t)x_0 = AT(t)x_0 = Ax$, i.e., $\dot{x} = Ax$. Since $x(0) = T(0)x_0 = x_0$, we conclude that x is a solution to the IVP (10.18).

To prove the uniqueness of x, assume that x_1 and x_2 are two different solutions of (10.18), and let $y(t) = x_1(t) - x_2(t)$. Define

$$z(t) = T(r-t)y(t), \quad 0 \leq t \leq r.$$

Using Theorem 10.5.1(vii), we compute

$$\frac{dz}{dt} = \frac{d}{dt}[T(r-t)y(t)] = -AT(r-t)y(t) + T(r-t)Ay(t) = 0.$$

Therefore, $z(t)$ is constant for $t \in [0,r]$, which implies $z(r) = z(0)$, i.e., $T(0)y(r) = T(r)y(0) = 0$, since $x_1(0) = x_2(0)$. Thus, $y(r) = 0$ for all $r \geq 0$, and consequently, $x_1 \equiv x_2$, as required.

(ii) If $x_0 \in D(A)$, then by differentiating both sides of (10.19) with respect to t and using the properties of the semigroup (Theorem 10.5.1), it can be verified that $x(t)$ satisfies (10.20). Therefore, in this case, $x(t)$ is referred to as a classical (or analytical) solution to (10.20). If $x_0 \notin D(A)$, then the function $x(t)$ given in (10.19) is called a mild solution of (10.20). Note that while every classical solution is also a mild solution, the converse does not hold. □

***Example* 10.5.7** Consider the following controlled heat equation

$$\begin{cases} \dfrac{\partial}{\partial t}z(\rho,t) = \dfrac{\partial^2}{\partial \rho^2}z(\rho,t) + w(\rho,t), \quad \rho \in [0,l], t \in [0,\tau], \\ z(0,t) = 0 = z(l,t), \\ z(\rho,0) = z_0(\rho), \quad z_0 \in L_2[0,l], \end{cases} \tag{10.21}$$

where $z(\rho,t)$ represents the temperature at ρ, at time t, $z_0(\rho)$ is the initial temperature profile, and $w(\rho,t)$ is the heat input along the rod. Let $x(t) = z(\cdot,t)$ and $u(t) = w(\cdot,t)$. Define the operators A and B as follows:

$$Ax = \frac{d^2x}{d\rho^2} \quad \text{with } D(A) = \left\{x \in L_2[0,l] : \frac{d^2x}{d\rho^2} \in L_2[0,l], x(0) = 0 = x(l)\right\},$$

$$B = I, \text{ the identity operator on } L_2[0,l].$$

Then (10.21) can be reformulated in abstract form as the following control system defined on $L_2[0,l]$

$$\begin{cases} \dot{x}(t) = Ax(t) + Bu(t), \quad t \in [0,\tau], \\ x(0) = z_0. \end{cases} \tag{10.22}$$

Note that the operator $A = \frac{d^2}{d\rho^2}$ with domain $D(A)$ defined above has eigenvalues $-\frac{j\pi}{l}$ and corresponding eigenfunctions $\sin(\frac{j\pi\rho}{l})$ for $\rho \in [0,l]$ and $j = 1,2,\ldots$. Let $\phi_j(\rho) = \sqrt{\frac{2}{l}}\sin(\frac{j\pi\rho}{l})$, $j = 1,2,\ldots$, which form an orthogonal basis for $L_2[0,l]$. Hence, every $x \in L_2[0,l]$ can be expressed as a Fourier series as follows.

$$x = \sum_{j=1}^{\infty} \langle x, \phi_j \rangle \phi_j \tag{10.23}$$

and

$$(Ax)(\rho) = \sum_{j=1}^{\infty} \left(\frac{-j^2\pi^2}{l^2} \right) \langle x, \phi_j \rangle \phi_j(\rho), \tag{10.24}$$

where $\langle x, \phi_j \rangle = \sqrt{\frac{2}{l}} \int_0^l z(\omega, t) \sin(\frac{j\pi\omega}{l}) d\omega$ for $j = 1, 2, \ldots$. Note that A is a self-adjoint and unbounded operator on $L_2[0, l]$. From Example 10.3.7, the C_0-semigroup generated by the operator A is

$$T(t)(x) = \sum_{j=1}^{\infty} e^{-\frac{j^2\pi^2}{l^2}t} \langle x, \phi_j \rangle \phi_j, \tag{10.25}$$

which is compact and self-adjoint for all $t > 0$. Substituting $T(t)$ from (10.25) and $w(t) = Bu(t)$ into (10.19), we obtain the mild solution to the controlled heat equation (10.21).

Exercise 10.5.1 Let A be the infinitesimal generator of a C_0-semigroup $T(t)$, and let (λ, ϕ) be an eigenpair of A. Show that $(e^{\lambda t}, \phi)$ is an eigenpair of $T(t)$.

Exercise 10.5.2 Let $A = \frac{d^3}{ds^3}$ be defined on $L_2[0, 2\pi]$ with domain $D(A) = \{z \in L_2[0, \pi] : z(0) = z(2\pi), \dot{z}(0) = \dot{z}(2\pi), \ddot{z}(0) = \ddot{z}(2\pi), z, \dot{z}$, and $\ddot{z}$ are continuous, and $\frac{d^3 z}{ds^3} \in L_2[0, 2\pi]\}$. Show that A is skew-symmetric and find the eigenpairs of A. If $T(t)$ is the C_0-semigroup generated by A, find the eigenpairs of $T(t)$. Also, show that $\|T(t)z\| = \|z\|$ for all $z \in L_2[0, 2\pi]$.

10.6 Controllability of Infinite Dimensional Systems

In this section, we shall briefly examine some controllability properties pertaining to infinite dimensional systems.

Let X and U be Hilbert spaces, and let $L_2([t_0, \tau], X)$ and $L_2([t_0, \tau], U)$ be the function spaces corresponding to X and U, respectively. Consider the linear control system

$$\begin{cases} \dot{x}(t) = Ax(t) + Bu(t), & t_0 \le t \le \tau, \\ x(t_0) = x_0, \end{cases} \tag{10.26}$$

where $x(t) \in X$, $u(t) \in U$, $A : D(A) \subseteq X \to X$ is a closed linear operator with domain $D(A)$ dense in X, and $B : U \to X$ is a bounded linear operator. Let $T(t) : t \ge t_0$ be the

C_0-semigroup generated by the operator A. Then, from (10.19), for a given control function u, the unique mild solution of the system (10.26) is

$$x(t) = T(t-t_0)x_0 + \int_{t_0}^{t} T(t-s)Bu(s)ds \quad \text{for } t_0 \leq t \leq \tau. \tag{10.27}$$

Definition 10.6.1 The control system (10.26) is said to be **exactly controllable** in the time interval $[t_0, \tau]$ if, for every x_0 and x_1 in X, there exists a control function $u(t)$ such that the mild solution $x(t)$ of the system corresponding to this control satisfies $x(t_0) = x_0$ and $x(\tau) = x_1$.

Definition 10.6.2 The control system (10.26) is said to be **approximately controllable** in the time interval $[t_0, \tau]$ if, for every x_0 and x_1 in X and for any given $\varepsilon > 0$, there exists a control function $u(t)$ such that the mild solution $x(t)$ of the system corresponding to this control satisfies $\|x(\tau) - x_1\| \leq \varepsilon$.

First, let us recall some relevant results on the controllability of finite dimensional systems discussed in Chapter 5. In a finite dimensional setting, the linear system (10.1) is controllable if and only if the $n \times n$ symmetric controllability Gramian matrix

$$W(t_0, \tau) = \int_{t_0}^{\tau} e^{A(t_0-s)}BB^*e^{A^*(t_0-s)}ds \tag{10.28}$$

is nonsingular. In this case, the control

$$u(t) = B^*e^{A^*(t_0-t)}W^{-1}(t_0, \tau)[e^{A(t_0-\tau)}x_1 - x_0],$$

steers the system from x_0 to x_τ (Theorem 5.3.3).

Now, let us formulate the solution of (10.1) in operator equation form. From (10.2), the solution of (10.1) is

$$x(t) = e^{A(t-t_0)}x_0 + \int_{t_0}^{t} e^{A(t-s)}Bu(s)ds.$$

Here, $x(t) \in \mathbb{R}^n$ and $u(t) \in \mathbb{R}^m$ for each t.

At the final time $t = \tau$,

$$x_1 - e^{A(\tau-t_0)}x_0 = \int_{t_0}^{\tau} e^{A(\tau-s)}Bu(s)ds. \tag{10.29}$$

Let $X = \mathbb{R}^n$, $U = \mathbb{R}^m$, $Z = L_2([t_0, \tau], X)$, and $Y = L_2([t_0, \tau], U)$. Define the operator $M : Y \to X$ as

$$Mu = \int_{t_0}^{\tau} e^{A(\tau-s)}Bu(s)ds,$$

where $u \in Y$, and $u(s) \in U$ for each $s \in [t_0, \tau]$. The adjoint M^* of M is defined as

$$M^* : X \to Y \text{ and } M^*x = B^* e^{A^*(\tau - s)} x.$$

In operator form, (10.29) can be written as

$$Mu = x_1 - e^{A(\tau - t_0)} x_0 = v \,(\text{say}). \tag{10.30}$$

From Definition 10.6.1, and the equations (10.28) and (10.30), it follows that the system (10.1) is exactly controllable if and only if, for every x_0 and x_1 in $\mathbb{R}^n$, there exists at least one $u \in Y$ that satisfies (10.30). In other words, the system (10.1) is controllable if and only if the range of M is $\mathbb{R}^n$. Note that

$$MM^*x = \int_{t_0}^{\tau} e^{A(\tau - s)} BB^* e^{A^*(\tau - s)} x ds,$$

which is the same as the controllability Gramian matrix $W[t_0, \tau]$ given in (10.28). Theorem 5.3.3 states that the controllability of (10.1) is equivalent to the nonsingularity of MM^* and the steering control is given by

$$u = M^*(MM^*)^{-1} v.$$

In the next section, we present some controllability results for control systems defined on infinite dimensional Hilbert spaces using an operator equation similar to (10.30).

Infinite Dimensional Control Systems

For the linear system (10.26), define the controlling operator $M : L_2([t_0, \tau], U) \to X$ by

$$Mu = \int_{t_0}^{\tau} T(\tau - s) Bu(s) ds. \tag{10.31}$$

Then, $M^* : X \to L_2([t_0, \tau], U)$ is given by $(M^*x)(t) = B^* T^*(\tau - t)x$. From (10.27) at $t = \tau$, we have

$$Mu = x_1 - T(\tau - t_0) x_0. \tag{10.32}$$

Let $\|T(t)\| \le C_\tau$ for $0 \le t \le \tau$ and $\|Bu\| \le b\|u\|$. The boundedness of the operators $T(t)$ and B implies that M is also bounded, and $\|Mu\| = \left\| \int_{t_0}^{\tau} T(\tau - s) Bu(s) ds \right\| \le \int_{t_0}^{\tau} \|T(\tau - s)\| \left\| Bu(s) \right\| ds \le C_\tau b \int_{t_0}^{\tau} \left\| u(s) \right\| ds \le C_\tau b \sqrt{\tau} \|u\|$ (by using Cauchy–Schwarz inequality).

In some applications, the associated semigroup $T(t)$ is compact for each $t > 0$, as in Example 10.5.7. For such problem, the next lemma shows that the operator M defined in (10.31) is compact.

Lemma 10.6.3 If $T(t)$ is compact for each $t > 0$ and B is bounded, then M defined in (10.31) is compact.

Proof. Since $T(t) : t \geq 0$ is a semigroup, it satisfies (a) $T(0) = I$ and (b) $T(t+s) = T(t)T(s)$ for all $t, s \geq 0$. As $T(t)$ is linear for each t, it can be verified that M is linear. Let

$$M_h u = \int_{t_0}^{\tau - h} T(\tau - s)Bu(s)ds,$$

then

$$M_h u = \int_{t_0}^{\tau - h} T(\tau - h + h - s)Bu(s)ds = T(h)\int_{t_0}^{\tau - h} T(\tau - h - s)Bu(s)ds.$$

Now, by using the Cauchy–Schwarz inequality, we obtain

$$||Mu - M_h u||_X = \left|\left|\int_{\tau - h}^{\tau} T(\tau - s)Bu(s)ds\right|\right| \leq C_\tau \int_{\tau - h}^{\tau} ||Bu(s)||ds \leq C_\tau b\sqrt{h}||u||.$$

Hence

$$||M - M_h||_{BL(Z,X)} = \sup_{||u||=1} ||(M - M_h)u|| \leq C_\tau b\sqrt{h},$$

which tends to 0 as $h \to 0^+$, implying the uniform convergence of M_h to M. Now

$$M_h u = T(h)(M^{\tau - h}u),$$

where

$$M^{\tau - h}u = \int_{t_0}^{\tau - h} T(\tau - h - s)Bu(s)ds.$$

As $T(t)$ and B are bounded linear operators, it can be seen that $M^{\tau - h}$ is a bounded linear operator. Since $T(t) : t > 0$ is compact and $M^{\tau - h}$ is bounded, by Property 3.6.14$(viii)$, M_h is compact. Hence, by Property 3.6.14(vii), we conclude that M is compact as M_h converges to M uniformly. □

In view of (10.27) and (10.32), Definitions 10.6.1 and 10.6.2 can be reworded in terms of M as given below.

Definition 10.6.4 The system (10.26) is exactly controllable on $[t_0, \tau]$ if any arbitrary point in X can be reached at the final time τ from any arbitrary initial state x_0, using a suitable control u, i.e., Range$(M) = X$.

Definition 10.6.5 The system (10.26) is approximately controllable on $[t_0, \tau]$ if, for any given $\varepsilon > 0$, it is possible to steer the system from any arbitrary initial state x_0 to the ε - neighborhood of any arbitrary final state x_τ using a control u, i.e., $\overline{\text{Range}(M)} = X$.

Let the controllability Gramian operator W of the system (10.26) be defined from X into itself by

$$Wx = \left(\int_{t_0}^{\tau} T(\tau - s)BB^*T^*(\tau - s)u(s)ds \right) x. \tag{10.33}$$

Then, $W = MM^*$.

For the linear system (10.26), the following results on exact controllability and approximate controllability hold (see [5]).

Theorem 10.6.6 The system (10.26) is exactly controllable on $[t_0, \tau]$ if and only if any one of the following conditions holds for some $\gamma > 0$ and all $x \in X$:

(i) $\langle Wx, x\rangle_X \geq \gamma \|x\|_X^2$,

(ii) $\int_{t_0}^{\tau} \|B^*T^*(s)x\|_U^2 ds \geq \gamma \|x\|_X^2$,

(iii) $\ker M^* = \{0\}$ and $\text{Range}(M)$ is closed.

Theorem 10.6.7 The system (10.26) is approximately controllable on $[t_0, \tau]$ if and only if any one of the following conditions holds:

(i) $W > 0$,

(ii) $\ker M^* = \{0\}$,

(iii) $B^*T^*(s)x = 0$ on $[t_0, \tau] \implies x = 0$.

The above two theorems can be applied to test the controllability properties of linear control systems.

Example **10.6.8** The controlled heat equation (10.21) is NOT exactly controllable.

Solution From Theorem 10.6.6(i), the condition for exact controllability is the existence of a scalar $\gamma > 0$ such that

$$\langle Wx, x\rangle \geq \gamma \|x\|^2. \tag{10.34}$$

From (10.23), we have

$$x(\cdot) = \frac{2}{l} \sum_{n=1}^{\infty} \int_0^l \sin\left(\frac{n\pi\omega}{l}\right) z(\omega, \cdot) d\omega \sin\left(\frac{n\pi\rho}{l}\right),$$

which gives

$$\|x(\cdot)\|^2 = \frac{2}{l}\sum_{n=1}^{\infty}\left|\int_0^l \sin\left(\frac{n\pi\omega}{l}\right) z(\omega,\cdot)d\omega\right|^2.$$

From (10.33), we obtain (by taking $B = I$, the identity operator)

$$MM^*x = \int_0^\tau T(2(\tau - s))xds = \sum_{i=1}^{\infty}\frac{1-e^{-\frac{2i^2\pi^2}{l^2}\tau}}{2i^2\pi^2}l^2\langle x,\phi_i\rangle\phi_i. \tag{10.35}$$

The condition (10.34) implies

$$\frac{2\gamma}{l}\sum_{n=1}^{\infty}\left|\int_0^l \sin\left(\frac{n\pi\omega}{l}\right) z(\omega,\cdot)d\omega\right|^2 \leq \frac{2}{l}\sum_{n=1}^{\infty}\left(\frac{1-e^{-\frac{2n^2\pi^2\tau}{l^2}}}{\frac{2n^2\pi^2}{l^2}}\right)\left|\int_0^l \sin\left(\frac{n\pi\omega}{l}\right) x(\omega,\cdot)d\omega\right|^2. \tag{10.36}$$

This shows that no such γ will exist. We can prove this by contradiction. Suppose there exists a fixed $\gamma > 0$ satisfying (10.36) for every $x \in L_2[0,l]$. Substituting $x(\cdot) = z(\omega,\cdot) = \sin(\frac{n\pi\omega}{l})$ into (10.36), we get $\gamma \leq \frac{1-e^{-\frac{2n^2\pi^2\tau}{l^2}}}{\frac{2n^2\pi^2}{l^2}}$ for all n. The RHS of this inequality tends to 0 as $n \to \infty$, which contradicts the assumption that γ is a fixed positive number. Hence, the system (10.21) is not exactly controllable on $[t_0,\tau]$.

***Example* 10.6.9** The system (10.21) is approximately controllable on $[t_0,\tau]$.

Solution We shall prove that the system (10.21) is approximately controllable using Theorem 10.6.7(ii), i.e., $M^*v = 0 \implies v = 0$.

Since $M^*v = T^*(\tau - s)v = T(\tau - s)v = \sum_{j=1}^{\infty} e^{-\frac{j^2\pi^2}{l^2}(\tau-s)}\langle v,\phi_j\rangle\phi_j$, $M^*v = 0$ implies $\langle M^*v,\phi_k\rangle = e^{-\frac{j^2\pi^2}{l^2}(\tau-s)}\langle v,\phi_k\rangle = 0$ for all $k = 1,2,\ldots$. Hence, $\langle v,\phi_k\rangle = 0$ for all $k = 1,2,\ldots$, which implies $v = 0$.

As the system (10.21) is approximately controllable, according to Definition 10.6.5, the Range(M) is dense in X. Hence, for a given $v \in X$, finding a solution $u \in L_2([t_0,\tau],X)$ satisfying $Mu = v$ is an ill-posed problem. A brief discussion on ill-posed problems and the computation of a regularized solution for the controlled heat equation (10.21) is presented in the next section.

Well-posed and Ill-posed Problems

Let $\mathscr{U}$ and $\mathscr{V}$ be normed linear spaces, and let $M : \mathscr{U} \to \mathscr{V}$ be a linear operator. The equation

$$Mu = v \tag{10.37}$$

is said to be **well-posed** if the following two conditions are satisfied.

(*i*) For every $v \in \mathscr{V}$, there exists a unique $u \in \mathscr{U}$ satisfying (10.37), i.e., $M^{-1} : \mathscr{V} \to \mathscr{U}$ exists.

(*ii*) The solution u depends continuously on the input v, i.e., for every $\varepsilon > 0$, there exists a $\delta > 0$ such that: If $v, v_1 \in \mathscr{V}$ satisfy $\|v_1 - v\| \leq \delta$, and $u, u_1 \in \mathscr{U}$ satisfy the equations

$$Mu = v \text{ and } Mu_1 = v_1,$$

then

$$\|u_1 - u\| \leq \varepsilon.$$

In other words, M^{-1} is continuous.

The equation (10.37) is said to be **ill-posed** if any one of the conditions for well-posedness is not satisfied.

From the open mapping theorem (Theorem 3.6.13), we infer that if $\mathscr{U}$ and $\mathscr{V}$ are Banach spaces, and $M : \mathscr{U} \to \mathscr{V}$ is a bijective bounded linear operator, then $M^{-1} : \mathscr{V} \to \mathscr{U}$ is continuous, and hence bounded. In this case, $M : \mathscr{U} \to \mathscr{V}$ is well-posed, and $M^{-1}v$ provides the unique solution.

Some ill-posed problems with finite dimensional range can be solved using the Moore–Penrose pseudoinverse. One such result is Theorem 5.3.3 on the (exact) controllability of a finite dimensional linear system. Let $\mathscr{U}$ be a Hilbert space, and $\mathscr{V}$ be a finite dimensional Hilbert space. Assume that the operator $M : \mathscr{U} \to \mathscr{V}$ is a bounded linear operator and onto; then, in Theorem 5.3.11, it has been shown that $MM^* : \mathscr{V} \to \mathscr{V}$ is nonsingular, and $u = M^*(MM^*)^{-1}v$ provides the minimum norm solution of the equation $Mu = v$. The same result holds for infinite dimensional systems, provided that the system is exactly controllable. However, this approach does not hold for approximately controllable systems, since Range(M) is dense in $\mathscr{V}$, implying that $(MM^*)^{-1}$ does not exist.

In the following, we shall assume that $\mathscr{U}$ and $\mathscr{V}$ are Hilbert spaces, and $M : \mathscr{U} \to \mathscr{V}$ is a bounded linear operator, whose range is dense in $\mathscr{V}$. For such ill-posed problems, approximate solution can be obtained through regularization methods. One such method is the **Tikhonov regularization**, which is stated in the following theorem.

Theorem 10.6.10 Let $\mathscr{U}$ and $\mathscr{V}$ be Hilbert spaces, and let $M : \mathscr{U} \to \mathscr{V}$ be a bounded linear operator. Then, for each $v \in \mathscr{V}$ and $\lambda > 0$, there exists a unique element $u_\lambda(v) \in \mathscr{U}$ that minimizes the functional

$$u \mapsto \|Mu - v\|^2 + \lambda \|u\|^2, \quad u \in \mathscr{U}.$$

Moreover, for each $\lambda > 0$, the map

$$R_\lambda : v \mapsto u_\lambda(v), \quad v \in \mathscr{V},$$

is a bounded linear operator from $\mathscr{V}$ to $\mathscr{U}$, and

$$R_\lambda v := (M^*M + \lambda I)^{-1} M^* v,$$

where M^* is the adjoint of M. For each $v \in \mathscr{V}$ and $\lambda > 0$, u_λ solves the operator equation

$$(M^*M + \lambda I)u_\lambda = M^*v \tag{10.38}$$

and satisfies

$$\|Mu_\lambda - v\| \to 0 \text{ as } \lambda \to 0. \tag{10.39}$$

For $v \in \mathscr{V}$ and $\lambda > 0$, the element $u_\lambda \in \mathscr{U}$, as in (10.38), is called the Tikhonov regularized solution of $Mu = v$. By the above theorem, $R_\lambda(v)$ is continuous with respect to v, and thus, for any given $\varepsilon > 0$, there exists $\lambda_\varepsilon > 0$ such that

$$\|Mu_\lambda - v\| \leq \varepsilon \quad \text{for all } \lambda \in (0, \lambda_\varepsilon].$$

However, this result does not prescribe any procedure to determine λ_ε. More details on the selection of the parameter λ, convergence results, and error estimate can be found in texts on regularization methods (see [16]). The above theorem provides a regularized solution for u, given by $u_\lambda = (M^*M + \lambda I)^{-1}M^*v$, where λ is a suitable parameter.

One can verify that $(M^*M + \lambda I_{\mathscr{U}})^{-1}M^* = M^*(MM^* + \lambda I_{\mathscr{V}})^{-1}$, by operating $M^*M + \lambda I_{\mathscr{U}}$ on both sides on the left. Hence, the regularized control u_λ can be computed using the following formula:

$$u_\lambda = M^*(MM^* + \lambda I_{\mathscr{V}})^{-1} \tag{10.40}$$

Exercise 10.6.1 Let $\mathscr{U}$ and $\mathscr{V}$ be real Hilbert spaces, and let $M : \mathscr{U} \to \mathscr{V}$ be a bounded linear operator. Show that:

(*i*) MM^* and M^*M are positive;

(*ii*) $MM^* + \lambda I_{\mathscr{V}}$ and $M^*M + \lambda I_{\mathscr{U}}$ are invertible for $\lambda > 0$;

(*iii*) $M^*(MM^* + \lambda I_{\mathscr{V}})^{-1} = (M^*M + \lambda I_{\mathscr{U}})^{-1}M^*$;

(*iv*) If MM^* is invertible, then $\bar{u} = M^*(MM^*)^{-1}v$ is a solution to $Mu = v$;

(v) If u_1 is any other solution of $Mu = v$, then $\langle \bar{u}, u_1 \rangle = \|\bar{u}\|^2$ and $\|\bar{u}\| \leq \|u_1\|$, i.e., $\bar{u}$ is the minimum norm solution of $Mu = v$.

10.7 Computation of Regularized Control

In this section, we shall demonstrate the computation of a regularized control function for the linear heat equation (10.21) provided with an initial and a final data. For $u \in L_2([0,\tau], L_2[0,l])$, the mild solution of (10.21) is obtained from (10.19) as

$$x(t) = T(t)z_0 + \int_0^\tau T(\tau - s)u(s)ds = T(t)z_0 + Mu,$$

where the operator $M : L_2([0,\tau], L_2[0,l]) \to L_2[0,l]$ is defined as

$$Mu = \int_0^\tau T(\tau - s)u(s)ds. \tag{10.41}$$

Then, $M^* : L_2[0,l] \to L_2([0,\tau], L_2[0,l])$ is defined by $M^*v = T^*(\tau - s)v = T(\tau - s)v$ for all $v \in L_2[0,l]$. From Lemma 10.6.9, it follows that the system (10.21) is approximately controllable. Now, let the final state to be reached be $x(\tau) = x_\tau \in L_2[0,l]$. To find a control corresponding to the final state x_τ, we need to solve the following equation:

$$Mu = x_\tau - T(\tau)x_0 = v \text{ (say)}. \tag{10.42}$$

Now, from (10.35)

$$MM^*v = \int_0^\tau T(2(\tau - s))vds = \sum_{i=1}^\infty \frac{1 - e^{-\frac{2i^2\pi^2}{l^2}\tau}}{2i^2\pi^2} l^2 \langle v, \phi_i \rangle \phi_i$$

and using (10.23) (by replacing x with v), we obtain

$$(MM^* + \lambda I)v = \sum_{i=1}^\infty \frac{l^2\left[1 - e^{-\frac{2i^2\pi^2}{l^2}\tau}\right] + 2\lambda i^2\pi^2}{2i^2\pi^2} \langle v, \phi_i \rangle \phi_i, \tag{10.43}$$

where

$$\phi_i(\rho) = \sqrt{\frac{2}{l}} \sin\left(\frac{i\pi\rho}{l}\right).$$

Let us denote the regularized control u by u_λ in order to indicate its dependence on λ. From (10.40),

$$u_\lambda = M^*(MM^* + \lambda I)^{-1}v.$$

Using (10.43), we obtain

$$u_\lambda(t) = M^* \sum_{i=1}^{\infty} \frac{2i^2\pi^2}{l^2\left[1 - e^{-\frac{2i^2\pi^2}{l^2}\tau}\right] + 2\lambda i^2\pi^2} \langle v, \phi_i\rangle \phi_i$$

$$= \sum_{j=1}^{\infty} e^{-\frac{j^2\pi^2}{l^2}(\tau - t)} \left\langle \sum_{i=1}^{\infty} \frac{2i^2\pi^2}{l^2\left[1 - e^{-\frac{2i^2\pi^2}{l^2}\tau}\right] + 2\lambda i^2\pi^2} \langle v, \phi_i\rangle \phi_i, \phi_j \right\rangle \phi_j$$

$$\text{(since } M^* = T^*(\tau - t))$$

$$= \sum_{j=1}^{\infty} \frac{2j^2\pi^2 e^{-\frac{j^2\pi^2}{l^2}(\tau - t)}}{l^2\left[1 - e^{-\frac{2j^2\pi^2}{l^2}\tau}\right] + 2\lambda j^2\pi^2} \langle v, \phi_j\rangle \phi_j.$$

Let

$$\beta_j = \langle v, \phi_j\rangle = \int_0^l v(\rho)\phi_j(\rho)d\rho \quad \text{for } j = 1, 2, 3, \ldots.$$

Then,

$$u_\lambda(t) = \sum_{j=1}^{\infty} \frac{e^{-\frac{j^2\pi^2}{l^2}(\tau - t)}}{\sigma_j^2 + \lambda} \beta_j \phi_j \tag{10.44}$$

where

$$\sigma_j^2 = \frac{l^2(1 - e^{-\frac{2j^2\pi^2}{l^2}\tau})}{2j^2\pi^2}.$$

Therefore,

$$\|u_\lambda(t)\|_X^2 = \sum_{j=1}^{\infty} \frac{e^{-\frac{2j^2\pi^2}{l^2}(\tau - t)}}{(\sigma_j^2 + \lambda)^2}.$$

Now,

$$Mu_\lambda = \int_0^\tau T(\tau - s)u_\lambda(s)ds = \int_0^\tau \sum_{i=1}^{\infty} e^{-\frac{i^2\pi^2}{l^2}(\tau - s)} \left\langle \sum_{j=1}^{\infty} \frac{e^{-\frac{j^2\pi^2}{l^2}(\tau - s)}}{\sigma_j^2 + \lambda} \beta_j \phi_j, \phi_i \right\rangle \phi_i ds$$

$$= \sum_{i=1}^{\infty} \int_0^\tau \frac{e^{-\frac{2i^2\pi^2}{l^2}(\tau - s)}}{\sigma_i^2 + \lambda} \beta_i \phi_i ds.$$

Hence,

$$Mu_\lambda = \sum_{i=1}^{\infty} \frac{l^2(1-e^{-\frac{2i^2\pi^2}{l^2}\tau})}{2i^2\pi^2(\sigma_i^2+\lambda)} \beta_i \phi_i.$$

Now,

$$v - Mu_\lambda = \sum_{i=1}^{\infty} \left(1 - \frac{l^2(1-e^{-\frac{2i^2\pi^2}{l^2}\tau})}{2i^2\pi^2(\sigma_i^2+\lambda)}\right) \beta_i \phi_i = \sum_{i=1}^{\infty} \frac{\lambda}{\sigma_i^2+\lambda} \beta_i \phi_i.$$

Hence, the square of the error E in the Tikhonov regularized control u_λ is given by

$$E^2 = \|v - Mu_\lambda\|_X^2 = \sum_{i=1}^{\infty} \frac{\lambda^2 \beta_i^2}{(\sigma_i^2+\lambda)^2}. \tag{10.45}$$

To illustrate the above result, let us take the parameters l, τ, and λ as: $l = \pi$, $\tau = 1$, and $\lambda = \frac{1}{n}$. Then, from (10.45),

$$E^2 = \|v - Mu_\lambda\|_X^2 = \sum_{i=1}^{\infty} \left(\frac{\beta_i}{n\sigma_i^2+1}\right)^2,$$

where

$$\sigma_i^2 = \frac{1-e^{-2i^2}}{2i^2}.$$

Now, let the initial condition be $x_0 = 0$ and the final state be $x_\tau = 2\sin\rho + 3\sin(3\rho)$, $0 \leq \rho \leq \pi$, which is to be reached using a suitable control function $u(t)$. Let $u_\lambda(t)$ be the regularized control function obtained from (10.44). Then

$$u_\lambda(t) = \frac{e^{t-1}}{\sigma_1^2+\lambda}\beta_1\phi_1 + \frac{e^{9(t-1)}}{\sigma_3^2+\lambda}\beta_3\phi_3,$$

where

$$v = x_\tau, \sigma_1^2 = \frac{1-e^{-2}}{2}, \sigma_3^2 = \frac{1-e^{-18}}{18}, \beta_1 = 2\sqrt{\frac{\pi}{2}}, \text{ and } \beta_3 = 3\sqrt{\frac{\pi}{2}}.$$

From (10.45), the error E is given by

$$E = \|v - Mu_\lambda\|_X = \sqrt{\frac{\lambda^2\beta_1^2}{(\sigma_1^2+\lambda)^2} + \frac{\lambda^2\beta_3^2}{(\sigma_3^2+\lambda)^2}}. \tag{10.46}$$

The error E is computed using (10.46) for different values of λ (specifically, $\lambda = \frac{1}{n}$, where $n \in \mathbb{N}$), and the results are shown in the following table.

n	50	100	200	1000	10000
E	1.0014	0.57634	0.3118	0.0667	0.0068

Thus, the table illustrates that the error approaches zero as λ decreases, as stated in Theorem 10.6.10.

10.8 Semilinear Control Systems

Consider the following semilinear control system

$$\begin{cases} \dot{x}(t) = Ax(t) + Bv(t) + f(t,x(t)) & \text{for } t_0 \le t \le \tau, \\ x(t_0) = x_0, \end{cases} \tag{10.47}$$

where A and B are operators, and X and U are spaces as described for the linear system (10.26). $f : [t_0,\tau] \times X \to X$ is a nonlinear function satisfying the **Lipschitz condition** in x, i.e.,

$$\|f(t,x_1) - f(t,x_2)\|_X \le \alpha \|x_1 - x_2\|_X \tag{10.48}$$

for some constant $\alpha > 0$, all $t \in [t_0,\tau]$ and $x_1, x_2 \in X$. Let $T(t) : t \ge 0$ be the C_0-semigroup generated by A with $\|T(t)\| \le C$ for $t_0 \le t \le \tau$.

Define $x(t)$ by the following integral equation for the above system

$$x(t) = T(t-t_0)x_0 + \int_{t_0}^{t} T(t-s)[Bv(s) + f(s,x(s))]ds \quad \text{for } t_0 \le t \le \tau. \tag{10.49}$$

For a given control $v(\cdot) \in L_2([t_0,\tau],U)$, the solution $x(t)$ (if it exists) of the integral equation (10.49) is called the mild solution of the semilinear system (10.47) on $[t_0,\tau]$.

Remark 10.8.1 If $x_0 \in D(A)$, then by differentiating both sides of (10.49) and using the properties of the semigroup $T(t)$, we can show that $x(t)$ satisfies the equation (10.47). If $x_0 \notin D(A)$, then $x(t)$ given in (10.49) is called the **mild solution** of (10.47). As stated in Remark 10.6.2, every mild solution of (10.47) can be approximated by a classical solution by suitably choosing the initial condition in $D(A)$.

Let the set of all possible final points be defined as

$$K_\tau(f) = \{x(\tau) \subset X : x(\cdot) \text{ is the mild solution of the system (10.47)} \\ \text{for } v(\cdot) \in L_2([t_0,\tau],U)\}$$

and referred to as the **reachable set** of the semilinear system (10.47). According to Definitions 10.6.1 and 10.6.2, the system (10.47) is exactly controllable if $K_\tau(f) = X$ and is approximately controllable if $\overline{K_\tau(f)} = X$.

Consider the following linear system

$$\begin{cases} \dot{x}(t) = Ax(t) + u(t), \\ x(0) = 0 \end{cases} \tag{10.50}$$

and the following semilinear system (10.47) with $B = I$, the identity operator,

$$\begin{cases} \dot{y}(t) = Ay(t) + v(t) + f(t, y(t)), \\ y(0) = 0. \end{cases} \tag{10.51}$$

For a given control function $u(t)$, the mild solution of system (10.50) is given by

$$x(t) = \int_0^t T(t-s)u(s)ds. \tag{10.52}$$

Now, consider

$$v(t) = u(t) - f(t, x(t)) \tag{10.53}$$

as the control function for the semilinear system (10.51). Then, the mild solution (10.49) can be expressed as

$$y(t) = \int_0^t T(t-s)f(s, y(s))ds + \int_0^t T(t-s)u(s)ds - \int_0^t T(t-s)f(s, x(s))ds. \tag{10.54}$$

Subtracting (10.54) from (10.52), we obtain

$$\begin{aligned} \|x(t) - y(t)\|_X &= \left\| \int_0^t T(t-s)f(s, x(s))ds - \int_0^t T(t-s)f(s, y(s))ds \right\| \\ &\leq \int_0^t \|T(t-s)\| \|f(s, x(s)) - f(s, y(s))\| ds \\ &\leq C\alpha \int_0^t \|x(s) - y(s)\|_X ds \text{ (using (10.48)).} \end{aligned}$$

Applying Gronwall's inequality (Section 4.10), it follows that $x(t) = y(t)$ for a.e. $t \in [0, \tau]$. Hence, every mild solution of the linear system with control u is also a mild solution of the semilinear system with the control v. Therefore, the semilinear system is approximately controllable whenever the linear system is approximately controllable.

In Section 10.6, computation of a regularized control $u_\lambda(t)$ for the linear heat control system (10.21) has been carried out using equation (10.44). Let $x_\lambda(t)$ be the regularized mild solution corresponding to $u_\lambda(t)$, which can be obtained from

(10.27) and (10.44). Note that $u_\lambda(t) = w_\lambda(\rho,t)$ and $x_\lambda(t) = z_\lambda(\rho,t)$ are the regularized control and the mild solution of the linear system (10.21), respectively. Now, using (10.53), the regularized control v_λ for the following semilinear heat equation can be obtained

$$\begin{cases} \dfrac{\partial}{\partial t}\tilde{z}(\rho,t) = \dfrac{\partial^2}{\partial \rho^2}\tilde{z}(\rho,t) + w(\rho,t) + f(t,\tilde{z}(\rho,t)), & \rho \in [0,l], t \in [0,\tau], \\ \tilde{z}(0,t) = 0 = \tilde{z}(l,t), \\ \tilde{z}(\rho,0) = \tilde{z}_0(\rho), \quad \tilde{z}_0 \in L_2[0,l]. \end{cases} \tag{10.55}$$

The relation (10.53) and the subsequent discussion imply that $\tilde{w}_\lambda(\rho,t) = w_\lambda(\rho,t) + f(t,z_\lambda(\rho,t))$ is a regularized control for the system (10.55), with the corresponding regularized mild solution $\tilde{z}_\lambda(\rho,t) = z_\lambda(\rho,t)$.

10.9 Summary

- Strongly continuous semigroup (C_0-semigroup) and uniformly continuous semigroup on a Banach space are defined along with examples in Section 10.2.

- If A is a bounded linear operator on a Banach space X, then the operators $e^{At}, t \geq 0$, form a uniformly continuous semigroup on X (Theorem 10.3.5). Moreover, A is the infinitesimal generator of the semigroup e^{At} (Theorem 10.4.2).

- An example of a self-adjoint and compact semigroup is given in Example 10.3.7.

- If $T(t), t \geq 0$, is a C_0-semigroup, then there exist constants $C > 1$ and $\alpha > 0$ such that $\|T(t)\| \leq Ce^{\alpha t}$ (Theorem 10.3.8).

- In Section 10.5, properties of a C_0-semigroup $T(t)$ and its infinitesimal generator A are discussed.

- From Theorem 10.5.1(vii), one can conclude the following: For every $x_0 \in D(A)$, the (classical) solution of the abstract Cauchy problem $\frac{dx}{dt} = Ax$, with the initial condition $x(0) = x_0$, is given by $x(t) = T(t)x_0$. If $x_0 \notin D(A)$, then $x(t)$ is referred to as the **mild solution** of the Cauchy problem. Since $D(A)$ is dense in X and $T(t)$ is bounded, every mild solution can be approximated by a classical solution. If $x_0 \notin D(A)$, then we can find a sequence $\{x_n\} \subset D(A)$ such that $x_n \to x_0$ as $n \to \infty$. Consequently, the sequence of classical solutions $\{T(t)x_n\}$ tends to the mild solution $T(t)x_0$.

- For the nonhomogeneous system $\dot{x} = Ax + w$ with the initial condition $x(0) = x_0$, the unique mild solution is given by $x(t) = T(t)x_0 + \int_0^t T(t-s)w(s)ds$ for $t \geq 0$ (Theorem 10.5.6).
- Approximate controllability of infinite dimensional linear control systems is discussed in Section 10.6. In addition, a brief discussion on well-posed and ill-posed problems is provided.
- Example 10.6.8 illustrates that the controlled heat equation is not exactly controllable, while Example 10.6.9 shows that it is approximately controllable.
- Computation of regularized control for the heat equation is carried out in Section 10.7 using the Tikhonov regularization method.

10.10 Exercise

1. Let $X = l^2$ be the space of sequences $\{x_n\}$ of complex numbers such that $\sum_{n=1}^{\infty} |x_n|^2 < \infty$, equipped with l^2 norm. Define the operator A on X as the left shift operator given by

   $$Ax = A(x_1, x_2, \ldots, x_n, \ldots) = (x_2, x_3, \ldots, x_{n+1}, \ldots) \quad \text{for every } x \in l^2.$$

   Show that A is a bounded linear operator with $\|A\| = 1$. Find the semigroup generated by A.

2. Show that the shift operator $T(t)$ (see Example 10.4.3) defined on $L_2[0,\infty)$ is a strongly continuous semigroup but not uniformly continuous.

3. Prove that $C_0[0,\infty) = \{x \in C[0,\infty) : \lim_{t\to\infty} x(t) = 0\}$ is complete and $C_c[0,\infty) = \{x \in C[0,\infty) : x \text{ is continuous with compact support}\}$ is dense in $C_0[0,\infty)$ under sup norm. Also, show that the shift operator on $C_0[0,\infty)$ is a C_0-semigroup.

4. On the Banach space $X = C_0[0,\infty)$ with sup norm, let $T(t) : X \to X$ be defined as

   $$T(t)f = e^{-t} f_t, \quad \text{where } f_t(x) = f(t+x).$$

   Show that $T(t) : t \geq 0$ is a C_0-semigroup.

5. Suppose $X = L_2[0,1]$. Define $T(t) : t \geq 0$ on X as

   $$(T(t)f)(\rho) = e^{at} f(\rho) \quad \text{for } 0 \leq \rho \leq 1,$$

   where $a \in \mathbb{R}$. Show that $T(t)$ is a uniformly continuous semigroup. Also, find its generator A.

6. Formulate an abstract IVP on $L_2(\mathbb{R})$ for the following first order partial differential equation

$$\begin{cases} \dfrac{\partial z}{\partial t} = \dfrac{\partial z}{\partial \rho}, & t > 0, \rho \in \mathbb{R}, \\ z(\rho, 0) = z_0(\rho), \end{cases}$$

by defining $A : D(A) \subset L_2(\mathbb{R}) \to L_2(\mathbb{R})$ as $A = \frac{d}{d\rho}$ with a suitable $D(A)$.

7. Let $A : L_2[0,\pi] \to L_2[0,\pi]$ be defined as $Af = \frac{d^2 f}{d\rho^2} + 2I$, where I is the identity operator, with domain

$$D(A) = \{f \in L_2[0,\pi] : f(0) = f(\pi) = 0 \text{ and } f'' \in L_2[0,\pi]\}.$$

Find the eigenvalues and the corresponding eigenfunctions of A. If $T(t) : t \geq 0$ is the semigroup generated by A, find $T(t)f$. Also, find a growth bound for $\|T(t)\|$.

8. Consider the following linear control system

$$\begin{cases} \dot{x} = Ax + u, & 0 < t \leq \tau, \\ x(0) = x_0, \end{cases}$$

where $A : D(A) \subset X \to X$ with $D(A)$ dense in X is the generator of a C_0-semigroup $T(t)$ on X. Show that if $x_0, x_1 \in D(A)$, then the control $u(t) = \frac{T(t)(x_1 - x_0) - \tau A x_1}{\tau}$ steers the mild solution of the system from $x(0) = x_0$ to $x(\tau) = x_1$.

9. (*i*) Consider the following linear heat equation

$$\begin{cases} \dfrac{\partial}{\partial t} z(t,\rho) = \dfrac{\partial^2}{\partial \rho^2} z(t,\rho) + w(t,\rho), & t \geq 0, \rho \in (0,\pi), \\ z(0,\rho) = \sin\rho, \\ z(t,0) = z(t,\pi) = 0. \end{cases} \tag{10.56}$$

Find a control function $w \in L_2([0,\tau], L_2[0,\pi])$ such that at $t = \tau$, the solution of the IVP (10.56) satisfies $z(\tau,\rho) = 3\sin\rho + \sin(2\rho)$.

(*ii*) For the following semilinear heat equation

$$\begin{cases} \dfrac{\partial}{\partial t} \tilde{z}(t,\rho) = \dfrac{\partial^2}{\partial \rho^2} \tilde{z}(t,\rho) + \tilde{z}(t,\rho) + |\tilde{z}(t,\rho)|, & t \geq 0, \rho \in (0,\pi), \\ \tilde{z}(0,\rho) = \sin\rho, \\ \tilde{z}(t,0) = \tilde{z}(t,\pi) = 0, \end{cases} \tag{10.57}$$

find a control function $\tilde{w} \in L_2([0,\tau], L_2[0,\pi])$ such that at $t = \tau$, the solution of the IVP (10.57) satisfies $\tilde{z}(t,\rho) = 3\sin\rho + \sin(2\rho)$.

Bibliography

[1] P.J. Antsaklis and A.N. Michel. *Linear Systems*. New York: McGraw-Hill, 1997.

[2] A.B. Bakushinsky, M.M. Kokurin, and M.Y. Kokurin. *Regularization Algorithms for Ill-Posed Problems*. Berlin: Walter de Gruyter, 2018.

[3] S. Barnett. *Introduction to Mathematical Control Theory*. London: Oxford University Press, 1975.

[4] E.A. Coddington and N. Levinson. *Theory of Ordinary Differential Equations*. New York: McGraw-Hill, 1955.

[5] R.F. Curtain and H. Zwart. *An Introduction to Infinite-Dimensional Linear Systems Theory*. New York: Springer-Verlag, 1995.

[6] M. Gopal. *Modern Control System Theory*. New Delhi: New Age International (P) Limited, 1993.

[7] W.M. Haddad and V.S. Chellaboina. *Dynamical Systems and Control: A Lyapunov-Based Approach*. Princeton: Princeton University Press, 2008.

[8] J.N. Kapur. *Mathematical Modelling*. New Delhi: New Age International (P) Limited, 1998.

[9] E. Kreyszig. *Advanced Engineering Mathematics*. Hoboken: John Wiley & Sons, 2006.

[10] E. Kreyszig. *Introductory Functional Analysis with Applications*. New York: John Wiley & Sons, 1989.

[11] J.P. LaSalle and S. Lefschetz. *Stability by Liapunov's Direct Method with Applications*. New York: Academic Press, 1961.

[12] F.L. Lewis, D.L. Vrabie, and V.L. Syrmos. *Optimal Control*. Hoboken: John Wiley & Sons, 2012.

[13] H. Logemann and E.P. Ryan. *Ordinary Differential Equations: Analysis, Qualitative Theory and Control.* London: Springer-Verlag, 2014.

[14] D.S. Naidu. *Optimal Control Systems.* Boca Raton: CRC Press, 2003.

[15] M.T. Nair. *Functional Analysis: A First Course.* New Delhi: Prentice-Hall of India, 2002.

[16] M.T. Nair. *Linear Operator Equations: Approximation and Regularization.* Singapore: World Scientific, 2009.

[17] B. Noble and J.W. Daniel. *Applied Linear Algebra.* Englewood Cliffs: Prentice-Hall, 1977.

[18] K. Ogata. *Modern Control Engineering.* Upper Saddle River: Prentice Hall, 1997.

[19] A. Pazy. *Semigroups of Linear Operators and Applications to Partial Differential Equations.* New York: Springer-Verlag, 1983.

[20] G. Strang. *Linear Algebra and Its Applications.* Belmont: Thomson, Brooks/ Cole, 2006.

Index